David Bashford.

David Bashford.

Engineers' Guide to Composite Materials

John W. Weeton

Consultant
Formerly Chief of Composite Materials Branch
at NASA

Dean M. Peters

Consultant

Karyn L. Thomas

ASM Editorial Staff

American Society for Metals

Metals Park, Ohio 44073

This book is a collective effort involving a large number of technical specialists. It brings together a wealth of information from world-wide sources to help scientists, engineers, and technicians solve current and long-range problems.

Great care is taken in the compilation and production of this book, but it should be made clear that no warranties, express or implied, are given in connection with the accuracy or completeness of this publication, and no responsibility can be taken for any claims that may arise.

Nothing contained in this book shall be construed as a grant of any right of manufacture, sale, use, or reproduction, in connection with any method, process, apparatus, product, composition, or system, whether or not covered by letters patent, copyright, or trademark, and nothing contained in this book shall be construed as a defense against any alleged infringement of letters patent, copyright, or trademark, or as a defense against any liability for such infringement.

Comments, criticisms, and suggestions are invited, and should be forwarded to the American Society for Metals, Metals Park, Ohio 44073.

Library of Congress Catalog Card Number: 86-72113
ISBN: 0-87170-226-6
SAN 204-7586

Editorial and production coordination by Carnes Publication Services, Inc.

PRINTED IN THE UNITED STATES OF AMERICA

FOREWORD

Composite materials, based on the principle that the sum is greater than the parts, are beginning to offer engineers new choices in materials selection for product design. With composites already in use for products ranging from golf-club shafts to turbine blades, many feel that the composites industry is poised for spectacular growth in the coming decades. To the majority of engineers, however, composite materials are viewed as uncharted waters: a vast expanse of technology sure to yield improved performance—but where does one begin?

With this book, ASM provides the compass. Herein, the reader will find both an ideal starting point and a lasting reference tool. The book presents information ranging from an explanation of what factors classify a material as a composite, to the Izod impact strength of glass-fiber reinforced polyester resins. Also presented are listings of important reference books, research institutes, consultants, and manufacturers—in short, this book is a carefully planned, concise guide to composite materials.

TIMOTHY L. GALL
ASM, October 1986

PREFACE

Many technological breakthroughs have occurred in the past 25 years in the field of composite materials. U.S. government agencies such as NASA, the Army, Navy, and Air Force, as well as industry, have sponsored innovative research that has led to the discovery of many new composite materials and their constituents. Through these joint efforts, important pioneering work has been accomplished. In the area of whisker and fiber technology, boron, silicon carbide, and graphite filaments are just a few of the materials that are beginning to have an impact on the composites field. Today, many filaments (both light and heavy) are so strong or stiff that they have permitted the fabrication of low- and high-temperature composites with strength-to-weight ratios or modulus-to-weight ratios greater than those of many of the best alloys. In short, the new types of filaments and whiskers developed through the research efforts of government and industry have created a new way of thinking about materials. Composites now give the engineer the freedom to design materials to suit the specific property needs of the structure.

Engineers' Guide to Composite Materials was developed to assist the engineer in designing and fabricating composites by presenting, in one convenient reference source, significant property data for polymer, metal, and ceramic matrix composites. Along with the data needed for designing composites, other helpful tools are included: a bibliography of books on composites, lists of R&D centers, names of relevant trade associations, names of consultants with specialties in composites, and an extensive guide to buyers and manufacturers. In addition, there is a glossary containing definitions of more than 1000 terms relating to the various composite disciplines, a textbook-type introduction to composites, a section on the economic aspects and future

of the composites industry, and extensive design sections including case histories and equations for designing composites.

Section 1, "Introduction to Composite Materials," was written as a supplement to the property data sections and covers ceramic, metal, and polymer matrix materials. Reinforcements and several types of composites are also discussed. The introduction should prove useful not only to the newcomer to composites, but also to engineers who want to examine an area that is outside their specialized field.

Composites—plastic, metal, and ceramic — are currently replacing many traditional materials. A section entitled "Economic Outlook for Composites and Reinforcements" provides a broad look at the short-term and long-term future of composites and some of their constituents. A brief look at the history of several composites as well as their current and future markets and applications is also presented.

The third and fourth sections of the book are included to assist the engineer with the design process. The section on case histories is a collection of 47 papers dealing with the use of composite materials in recreational, aerospace, automotive, marine, electronic, and medical applications. These papers have been condensed and are carefully referenced. This part of the book should provide a good source of ideas for design and material selection. In conjunction with the case history section, there is also a design-equation section that gives basic and advanced equations for ceramic, polymer, and metal matrix composite design.

The largest portion of the book consists of 225 pages of tables and illustrations dealing with property data, fabrication methods, applications, and testing methods for reinforcements, polymer, metal, and ceramic matrix composites. The hard data presented here have been carefully selected and organized to ensure ease of use and to cover as many areas as possible. Early research results have been included, along with current data where appropriate. Most tables and illustrations are referenced for further investigation of a particular material or subject.

This first edition of *Engineers' Guide to Composite Materials* is, as one consulting engineer on the project says, "jam-packed with information."

Data and information from the last 25 years in composites technology have been collected and organized under one "roof." Whether you are new to composite materials or a seasoned composites engineer, the editors are certain that you will find this book a valuable reference tool for many years to come.

J. W. WEETON
D. M. PETERS
K. L. THOMAS

CONTENTS

SECTION 1
Introduction to Composite Materials

Ever since it was recognized that combinations of different materials often resulted in superior products, materials have been combined to produce composites. Mud bricks reinforced with straw were known to have been made hundreds of years B.C., as were laminated woods. Early history reports Mongol bows made from cattle tendons, wood, and silk bonded together with adhesives. Other examples include Japanese ceremonial swords and Damascus gun barrels fabricated from iron and steel laminates. This introduction, however, deals with more recent composites, dating primarily from after World War II. Contemporary composites range from glass-fiber-reinforced automobile bodies to SiC-particulate-reinforced aluminum for lightweight space and military applications. Other modern-day composites include plywoods, plasterboard, concrete, fiber-reinforced pneumatic tires, and many other important materials.

At present, there is no universally accepted definition of the term "composite material." One definition might be "any material that consists of two or more identifiable constituents." Natural as well as man-made composites could be considered composites by this definition and might include plants such as wood, animals (a composite of bone and tissue), or even many rocks and minerals. Since this definition could include almost everything except single-phase or homogeneous materials, this book will consider only composites that consist of two or more materials deliberately combined to form heterogeneous structures with desired or intended properties. Although this considerably narrows the scope of materials that may be considered composites, it would still be too all-inclusive for this introduction. Metal alloys, dispersion-strengthened materials, copolymers, conventional cements, rubber tire materials, plywood, asphalt, some glasses, nuclear materials, and many other such materials might technically be considered composites, but will not be discussed. However, the following types of composites and some of their constituents will be included in this introduction: fiber- and whisker-reinforced, directionally solidified, filled, flake, particulate, and laminar composites. Five of these types are illustrated in Fig. 1.

Perhaps the most typical types of composites are composed of an additive constituent such as fibers or particles embedded in a matrix. Usually these are structural materials, but they can also be special materials such as electrical con-

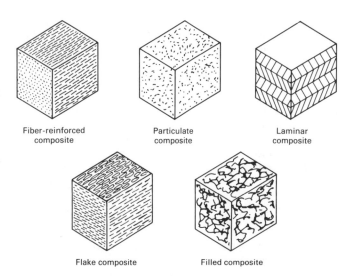

Fiber-reinforced composite · Particulate composite · Laminar composite

Flake composite · Filled composite

Fig. 1. Some classes of composites.

ductors. Some composites have no matrix and are composed of one or more constituent forms consisting of two or more different materials. Sandwiches and laminates, for example, are composed entirely of layers which, taken together, give the composite its form. Many felts and fabrics have no body matrix, but consist entirely of fibers of several compositions, with or without a bonding phase. Reinforcing or additive constituents used for structural materials usually carry most of the load, or furnish the dominant properties. Figure 2 illustrates the constituent forms of some reinforcing or additive phases or materials used in composites.

Composite types are sometimes difficult to distinguish from one another. The differences between one composite type and another may seem fairly clear, but in practice it is not always easy to draw the line. Consider a product such as a fiberglass boat hull. The hull is made of several layers containing fibers bonded together. According to some composite fabricators, this is called a laminar composite, or laminate. The individual layers are also strengthened by glass fibers. Individual layers containing fibers can also be termed plies or tapes,

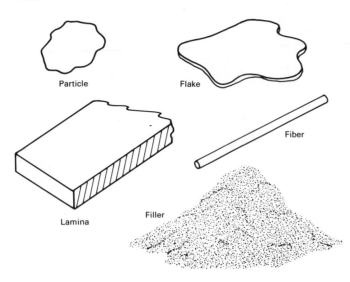

Fig. 2. Constituent forms in composites.

and may be considered "building blocks" used to form bulk fiber composites. This book will consider such a material to be a fiber-reinforced composite.

Matrices

In those structural composites that consist in part of matrices, the matrix serves two very important functions: (1) it holds the reinforcement phase in place, and (2) under an applied force it deforms and distributes the stress to the reinforcement constituents. Sometimes the matrix itself is a key strengthening element. This occurs in certain metal-matrix composites. In other cases, a matrix may have to stand up to heat or cold. It may conduct or resist electricity, keep out moisture, or protect against corrosion. It may be chosen for its weight, ease of handling, or any of many other properties.

Any solid that can be processed so as to embed and adherently grip a reinforcing phase is a potential matrix material. Polymers and metals have been very successful in this role; inorganic materials, such as glass, plaster, portland cement, carbon, and silicon, have also been used as matrix materials with varying success. These latter materials remain elastic up to their points of failure and characteristically exhibit low failure strains under tensile loading, but are strong under compression. Prestressed reinforced concrete is one example of this type of composite, but these materials fall outside the scope of this volume and will not be discussed further.

One important consideration in composite production is how the constituents of a composite interact during fabrication and/or use. They should not react chemically or metallurgically in a way that harms either. In general, they should not have greatly different coefficients of linear expansion. The area of contiguous contact between the matrix and reinforcing material is called the interface, which in some ways is analogous to the grain boundaries in monolithic materials. In certain cases, however, the contiguous region is a distinct added phase, called an interphase. Examples are the coating on the glass fibers in reinforced plastics, and the adhesive that bonds the layers of a laminate together. When such an interphase is present, there are two interfaces—one between each surface of the interphase and its adjoining constituent. In still other composites, the surfaces of the dissimilar constituents interact to produce an interphase.

Fabrication methods depend to a great degree on the matrix properties, and how the matrix affects the properties of the reinforcements. For this reason, the next few sections of this discussion will look into the general characteristics of different types of matrix materials: metals, polymers, ceramics and glasses, and carbon/graphite.

Metal Matrix Materials

Metal-matrix composites, although not as widely used as plastic ones, are currently a subject of great interest. Metal matrices offer greater strength and stiffness than those provided by polymers. Fracture toughness is superior, and metal-matrix composites offer less-pronounced anisotropy and greater temperature capability in oxidizing environments than do their polymeric counterparts. Although most metals and alloys could serve as matrices, in practice the choices for low-temperature applications are sharply limited to the light metals because of the penalty in weight that would result from use of heavier metals.

The metals most used as matrices are aluminum, titanium, and magnesium, which are particularly favorable for aircraft applications. These metallic matrix materials require high-modulus fibers or reinforcements in order to operate effectively, because their modulus values—69,110, and 45 GPa, or 10×10^6, 16×10^6, and 6.5×10^6 psi, respectively—are high relative to those of most polymers. The resulting composites can have strength-to-weight ratios higher than those of almost all known bulk metal alloys. Many of them perform well at elevated temperatures (again, relative to polymers).

The service temperature of a composite depends on the melting point of the matrix material, the reactivity of the fiber and matrix materials, and the mechanical and physical properties of the composite at various temperatures. Various fiber/metal combinations can be used in a wide variety of temperature ranges, from the cryogenic range to 2760 °C (5000 °F).

The choice of fibers for use with lower-melting-point matrices is relatively broad—most metals, ceramics, and compounds can be used in a magnesium or aluminum matrix. The choice becomes more and more limited as the melting point of the matrix material increases. At the upper end of the service-temperature range, the superalloys and refractory metals are the principal matrix materials for consideration, and the choice of reinforcements is reduced to oxides, intermetallic compounds, and refractory-metal fibers.

Polymer Matrix Materials

Polymers are particularly attractive as matrix materials because of their relatively easy processibility, low density, and good mechanical and dielectric properties. High-temperature resins for use in composites are of particular interest to the high-speed aircraft, rocket, space, and electronic fields. The temperature capability of a matrix resin is determined by its softening temperature, its oxidate resistance, or its intrinsic thermal breakdown.

There are two principal types of polymers: the thermoplastics and the thermosets. The thermoplastics are largely one- or two-dimensional in molecular structure. At high temperatures they soften and show a discrete melting point. The process of softening at temperature and regaining rigidity upon cooling is a reversible one. These molding compounds are processed by use of conventional techniques of compression, transfer, or injection molding or extrusion. Polyethylene, polystyrene, polypropylenes, polyamides, and nylon are examples of thermoplastics.

Thermosets, on the other hand, develop a well-bonded three-dimensional molecular structure upon curing. Once crosslinked or hardened, these polymers will decompose rather than melt. The conditions necessary to effect cure and the characteristics of the uncured resin can be varied by changing the basic formulation. Moreover, some thermosets can be held in a partially cured condition (B-stage) for prolonged periods of time. This inherent flexibility makes the thermosets ideally suited for use as matrix materials for the newer continuous fiber composites. Most often, however, the thermosets are used in chopped fiber composites in which the starting material is a premix or molding compound containing fibers of a desired quantity and aspect ratio. Epoxy, polyester, and phenolic polyimide resins are examples of thermosets.

Thermoplastics. Thermoplastic-reinforced resin systems are the most rapidly growing class of composites. In this class the focus is on improving the base properties of the resin to allow these materials to perform functionally in new applications or in those previously requiring metals, such as die casting. Thermoplastics may be crystalline or amorphous. For crystalline thermoplastics, morphology may be significantly influenced by the reinforcement, which can act as a nucleation catalyst. In both types, there is a range of temperature over which creep of the resins increases to the point where it limits usage. The reinforcements in these systems can increase failure load as well as creep resistance. Some shrinkage also occurs during processing, plus a tendency of a shape to remember its original form. The reinforcement can modify this response as well. Thermoplastic systems have advantages over thermosets in that there are no chemical reactions involved that cause release of gas products or exothermal heat. Processing is limited only by the time needed to heat, shape, and cool the structure, and the material can be salvaged or otherwise reworked. However, solvent resistance, heat resistance, and absolute performance are not likely to be as good as those of thermosets.

Generally, thermoplastic resins are marketed in the form of molding compounds. Usually, these compounds are reinforced by short glass fibers. Graphite and other fiber materials may also be used. In most cases, the fibers are dispersed randomly, making the reinforcement nearly isotropic. Molding processes, however, may cause the fibers to lie directionally.

Fillers can be added to raise the heat resistance of thermoplastics, but all thermoplastic composites tend to lose strength dramatically at high temperatures. At normal temperatures, however, thermoplastic composites offer toughness, rigidity, and resistance to creep.

The auto industry uses about 50% of the thermoplastics produced in the United States. Appliance and business-machine makers use another 30%. Automobile instrument panels and pump bodies, computer and appliance housings, and washing-machine agitators are typical thermoplastic composite products.

Several recent developments promise to widen the use of thermoplastics. For example, reinforced thermoplastics are now available in large sheets, which can be stamped or heated and formed in molds. Large parts can be more easily fabricated from this sheet material than from molding compounds. Newly developed reinforced thermoplastic foams are also suited to the molding of large parts, and filament winding of thermoplastic materials is in development.

Thermosets. A list of all the composite products with thermoset matrices would fill a book. Some of the most important uses involve fiber composites designed for structural use. Many aircraft, space, military, and automotive parts fall under this heading. The electrical industry uses epoxy-matrix materials in such parts as printed circuit boards.

Thermoset resins are generally produced either by condensation polymerization directly or by additional polymerization followed by a condensation rearrangement reaction to form the heterocyclic entities. That is, H_2O is a reaction product in either case and creates inherent difficulties in producing void-free composites. Voids have a deleterious effect on shear strength, flexural strength, and dielectric properties, among others. As stated earlier, there are several types of thermoset resins. Two of the most prominent are epoxies and polyesters.

Epoxy resins play a major role in filament-wound composites, and are well suited for use in molding compounds and prepregs. They also have good chemical stability and flow properties, and exhibit excellent adherence and water resistance, slow shrinkage during cure (about 3%), and freedom from gas evolution. However, the epoxies are relatively expensive and are generally limited to service temperatures below 150 °C (300 °F). This restricts many aerospace applications, where higher use temperatures are required.

Polyester resins are cheap and versatile resins and are widely used. Liquid polyesters can be stored at room temperature for months or even years, and can be brought to cure in a matter of minutes simply by adding a catalyst. About 80% of the polyester resins produced in the United States are used with reinforcing materials, mainly fiberglass. Reinforced polyesters are used in boat hulls and building panels, and in structural parts in aircraft, automobiles, and appliances.

Although many types of plastic resins, both thermoset and thermoplastic, are being reinforced with glass fibers, polyester resins are the most widely used, especially for low-performance applications. In the cured state, polyesters are hard, light-colored, transparent materials, which may be rigid or flexible. They are resistant to water, weather, aging, and a variety of chemicals. They can be used at temperatures up to about 80 °C (175 °F) or higher, depending on the formulation of the resin or service requirements of the application. Other principal advantages of polyesters are that they combine easily with glass-fiber reinforcements and can be used with all types of reinforced plastic fabrication equipment. Disadvantages of polyesters include vulnerability to some chemicals and a tendency to burn easily.

Ceramic and Glass Matrix Materials

Ceramic materials in general are characterized by high melting points, high compressive strength, good strength retention at high temperatures, and excellent resistance to oxidation. These are very desirable properties from the standpoint of selecting a structural material that will operate at 1650 °C (3000 °F) and higher. In fact, these properties have been responsible for the use of ceramic materials for furnace, kiln, and other refractory applications for centuries. Many modern-day applications for ceramics, however, require materials with high tensile strengths and resistance to impact, vibration, and thermal shock. Therefore, before the desirable properties of ceramics can be utilized for today's high-temperature applications, several major weaknesses that also characterize these materials must be overcome. These weaknesses are relatively low tensile strength, poor impact resistance, and poor thermal-shock resistance. It is toward overcoming these deficiencies that the research into the reinforcement of ceramics has been directed.

Ceramic Matrices. Fibrous reinforcements of ceramic matrices present some unusual problems. It is generally understood that reinforcement of a material is undertaken primarily for the sake of utilizing the higher tensile strength of the fiber so as to produce an increase in the load-bearing capacity of the matrix. However, the addition of a high-strength fiber to a relatively weak ceramic does not always result in a composite with a tensile strength greater than that of the ceramic alone. In many cases it will actually result in a weaker composite.

The combination of low tensile strain and high modulus of elasticity exhibited by most ceramics is the primary reason for failure to obtain strength improvement by the addition of a reinforcement. That is, at stress levels sufficient to rupture the ceramic, the elongation of the matrix is insufficient to transfer a significant amount of the load to the reinforcement, and the composite will fail unless the volume percentage of the fiber is extraordinarily high. This difficulty can be mitigated to some extent, of course, by using a reinforcement with an unusually high modulus of elasticity. Another, and perhaps a more satisfactory, solution is to prestress the fiber in the ceramic matrix. This may be accomplished by utilizing a reinforcement with a coefficient of thermal expansion higher than that of the matrix. If such a system can be brought to a "no-stress" condition at high temperature, the prestressing will occur during cooling. Both of these solutions require, of course, that a sufficient bond be established between the ceramic and the reinforcement to effect the necessary bond transfer.

The tensile strength of a composite in which the ceramic has a coefficient of thermal expansion higher than that of the reinforcement may not be superior to the strength of the ceramic alone. Such a composite will develop tensile stresses within the ceramic on cooling. This usually produces microcracks, extending from fiber to fiber, within the matrix. While microcracking does not destroy the integrity of the composite, it can result in a composite tensile strength that is lower than that of the ceramic alone. Matrix brittleness, shrinkage associated with sintering, reactivity, and the generation of internal stresses due to thermal mismatch have proven to be only a few of the problems associated with the development of ceramic-matrix composites.[1]

Glass Matrices. Glass is a convenient matrix material because it is a relatively inert, inorganic thermoplastic material. Glass, in fact, lends itself to some of the composite processing methods applicable to polymers, such as melt infiltration and compression molding. Glass can also provide several important advantages over crystalline ceramics.

In some cases, glass-matrix composites can be produced with both high elastic modulus and strength that can be maintained at temperatures as high as 600 °C (1110 °F). Also, the resulting composites can be dimensionally superior to resin or metal systems because of the low-thermal-expansion behavior of the glass matrix. The thermoplastic nature of glass matrices also permits fabrication of a wide range of articles. Complex shapes such as tubes and hat sections have been made to final shape using procedures similar in format to those currently practiced for resin-matrix composites. In addition, glass matrices also offer unique wear characteristics.

In the case of fiber-reinforced glass composites, the flow characteristics of glass enable it to be readily densified with fibers in a composite structure by the application of pressure at elevated temperatures. Unlike crystalline ceramics, this densification can be carried out to achieve very high fiber contents without causing damage to the reinforcing fibers. In further contrast to crystalline ceramics, glass has an elastic modulus well below those of many intended fiber reinforcements. In fact, glass has an elastic modulus of about 69 GPa $(10 \times 10^6$ psi), which is comparable to that of aluminum, whereas crystalline ceramics of interest for high-temperature applications have much higher moduli and are difficult to reinforce, as noted earlier.

Thus, during the application of stress to the composite, high-modulus fibers provide the main load-bearing composite constituent, and strengths in excess of those of the parent glass matrix can be achieved. Further advantages include the availability of numerous glass compositions that provide a broad selection of chemically compatible matrices that can be combined with ceramic fibers.

Carbon/Graphite Matrix Materials

Carbon and graphite are superior high-temperature materials with strength and stiffness properties maintainable at temperatures up to 2500 K. The combination of carbon or graphite fibers in a carbon or graphite matrix results in a most advanced material with unusual property characteristics. This material is referred to as a carbon/carbon composite. Briefly, carbon/carbon composites are fabricated by the multiple impregnation of porous "all-carbon" frames or configurations. These frames of fibers are impregnated with a liquid carbonizable precursor (for example, pitch) which is subsequently pyrolized. Carbon/carbon composites can also be produced through a process of chemical vapor deposition of pyrolytic carbon. The gas-impregnation method is difficult to perform because of the tendency for pore closure instead of pore filling in the fiber skeleton.

The resultant composites are not particularly impressive at ambient temperatures. In fact, many conventional structural materials are stronger. The main feature of carbon/carbon composites, however, is their retention of usable properties at temperatures from room temperature to 2760 °C (5000 °F). These composites also remain dimensionally stable over a comparable temperature range.

Over the past 20 years, carbon/carbon composites have been used for various aeronautical, biomedical, defense, industrial, and space applications. Originally, these materials were produced for applications where hardware was exposed to extreme temperatures, requiring high performance standards, such as solid rocket motors. Today, carbon/carbon composites are used in commercial as well as military applications. The combination of high-temperature resistance, light weight, and stiffness makes them ideal candidates for use in missiles and space vehicles.

Composites and Reinforcements

Reinforcing materials supply the basic strength of the composite. They can, however, contribute much more than strength: they can conduct heat or resist chemical corrosion; they can resist or conduct electricity; and they may be chosen for their stiffness (modulus of elasticity) or for many other properties.

If a reinforcement is to improve the strength of a given matrix, it must be both stronger and stiffer than the matrix, and it must significantly modify the failure mechanism in an advantageous way. The requirement of high strength and high stiffness implies little or no ductility and, thus, relatively brittle behavior. Fibers are probably the most important class of reinforcements due to their ability to transfer strength to matrix materials and greatly influence their properties. Other types of reinforcements include fillers, sometimes known as particles, and flakes.

Before examining fibers, we will first look at the products (fiber-reinforced composites) that result from adding fibers to a matrix. Continuing on in this fashion, we will examine fillers as well as filled composites and flake composites.

Fiber-Reinforced Composites

Glass-fiber-reinforced resins were among the earliest fiber-reinforced composites. These initial composites were fabricated from solid glass fibers of circular cross section and flexible plastic matrices. As work with composites continued, it was discovered that other materials besides glass fibers and plastics could be combined to form useful materials. Metal and ceramic fibers, hollow fibers, and fibers of noncircular cross section were combined with stiffer and more heat-resistant matrices. Today, fiber-reinforced composites are produced from a wide range of constituents. Many types of fibers are combined with metal, resin, and ceramic matrices to form extremely useful materials. Of all the composites, fiber-reinforced composites have evoked the most interest among engineers concerned with structural applications.

Factors Affecting Fiber Performance. The engineering performance of a fiber composite is determined by the orientation, length, shape, and composition of the fibers, the mechanical properties of the matrix, and the integrity of the bond between fibers and matrix. Fiber orientation (how the individual strands are positioned) determines the mechanical strength of the composite and the direction in which that strength will be the greatest. There are several ways of thinking about fiber orientation and how orientation affects the strength of a composite. First, however, it must be realized that unidirectionally oriented fibers (or longitudinal fibers) provide maximum composite strength and modulus when loads are applied in the direction of the fibers. In contrast, when loads are applied at even very small angles from the fiber direction, composite strength is drastically lowered. (For more insight into the effects of "off-angle" loads, see the section of this book entitled "Design Equations.") Since very few structures are loaded unidirectionally, it is necessary in most cases to mix orientations in a given part or structure.

Some of the types of possible orientations are shown in Fig. 3. Both continuous and discontinuous fibers can be unidirectionally oriented in thin layers (monolayer tapes). These tapes can be stacked and consolidated into plies containing many or few layers of filaments oriented in the same direction. Plies can also be made by direct means by many different methods.

Plies, however, are planar structures. Layers of tapes or plies with longitudinally oriented fibers can be stacked together so that the fibers in each layer are at different orientations. Thus, plies can be stacked so that the fibers are oriented at right angles to each other (cross plied), or so that the fibers are oriented at other angles to each other (angle plied). Plies can be stacked in very complicated ways, and computers frequently are used to determine the orientations needed to achieve desired properties. Thus, strengths in planar composites (sheets) can be varied from those of unidirectionally

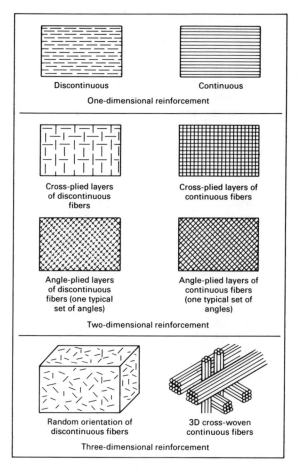

Fig. 3. Several possible ways to orient fibers in fiber-reinforced composites.

oriented fiber composites to produce composites with almost isotropic properties (quasi-isotropic composites).

As an example of possible orientations, the quasi-isotropic material has plies with fibers at $0°/90°/+45°/-45°/-45°/+45°/90°/0°$. Rule-of-thumb strengths and moduli for the latter type of composite are about one-quarter and one-third, respectively, of those of longitudinally oriented composites. For angle-plied composites other than quasi-isotropic ones, properties can vary with the number of plies and their respective orientations. These ratios assume that the composite variables are constant and that the matrices (e.g., plastics) are very weak relative to the fibers. To go to a three-dimensional orientation of continuous fibers, it is usually necessary to weave fibers in 3D and infiltrate materials into their interstices later. Assuming that fiber volume percentages would be equal in all three axes of such composites, the strengths in any axis would be about one-third that of unidirectional fiber composites.

It is possible to orient short-length fibers differently than by the methods noted above. For example, short fibers can be given random orientations by sprinkling or shaking them onto a given plane (to achieve random orientation in two dimensions). The matrix could be added either before or after deposition of the fibers using liquid or solid-state methods. Three-dimensional random orientation can be achieved by mixing solids or liquids with the short fibers by several methods. Randomizing fiber orientations in 2D would yield a composite with a strength about one-third that of a composite with unidirectional fibers stressed in the direction of the fibers, while randomizing the fibers in 3D would yield a composite with a comparable ratio of somewhat less than one-fifth.

Extremely strong matrices (e.g., titanium alloys) would not be reduced to such low strengths or moduli as those noted in all examples above. One can approximate the strength values of the composites with such matrices with various orientations. By assuming that the fibers have been reduced to "effective strengths" by the ratios given above, one can calculate the longitudinal strength of the composite (which is an approximation of the strength of the strong matrix composite with fibers that are not oriented longitudinally) using these values.

Fiber composites may be constructed with either continuous or short fibers. Although continuous fibers (filaments) are more efficiently oriented than short fibers, they are not necessarily better. One of the main characteristics of a fiber is that its length is considerably greater than its effective diameter—that is, it has a high aspect ratio. Filaments are fibers with very high aspect ratios. Manufacturing processes for filaments are often continuous and the filaments spoolable. Some filaments are subsequently chopped for certain uses. The principal advantage of filaments is that they can be mass-produced. They can be wound (often in a resin matrix), laid in strata, twisted, woven, knit, and in general can perform similarly to conventional fabric or textile materials. Filaments have high strengths, and many have low densities.

If the fibers could be properly oriented, composites made from some specific types of shorter fibers such as glass, ceramics, and multiphase fibers could have substantially greater strengths than those made from continuous fibers. Some short fibers can be produced with few surface flaws and come extremely close to achieving their theoretical strength.

The length of a fiber impacts not only the mechanical properties of a composite, but also its processibility. In general, continuous fibers are easier to handle than short fibers. Continuous filaments can be incorporated by the filament-winding process, which wraps a continuous fiber (impreg-nated with a matrix material) around a mandrel having the shape of the part, ensuring good distribution and favorable orientation of the fiber in the finished article. Filament winding is limited chiefly to the fabrication of bodies of revolution, but flat and irregular shapes can also be made. Short-fiber composites, on the other hand, can be made by the numerous open- and closed-mold processes. These composites are less efficient in their use of fiber, but generally have higher output and lower cost than filament winding.

Most fibers in use today are solid (easy to produce and handle) and have a circular cross section, but other shapes and hollow fibers show promise of leading to improved mechanical properties in composites. Some shapes that show potential are hexagonal, rectangular, polygonal, and irregular. Another element to consider is the fiber diameter. Generally, the smaller the diameter the greater the strength of the fiber, due to the elimination of surface flaws during production. Figure 4 compares cross-sectional dimensions of various fibers.

The development of thin, flat-sided filaments has led to fibers with rectangular cross sections, which show potential for use in high-strength cylindrical structures. The shape of these fibers makes it possible to obtain almost perfect packing. Hollow glass fibers demonstrate improved structural efficiency for applications where stiffness and compressive strength are the governing criteria. The transverse compressive strength of a hollow-fiber composite is lower than that of a solid-fiber composite when the hollow portion of the fiber is more than half the total fiber diameter. Hollow fibers, on the other hand, are quite difficult to handle and to incorporate into a composite.

Types of Fibers

Composite materials are reinforced with both organic fibers* and inorganic fibers. Organic fibers made from polypropylene, nylon, and graphite can be characterized in gen-

---

*Compounds of carbon that often contain such elements as hydrogen, nitrogen, and oxygen.

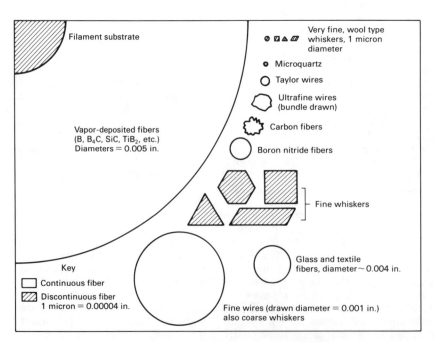

Fig. 4. Comparison of cross-sectional dimensions for various fiber reinforcements.

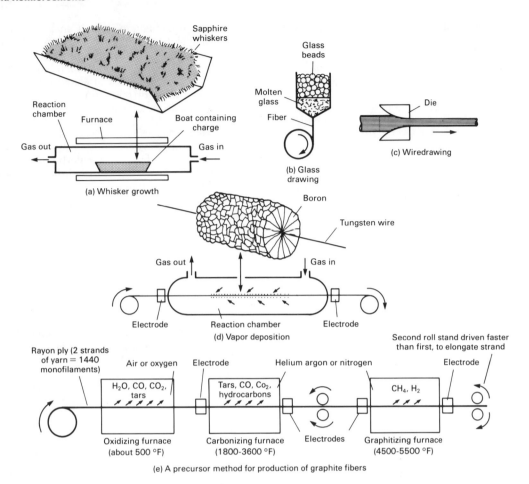

Fig. 5. **Methods of fiber production.**

eral as lightweight, flexible, elastic, and (except for graphite) heat-sensitive. Inorganic fibers such as steels, glass, tungsten, and ceramics in general can be described as very high in strength, heat resistant, more rigid or stiff than organic fibers, fatigue resistant, and low in energy absorption. While many organic fibers satisfy both strength and elasticity requirements for some structural composites, graphite (which is more often than not considered a ceramic) has become very popular in recent years. Organics are marketed in far greater quantities than inorganics. Of the latter, glass fibers are produced in the greatest quantities. Where extremely high strength or stiffness is required, graphite, silicon carbide, and boron have received the most attention; but for stringent applications such as high-temperature use, steels, superalloys, refractory metals, and several ceramics (including whiskers) are under investigation for their ultimate potential. Methods of fiber production are illustrated in Fig. 5

Glass Fibers. Glass fibers comprise well over 90% of the fibers used in reinforced plastics because they are inexpensive to produce and possess high strength, high stiffness (relative to plastics), low specific gravity, chemical resistance, and good insulating characteristics. One problem with glass, however, is that when subjected to tensile loads for prolonged periods of time, glass breaks at stress levels much below those measured in short-time (2 to 5 minutes) laboratory tests. This behavior, known as static fatigue (actually it is more like a stress-rupture phenomenon), effectively reduces the useful strength of glass if it is intended to sustain such

loads for months or years in service. The reduction can be as much as 70 to 80% depending on the load duration, temperature, moisture conditions, and other factors, and if fracture does occur, it gives little or no prior warning since glass is a brittle material. Conservative design practices and periodic inspections are recommended preventive measures in such applications where failure could have critical consequences.

Because they are available in so many different forms, glass fibers lend themselves to a variety of fabrication processes—e.g., filament winding, matched-die molding, lay-up, etc. Glass fibers come in the forms of cloths, mats, tapes, continuous or chopped filaments, rovings, and yarns. Common applications for glass-reinforced plastics are automobile bodies and large boat hulls. Glass-reinforced fishing poles have taken the place of bamboo and steel poles because of their light weight and flexibility.

Glass Fiber Types. Several chemicals may be added to silica sand in preparation for glassmaking. Different combinations give rise to different types of glass. At present there are four major types of glass used to make glass fibers:

1. *A-glass* is a high-alkali glass. An alkali is a chemical that might be called the opposite of an acid. Alkalis neutralize acids in chemical reactions. The important alkalis in A-glass are soda (sodium oxide) and lime (calcium oxide). Together they make up about 25% of A-glass by weight. A-glass offers very good resistance to

chemicals. However, its high alkali content lowers its electrical properties.

2. *C-glass* (chemical glass) is a special mixture with extremely high resistance to chemicals. It is intended for use in situations where even the chemical resistance of A-glass would not be good enough.

3. *E-glass* is named for its electrical properties. With a low alkali content, it offers much better electrical insulation than A-glass. As a fiber, it has good all-around strength, and it strongly resists attack by water. This is important in many glass fiber/resin composites that require high performance under moist conditions. At present, E-glass accounts for more than half of the glass-fiber reinforcements made.

4. *S-glass* is a high-strength material. For example, its tensile strength is about 33% higher than that of E-glass. It also holds its strength better at high temperatures and resists fatigue well. It has found wide use in such high-performance materials as rocket-motor casings and aircraft parts.

A fifth type, D-glass, has also been developed. It has even better electrical properties than E-glass. (The D stands for low dielectric constant.) Although its mechanical properties are not as good as those of E- or S-glass, applications for high-performance radomes have been explored. To date, however, the use of D-glass has been limited.

High Silica and Quartz Fibers. The glasses just examined typically contain about 55 to 75% silica (SiO_2). Silica glass is a much purer form and is made by treating ordinary fiberglass in a hot-acid bath. The acid removes almost all impurities but leaves the silica intact. The glass that results contains from 95 to 99.4% SiO_2.

An even purer substance is quartz. Quartz fibers are made from natural quartz crystals containing 99.95% SiO_2. These fibers have nearly all the properties of pure, solid quartz. (Note: Trade names often use the word "quartz" for any high-purity glass. For our purposes, however, quartz is the very pure substance described here.)

High silica and quartz fibers have much in common with ordinary fiberglass. Both are produced in a variety of fiber diameters. Both can be made into rovings or yarns, woven into fabrics, and so on. Both can be used with most of the same matrix materials with which ordinary glass is used. However, high silica and quartz also differ from ordinary glass, and this is especially true in terms of heat properties.

There are two major differences between high silica and quartz. The first is cost. Quartz crystals are found throughout the world, but the high-purity kind (mined mainly in Brazil) are fairly rare. The raw material for high silica, on the other hand, is the same as for glass fibers — common silica sand. As a result, quartz costs considerably more than high silica. The second major difference is strength. At high temperatures, quartz can be considerably stronger than high silica.

In most other ways, high silica and quartz fibers are very much alike. Both can be stretched to about 1% of their length before breaking and both are perfectly elastic. (That is, they always return to their original shape after deforming under stress. They do not exhibit a yield strength, and the elastic limit coincides with the fracture stress.) Chemically, high silica and quartz are not affected by most acids and resist moisture well. Both, however, react to alkalis. High silica and quartz also have low dielectric constants.

The thermal properties of high silica and quartz make them attractive for use as fibers. Both are excellent insulators, and neither melts or vaporizes at temperatures below 1650 °C (3000 °F). The two fibers also have very low coefficients of linear thermal expansion, which enable them to withstand extreme temperature changes. For example, they can be heated to 1095 °C (2000 °F) and then plunged into water without damage. These thermal properties lend high silica and quartz fiber reinforced composites to aerospace applications. They are used in rocket nozzles, nose cones, and re-entry heat shields for spacecraft.

Metal Fibers and Wires. Metal filaments, or wires, have several advantages as reinforcing materials. They are relatively easy to produce, particularly by wiredrawing processes, and are less sensitive to surface damage than other fibers. Thus, metal fibers can be made by a wider range of fabrication processes than can many other fibers. Metal fibers are also inherently more ductile than ceramic, polycrystalline, and multiphase materials. They are extremely strong, and some have excellent resistance to high temperatures.

Metal filaments also have some drawbacks. Most are heavy compared with most ceramic and multiphase materials, and they tend to react by various alloying mechanisms with metal matrices.

Steel wire used for tire reinforcement comprises the major large-scale application for metal filaments. Wire is also used for enhancing the tensile strength of concrete. In addition, continuous metal fibers are used to reinforce metals and ceramic materials. The incorporation of a fine metal skeleton in a refractory ceramic can almost certainly improve the ceramic's impact resistance and thermal-shock properties.

Continuous metal wires have also been used to reinforce plastics—i.e., epoxy or polyethylene reinforced with steel wire. The advantages of such a combination of materials are high strength, light weight, and good chemical and fatigue resistance. One further advantage is that continuous metal fibers are easier to handle than glass fibers. In addition to better flexural properties, some metal fiber/plastic composites offer better modulus-to-weight ratios than glass-fiber composites. Some of the disadvantages of metal-fiber-reinforced plastics are poor high-temperature performance, and problems resulting from the great difference between the coefficients of thermal expansion of the metal fibers and the resins.

Alumina Fibers. Alumina is the compound aluminum oxide, Al_2O_3. One commercial experimental alumina fiber produced by Du Pont is known as fiber FP. These fibers were originally developed for use in metal matrices, but they also show promise as reinforcements for resin-matrix composites.

The tensile strength of FP is not especially high, but it does supply good compressive strength when set in a matrix. Typical compressive strengths (in the direction of the fibers) for FP/epoxy composites range from 2275 to 2413 MPa (330 to 350 ksi).

An outstanding feature of FP is its high melting point, 2045 °C (3713 °F). The fiber can also withstand temperatures of up to 1000 °C (1830 °F) with no loss of strength or modulus of elasticity. These properties qualify fiber FP for use in high-temperature situations. It has been successfully used in magnesium and aluminum matrices. Even when molten, these metals do not damage the alumina fibers.

Graphite Fibers. Although the names "carbon" and "graphite" are used interchangeably when relating to fibers, there is a difference. Typically, PAN-base carbon fibers are 93 to 95% carbon by elemental analysis, whereas graphite fibers are usually 99+% carbon. The basic difference is the temperature at which the fibers are made or heat treated. PAN-base carbon is produced at about 1315 °C (2400 °F), while higher-modulus graphite fibers are graphitized at 1900 or 3010 °C (3450 or 5450 °F). This also applies to carbon and graphite cloths. Unfortunately, with only rare exceptions, none of the carbon fibers are ever converted into classic graphite regardless of heat treatment.[2]

Graphite filaments have very high strengths and moduli, and these properties remain constant at high temperatures. Unfortunately, graphite reacts readily with many metals during fabrication and high-temperature use. In some metals (such as magnesium) and in plastic matrices, it furnishes high strength and stiffness to composites, and in such cases reactions are not a problem. In other metals (such as aluminum), matrix reaction can produce carbides at the interface, which in turn react with moisture.

Graphite fibers are the stiffest fibers known. Theoretically they can be almost five times more rigid than steel. Practically speaking, those now produced commercially are 1.5 to 2 times stiffer, and laboratory specimens a bit more than 3 times stiffer have been made. The fibers themselves are composites: only part of the carbon present has been converted to graphite in tiny crystalline platelets specially oriented with respect to the fiber axis. The higher the graphite content, the stiffer the fiber. Unfortunately, stiffness and strength are inversely related.

A basic disadvantage of the use of graphite fibers is their cost: the cheapest and lowest-quality graphite fibers still are many times more expensive than glass fibers. The raw materials, or precursors, for PAN-base fibers are expensive, and the processes of carbonization and graphitization consume much time, energy, and materials, and require close control throughout. As a result, their use in composites is limited to applications that place a premium on saving weight — aircraft, missiles, some sports equipment, and certain specialized hardware. Less-expensive pitch-base fibers with greatly improved properties, however, are now being developed and will ultimately lead to many new applications for graphite fibers.

Multiphase Fibers. Multiphase fibers are usually spoolable filaments produced by chemical vapor deposition (CVD) processes. Typical multiphase fibers consist of materials such as boron, boron carbide, silicon carbide, and others formed on the surface of a very fine filament substrate of tungsten or carbon. These fibers offer good potential for use at high temperatures because they are less reactive with higher-melting-temperature metals than are graphite and many metallic fibers. Boron filaments are good candidate materials for structural and intermediate-temperature composites.

Another type of multiphase, or polyphase, fiber is a core-sheath fiber, which consists of a polycrystalline or crystallizable core material enclosed in a vitreous or glasslike sheath. The sheath configuration enables high-modulus oxide materials to be drawn into fibers or filaments 25 to 200 μm (1 to 8 mils) in diameter.

Boron Fibers. It is not strictly accurate to speak of "boron fibers." Actually, these fibers are themselves composites. The metal boron is coated on a thin filament of another substance called a *substrate*. Substrates usually are tungsten (W) or carbon.

Boron-tungsten fibers are made by passing a hot tungsten filament through a mixture of hydrogen and boron trichloride BCl$_3$) gas. The boron deposits out on the tungsten, and continued processing produces the desired fiber thickness. At present, boron fibers are made with diameters of 0.05, 0.1, 0.14, and 0.2 mm (0.002, 0.004, 0.0056, and 0.008 in.). Other diameters as high as 0.4 mm (0.016 in.) are under test. The tungsten substrate always has the same thickness, 0.01 mm (0.0005 in.).

About 90% of the boron-tungsten fiber marketed is 0.05 mm (0.002 in.) in diameter and is used primarily in boron/epoxy prepregs. Almost all of these prepregs are sold as tapes or broadgoods. (Broadgoods is the industry term for any fiber material over 45.6 cm, or 18 in., wide.) A popular width for boron/epoxy broadgoods is 121.9 cm (48 in.). Both tapes and broadgoods are shipped in continuous lengths up to 305 m (1000 ft). Other boron-tungsten fibers are 0.14 mm (0.0056 in.) in diameter and are used in aluminum matrices.

The properties of boron fibers vary somewhat with diameter. This is partly due to the changing ratio of boron to tungsten and to surface defects that vary with size. However, all boron fibers exhibit good strength and outstanding stiffness. Tensile strengths range from about 2758 MPa (400 ksi) to over 3447 MPa (500 ksi). These strengths compare with the strengths of many glass fibers. Tensile modulus, on the other hand, is much higher for boron. A typical tensile modulus for a boron-tungsten fiber is 400 GPa (58×10^6 psi). This is nearly five times the tensile modulus of S-glass.

Boron coated on a carbon substrate is another type of boron fiber. Carbon is cheaper and lighter than tungsten, but it also has a lower modulus of elasticity. The tensile modulus of boron-carbon (B-C) is more than 10% lower than that of boron-tungsten.

Silicon Carbide Fibers. Like boron, silicon carbide (SiC) can be coated onto tungsten (W) or carbon substrates. SiC-W fibers are made in diameters of 0.1 and 0.14 mm (0.004 and 0.0056 in.). These fibers have room-temperature tensile strengths and tensile moduli that are very similar to those of B-W.

SiC-W has several advantages over uncoated B-W fibers. The most important is its high-temperature performance. In addition, SiC-W fibers have been tested at temperatures up to 1370 °C (2500 °F) with only a 30% loss of strength. Both SiC-W and SiC-C have been shown to have very high stress-rupture strengths at 1095 °C (2000 °F) and 1315 °C (2400 °F). Uncoated B-W fibers lose *all* their strength at temperatures of about 650 °C (1200 °F). Another feature of SiC fibers, compared with uncoated boron, is that SiC does not react with molten aluminum. The fiber also stands up well to the high temperatures used in hot pressing titanium matrices.

SiC-W fibers have been tested in several other metal-alloy matrices, and much current research focuses on the titanium-aluminum-vanadium alloy known as Ti-6Al-4V. The objective is to develop composite compressor blades and other jet-engine parts. Such parts must withstand great stresses at temperatures of up to 480 °C (900 °F).

On the negative side, SiC-W fibers are somewhat heavier than B-W fibers of the same diameter. Also, SiC is highly subject to surface damage, and care must be taken in handling the fibers, especially when incorporating them in a matrix. Another difficulty arises at temperatures above 925 °C (1700 °F): weakening reactions begin to occur between the tungsten and SiC. This can be a problem during high-temperature metal-matrix formation.

SiC on a carbon substrate is a very promising fiber material. Advantages are: no high-temperature reaction occurs between the carbon and the SiC; SiC-carbon is lighter than SiC-W; and the tensile strength and tensile modulus rival or better those of SiC-W and boron fibers. SiC-carbon is also considerably cheaper than either SiC-W or boron and, as noted above, has excellent stress-rupture strength. This fiber shows promise of eventually taking the place of SiC-W and boron fibers.

Aramid Fibers. Aramid is the generic term for fibers produced from "aromatic polyamides." Polyamides are long-chain polymers and *aromatic rings* (commonly called "benzene rings") are molecular structures in which six carbon atoms are bonded to each other and to different combinations of hydrogen atoms. These rings recur again and again in the structure of aramid fibers.

Du Pont is the largest commercial producer of aramid fibers. The company's fibers are sold under the Kevlar 29 and Kevlar 49 trade names. These fibers were introduced in the early 1970's and were originally developed to replace steel used in radial tires. They still fulfill this function and have found applications in many other areas as well.

The outstanding features of the aramids are low weight, high tensile strength, and high modulus. They produce tough, impact-resistant structures with about half the stiffness of graphite. Both commercially produced Kevlars have virgin tensile strengths of about 2344 MPa (340 ksi). Strain to failure is about 1.8%. The tensile modulus of Kevlar 49 is about twice that of Kevlar 29 and above that of titanium alloys. The density of Kevlar fibers is much less than that of glass fibers, and somewhat less than that of carbon graphite.

Aramid fibers resist flame and high temperatures, as well as organic solvents, fuels, and lubricants. They also resist neutral chemicals, but are attacked by acids and alkalis—especially strong acids. In addition, aramid fibers are very pliable and easily woven by ordinary means. (Glass and graphite, by comparison, are brittle and relatively difficult to handle in weaving.)

Directionally Solidified Eutectics

In situ fibers, or laminae, are often produced by directional solidification of cast alloys. They are actually a part of the alloy that is being precipitated from the melt as the alloy solidifies. This type of reaction usually involves eutectic alloys in which the molten material decomposes into two or more phases at a constant temperature. When this reaction is carried out so that the solidifying phases are unidirectional, the product is termed a "directionally solidified eutectic."

Normally, as an alloy solidifies, crystals nucleate from the edges, top, or bottom of the mold or from any cool spot. The resulting structure has many crystalline particles, or grains, that grow into each other. In directional solidification, this random form of structure is not allowed to occur.

Early work in this area produced Al_3Ni crystals from an aluminum-nickel alloy system (presumed to be whiskers) having very high strengths. Similarly, chromium "whiskers" with strengths over 6.9 GPa (10^6 psi) have been extracted from a copper-chromium eutectic. More recently, tantalum carbide rods in tantalum, and niobium carbide in niobium, have been produced.

Fiber-reinforced ceramic-matrix composites can also be grown *in situ*. Structures have been achieved very much like those in the case of all-metallic systems. More than thirty eutectic combinations with such metals as Cr, Nb, Ta, Mo, and W have been identified as providing *reinforcement* capabilities in conjunction with ceramic major phases such as $(Al,Cr)_2O_3$, CeO_2, Cr_2O_3, UO_2, and ZrO_2. In addition, various all-ceramic systems, such as ZrO_2-MgO, NiO-CaO, and Al_2O_3-ZrO_2, have been shown to lead to similar structures. Among the benefits of such composites are improvements in fracture toughness, greater damage tolerance, and improved high-temperature strength.

Whiskers

Whiskers are single crystals that are grown with almost no defects. They are short, discontinuous fibers of polygonal cross sections that can be made from many (over 100) different materials. Examples are copper, iron, graphite, silicon carbide, aluminum oxide, silicon nitride, boron carbide, and beryllium oxide.

Whiskers have been grown inadvertently in laboratories, and nature has produced some geological structures. In 1952, Herring and Galt of the Bell Telephone Laboratories observed that the strength of metallic tin whiskers was 6.9 GPa (10^6 psi), yield strength was very high, and elastic strains were greater than 1%. For some time, whiskers were little more than a laboratory curiosity, but their discovery stimulated the study of crystal structures and growth in metals, and the study of the effect of defects on the strength of materials. Compared with other material forms, whiskers are relatively free from defects. Whiskers can be incorporated into composites by several techniques. Slip-casting and powder metallurgy techniques have been used successfully to prepare metal/whisker systems.

Strengthening of metals at elevated temperatures with whiskers has been demonstrated in laboratories and limited prototype operations. If all or part of the strength of whiskers could be retained when they are incorporated into an engineering material, a superior product would result, but their very fine size makes whiskers extremely difficult to handle and fabricate into fiber composites.

Some of the earliest research on whiskers showed, in general, that whisker strength varied inversely with effective diameter or cross-sectional size. For R&D purposes where whiskers were embedded in matrices, it was generally felt that whiskers about 2 to 10 μm in diameter would yield reasonably good composites. Examinations of composite specimens with optical microscopes permitted measurements of diameters, volume percentages, etc. Today, composites are being made with much finer whiskers. In fact, one Japanese producer has made, and is producing, whiskers of Si_3N_4 as fine as 0.2 to 0.5 μm in diameter and SiC whiskers as fine as 0.05 to 0.2 μm. To appreciate how small these diameters are, consider that two points 0.05 μm in size are barely (if at all) resolvable with the best optical microscopes at magnifications of about 2000 diameters. In other words, some of the whiskers being seriously considered for use require a scanning electron microscope to reveal their dimensions.

Whiskers of ceramic materials have low densities, high moduli, and useful strengths ranging from 2.8 to 24 GPa (400 to 3500 ksi). Specific strength (strength/density) and specific modulus (modulus/density) are high, making ceramic whiskers suitable for structures requiring minimum weight. Ceramic whiskers resist heat, mechanical damage, and oxidation to a greater degree than do metallic whiskers. Also, metallic whiskers are more dense than ceramic whiskers. The principal reason that metallic whiskers are not commercially available is that they are easily damaged by handling. Oxide, carbide, and nitride whiskers are those of primary interest in composite technology.

Many whiskers are grown from materials in a boat or disk which is placed in a heated, controlled-atmosphere reaction chamber. For example, aluminum in a ceramic boat, in the presence of hydrogen, forms vapors of Al_2O or AlO. These vapors, in turn, react with oxygen (present in the form of water vapor or SiO vapor that forms from the boat material), and the result is solid, needlelike, single crystals of alumina that grow on the surfaces of the boat or its contents.

One method of continuous whisker production is by controlling the hydrolysis of aluminum chloride in a fluidized bed on an aluminum or alumina nucleus. By passing chlorine over molten aluminum, aluminum chloride is prepared and is then passed into a fluidized bed reaction chamber with Co, H_2 and CO_2. From this production technique a variety of whiskers can be produced. One disadvantage of this technique, however, is the amount of debris that also forms, especially when the growth occurs on the alumina nuclei.

The usual method for forming silicon carbide whiskers is by the pyrolysis of chlorosilanes at temperatures above 1400 °C (2550 °F) in hydrogen. Production of silicon carbide whiskers has improved greatly in recent years through the use of proper silanes and suitable reaction chambers, and by ensuring optimum reaction conditions during formation.

Types of Filled Composites

Two types of composites can be thought of as being filled. In one, filler materials are added to a plastic matrix to replace part of the matrix, or to add to or change the over-all properties of the composite. In some cases, the fillers actually help to strengthen the composite. In others, they are added to reduce weight, reduce the quantity of plastic used, or change such properties as electrical conductivity and color.

The second type of filled composite consists of a continuous three-dimensional skeletal structure infiltrated or impregnated with a second-phase filler material. The skeleton itself may be a group of cells, honeycomb structures, or a random spongelike network of open pores. In some cases, the infiltrant can separate but bind the skeletal components (originally powders or fibers), while in others it can simply fill the voids. When made from powders, this type may also be considered a particulate composite.

The open matrix in a pore or spongelike type of skeletal composite is formed naturally during processing and can be filled with a wide range of materials, including metals, plastics, and lubricants. Metal impregnates, for instance, might be used to improve the strength of a matrix or provide better bearing properties. Examples of this type of filled composite include metal castings, powder metallurgy parts, ceramics, carbides, graphite, and foams. Another form of skeletal composite is the honeycomb structure. In this case the matrix is not formed naturally, but is specifically designed to a given shape. The typical honeycomb is a hexagonal-shape sheet material (paper, glass fabric, aluminum, etc.) that is impregnated with a resin or foam. It is then used as a core material in a sandwich constructed composite. More about this type of filled composite can be found in Section 6 of this volume.

Particle-Filled Composites. Fillers can constitute either a major or a minor part of the composition of a composite. The structure of filler particles can range from irregular masses to precise geometrical forms such as spheres, polyhedrons, or short fibers. Fillers are used for nondecorative purposes, although they may incidentally impart color or opacity to a composition. The use of fillers to modify the properties of compositions can be traced at least as far back as early Roman times when artisans used ground marble in lime plaster, frescoes, and pozzolanic mortar. Early in the 20th century, Goodrich added carbon black to rubber, and Baekeland formulated phenol-formaldehyde plastics with wood flour.

Fillers are used extensively in reinforced plastics. Typical fillers include clay, sand or silica, calcium carbonate, diatomaceous earth, alumina, calcium silicate, carbon black, and titanium dioxide. As inert additives, fillers are capable of modifying nearly any basic resin property in the direction desired, thereby overcoming many of the limitations of the basic resins. Particle size, size distribution, shape, surface treatment, and blends of particle types all affect the final composite properties.

When a plastic is combined with a filler, the two act as two separate constituents. The materials do not alloy, and, except for the necessary bonding, do not combine chemically to any significant extent. As in the case of fiber-reinforced materials, it is also important that the constituents be compatible and not degrade or destroy one another's inherent properties.

In some filled composites, the matrix provides the framework and the filler provides the desired engineering properties. For example, although the matrix usually makes up the bulk of the composite, the filler material is often used to such a large extent that it becomes the dominant material and makes a significant contribution to the over-all strength and structure of the composite.

Fillers offer a variety of benefits: increased strength and stiffness, heat resistance, heat conductivity, stability, wet strength, fabrication mobility, viscosity, abrasion resistance, and impact strength; reduced cost, shrinkage, exothermic heat, thermal-expansion coefficient, porosity, and crazing; and improved surface appearance. However, fillers also possess disadvantages: they may limit the method of fabrication, inhibit curing of certain resins, shorten the pot life of some resins, and even weaken some composites. Table 1 shows effects of commonly used fillers on resin properties.

Microspheres. Microspheres are one of the most useful fillers available to the plastics industry. When the need arises, the plastics manufacturer can take advantage of the microspheres' sphericity, controlled particle size, strength, and controlled density to modify a product without sacrificing profitability or physical properties. There are many types of microspheres, including siliceous, ceramic, glass, polymeric, and mineral.

Solid Microspheres. Solid glass microspheres, which are manufactured from A-glass, are commercially available in graded sizes from 5 to 5000 μm. The size most often used for plastics is less than 325 mesh, and the average sphere is 30 μm in diameter. Most of the solid glass microspheres marketed for use in plastics are coated with a coupling agent. This coupling agent bonds both to the sphere's surface and to the resin. It maximizes the strength of the bond between the two and essentially eliminates absorption of liquids into the separations around the spheres.

The relatively low specific gravity of solid microspheres directly influences the weight and economics of the finished product. Only in comparison with carbon at 1.8 g/cm^3 would the solid glass microspheres at 2.48 g/cm^3 represent an increase in the weight of the system.

An important aspect of solid glass microspheres is their inherent strength. Physical data on the strength of the spheres are not available, but research has shown that their inherent strength carries over to the finished molded part. In an epoxy/glass-fiber system in which a part of the epoxy was replaced with solid microspheres, both the flexural strength and flexural modulus were increased. Tests conducted on a 40 wt% microsphere-filled nylon 6/6 system show that tensile strength,

Table 1. Effects of commonly used fillers on resin properties

Type of filler[a]	Property increase							Property decrease					
	Thermal conductivity	Thermal-shock resistance	Impact resistance	Compressive strength	Arc resistance	Machinability	Electrical conductivity	Cost	Cracking	Exotherm	Coefficient of expansion	Density	Shrinkage
Bulk													
Sand	X			X				(b)		X	X		X
Silica	X			X				(b)		X	X		X
Talc						(b)		X		X			X
Clay	X			X		(b)		X	X	X	X		X
Calcium carbonate						(b)		X		X			X
Calcium sulfate (anhydrous)	X				X	(b)		X		X	X		X
Reinforcing													
Mica		X	(b)					X	X				
Asbestos		X	(b)						X				
Wollastonite	X	X	(b)					X	X				
Chopped glass		X	(b)						X				
Wood flour		X	X			X	(b)	X				(b)	
Sawdust		X	X			X	(b)	X				(b)	
Specialty													
Quartz	X		X	X				X			(b)		X
Aluminum	X	X		X						X	(b)		X
Hydrate alumina					(b)								
Li-Al silicate	X			X					X	X	(b)		X
Beryl	X			X				X	X		(b)	X	
Graphite						X	(b)						
Powder metals	X	X	X	X		X	X	X	X	X	X		X
Low-density spheres												(b)	

[a]Particle size of fillers is 74 μm (200-mesh) or finer, except for sand, hollow spheres, and reinforcing fillers that depend on particle configuration for the desired effect. [b]Denotes most significant property of each filler listed.

flexural modulus, and compressive strength are greater than for an unfilled system.

Hollow Microspheres. The silicate-base (siliceous) microspheres are manufactured at controlled densities. Commercial products are available in densities from 0.9 to 0.22 g/cm^3 and in a particle-size range from 10 to 200 μm. The average microsphere diameter ranges from 65 to 75 μm. These products are somewhat larger in size than the solid glass spheres used in plastics, and are supplied in a broader particle-size range.

Silicate-base hollow microspheres whose compositions have been modified with organic compounds are also available. The modification lowers the microsphere's sensitivity to moisture and reduces the attraction between particles. The reduced particle attraction is particularly important in highly filled liquid polymer composites where viscosity buildup limits the amount of filler loading.

When hollow microspheres were first introduced, their major application areas were limited to thermosetting resin systems. Primarily, this was a function of the low molding pressures required for thermosets. However, newer, stronger spheres are available. These spheres are up to five times stronger than the standard grades of hollow microspheres in static crush strength and four times as durable in shear. These stronger spheres are designed for certain thermoset, compression-molded plastics.

Ceramic aluminosilicate microspheres have recently been introduced for use in thermoplastic systems. These hollow microspheres are available in densities from 0.5 to 0.7 g/cm^3 and particle sizes from 5 to 300 μm. The ceramic spheres are generally higher in density and considerably stronger than siliceous microspheres. On a per pound basis, they are approximately one-third the price of silicate-base spheres. Because ceramic microspheres are strong and abrasion resistant, they are recommended for applications that require either high-shear or high-pressure conditions.

The density of available hollow glass microspheres is 0.15 to 0.38 g/cm.^{3} If a hollow microsphere with a specific gravity of 0.21 g/cm^3 is compared with the denser fillers with densities of approximately 2.4 to 2.6 g/cm,^{3} it is apparent that a factor of about 12 exists in their relative contributions to the weight of a given plastic item. For this reason, the primary use of hollow microspheres is to reduce the weight of plastics systems.

It should also be noted, however, that hollow microspheres have approximately one-fifth the density of pure resin. They can, therefore, be used to lighten even those compounds consisting mostly of resin. This is particularly significant in hydrospace, aerospace, and automotive applications where weight reduction for energy conservation is of prime importance.

Hollow microspheres do not have the crush resistance exhibited by solid spheres. For this reason, their use in systems requiring extremely high-shear mixing or high-pressure molding is prohibited. However, when used properly, hollow microspheres can improve the stiffness of polyester-laminate construction. They can also improve the impact resistance of acrylic sheet rigidized with a polyester/glass-fiber laminate. For PVC plastisol applications in which the finished part is subjected to considerable bending, the presence of hollow microspheres reduces or eliminates crazing, or crease whitening, at the bend. They can also improve compressive strength, and the crack-arresting properties of microspheres result in easier machining, sanding, and nailing of finished goods.

Microsphere Properties. Many of the advantages of solid and hollow microspheres are directly related to their spherical form. Their spherical shape allows them to act as tiny ball bearings, providing better flow properties than platelet, needle, or randomly shaped filler particles. The spherical particles also distribute stress more uniformly throughout a resin matrix, and their isotropic behavior yields better stress distribution within a finished part.

Another positive benefit of the spherical shape of microspheres is derived from their surface area. A spherical particle has a minimum ratio of surface area to volume. Therefore, the amount of viscous drag between the surface of the filler particles and the resin is smaller with spheres than with other filler shapes.

To obtain a smooth surface with a filled or reinforced system, it is necessary to have a resin-rich surface to hide the fiber or filler pattern. Microspheres, which are free of orientation and have no sharp edges, produce a smoother surface than fibers or randomly shaped fillers.

Flake Composites

Composites made from flakes represent another type of composite. Flakes are used for a wide variety of reasons and may be used in place of fibers because they can be more tightly packed. Metal flakes touching each other in a polymer matrix can conduct heat or electricity. Glass or mica flakes, in contrast, can resist heat or electricity. Flakes can also be easier and less expensive to produce than fibers. Among the disadvantages that flakes have are the quality control of the size, shape, and degree of flaws in the final product. Glass flakes, for example, may have notches or cracks around the edges which produce a weakening effect. Flakes can also be difficult to line up parallel to one another in a matrix, resulting in uneven strength.

Flake composites have several familiar uses. Aluminum flakes are used in metallic automobile paints and in molded plastics to provide decorative color effects and various degrees of transparency. Glass flakes are found in printed circuit boards for computers and calculators. Mica flakes are also used as insulation in home heaters.

Flakes may be set in a matrix or simply held together by a gluelike binder — which is also a matrix. Depending on the material's end use, the flakes can be present in a small amount or constitute almost the entire composite.

In structural applications, flakes offer several advantages over fibers. For example, as long as the flakes are parallel, flake composites can provide uniform mechanical properties in the plane of the flakes. Although properties approaching isotropic can be obtained in continuous fiber composites, an-

gle plying is required to produce nearly isotropic properties (i.e., quasi-isotropic properties).

Flake composites also have a higher theoretical modulus than fiber composites. Compared with fibers, flakes are relatively inexpensive to produce and can be handled in batch quantities. Although they are relatively easy to incorporate into composites, obtaining parallel orientation is not easy.

Particulate Composites – Inorganic

Ceramic and metal composites that consist of a microstructure that is made up of particles with one phase dispersed in the other phase, or whose microstructure is made up of two interpenetrating skeletons, are known as particle-reinforced composites.

This class of composites is distinguished from the filamentary type by the fact that the reinforcing phase has no long dimension. It may be round, square, or even triangular, but the dimensions of its sides are approximately equal. It is distinguished from dispersion-hardened materials by the size and volume concentration of the dispersoid. In particulate composites, the dispersoid size is at least several microns (and may go up to several hundred microns), and its volume concentration is greater than 25% (and usually 60 to 90%). The distinction between particulate composites and dispersion-strengthened materials, whose dispersoids are 0.01 to 0.1 µm in size and are present in volume concentrations of less than 15%, is readily apparent. A further distinction between dispersion-strengthened materials and particulate composites lies in the mechanism of strengthening. The fine dispersoid present in dispersion-strengthened materials reinforces the matrix alloy by restricting the motion of dislocations, requiring larger stresses to break through or fracture the roadblock imposed by the dispersion. The particles of particulate composites, on the other hand, strengthen by hydrostatically constraining the fillers of matrix between them as well as by their inherent hardness relative to the matrix phase.[3]

The three-dimensional reinforcement in a particulate composite can lead to isotropic properties, since the material is symmetrical across the three orthogonal planes. Because the particulate composite, however, is not homogeneous, the material properties are sensitive not only to the constituent properties, but also to the interfacial properties and geometric shapes of the array. The strength of the composite normally depends on the diameter of the particles, the interparticle spacing, and the volume fraction of the reinforcement. Matrix properties are also important.

Cermets. Cermets are one important class of particle-strengthened composites. A cermet consists of ceramic grains (greater than 1 µm), either borides, carbides, or oxides (nitrides, carbonitrides, and silicides are sometimes included), dispersed in a refractory ductile metal matrix, which accounts for 25 to 90% of the total volume. The bonding between the metal and ceramic constituents results from a small amount of mutual or partial solubility. Metal oxide systems, however, exhibit poor bonding between phases and require additions to serve as bonding agents. Structures made from cermets are produced by powder metallurgy (PM) techniques. The parts have a wide range of potential properties, depending on the compositions and relative volumes of the metal and ceramic constituents. Another method of producing cermets involves impregnating a porous ceramic structure with a metallic matrix binder (skeletal composite). Cermets can also be used in powder form as coatings. The powdered mixture is sprayed through an acetylene flame, and fuses to a base material.

Although a great variety of cermets have been produced on a small scale, only a few types have significant commercial use. These fall into two main groups: oxide-base and carbide-base cermets. The most common type of oxide-base cermet contains aluminum oxide ceramic particles (ranging from 30 to 70 vol %) and a chromium or chromium alloy matrix. In general, oxide-base cermets have specific gravities between 4.5 and 9.0, and tensile strengths ranging from 145 to 269 MPa (21 to 39 ksi). Tungsten, chromium, and titanium carbides comprise the three major types of carbide-base cermets. Each one of these groups is made up of a variety of compositional types or grades. Tungsten carbide cermets (Fig. 6a) contain up to about 30% cobalt as the matrix binder and are the heaviest type of cermet (specific gravity of 11 to 15). Titanium carbide cermets (Fig. 6b) have nickel or nickel alloys as the metallic matrix, which results in high-temperature resistance. Cermets in this group have relatively low density combined with high stiffness and strength at high temperatures (above 2200 °C).

Laminar Composites or Laminates

There are as many possible laminar composites as there are possible combinations of materials. In the immediately ensuing pages, some types of laminates will be mentioned briefly and described to some extent to help illustrate the wide versatility that a materials designer may have in thinking "composites." However, the greatest emphasis in the presentations of properties from the literature will be on structural laminates or laminates with useful mechanical properties.

For this book, laminates or laminar composites will be defined as materials consisting of two or more layers of monolithic materials bonded together; clad and precoated materials will also, in some cases, be included. Excluded will be information on what may be termed "composite structures." Such structures as honeycombs enclosed between sheet materials and syntactic foams or conventional foams centered between layers of sheet materials are believed to be structures. Admittedly, defining such "materials" as "structures" is arbitrary, but definitions of materials that are too inclusive are not very useful. However, more information on this form of composite may be found in Section 6 of this book. Another exclusion from the category of laminates or laminar composites is materials made by layering of tapes or sheets of fiber-reinforced plastics or other matrix composites. Such layering is termed "plying," "cross plying," and "angle plying," and is a means of producing fiber-reinforced composites with designed-in properties.

Metal/Metal Laminates

Metal/metal laminates may be better understood with the following letter designations. Consider first that letters A, B, C, etc., represent different metals stacked or bonded together to form laminar composites. Thus some of many possible stacking combinations follow:

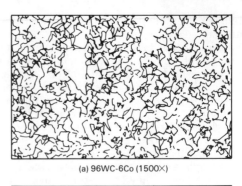

(a) 96WC-6Co (1500×)

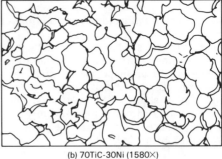

(b) 70TiC-30Ni (1580×)

Fig. 6. Metal/ceramic composites (illustrated from a micrograph)

AB	A could represent a clad or protective layer, and B the base material
ABA	Same as above; or this structure could be considered a sandwich; or a clad material
ABABABA	Typical of a high-strength composite where A is strong and B is ductile, for example; or where A is weak but corrosion resistant and B is corrosion poor
ABCABCABCBACBACBA	Just another possibility where there are three different metals
ADBDADBDA	Where A and B are separated by a diffusion barrier D, or a bonding agent, depending on the case

Laminar clad and sandwich materials have been successfully used for many applications. It is surprising, though, that the multilayer approach has not been exploited more than it has. Such materials have been shown to obey the rule of mixtures from a strength and modulus standpoint and, in addition, have several intrinsic advantages relative to metal/metal fiber composites.

First of all, it should be noted that the laminae can be made from controlled-thickness materials. For example, very thin high-strength foils can be used as part of a laminar composite, as can sheet materials. Powders or sprayed metals can also be layered and formed into laminar products.

Fabrication of different alloys of sheet, foil, powder, or sprayed materials can be accomplished by a wide variety of metallurgical and powder metallurgical processes such as diffusion bonding, hot pressing, roll bonding, liquid phase sintering, brazing, etc., depending on the nature of the materials being joined together. In general, a given alloy cannot be made as strong in sheet or foil form as in fiber or wire form. However, a composite made from foil or sheet can be made isotropic in two dimensions far more readily than can fiber composites. Also, in some cases, foil or sheet laminar composites can be made to have very high percentages of the phase that contributes the desired main properties to a composite. For example, a strong sheet may be used in percentages of over 90% in a laminar structure, whereas it is very

Table 2. Composite metal combinations[a]

Cladding	Aluminum and alloys	Beryllium copper	Brass	Bronze	Copper	Cupro-nickel	Gold	Gold-nickel alloy	Indium	Fe/Fe-Ni alloys	Magnesium	Molybdenum	Nickel	Ni-Fe alloys	Ni-Cu alloys	Ni-Cr-Fe alloys	Nickel-silver	Platinum	Silver	Stainless steels	Steel, carbon	Tantalum	Titanium	Tungsten	Zinc	Zirconium
Aluminum	X	X	X	X	X		X		X	X	X	X	X	X				X	X		X	X	X	X	X	
Aluminum alloys	X				X					X		X						X	X						X	
Bismuth alloys					X																					
Brass	X																				X					
Bronze					X																					
Cadmium	X				X					X									X							
Calcium													X						X		X					
Copper and alloys	X									X	X	X	X	X	X	X			X	X	X	X				
Gold and alloys	X	X	X	X	X					X						X	X	X	X			X	X	X		
Indium	X				X					X		X	X	X								X	X			
Fe/Fe-Ni low-expansion alloys	X		X	X	X					X																
Lead and alloys	X		X	X	X					X		X	X	X					X		X	X				
Magnesium										X											X					
Nickel and alloys	X		X		X								X			X			X	X				X		
Palladium and alloys		X	X	X	X	X							X	X	X	X			X	X	X					X
Platinum and alloys		X	X	X		X				X			X	X	X	X			X	X	X					X
Silver and alloys	X	X	X	X	X	X				X	X	X	X	X	X	X		X	X	X	X	X	X	X	X	X
Stainless steels	X				X																X					
Steel, low-carbon	X				X					X																
Tantalum					X												X									
Tin and alloys	X	X	X	X	X	X	X			X		X	X	X			X		X	X	X	X	X			
Titanium					X															X	X					
Zinc	X							X																		

[a]Courtesy of *Stamping/Diemaking*.

difficult to make fiber composites containing more than 70% strong fibers.

There are two major functional categories of metal/metal laminates other than those that are intended to have high strength or stiffness: (1) those with one layer or face that is decorative, and (2) those with one or more surface layers that give the composite special properties, or that lessen the cost of the material. For the above types of materials there are two major methods of production: precoating and cladding.

In precoated metals, the face is formed by building up the second constituent on a substrate to form a thin, essentially continuous film. This is usually done by electroplating or hot dipping, although chemical plating is also used. Clad metals, however, are faced in the fashion of a solid wrought material and are suitable for more severe environments where a thicker face is needed. Rolling, hot pressing, casting, extruding, brazing, and welding are used as cladding methods.

Many electrical contact materials are made from silver, platinum, and platinum-group metals clad to copper, bronze, and nickel. Clad metals are also used where heat reflectivity or radiation shielding is desired. Other interesting applications include those where more than one special property has been gained through the combination of the materials—e.g., heat transfer and corrosion resistance in stainless-clad copper or aluminum for pots and pans. Table 2 shows composite metal combinations.

Metal/Plastic Laminates

There are many combinations of sheet or foil that can be combined for low-temperature applications by using adhesives. Such materials (either metals or plastics) with letter designations A, B, C, etc., could be stacked and glued together with G as shown below. Under some circumstances, G could be replaced by P for plastics. Thus, a few of many possible composites, such as the following, could be made:

AGBGCGBGA	This would be a glued-together laminate that could have high tensile strength or high stiffness (modulus). This would be one possibility of many.
APAPAPA	This would be a metal/plastic laminate; it would be one of many possibilities.
PA	This could represent a metal coated or covered with a protective or decorative plastic layer.

The best-known metal/organic laminate is probably prefinished or prepainted metal, whose primary advantage is the elimination of final finishing by the user. Another is the metal/plastic laminate. Although there are many possible combinations of solid organic films and metals, metal/plastic combinations are probably the most familiar. Vinyl/metal laminates probably account for 90 to 95% of all metal/plastic laminates now used. These laminates are commonly made by adhesive bonding of preprocessed vinyl sheet to metal. Other metal/organic laminates include such items as jacketed wire, metal-faced plywood, and a great variety of rubber/metal combinations, such as rubber-covered steel rolls, rubber-lined tanks, etc.

Metal/Ceramic Laminates

Since ceramics are most useful at high temperatures, one of the most useful composites of this type is a metal sheet coated on one or two sides with an oxidation-resistant or corrosion-resistant layer. Glass-coated hot water tanks, ceramic-lined mufflers, and thermal barrier layers on superalloy components of gas turbines are typical examples. Porcelain enameled steel or copper are composites of the type under consideration, but are not part of this survey. Coated and many-layered composites where A is a metal and C is a ceramic, or inorganic material, are schematically indicated below:

CA	Where C is a ceramic or inorganic coat
CAC	Where C is a ceramic or inorganic coat
CACACAC	A multilayered composite (an example)
ACACACA	A multilayered composite (an example)

The above types of composites — again, a few of many possible combinations of materials — would have to be made by methods suitable for the specific combinations of materials desired. In addition to some of the fabrication methods noted for the metal/metal types of composites, some of the ceramic/metal composites could be made by spraying or dipping metal sheets in slurries of some inorganics or ceramics. These would then be given appropriate thermal treatments to densify the inorganic.

Other Laminates

Other possible combinations of materials that form laminates might include organic/organic (including plastic/plastic), ceramic/plastic, or ceramic/ceramic laminates. The most common type of ceramic/ceramic laminate would be made of glass layers. Structural glass/glass laminates, for example, find wide applications as partitions, table tops, countertops, and signs. These laminates are also used for special lenses. Laminated safety glass is a type of ceramic/plastic laminate. Safety glass is generally made from two or more layers of glass sheet and one or more layers of plastic, although sometimes wood or fabric is used. Bulletproof glass is a composite structure consisting of multiple layers of plate glass bonded together with alternate layers of vinyl to provide even greater thickness and energy absorption. Another application for glass/plastic laminates is light filters made from colored glasses with interposed transparent plastic layers.

One other category of laminates would be organic/organic. This group includes laminated glued wood, plastic-faced wood, laminated paper, and rubber fabric industrial belting, to name a few. One important class of plastic/organic laminates includes high-temperature thermosetting plastics. These plastic laminates, consisting of layers of resin and impregnated paper, fabric, or mat, are used for a wide variety of mechanical and electrical applications.

References

1. J. D. Walton, Jr., and W. J. Corbett, "Metal Fiber Reinforced Ceramics," Fiber Composite Materials, Papers presented at the American Society for Metals, Metals Park, OH: ASM, October 17, 18, 1964
2. *Composite Materials Handbook*, Mel M. Schwartz; New York: McGraw-Hill, 1984
3. *Modern Composite Materials*, L. J. Broutman, R. H. Krock, eds.; Addison-Wesley Publishing Company, Reading, MA, 1967

Additional Sources of This Article

Advanced Fibers and Composites for Elevated Temperatures, I. Ahmad and B. R. Noton, eds.; New York: AIME, 1980

Inorganic Fibres and Composite Materials, P. Bracke, H. Schurmans, and J. Verhoest; Oxford: Pergamon, 1984

Encyclopedia of Composite Materials and Components, Martin Grayson, ed.; New York: John Wiley, 1983

Summary of the Eighth Refractory Composites Working Group Meeting: Vol III, prepared by Lt. D. R. James and L. N. Hjelm; Dayton: Wright-Patterson AFB, January 1964

Composite Materials: Vol 4–Metal Matrix Composites, L. J. Broutman and R. H. Krock; K. G. Kreider, ed.; New York: Academic Press, 1974

Fabrication of Composite Materials: Source Book, Mel M. Schwartz, ed.; Metals Park, OH: ASM, 1985

Composites: State of the Art, John W. Weeton and E. Scala, eds.; New York: AIME, 1974

Machine Design, February 10, 1969, pp 141-156; "Design Guide: Fiber-Metal Matrix Composites," by John W. Weeton

P.Q. Corporation Technical Bulletin PA-101

J. P. Halpin, "Primer on Composite Materials: Analysis," Technomic Publishing Co., Lancaster, PA, 1984

SECTION 2
Economic Outlook for Composites and Reinforcements

Introduction

The origins of commercial markets for composite materials vary widely from one material to another. The Society of the Plastics Industry announces, in 1986, that fiber-reinforced plastics are 41 years old, dating from the first widely known use of a reinforced plastic in a commercial application — a semistructural, tubular, glass-reinforced phenolic-nylon fishing rod. The metal-matrix specialists report a 20-year history, starting with the initial development by the aeronautical and nuclear industries of lightweight, high-temperature-resistant composite materials and corrosion-resistant refractory materials. A survey of known materials reports the existence of ceramic-matrix composites as early as 1927, consisting of an alumina matrix reinforced with oxidation-resistant metal wires. As impressive as this sounds for ceramics, no further developments occurred until the 1960's.

Each composite material's status in the marketplace is defined by a unique set of characteristics. Industry age, material cost and availability, demands by government and consumers for new or enhanced products, and current technology all affect present and future markets for composites. The following discussion will examine these areas for polymer-, metal-, and ceramic-matrix composites. Reinforcement materials applicable to the production of composites will also be investigated. Asphalt, concrete, and other composite specialties are excluded—not because of a lack of importance or impact of these materials on the marketplace, but simply because their treatment does not lie within the scope of this book.

Polymer-Matrix Composites

Because plastic composites are a much more mature industry than ceramic- or metal-matrix composites, their economic growth forecasts are much more detailed and plentiful. Various sources report different figures, but the main consensus is one of strong economic growth. This is in contrast to the market decline following the 1979/1980 oil crisis. The market for nearly all major petrochemical products, especially plastics, became depressed because of high prices for energy and raw materials and the subsequent worldwide economic recession. While there still may be some problems for the raw-material-producing segment of the plastics industry, with offshore chemical and plastics plants still not up to capacity, the outlook for plastics is promising. Recent announcements of phenomenal decreases in crude-oil prices should ensure an extremely bright economic outlook, at least in the near-term.

The major industries that support the reinforced plastics market are land transporation, construction, corrosion-resistant equipment and applications, marine and marine accessories, electrical and electronic consumer products, appliances and business equipment, aircraft/aerospace/military, and the specialties markets. The reinforced plastics/composites industry reports 1985 total sales of more than 998,000 kg (2.2 billion pounds)—an increase of nearly 3% over the 976,600 kg (2.153 billion pounds) sold in 1984. The many previous years of developing new resins, reinforcements, and processing methods and machinery are beginning to pay off for the industry, with record sales in the last couple of years.

Of the more than $2 billion that comprises the total reinforced plastics market, about half is accounted for by advanced polymer composites. Advanced composites are generally understood to be high-performance materials consisting of a polymer matrix reinforced with S-glass, carbon/graphite, boron, aramid, or other developmental fiber. The fibers in advanced composites are often continuous and oriented in the matrices. The advanced composites field is probably the fastest-growing and most exciting segment of the fiber-reinforced plastics market. It is projected to grow 16% annually through the year 2000, reaching an estimated $12 billion. An annual growth of 22% is projected for the aerospace industry, as well as a healthy 10% for industrial applications. Although this is one of the more optimistic reports, other analysts concur that at least a 10% annual growth rate through 1995 may be expected.

The majority of advanced composites—presently 99%—use one of three commercial fibers: S-glass, carbon/graphite, or aramid. They are also primarily made from epoxy resins. This is expected to change over the next 15 years with the development of new macromolecular materials made into fibers and the increased use of thermoplastic matrices. Currently, epoxy is used in approximately 80% of the advanced composites, while the remaining thermosets and thermoplastics each account for about 10% of the market. Although the thermoset resins will maintain a dominant position in this market, an increased market share for thermoplastic resins is expected.

Some applications for advanced polymer composites

Space industry	Antenna parts, carriers for solar cells, robot arms
Aviation	Fairings, ailerons, spoilers, rudders, stabilizers, blades, cabin interiors
Automobile construction	Drive shafts, connecting rods, piston pins, motor fastening, coach springs
Leisure time	Skis, fishing rods, sailboat and surfboard masts, golf clubs, tennis racquets

One of the main reasons for the expected growth of thermoplastic matrices in advanced and traditional fiber-reinforced plastics is their ease of fabrication. Because they are processed by melting and solidification, they can be processed at least ten times faster than the thermoset systems, which must be polymerized and cured as part of the fabrication cycle. The combination of strength and toughness, improved low-temperature impact properties, reduced wear, lower mold shrinkage, and high tensile and flexural properties are the most-mentioned advantages of reinforced thermoplastics. Polyetheretherketone (PEEK), polyphenylene sulfide, and polysulfones in particular are expected to penetrate the thermoset market.

Today, the major matrix polymer is epoxy resin. Combined with carbon fibers, it produces an advanced composite that displays an impressive spectrum of properties. Its stiffness, tensile strength, and resistance to vibration and wear are superior to those of steel. These properties, along with low weight, make this material particularly attractive for the aircraft and aerospace industries. It has been estimated that every kilogram of weight reduction in an aircraft means lower fuel consumption or higher load capacity, and thus this material could represent eventual cost reductions.

The auto industry is expected to have a strong impact on the increased demand for epoxy-matrix composites and thermosets in general. Exterior body panels and other related auto components are predicted to be the fastest-growing applications of engineering plastics over the next five years. Reinforced thermosets, a part of this market, are forecast to grow at a faster rate than auto production due to the increases in applications. In fact, 50% of the growth will come from new applications such as closure panels, springs, and bumper systems. Another change for the industry will be the strong growth of reinforced reaction-injection molding (RRIM) as a manufacturing process. Although compression molding will remain the dominant process, constituting around 70% of production processes, RRIM is expected to increase from an estimated 203.2 kg (20 million lb) in 1985 to more than 812.8 kg (80 million lb) by 1990.

Epoxy composite materials are also being studied as potential tooling materials in the auto industry. General Motors Corporation is conducting these studies for the Saturn project as well as other possible new car programs. So far, impressive results have been reported. These composites, consisting of an epoxy resin filled with either 61 or 70% aluminum, have been made into reaction-injection molding, pressure molding, injection molding, and drop hammer tools. In three months of testing, the GM engineers poured 210 die sets at total labor and material costs of $1.5 million, compared with an $8 million outlay during 12 months for the same number of tools produced with hard-tooling methods. Whether this will remain a prototype program or become part of the GM manufacturing process is not known at this time, but the long-term outlook for another new application for thermosetting composites in the auto industry appears to be excellent.

The market for FRP products in the construction industry has withstood the rise and fall of the housing crisis over the last several years, mainly because of the introduction of new products and applications of reinforced plastics. Tub/shower and paneling areas have always represented the majority of the construction market in plastics. New products, however, are opening new markets for reinforced plastics. New products include cushioned bathtubs, pultruded sashes and frames, reinforced plastic doors with one-piece thermoplastic door sills that have the look of wood, and structural applications of C-glass-reinforced vinyl ester (with an outer gel coat) molded into concrete-type rebars. The obvious advantages of these products are resistance to corrosion, lower weight, and electrical nonconductivity.

The electrical/electronics segment of the reinforced plastics industry is also contributing to the recent increase in the production of plastics. One new application that has been seen as a potential growth area is the consumer satellite dish market. Several new reinforced thermoplastic composites have been introduced in the last several years with the needs of this market in mind. One such formulation is a reinforced polycarbonate, which has been put on the market for electromagnetic interference (EMI) shielded parts. This once very promising satellite-dish market is being threatened, however, by the scrambling of television transmissions by pay-TV companies such as Home Box Office, resulting in less access to programming for owners without the necessary expensive "descrambling" devices. It will be some time before the effects of these actions on this market will be known.

Pultruded reinforced plastics have been serving the electrical industry for years through products such as light poles, electrical conduit, bus-bar insulation, and third-rail protective coverboards. Other electrical applications include a new reinforced plastic extension ladder with both rungs and rails made of plastic, replacing the previously used "composite" ladders of aluminum rungs and plastic rails. Wind turbine blades also show promise and should ultimately replace aluminum blades. These blades are pultruded from glass-mat and roving reinforced vinyl ester, and are reported to feature advantages such as extended service life, increased aerodynamic characteristics, and a 20% electrical output.

The reinforced plastic composites market projected to grow the fastest, particularly in the advanced composites segment, is the aircraft and aerospace market. Most of the new innovations are, however, being advanced by the military/aerospace segment rather than the commercial aircraft industry. Little, in fact, is being done in the way of design or development of new commercial planes, even though some forecasts predict that plastic composites will comprise up to 65% of the material composition of planes within the next decade. In contrast, however, it is the military planes that are using virtually every new technology in composites. Although this market represents one of the smallest segments, it also represents some of the most sophisticated and highly engineered applications in the industry. It is truly on the leading edge of technology in material development.

Because much of this activity is developmental, poundage shipments to the aerospace market are not being felt in any great magnitude. The ultimate dollar figures will depend on how well some military contracts move along now and into the future. The new technology that is being developed, however, will eventually be felt in the commercial markets for many years to come.

Compared with other composites, fiber-reinforced plastics constitute a relatively mature industry. Through increased new applications and the new technology of the advanced polymer composites, their market share continues to grow. As the automotive, aerospace, and many other industries expand their uses for plastics, we should expect to see more and more of these materials.

Metal-Matrix Composites

Among the many reasons for the recent resurgence of interest in metal-matrix composites (MMC's) are the engineering properties of these composites. MMC's are light in weight and have good stiffness and strength, low density, good high-temperature capabilities, and low thermal expansion. Indications are that these materials will provide up to 60% savings in weight while still retaining key properties. Another reason is the apparent temperature and strength limitations of other composite materials, namely the polymer-matrix materials, which are sensitive to moisture and which in some cases also outgas or release moisture. These are serious problems in space applications. Other important advantages that MMC's have over polymer-matrix materials include greater transverse stiffness and strength, greater thermal stability, and, in some cases, improved fracture toughness and ductility, greater radiation resistance, and improved high-temperature performance.

Cost is the major concern regarding future commercial applications for many MMC's. Over the past 20 years, the MMC's have undergone rapid development where cost was secondary and high strength and minimum weight were critical. Recently, reductions in raw-material and processing/fabricating costs, and the desirability of the special properties of MMC's, have revived interest.

The special properties of metal-matrix composites can sometimes offset extremely high material costs. For example, a significant weight reduction in an aircraft engine shaft can have many cumulative engineering effects. Use of the lower-weight metal composite can increase load-bearing capabilities, reduce wear, reduce lubrication needs, and also increase output speed, all of which result in cost savings.

Not all MMC materials are affected by the problems of high cost. Some manufacturers using particulate and discontinuous fiber materials report that their costs are already competitive with those of conventional materials. The success of these products could make metal composites standard materials for many industrial applications. One such application might be high-speed machinery. The combination of great stiffness and strength, along with low density, can significantly increase the productivity of the machines. If costs are as competitive as claimed, this will be a prime market for this particular type of MMC.

Although a considerable number of companies are engaged in MMC activities, especially in conjunction with the aerospace industry, there are basically only about five major commercial producers of MMC's in the United States: DWA, Amercon, Materials Concept, Avco, and Arco Metals. These companies also are prime contractors to the aerospace industry and are still not totally commercially independent of government support. Government funding through military and aerospace programs has contributed greatly to these companies for materials testing, prototype development, and full-scale production of component systems. For example, silicon carbide development is currently sponsored by numerous government agencies through private industry.

A report by Du Pont tells of growing interest in its developmental fiber FP. This is a sintered aluminum oxide fiber originally developed for aluminum-matrix composites, but also used in lead and magnesium matrices. Applications for aluminum- and magnesium-matrix composites are of particular interest to the automotive industry. Costs are still too high, but potential products include pistons, wrist pins, connecting rods, brake calipers, transmission housings, gearshift and clutch components, and differential housings.

Similarly, Honda Motor Co., Ltd., of Japan, has reported success in making connecting rods on a mass-production basis from fiber-reinforced metals. The company has announced that it will use these rods in the production of small passenger cars. By making the connecting rods out of fiber-

reinforced light alloys of stainless steel or aluminum, fuel efficiency and engine output will be improved because of the 30% weight reduction. Another new example of MMC development is the recent announcement by a Japanese manufacturer that it can produce pistons from aluminum reinforced with alumina silica fibers at a price 15% below that of domestically produced conventional pistons.

Besides aerospace/military and automotive applications, MMC composites have potential applications in the medical and sports-equipment industries. Table 1 is a list of possible MMC commercial products and suggested composite systems.

As more and more corporate giants expand their development of technological capabilities, prices should continue to decline. One industry leader is already predicting missile-application cost reductions of anywhere from 25 to 300% in the not too distant future. Due to new automated processes and increased demands for products, other MMC companies are also reporting substantial, if not totally revolutionary, price reductions. In addition, the special properties (high specific strength and modulus) of most MMC's make them attractive to designers. With the promise of cost reductions, aided by large metal resources, increasing automation, and rapidly advancing technology, future commercial development of MMC's seems assured.

Ceramic-Matrix Composites

Overcoming brittleness has been a major stumbling block in the commercial development of new ceramic products and applications. Now, with new technology and the demand for high-temperature-performance products, the advanced ceramics market is growing dramatically.

The Japanese, with low natural supplies of metal and energy sources, are aggressively pursuing, as they call it, the fine ceramics market. They are currently ahead of the United States in terms of actual structural products on the market, but are still lagging in technology.

A number of important technological barriers need to be overcome before advanced ceramics can achieve their full potential. These lie in the areas of production costs, reliability in service, and reproducibility in manufacture. Some of these barriers are currently retarding commercialization (e.g., ceramic heat engines). In other cases, commercialization has occurred but the aforementioned barriers are limiting market penetration. The severity of these technological barriers, and the likely timing of their removal, vary considerably among the different ceramic application areas. This is indeed the case relative to the importance of cost, reliability and reproducibility for the different areas.

Typical established ceramic industries compared with new ceramic industries

Established and evolving	New and rapidly changing
Refractories	Structural heat-engine ceramics
Whitewares	Wear-resistant specialty ceramics
Electrical ceramics	Electronic ceramics
Flat and container glass	Optical communication glass
Cement and concrete	Composite involving cement
Mineral resources	Synthetic powders and fibers
Enameled metals	Ceramic fiber reinforced metals
Co-WC cutting tools	Ceramic cutting tools
Ceramic nuclear fuels	Nuclear waste disposal ceramics

Table 1. Potential commercial applications of metal-matrix composites
(From *Commercial Opportunities for Advanced Composites,* edited by A. A. Watts, STP 704, ASTM, Philadelphia, 1980, p 114–115)

Application	Desired properties	Suggested composite systems
Aerospace		
Space structures	Light weight, stiffness, or resistance	B/Al, B/Mg, Gr/Mg
Antennae	Light weight, stiffness, or resistance	B/Al, B/Mg, Gr/Mg
Aircraft		
Airplanes:		
Pylons	Light weight, stiffness, heat resistance	B/Al, SiC[a]/Al
Struts	Light weight, stiffness, strength	B/Al, SiC[a]/Al
Fairings	Light weight, stiffness	B/Al, SiC[a]/Al[a], Gr/Al
Access doors	Light weight, stiffness, strength	B/Al, SiC[a]/Al
Wing boxes	Light weight, stiffness, strength	B/Al, SiC[a]/Al
Frames	Light weight, stiffness, strength	B/Al, SiC[a]/Al, Gr/Al
Stiffeners	Light weight, stiffness, strength	B/Al, SiC[a]/Al, Gr/Al
Floor beams	Light weight, stiffness, strength	B/Al, SiC[a]/Al, Gr/Al
Fan and compressor blades	Strength, stiffness, heat resistance, impact resistance	B/Al, SiC[a]/Al, Gr/Al
Turbine blades	Strength, stiffness, heat resistance, impact resistance	Tungsten or tantalum fiber-reinforced superalloys. Directionally solidified eutectics Ni_3Al-Ni_3Cb, Ni_3Al-Ni_3Cb-Ni, Ni-Mo wire
Helicopters:		
Transmission cases	Light weight, stiffness, strength	Al_2O_3/Mg, Gr/Al, Gr/Mg, Al_2O_3/Al
Truss structures	Light weight, strength, stiffness	B/Al, SiC[a]/Al, Al_2O_3/Al
Swash plates	Light weight, strength, stiffness	Al_2O_3/Al, SiC[a]/Al
Push rods	Light weight, strength, stiffness	SiC[a]/Al, B/Al
Trailing edges of tail rotor blades	Light weight, stiffness, strength	Gr/Al, SiC[a]/Al
Landing-gear steps		SiC[b]/Al, Al_2O_3/Al
Automotive		
Engine blocks	Light weight, heat resistance, strength, stiffness	SiC[b]/Al
Push rods	Light weight, heat resistance, strength, stiffness	SiC[a]/Al, B/Al
Frames	Light weight, strength, stiffness	SiC[b]/Al
Piston rods	Light weight, strength, stiffness	SiC[b]/Al
Battery plates	Stiffness	Gr/Pb
Electrical		
Motor brushes	Electrical conductivity, wear resistance	Gr/Cu
Cable, electrical contacts	Electrical conductivity, strength	Gr/Cu
Utility battery plates	Stiffness, strength, corrosion resistance	Al_2O_3/Pb, Gr/Pb, fiberglass/Pb
Medical		
Prostheses	Light weight, stiffness, strength	B/Al, SiC[a]/Al
Wheel chairs	Light weight, stiffness, strength	B/Al, SiC[a]/Al
Orthofies	Light weight, stiffness, strength	B/Al, SiC[a]/Al
Sports Equipment		
Tennis racquets	Light weight, stiffness, strength	B/Al, Gr/Al, SiC[b]/Al
Ski poles	Light weight, stiffness, strength	B/Al, Gr/Al, SiC[b]/Al
Skis	Light weight, stiffness, strength	B/Al, Gr/Al, SiC[b]/Al
Fishing rods	Light weight, strength, flexibility	B/Al, Gr/Al, SiC[b]/Al
Golf clubs	Light weight, strength, flexibility	B/Al, Gr/Al, SiC[b]/Al
Bicycle frames	Light weight, strength, stiffness	B/Al, Gr/Al, SiC[b]/Al
Motorcycle frames	Light weight, strength, stiffness	B/Al, Gr/Al, SiC[b]/Al
Textile industry		
Shuttles	Light weight, wear resistance	B/Al, Gr/Al, SiC[a]/Al
Other		
Bearings	. . .	Gr/Pb
Chemical process equipment	. . .	Al_2O_3/Pb
Abrasive tools	. . .	B/Al_2O_3, SiC/Al_2O_3

[a]SiC whisker and/or continuous SiC filament.
[b]SiC whiskers only.

A segment of the advanced ceramics market is ceramic composites. Cermets, an established but still evolving field, have been stimulated in the last few years by increased automobile production and the discovery that cermets have greater performance characteristics than traditional ceramic cutting tools. The other segment, fiber-reinforced ceramic-matrix composites, is part of the evolving ceramics industry and includes ceramic-ceramic, carbon-carbon, and whisker- and fiber-reinforced ceramic composites, to mention a few.

As already noted, the automotive industry is contributing to the changing ceramic industry. Influencing cermet production through end-use of cutting tool products, the industry is also having an impact through its need for advanced structural parts and designs. The automotive industry continues to seek greater and greater fuel savings, principally through the weight reduction of cars. Ceramic composite engines offer the potential of weight savings by allowing full combustion at higher temperatures than metals, and auto designs that do not need cooling systems (radiators, pumps, fans). Some of this latter savings is taken up, however, by the need for other design considerations. In order for the auto industry to take advantage of the up to 30% fuel savings that can be realized by using boron and carbon fiber-reinforced ceramic materials, the makers must first invest in new equipment and remodel production lines. Automakers are already seeking sources of supply from abroad.

New investments in technological developments are opening the future for other commercial ceramic composite opportunities. For example, a recent development has been reported by Showa Denko and Tateho Chemical Industry of Japan. They have jointly developed a new ceramic composite coating process that greatly improves thermal-shock and peeling resistance. The material is prepared by the addition of whiskers to a few percent by weight of alumina or zirconia ceramic material. The mixed ceramic substance is then plasma-spray coated on a substrate to form a reinforcing oxide coating. By this method, there is a two- to three-fold improvement in thermal-shock resistance over that of conventional ceramic spray coatings, while adhesion is improved by 130 to 140%. As a result, the coating can be applied directly to engine parts and other heat-resistant and high-temperature corrosion-resistant parts.

Carbon-carbon composites are probably the most highly developed of the ceramic-ceramic composites. These materials were originally developed for use as brake materials for the aerospace industry. The composites are generally produced by impregnating shaped carbon or graphite fibers with a carbonized precursor which is subsequently pyrolyzed, or by chemical vapor deposition of carbon or graphite. The resulting materials are strong and lightweight, and have excellent high-temperature properties, good corrosion resistance, and good thermal-shock, electrical, and wear characteristics. Their ease of forming makes them useful for widespread applications.

One of the exciting new areas for carbon-carbon composites is the biomedical field. Carbon fibers are not rejected by the human body in blood, bone, or soft tissue. In addition, tissue grows around the material within a few weeks after implantation. Another promising application is the use of carbon-carbon composites as bone plates. The material has already been used successfully in human arms and legs.

The mass production, however, of carbon-carbon composite structures is not expected in the near future. Developments will probably be more extensive in the area of fabrication of multidirectional structures in direct relationship to specific applications.

Carbon-carbon composite applications

Industrial and commercial applications	Medical applications	Aerospace applications
Thermal insulation	Internal bone plates	Brakes
Electrodes	Prostheses	Re-entry systems
Bearings	Implants	Rocket nozzles
Brake linings, disks		Turbine blades
Nuclear reactors		Rocket exhaust systems
Seals		Heat shields
Screws and gaskets		
Pistons for engines		
Foundry molds		
Gas turbines		

Another area of commercial interest for ceramic composites is fiber-reinforced glass-matrix composites. Information was first published in the early 1970's about the reinforcing of low-modulus glass with high-strength ceramic fibers. It was found that boron, silicon carbide, or carbon or graphite fibers in a glass matrix would produce extremely tough materials. Some of the most advanced work to date has been performed by United Technologies. New silicon carbide fibers available from such companies as Avco and Yajima have been placed in glass matrices to produce composites with high levels of mechanical performance and excellent oxidation resistance.

The present methods of production, however, still present some major hurdles that need to be overcome. The current hot-pressing method is expensive and shape limiting. Other problems exist with matrix cracking due to oxidation of the glass matrix. Important specialty uses, however, of aerospace related products, internal combustion engines, and armor; as well as some domestic cookware, should help these materials find a place in the advanced structural ceramics market.

The new ceramics industry, including fiber-reinforced ceramic composites, represents a commercially important set of materials. It is obvious that market growth depends primarily on new technological developments. Currently the industry is producing specialized, rather than general-purpose, materials. Production is multikind and small-lot, and products are of high added value. Over the long run, other considerations arise. As ceramic technology advances, giving rise to products with higher performance and longer life, the value of these products per unit weight will increase. This will make shipping over long distances economically feasible and will increasingly make ceramic industries subject to worldwide competition. Because there is such a gap at present between the technologies of the new ceramics and the already established materials, it is difficult to forecast the impact that these new products will have. It is believed that, as adaptation of traditional products occurs, the new ceramics will have a significant impact on the marketplace.

Because of the intense competition among different families of new ceramics, between ceramics and other materials, and among firms in many developed countries competing for market share, it is difficult to predict who will emerge as the leaders in the new advanced ceramic fields. As stated previously, the Japanese, with a large abundance of natural resources and the desire to develop products that are not dependent on foreign resources, are currently in the lead. The U.S. position in the international market is favored by its ability to gain the necessary technological breakthroughs, but the low labor costs of developing countries suggest that joint ventures might be the more favorable approach to commercial development of ceramics composites production.

Reinforcements – Fibers and Fillers

Having already discussed the various economic aspects of polymer-, metal-, and ceramic-matrix composites, it would seem academic to point out that as the markets for these products grow, so will the need for many reinforcements. Many of today's composites, disregarding particulate and in-situ composites, are fabricated with fibers or fillers. Some of these reinforcements are already mature products—i.e., fiberglass, carbonates, clays, and talcs—but still new technological breakthroughs are increasing the uses for these old workhorses, or, in many cases, making them easier to handle. Silicon carbide, boron, and FP alumina fibers are infants in the commercial arena, but as metal- and ceramic-matrix composites grow, so will the production of these reinforcements.

The big producers are still adding to their existing lines of glass fibers, and new producers are still entering the marketplace. New chopped strand grades of fibers are in development for polypropylene and other high-temperature resins such as modified polyphenylene oxide (PPO) and polysulfone. Additional new fibers for thermoset injection molding compounds, which offer improved impact strength, are also not far from the assembly line. Gun rovings for spray-up and lay-up with fast wettability and wetout, new chopped strand rovings for bulk molding compounds (BMC's), and a new electric melt process for producers of large-volume glass fibers are some of the new developments.

In the early 1970's, Du Pont first introduced an aramid (aromatic polyamide) fiber for industrial applications. That material, which today is known as Kevlar, is produced in continuous fibers and fabrics for commercial uses with polymer matrices for aircraft, aerospace, and marine markets. This fiber is also found in high-performance ropes and cable, friction products, gaskets, ballistic fabrics, tires, and mechanical rubber goods.

Kevlar, in some cases, is replacing fiberglass, metal, and wood in boat structures such as hulls, decks, and bulkheads. Cost, however, is still a factor in many applications where aramid fibers are used as a replacement for glass fibers. Some of the world's leading tire manufacturers are now using Kevlar fiber for added strength in their automobile radial tires. Kevlar is also found in commercial aircraft and military helicopters. In spite of all the important features of some of the potential new products and uses for Kevlar, cost and adaptation of old products still remain issues. The over-all outlook for the continued growth of this fiber, however, remains excellent.

Increased research on production of composites, and the need for high-temperature insulation in industrial furnaces to alleviate energy costs, have resulted in an increased interest by the industry in the development of suitable ceramic fibers. Although these fibers are currently being used in the fabrication of metal- and ceramic-matrix composites, little has been pursued regarding resins. This may be changing, however, as evidenced by reports from one producer that an experimental sintered aluminum oxide fiber is being studied for potential plastic applications. Ceramic whiskers are also attracting more and more attention. Compared with ceramic fibers, they are thinner and shorter, have greater fracture toughness and stiffness, and maintain their physical properties at high temperatures.

One ceramic whisker manufacturer, the J. M. Huber Corporation, has recently announced the availability of a new submicron ceramic whisker. This product is being tested as a reinforcement for adhesives, ceramics, and coatings. The-

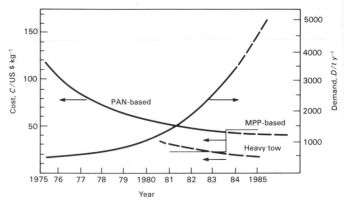

Fig. 1. World demand and price of carbon fibers. (From E. Fitzer et al., Carbon Fibres and Their Composites, *High Temperatures-High Pressures*, Vol 16, 1984, p 370)

ories are that the short fibers should be especially effective as reinforcements in matrices with high elastic moduli.

Scarcity of whiskers has prevented widespread investigation of the practical incorporation of whiskers in composites by many manufacturers. It is anticipated, however, that as ceramic whiskers become more readily available, the development of processing expertise will increase and the demand for whiskers will spur the commercialization of other whiskers that are now just in the experimental stages.

Today only two raw materials are being widely used for the production of carbon fibers: polyacrylontrile (PAN), which has been commercially available in fiber form for years because of its use as a textile fiber, and mesophase pitch (MPP). World demand and price of PAN-based and MPP-based carbon fibers, 1975–1985, are plotted in Fig. 1. The pitch-based carbon fibers were originally developed to reduce costs because of the lower price of pitch as a primary raw material. Although they effectively reduced costs at the outset, certain production differences between the two fibers have limited much of this initial cost savings. In addition, mesophase pitch fibers do not have the same high performance levels as the PAN-based fibers, and current applications for pitch fibers are primarily focused on carbon/carbon composites. The PAN-based fibers are used exclusively for reinforcements of polymers, but research is under way to develop high-quality pitch-based reinforcements in plastics for automotive and industrial uses at competitive prices. One major producer predicts that sales of the pitch fibers could increase 20 to 30% by the year 1990 because of new technological developments that will broaden their uses.

Another rather recent commercial development in the fiber industry is a high-strength polyethylene fiber manufactured by Allied Corporation. With a tensile strength of 2585 MPa (375,000 psi) and a modulus of 117,210 MPa (17 million psi), this fiber (Spectra 900) should ultimately be found in a large variety of applications. Plans are under way to advance its use in the marine industry, where it is already found in sails, braided rope, and cordage. Predictions are that the applications of Spectra 900 will eventually spread to areas such as ballistic protection products, sporting equipment, medical prostheses, and other industrial uses, and a 101.6-kg (10-million-lb) annual market for this new fiber is being forecast by its producer.

Other areas of economic interest in reinforcements are the nonfibrous enhancers, or fillers, called Microspheres—reportedly the fastest-growing segment of filler reinforcements. Microspheres come in hollow and solid spheres and irregular shapes, and are made from glass, ceramics, and resins. New grades and modifications in the last few years have broadened the range of applications, and use of solid glass spheres to replace parts of glass reinforcements in resins is not only producing good end-product results, but also helping to lower costs. In a new and unanticipated area of growth, Microspheres are being used as antiblocking agents in polyolefin film. At present, they are already beginning to replace traditional slip agents such as silica and diatomaceous earth.

Mineral fillers are also changing, and surface modification is making them more useful. Today's fillers are doing an overall better job of meeting the specific requirements demanded of them by the rapidly growing reinforced plastics industry. The surface treatment of minerals in many cases has helped extend their reinforcing abilities and uses. New technology leading to color improvements in mica has given this mineral a wider range of applications. The use of minerals, in general, has grown an estimated 25% in the last few years. As more and more producers continue to create new products and enhance and modify their old standbys, this growth is expected to continue well into the future.

Finally, carbonates, clays, talcs, and conductives are making news of their own. Here again, technological upgrading is making the difference. Calcium carbonates have been made easier to handle and, therefore, have improved productivity in fabrication of materials. Companies are also expanding their products as well as their production lines. In addition, metallics and carbon black fillers that provide electromagnetic interference (EMI) protection are becoming a new and important market segment.

Fillers and other nonfibrous reinforcements appear to be meeting the challenge of providing new products and technological developments for the expanding composites industry. Growth of close to 18% per year has occurred during the past few years and is expected to follow the tide of demand generated by the end users of these products.

Conclusion

The reasons for industry turning more and more to composite materials are multifaceted. Economic factors as well as new consumer demands are two obvious reasons, but many of the factors are complex and interwoven. The oil embargo in the 1970's established a chain of events that has led industry to explore many alternatives to high fuel costs, shortage of petroleum as raw material for polymers, and the dependence on foreign countries for these resources. Foreign competition for many of our established markets has placed strong demands on industry to face new challenges in developing composite materials to help cut costs—of both material and labor—through greater ease of fabrication. Although the recent decline in crude oil prices and the value of the dollar should eventually ease some of the foreign trade deficit, competition from abroad will continue.

Many of the new composite materials—especially the metal and ceramic varieties—are still developmental and in the "cottage industry" stage, but the specialized markets that they are now serving will increase as new technology facilitates fabrication and reduces costs. It is up to industry at this point to continue to redirect its resources and modernize its plants and thinking. International competition is making this imperative.

Bibliography

Sam L. Jones, Epoxy Resin Tooling for Saturn Auto Projects Demonstrated Savings in Test, *American Metal Market Metalworking News,* January 20, 1986, p 9

Fiber Reinforced Plastics Move Inside Engine, *Automotive Engineering,* Vol 93, No. 5, May 1985, p 34–36

Alan Gwinn, Reinforced Plastics Surge Ahead by 12 Percent in '84, *Plastics Engineering,* April 1985, p 61–65

SAMPE Journal, May/June 1985

The Outlook for Glass Fiber Reinforced Thermoset Composites in Automobiles and Trucks: 1984–1989, product literature, Owens/Corning Fiberglas Corporation, Toledo, Ohio

A. Stuart Wood, Fibrous Reinforcements, *Modern Plastics,* July 1984, p 52–54

Rosalind Juran, Nonfibrous/Enhancers, *Modern Plastics,* July 1984, p 55–56

Industry/News Focus, *Plastics Technology,* May 1985, p 195, 196

J. M. Huber Corporation, Press Release, January 24, 1985

BASF Aktiengesellschaft, Press Release, December 12, 1985

Erich Fitzer, Antonios Gkogkidis, and Michael Heine, Carbon Fibres and Their Composites, *High Temperatures-High Pressures,* Vol 16, 1984, p 363–392

P. Bracke, H. Schurmans, and J. Verhoest, *Inorganic Fibers and Composite Materials,* Pergamon Press, Oxford, U.K., 1984

Honda Motor Co. Ltd., *Trading Times,* January 25, 1985, p 5

Robert B. Aronson, Metal-Matrix Composites—Materials of the Future, *Machine Design,* August 8, 1985, p 68–73

Super Fibers: New Role for PE, *Plastics World,* May 1985, p 21

"Advances in Materials Technology: Monitor," United Nations Industrial Development Organization, Vienna, Austria, August 1984

SECTION 3
Design Equations

Introduction and Comments on Composite Design Equations

John W. Weeton

Equations for calculating the properties of composites for design purposes will be presented in two separate articles. The first article, "Calculations of Composite Properties," deals primarily with fiber-reinforced metal and polymer matrix composites. They are relatively simple equations that can be used to calculate and predict in advance the properties of many different types of composites. The second article, "Simplified Composite Micromechanics Equations for Mechanical, Thermal, and Moisture-Related Properties," is a combination or merging of two papers by C. C. Chamis (Ref 1 and 2). That article deals largely with the micromechanics of plastic matrix composite plies and how they can be combined to form larger structures. Both articles contain notations on how some equations can be used to calculate the properties of systems other than metal or plastic.

Some of the equations presented in the first article were developed subsequent to conducting experiments with actual composites. Some are the equivalents of equations in the second article but have different notations or symbols. In both articles, many equations have been verified experimentally with both practical and model systems, but in some cases, as Chamis points out, additional data and research to verify equations should be obtained.

The main assumption that should be noted in the second article is that "the ply and its constituents behave linearly elastic to fracture." This is different from cases described for the metal matrix composites noted in the first article, where the matrix may be in plastic deformation in tension. In some cases the fiber may also be in plastic deformation (see Fig. 1). The elastic/elastic assumption for plastic matrix composites in the second article should be sound, and results obtained using the equations should be close enough to those that would be obtained if the matrix were in plastic deformation (since both strengths and moduli are low in any case). If the fiber used in the plastic matrix were metallic and had a low proportional limit, results relating to moduli of elasticity would be incorrect. But, in most cases, the reinforcements are brittle materials such as graphite or ceramics.

The equations presented in the second article apply mainly to plies or sheets with single or multiple layers of unidirectionally oriented filaments. Some of the equations can also apply to discontinuous filament composites, and even to bulk or shaped composites, as long as the filaments are oriented longitudinally. The advantage of plies over other preliminary shapes is that they can be considered "building blocks" from which many shaped products (including finished products) can be constructed. Plies can be stacked or layered, thus orienting the unidirectionally oriented plies at different angles to each other. This enables the designer to control properties such as moduli and strengths so that the product can withstand stresses in different areas or directions. In two-dimensional planes, for example, it is possible to produce angle-plied structures with almost isotropic properties (quasi-isotropic properties). Or, as another example, a structure can have high strengths in one direction and simultaneously resist shear stresses at angles to the main applied stresses.

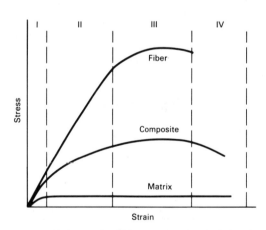

Fig. 1. Schematic illustration of four stages of deformation of fibers, matrix, and composite. Stage I: elastic deformation of both fibers and matrix. Stage II: elastic deformation of fibers; plastic deformation of matrix. Stage III: plastic deformation of both fibers and matrix. Stage IV: failure of both fibers and matrix (successive failure of fibers).

Degradation of Plastic Properties by Moisture

Some of the terminology used in the second article will allude to problems specifically associated with plastics. The terms hygral,* hygrothermal,† and hygrothermomechanical† are associated with problems of moisture and heat as they affect properties of the plastic matrices. Void formations, and diffusivity of moisture through bulk plastics and along interfaces of filaments, are represented by various equations. The author notes that some of the equations relating moisture and related theories have not been verified by research or tests. They do, however, permit estimations of possible damage to plastics during use and give limiting properties for the materials. Analysis of plastic matrix composite test results can show how accurately the calculated values represent the true properties.

Degradation of Metal and Ceramics by Fabrication and Use

The same approach has been and can be used for some fiber/metal or fiber/ceramic compositions where simple degradation mechanisms can be anticipated or where no degradation of an unusual nature will exist. Property reductions in matrices, however, can occur by so many mechanisms and from so many starting conditions of the constituents, both prior to and subsequent to fabrication, that very few metallurgists would attempt to approach many problems with generalized equations. The approach used by some has consisted of judicious preselection of materials that should be compatible, fabrication of these materials by one or more methods, and study of their microscopic structures prior to and subsequent to fabrication. Analyses of optical, electron microscopic, x-ray, diffusivity, and other metallurgical studies often reveal mechanisms of failure or can account for adequate or outstanding properties. Equations predicting the lives of such metal/metal or ceramic/metal combinations prior to producing them or even testing them are usually based on assumptions that no serious degradation of constituents will occur during fabrication or use. If either fiber or matrix properties are reduced by mechanical processes or heat treatments that cause metallurgical reactions, the properties of individual constituents may, after simulated treatments, be used to calculate more accurately the properties of the composites. If degradation problems are insurmountable for a given metal/metal, metal/ceramic, or ceramic/ceramic composite, the fiber or matrix, or even the fabrication method, may be changed. Coatings sometimes may be applied to the fibers to prevent reactions with the matrix.

Symbols Used

The symbols used in the first article are defined in conjunction with the presentation of the equations. Because they are more complicated and relate to plies and to unidirectionally oriented (or longitudinally oriented) fibers, the sym-

bols in the second article are presented in the text in a single listing. Since the symbol notations are part of a larger set of symbols for different types of composites used by Chamis, they are printed exactly as given in Chamis' original papers (Ref 1 and 2). However, for those not familiar with micromechanics, some additional descriptions relative to the symbols, other than those given in the text, should be helpful. Three such descriptions follow.

First, consider the tensile-strength designation "S_{l11T}" used in the second article (refer to Fig. 3 in that article). The "S" represents strength. The "l" indicates that we are dealing with a ply (or a layer with fibers oriented unidirectionally). If we were working with completed composites, the "l" would be replaced with a "c." Now the axes that would be used would be structural axes relating to the composite and would be designated as XX, YY, or ZZ. The first "1" designates the face of the plane that is under consideration. The second "1" designates the direction of the stress (in the direction of the fibers). The "T" represents a tensile stress.

Next, consider a compressive stress acting on a different plane — the plane perpendicular to the direction of the stress (the transverse stress). Again refer to Fig. 3 in the second article, but replace the "T" with a "C" for compressive stress. Now the stress is expressed by "S_{l22C}." In this designation, the "S" and the "l" have the same definitions as those in the first example. The first "2" represents the face of the plane under consideration, and the second "2" the direction of the stress. The "C" represents a compressive rather than a tensile stress.

Finally, consider a shear stress represented by "S_{l12S}." Here, the "S" and the "l" again have the same definitions as in the first example. The "1" indicates the face acted upon by the shear stress. The "2" represents the direction of the shear stress (the transverse direction). The "S" indicates a shear stress.

Angle-Plied Composites

In the first article it is graphically shown how a unidirectionally reinforced composite will have a far lower strength in transverse tension than one loaded exactly in the direction of the fibers (i.e., the rule-of-mixtures value in many cases). The reduced properties result from the shearing of the weak matrix. In fact, both the strength and moduli of a composite in a ply are reduced considerably (if the matrix is weak and/or has a low modulus) when the angle of the applied stress deviates from the direction of the filaments in the composite. Thus, because of their low mechanical properties relative to fibers, plastic matrix composites and some metal matrix composites will need to be strengthened or stiffened by laying up plies in different directions (see Fig. 2 in the second article). Such lamination will be necessary because stresses in a component can vary in both direction and type of stress. Properties of various combinations of various plies oriented at different angles can be calculated by resolution-of-forces methods or through the use of computer programs that can calculate various properties of any combination of oriented plies.

Hybrid Composites

A hybrid composite is one containing more than one type of fiber reinforcement, more than one matrix, or both. There are also various combinations of monolithic sheets or foils, or combinations of sheets or foils plus fiber composites, that may be considered hybrid composites. In the broadest sense,

---

*"Hygral" is a term which denotes the effect of water molecules on the plastic (polymer) network of molecules. The water molecules apparently distort the network, thereby weakening or softening the plastic.

†The prefix "hygro" applies to effects of moisture. This moisture alone can weaken some plastics or, in conjunction with heat, can be even more deleterious to the properties of the plastic. Thus, the words "hygrothermal" and "hygrothermomechanical" are compound words relating heat and/or mechanical properties to moisture. The prefix "hydro," although not used here, refers to liquid water.

hybrid composites could even include combinations of fibers and particulate composites or combinations of different particulate composites. However, composites with particulates will not be considered in this discussion.

There are both technical and economic reasons for considering the use of hybrid composites. One example of the many technical objectives might be improving impact resistance by adding some less-brittle (or even ductile) fibers to a composite strengthened by brittle fibers (such as graphite or boron). Such additions would tend to delay fracture on impact, allow absorption of more energy, and prevent structural parts from shattering in a brittle fashion. On the other hand, by combining stiff (high-modulus) plies with ductile (low-modulus) metallic sheet or foil (e.g., aluminum alloys), a higher-modulus, lighter-weight hybrid composite with the surface characteristics of aluminum can be obtained. Depending on the type of composite used to stiffen the aluminum, one might diffusion bond the materials together (in the case of a metal matrix composite) or glue the materials together (in the case of a plastic matrix composite). One could also make a sandwich structure or a multilayered laminar composite.

Weight reductions resulting from the use of such structures as those noted above, or from the use of lighter-weight fibers as a partial replacement for heavier fibers, can often yield cost-effective results. In space, aeronautical, and automotive vehicles, considerable fuel savings can be realized, for example. Other savings can be obtained simply by substituting some low-cost filaments for high-cost ones, although in such cases some properties usually have to be sacrificed. Because of the great number of possible combinations of materials, the potential number of variations in properties is also huge, as is the number of ways in which to achieve cost savings.

The hybrid composites that make use of plies or layers of different fiber-reinforced composites or plies of fiber-reinforced composites and monolithic sheet or foil have been termed by some as "interply hybrid" and "superhybrid" composites, respectively. Laminate theories apply to these composites. In the fiber-reinforced composites noted immediately above, either fibers or matrices, or both, can be varied.

Most of the research, development, and analytical work on hybrids has been done with composites fabricated from plies that contain unidirectionally oriented fibers of more than one type (usually two types of fibers). Such composites are termed "intraply" composites or "unidirectional intraply hybrid" composites. This type of composite can be angle plied to form intraply hybrid angle-plied composites.

Strengths and moduli, and some thermal, physical, and hygral properties, can be predicted from equations based on rule-of-mixtures equations and on several assumptions that will not be described herein (Ref 5). At this point, it is suggested that tensile, compressive, and stress-rupture strengths (and moduli) of unidirectionally oriented composites with more than one type of fiber can be approximated by rule-of-mixtures equations such as those shown in both of the two articles that follow provided that properties representing each fiber and the volume percentages of each fiber are included in the equations. It should be noted that this approach is based on the assumption that each fiber will fracture at the same strain.

References

1. C. C. Chamis, "Simplified Composite Micromechanics Equations For Hygral, Thermal, and Mechanical Properties," NASA TM 83320, 1983
2. C. C. Chamis, "Simplified Composite Micromechanics Equations For Strength, Fracture Toughness, Impact Resistance, and Environmental Effects," NASA TM 83696
3. S. W. Tsai, *Composites Design–1985*, United States Air Force Materials Laboratory, THINK COMPOSITES: Dayton, Paris, Tokyo, 1985
4. J. C. Halpin, *Primer on Composite Materials: Analyses*, Air Force Materials Center, TECHNOMIC Publishing Co., Inc., 1984
5. C. C. Chamis and J. H. Sinclair, Mechanics of Intraply Composites — Properties, Analysis, and Design, Polymer Composites, Vol 1, No. 1, September 1980

Calculations of Composite Properties

John W. Weeton

This article pertains only to some types of fiber-reinforced composites. The equations presented here apply largely to metal matrix or plastic matrix composites where composite fracture is fiber-controlled, where the modulus of elasticity of the fiber is greater than that of the matrix, and where the strain to failure of the matrix is greater than that of the fiber. Most equations will be for unidirectionally oriented fiber composites, but some will treat off-angle loading of composites.

References to publications containing detailed equations that define the behavior of composites of different configurations, and particularly cross-plied composites, will be presented at the end of this article. To gain an understanding of the strengths of cross-plied fiber-reinforced composites, both metallic and plastic, one can refer to the various data presented elsewhere in this book. Values for pseudo-isotropic or quasi-isotropic materials, for angled plies, etc., are presented in several tables and may be compared with unidirectional properties to determine the degree of reduction of their strengths relative to those of unidirectional composites.

The equations presented here are relatively simple compared with some used by those versed in the micromechanics of composites. They should permit calculations, or at least approximations, of some of the properties that are of interest to most engineers. In addition, they should enable one to determine whether or not a composite may be superior to, or can be used in place of, conventional monolithic materials such as metals, plastics or polymers, or ceramics. These equations will also permit the developer of new types of materials to calculate properties attainable from different combinations of fibers and matrix materials. If what look like potential materials are then fabricated and the calculated values are not achieved, it will be obvious that there are problems with the fabrication procedure or with the compatibility of the fiber and matrix materials.

Actually, any manufacturer seriously considering the substitution of composites for monolithic materials should realize that some companies are well advanced in techniques of composite manufacture, lay-ups of plies, fabrication of panels, tubes, and other components with complex configurations, and fabrication of finished products. The equipment needed to produce some of these products is complex and includes computerized filament winders, weaving equipment, plastic molding and fabrication equipment, vacuum or atmosphere hot pressing equipment (for metal matrix materials), and casting equipment, just to name a few examples. Not only can the materials technologies be complex and require specific knowledge relative to fabrication of composites, but the micromechanics and mathematics behind choices and designs of components can be most complex. Reiterating then, this article will present some simple equations that will enable a manufacturer to determine on a preliminary basis whether or not composites are worth buying or manufacturing. If composite manufacture appears warranted, personnel will have to be hired for or reassigned to such an effort, and the necessary equipment will have to be obtained.

Composites may have fibers or filaments that are continuous over a given region or may have separated, segmented (i.e., discontinuous) filaments of small size relative to the region in which they are embedded. The filaments or fibers may be oriented unidirectionally, at angles, and at random. Unidirectional composites in the form of plies may be combined with other plies and bonded together to yield a composite with layers of differently oriented filaments designed to carry loads in different directions.

Fibers in the form of glass filaments have been used for years to reinforce plastic automobile bodies, boat structures, and high-pressure tanks. The strengths and moduli of these composites could be predicted to some degree from properties of the glass alone, because the plastic matrices had very little strength and stiffness relative to the fibers.

Although glass fibers have very high strengths, particularly at room, cryogenic, and reasonably high temperatures, they have low moduli of elasticities — nominally 10×10 psi. As reinforcements for metals, they could be expected to have little use because of their potential for reaction with metal matrices and because their moduli are low relative to those of many metals. Since many designs of aerospace components are based on moduli of elasticities, and since weight savings are of prime importance, many research engineers during the 1950's noted with great interest the properties of whiskers, one of the strongest known manmade materials.

Results of research on whiskers showed that both metallic and ceramic whiskers could be made that had strengths approaching theoretical values (as much as 2 to 3 million psi) and moduli many times greater than those of glass filaments or metal wires. Also, some investigators were working on high-strength, high-modulus filaments. A great impetus to the

composite field occurred when the results of a study of a model system — a tungsten-fiber/copper-matrix composite — showed that the strengths of both continuous and discontinuous fiber composites (unidirectionally oriented) could be represented by a rule-of-mixtures strength equation (Ref 1). Further work published in Ref 2 and 3 refined concepts relative to composites and also showed that moduli of elasticities of composites could be represented by the same type of equation. Many equations evolved during the 1960's or later, and some will be presented and referenced here.

Composites with Unidirectional (or Longitudinal) Fibers (Either Continuous or Discontinuous Fibers)

At this point let us consider a composite bar (e.g., Fig. 1) with filaments or fibers extended from end to end, parallel to the axis of the bar but separated by a matrix — i.e., not touching each other (Ref 3). In this composite, the load is transferred from the matrix to the filaments by a shear mechanism. When stressed in tension, the fibers and matrix of the composite elongate equally (the principle of combined action). The stresses on the composite constituents are determined by the moduli of elasticities or elongations of constituents comprising the composite. Thus, both constituents may be in elastic deformation, the fiber may be in elastic deformation while the matrix is in plastic deformation, or both constituents may be in plastic deformation at a given elongation. Equations relating to orientation effects will be presented later.

Rule-of-Mixtures Equations Relating to Strengths of Composites

Tensile Strength

The generalized equation for predicting stresses carried by composites for all values of strain is

$$\sigma_c^* = \sigma_f^* V_f + \sigma_m^* V_m$$

or

$$\sigma_c^* = \sigma_f^* V_f + \sigma_m^* (1 - V_f)$$

where the σ's represent stresses at those particular values of strain taken from the stress-strain curves of the components of the composite, in the conditions in which they exist in the composite (Ref 3): σ_c^* is the stress carried by the composite at a particular strain, σ_f^* is the stress carried by the fiber at a particular strain, V_f is the volume fraction of the fiber, σ_m^* is the stress carried by the matrix at a particular strain, and V_m is the volume fraction of the matrix.

Ultimate Tensile Strength

$$\sigma_c = \sigma_f V_f + \sigma_m^* V_m$$

where σ_m^* is the stress on the matrix taken from the stress-strain curve at an equivalent strain to that at which the ultimate tensile strength of the fiber is achieved, σ_f is the ultimate strength of the fiber, and V_f and V_m are the same as for the generalized equation.

Yield Strength

See generalized equation under "Tensile Strength."

Critical Volume Fraction

The above equations are valid provided that a critical volume fraction of filaments has been added to the composite.

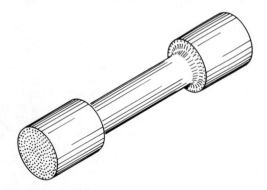

Fig. 1. Composite bar with fibers oriented parallel to the axis of the bar (Ref 3)

In other words, there must be enough fibers added to the composite to give the composite a strength above that at which the fibers first begin to strengthen rather than weaken the matrix. This approximates the fiber content needed to raise the critical volume or V_{crit} and may be calculated from the following equation (Ref 4):

$$V_{crit} = \frac{\sigma_m - \sigma_m^*}{\sigma_f - \sigma_m^*}$$

where, except for σ_m (which equals the UTS of the matrix), all terms are the same as for the ultimate tensile strength equation. Typical values for different matrix materials and for fibers of different strengths are given in Table 1 (Ref 4). Note that values of V_{crit} decrease for fibers of increasing strength and that the values decrease as the difference between the matrix UTS (σ_m) and the σ_m^* increases. At first glance, it would appear that the stronger the matrix, the more fibers are needed to achieve V_{crit}, but this is not true. If one takes the titanium alloy Ti-6Al-4V, which has a UTS of about 130 to 200 ksi depending on the condition of anneal or heat treatment, a V_{crit} of 0.03 (Ref 5) can be obtained since the σ_m^* of Ti is very high and close to the yield strength of about 124 ksi. V_{crit} may be observed in the curve of Fig. 2 (Ref 3), which represents actual data. Note that the data curve extrapolates to σ_m^* and that it is below V_{crit} and less than σ_m. Almost always, one would not utilize anything but a high-strength filament to strengthen a material. This would mean that one would not have to be overly concerned with V_{crit}, but it cannot be ignored.

Moduli of Elasticities

Case I–fiber and matrix are in elastic deformation:

$$E_c = E_f V_f + E_m V_m$$

or

$$E_c = E_f V_f + E_m(1 - V_f)$$

or (for plastic matrix composites)

$$E_c = E_f V_f$$

where E_c is the modulus of elasticity of the composite, E_m is the modulus of elasticity of the matrix, V_f is the volume fraction of the fiber, V_m is the volume fraction of the matrix, and E_f is the modulus of elasticity of the fiber.

Case II–fiber is in elastic deformation and matrix is in plastic deformation (or a stage of deformation in some metal matrix composites where the modulus is termed a "secondary modulus") (Ref 3):

$$E_c = E_f V_f + (\sigma_m^*/\epsilon) V_m$$

where σ_m^*/ϵ is the slope of the stress-strain curve of the matrix at a given strain beyond the proportional limit of the matrix.

Stress-Rupture Strength

(High-temperature strengths at a given time period — pertains to metal matrix and some ceramic and glass matrix composites) (Ref 6):

$$(\sigma_c)_t = [(\sigma_f)_t V_f + (\sigma_m)_t V_m]$$

where t is a notation that signifies a constant rupture time, σ_c is the stress to cause rupture at a given temperature and a given time (stress-rupture strength), σ_f is the stress-rupture strength of the fiber, V_f is the volume fraction of the fiber, and V_m is the volume fraction of the matrix.

Table 1. Values of V_{crit} for fibers of various strengths in ductile metals(a)

Matrix material and properties	Critical fiber volume fraction, V_{crit}, for fiber UTS, σ_{f}, of:			
	100 ksi	250 ksi	500 ksi	1000 ksi
Aluminum ($\sigma_m^* = 4$ ksi; $\sigma_m = 12$ ksi)	0.083	0.033	0.016	0.008
Copper ($\sigma_m^* = 6$ ksi; $\sigma_m = 30$ ksi)	0.255	0.098	0.047	0.024
Nickel ($\sigma_m^* = 9$ ksi; $\sigma_m = 45$ ksi)	0.396	0.150	0.073	0.036
18-8 stainless steel ($\sigma_m^* = 25$ ksi; $\sigma_m = 65$ ksi) ...	0.534	0.178	0.084	0.041

(a) From Kelly and Davies (Ref 4). Symbols have been changed slightly for conformance with the equations presented in this article.

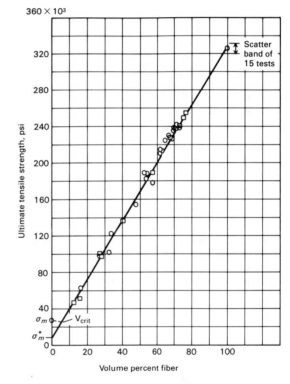

Fig. 2. Tensile strengths of tungsten-fiber-reinforced copper composites (for continuous reinforcement with 5-mil-diam tungsten fibers) (Ref 3)

Compressive Strengths

Same equations as for tensile strengths except that all values of σ are for compressive stresses.

Poisson's Ratio

$$\nu_c = \nu_f V_f + \nu_m V_m$$

where ν is Poisson's ratio and the subscripts c, f, and m refer to composite, fiber, and matrix, respectively (Ref 7).

Density

$$\rho_c = \rho_f V_f + \rho_m V_m$$

where ρ is density and the subscripts c, f, and m represent composite, fiber, and matrix, respectively.

Electrical Conductivity

$$K_c = K_f V_f + K_m V_m$$

(Ref 8) where K is conductivity and the subscripts c, f, and m represent composite, fiber, and matrix, respectively.

Equations and Concepts Relating to Discontinuous Fiber Composites

Background

Shear transfer of loads between discontinuous fiber composites may be better understood with the aid of Fig. 3 (Ref 9). By equating the tensile fracture strength of the fiber to the

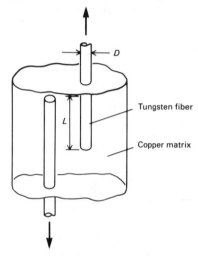

Fig. 3. Schematic diagram of shear load transfer mechanism in fiber-reinforced composites reinforced with discontinuous fibers (Ref 9)

shear load of the overlapped portion of the fiber shown, a minimum or critical length-to-diameter ratio necessary to fracture the fiber can readily be calculated by

$$\frac{L_c}{D} = \frac{1}{4}\frac{\sigma_f}{\tau}$$

but this becomes the following when both ends of the fiber are embedded in the matrix.

Critical Length-to-Diameter Ratio (or Aspect Ratio)

$$\frac{L_c}{D} = \frac{1}{2}\frac{\sigma_f}{\tau}$$

where L_c is the critical fiber length, D is the diameter of the fiber, σ_f is the tensile strength of the fiber, and τ is the shear strength of the fiber-matrix interfacial bond or the matrix shear strength, whichever is less.

Modified Rule-of-Mixtures Equation for Discontinuous Fiber Composites

Background

The ends of a short-length fiber embedded in a matrix cannot bear any load since the fiber has no shear length to receive a load from the matrix. As one progresses along the fiber from the end, the shear length becomes greater and more load is transmitted to the fiber. The shaded areas in Fig. 4 represent the magnitudes of the stresses that the fiber can

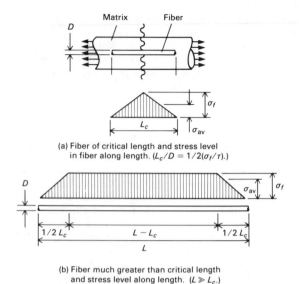

(a) Fiber of critical length and stress level in fiber along length. ($L_c/D = 1/2(\sigma_f/\tau)$.)

(b) Fiber much greater than critical length and stress level along length. ($L \gg L_c$.)

Fig. 4. Schematic illustration of stresses in critical length of fiber in matrix and effect of increasing length on average fiber stress (Ref 9)

carry. When the fiber is exactly as long as the critical length (Fig. 4a), the average load is the maximum load. As the fiber length is increased above the critical length, the stresses in the fiber appear as represented in Fig. 4(b). The greater the length of the fiber relative to the critical length, the more nearly the strength of the composite will approach the rule-of-mixtures value.

Theoretically, the short-length fiber composites can never equal the rule-of-mixtures value (assuming perfect bonding and mutual insolubility between fiber and matrix). Practically, the results can be so close to rule-of-mixtures values that one cannot detect differences because they can be within experimental error. As an example, using the next equation that will be presented, it can be shown that a fiber with an actual length only 200 times the critical length can carry a load that is 99.5% of its tensile strength. The equation is from Ref 10.

Ultimate Tensile Strength of Discontinuous Fiber Composite

$$\sigma_c = \sigma_f\left(1 - \frac{L_c}{2L}\right)V_f + \sigma_m^* V_m$$

where σ_c is the ultimate tensile strength of the composite, σ_m^* is the stress on the matrix at which the fiber reaches its ultimate tensile strength, V_f is the volume fraction of the fiber, V_m is the volume fraction of the matrix, L is the actual fiber length, and L_c is the critical fiber length.

Orientation Effects on Strengths and Failure Mechanisms of Composites (Unidirectionally Alined Fibers)

To appreciate the significance of the orientation of fibers relative to the direction of loading, consider a sheet material in which all fibers are oriented in a single direction (Fig. 5). A thin sheet could also be a ply with one or more layers of fibers bonded together. We know that the full fiber strength can be utilized by a composite if the tension axis is in the direction of the fibers. Turn the sheet or ply 90° and it becomes obvious that tension applied perpendicular to the fi-

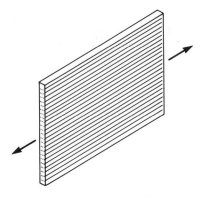

Fig. 5. Sheet material in which all fibers are oriented in the same direction

ber orientation could pull apart the matrix, or the bonds between fibers and matrices, or the fibers themselves. If any of these latter zones in the composite are weak in tension, they may be even weaker in shear. A bulk specimen other than a sheet with fibers oriented in a single direction would also behave in the same way. The curve in Fig. 6 (Ref 9) shows how composite strength drops off drastically as the angle between stress application and fiber direction is increased. Also shown are three equations that represent different portions of the curve where failure mechanisms occur. The lowest strength is at 45° in shear (usually of the matrix). This points to some very important considerations. For example, in a specimen stressed in tension, an angle of misorientation of fibers relative to the direction of tension could cause a drastic reduction in strength. Or, to fabricate a composite sheet that would compete with a metal sheet alloy which has more or less isotropic properties, it would be necessary to cross ply or orient plies containing unidirectional fibers in order to permit the sheet to carry large stresses in different directions. More will be said of plies and the construction of composites by cross plying later. For now, referring to Fig 6, consider the following equations for three different failure mechanisms (Ref 9):

Tensile failure of fiber: $\sigma_c = \sigma$ (1)

Matrix or interfacial shear: $\tau = \sigma \sin \phi \cos \phi$ (2)

Tensile failure $\sigma_u = \sigma \sin \phi$ (3)

where σ_c is the composite tensile strength (assuming axially alined fibers), σ is the applied stress on the composite, ϕ is the angle between the fibers and the tensile axis of the specimen, τ is the shear strength of the matrix or interface, and σ_u is the tensile strength of the matrix.

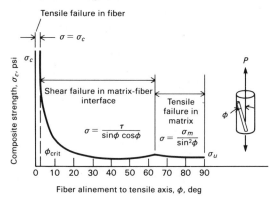

Fig. 6. Failure mode as function of alinement of fibers in discontinuous fiber-reinforced composite (Ref 9)

Critical Angle of Misalinement

Combining Equations 2 and 3 above,

$$\phi_{crit} = 1/2 \text{ arc sin } (2\tau/\sigma_c)$$ (4)

where ϕ_{crit} is the critical angle of misalinement and the other symbols are the same as for Equations 1, 2, and 3 above.

Note: the above equation has considerable significance. For example, stronger fibers and greater volume fractions of fibers will increase composite strength and lower the critical angle of misalinement, and high-temperature applications will lower the shear strengths of the matrix and thus lower the critical angle. Therefore, a composite with a low angle of misalinement may be far weaker than might be expected from rule-of-mixtures calculations. Figure 7 shows how the fracture mechanism changes drastically, from tension to shear, with misalinement (Ref 9).

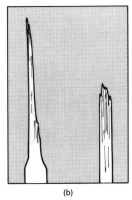

Fig. 7. Test specimens that failed (a) in tension ($\sigma_c = \sigma$) and (b) in shear ($\tau = \sigma \sin \phi \cos \phi$) (Ref 9)

A similar change in the mechanism of failure would occur in a single ply such as that shown in Fig. 4, and, in fact, a polymer matrix composite ply would have very low strengths at off angles. Thus, to obtain more uniform strength in a sheet, for example, and also to increase strengths in any direction, it is necessary to cross orient such plies as that shown in Fig. 4. By doing this, it is possible to obtain nearly isotropic properties in a sheet (pseudo-isotropic or quasi-isotropic properties), or to give a sheet major strength in one direction and lesser strengths in other directions. Similarly, in tubes or structures, by varying plies or laminates, necessary strengths and stiffness can be designed into materials. In all cases, however, the strengths of the cross-plied composites of a given composition would be less than those of perfectly alined, unidirectionally oriented fiber composites of the same composition. Actually, some of the equations and approaches treat not only two-dimensional materials but also materials with three-dimensional fiber configurations. Additional equations for composite design may be found in the publication *Composites Design–1985,* by Think Composites, Dayton, OH, 1985, and *Primer on Composite Materials: Analyses,* by Technomic Publishing Co., Inc., Lancaster, PA, 1984.

References

1. Jech, R. W., McDanels, D. L., and Weeton, J. W., Fiber Reinforced Metallic Composites, in *Composite Materials and Composite Structures,* Proceedings of the Sixth Sagamore Ordnance Materials Research Conference, Rep. MET 661-601, Syracuse Univ. Res. Inst., 1959, p 116–143

2. McDanels, D. L., Jech, R. W., and Weeton, J. W., Metals Reinforced with Fibers, *Metal Progr.,* Vol 78, No. 6, Dec 1960, p 118–121

3. McDanels, D. L., Jech, R. W., and Weeton, J. W., "Stress-Strain Behavior of Tungsten-Fiber-Reinforced Copper Composites," NASA TN D-1881, 1963; also, McDanels, D. L., Jech, R. W., and Weeton, J. W., Analysis of Stress-Strain Behavior of Tungsten-Fiber-Reinforced Copper Composites, *Trans. AIME,* Vol 233, No. 4, April 1965, p 636–642

4. Kelly, A., and Davies, G. J., The Principles of the Fibre Reinforcement of Metals, *Met. Rev.,* Vol 10, No. 37, 1965, p 1–77

5. Schoutens, J. E., "Introduction to Metal Matrix Composite Materials," MMC No. 272, DOD Metal Matrix Composites Information Analysis Center, Santa Barbara, CA, June 1982, p 7–18

6. McDanels, D. L., Signorelli, R. A., and Weeton, J. W., "Analysis of Stress-Rupture and Creep Properties of Tungsten-Fiber-Reinforced Copper Composites," NASA TN D-4173, 1967; also, McDanels, D. L., Signorelli, R. A., and Weeton, J. W., Analysis of Stress-Rupture and Creep Properties of Tungsten-Fiber-Reinforced Copper Composites, in *Fiber Strengthened Metallic Composites,* STP 427, ASTM, 1967

7. Chamis, C. C., Simplified Composite Micromechanics Equations for Hygral, Thermal, and Mechanical Properties, *SAMPE Quarterly,* April 1984

8. McDanels, D. L., The Electrical Resistivity and Conductivity of Tungsten-Fiber-Reinforced Copper Composites, *ASM Transactions Quarterly,* Vol 59, No. 4, Dec 1966, p 994–997

9. Petrasek, D. W., Signorelli, R. A., and Weeton, J. W., "Metallurgical and Geometrical Factors Affecting Elevated-Temperature Tensile Properties of Discontinuous-Fiber Composites," NASA TN D3886, March 1967

10. Kelly, A., and Tyson, W. R., Fiber-Strengthened Materials, in *High-Strength Materials,* 2nd International Materials Conference, Berkeley, CA, V. F. Zackay, ed., John Wiley & Sons, 1965, p 578–602

Simplified Composite Micromechanics Equations for Mechanical, Thermal, and Moisture-Related Properties*

C. C. Chamis†

Summary

A unified set of composite micromechanics equations of simple form is summarized and described. This unified set can be used to predict unidirectional (ply) geometric, mechanical, thermal, and hygral properties using constituent material (fiber/matrix) properties. The set of equations can be used for predicting (1) ply in-plane uniaxial strengths; (2) through-the-thickness strength (interlaminar and flexural); (3) in-plane fracture toughness; (4) in-plane impact resistance; and (5) through-the-thickness (interlaminar and flexural) impact resistance. Equations are also included for predicting the hygrothermal effects on strengths, fracture toughness, and impact resistance. And finally, the set also includes approximate equations for predicting (1) moisture absorption; (2) glass transition temperature of wet resins; and (3) hygrothermal degradation effects. Several numerical examples are worked out to illustrate ease of use and versatility of these equations. These numerical examples also demonstrate the interrelationship of the various factors (geometric to environmental) and help provide insight into composite behavior at the micromechanistic level.

Introduction

Mechanical, thermal, and moisture-related properties (in the case of plastic matrix composites) of unidirectional composites are fundamental to analysis and design of fiber composite structures. Though some of these properties are determined by physical experiments, several of them are not readily amenable to direct measurement by testing. In addition, testing is usually time-consuming and costly, and composites of specific configurations must have been made prior to testing. Furthermore, parametric studies of the effect of fiber volume ratio on properties such as impact resistance and fracture toughness can be made only by conducting an extensive series of tests. Another approach is to use composite micromechanics to derive equations for predicting composite properties based on constituent (fiber and matrix) properties. Over the last 20 years, composite micromechanics has been used to derive equations for predicting selected composite strengths (Ref 1). However, many of these equations are not readily available since equations for different properties are scattered throughout the literature (Ref 2–4). In addition, recent developments on hygrothermal effects (Ref 5–7) provide simple equations for estimating moisture absorption, glass transition temperature of wet resins, and hygrothermal degradation. It is timely, therefore, to provide a summary of the various micromechanics equations.

Herein, a unified set of composite micromechanics equations is summarized and described. The set includes simple equations for predicting ply (unidirectional composite) properties using constituent properties. Equations are for: (1) tensile strengths (in-plane and through-the-thickness), (2) compression strengths, (3) flexural strength, (4) impact resistance, (5) fracture toughness, (6) ply geometry, (7) thermal properties, (8) moisture-related properties, and (9) hygral properties. The latter three include simple equations for pre-

---

*This article is a combination of two reports that are referenced and described in the introduction to this section. Before reading this presentation, it is suggested that the introduction be read as an aid to understanding the scope, symbolism, and some of the terminology used herein.

†Dr. Chamis is an Aerospace Structures and Composites Engineer at the National Aeronautics and Space Administration, Lewis Research Center, Cleveland, Ohio.

SYMBOLS

c	Heat capacity	θ	Ply-orientation angle
D	Diffusivity	λ	Weight percent
d	Diameter	ρ	Density
E	Modulus of elasticity	σ	Stress
G	Shear modulus		
$\mathscr{E}_\ell$	Impact energy density	Subscripts:	
K	Heat conductivity	F	Fiber property
k	Volume ratio	C	Compression property
M	Moisture (percent by weight)	D	Dry property
N_f	Number of filaments per roving end	G	Glass-transition
P	Property	F	Flexural
RHR	Relative humidity ratio	ℓ	Ply property
S	Strength	m	Matrix property
$\mathscr{S}$	Fracture "toughness"	S	Shear
T	Temperature	SB	Short-beam shear
t	Thickness	T	Tension
x,y,z	Structural reference axes	v	Void
1,2,3	Ply material axes	W	Wet
α	Thermal expansion coefficient	0	Reference property/temperature
β	Moisture expansion coefficient	∞	Saturation
δ	Interfiber/interply spacing	1,2,3	Direction corresponding to 1,2,3 ply
ξ	Fracture strain, strain		material axes

dicting the moisture absorption, the glass transition temperature of the wet resin, and the degradation effects due to hygrothermal environments. Results predicted by these equations are compared with available experimental data. These data are primarily from Ref 1 and 8–11. The equations are summarized in subsets corresponding to related strengths such as in-plane, through-the-thickness, etc. The description consists of the significance of the participating variables in the equations of each subset, several numerical examples, and possible implications.

The equations of each subset (strengths, fracture toughness, impact resistance, and hygrothermal degradation effects) are summarized in chart form (labeled figures). This allows the equations for each subset to be on one page for convenience of use and identification of interrelationships. Constituent material properties used in the numerical examples are tabulated and identified with the same symbols used in the equations. The numerical examples are presented in narrative, rather than tabular, form.

Many of the equations included in this composite micromechanics unified set appear in their present simplified forms for the first time. These equations evolved from continuing research on composite micromechanics and composite computational mechanics at Lewis Research Center. In addition, this is the first unified set which provides a quantified description of composite strength and strength-related behavior (fracture toughness and impact resistance) at the micromechanistic level.

Composite Mechanics—Definitions and Constituent Materials

The branch of composite mechanics which provides the formal structure by which ply uniaxial strengths are related to constituent properties is called composite micromechanics. Composite micromechanics for uniaxial strengths is identified concisely in the schematic in Fig. 1. The schematic in this figure defines the inputs to composite micromechanics and the resulting outputs. The inputs consist of constituent material (fiber/matrix) properties, geometric configurations, environmental conditions, and the fabrication process. The outputs consist of ply uniaxial strengths,

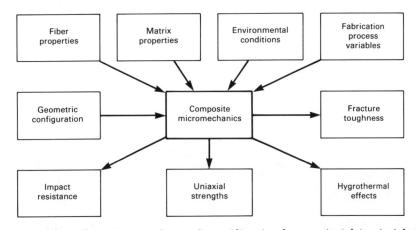

Fig. 1. Concepts, math-models, and equations used to predict unidirectional composite (ply) uniaxial strengths from constituent material properties, geometric configuration, fabrication process variables, and environmental conditions

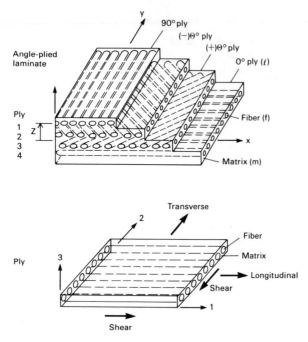

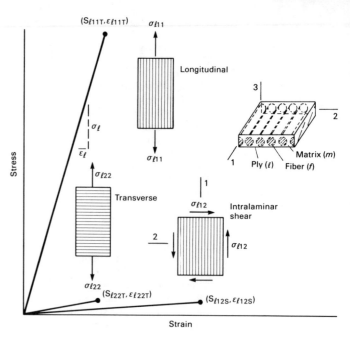

Fig. 2. Typical fiber composite geometry

Fig. 3. Typical stress-strain behavior of unidirectional fiber composites

impact resistance, fracture toughness, and hygrothermal effects.

The formal structure of composite micromechanics (concepts, math-models and equations) is developed based on certain assumptions (consistent with the physical situation) and the principles of solid mechanics. The four main assumptions made in deriving the equations described herein are: (1) the ply resists loads as depicted schematically in Fig. 2; (2) the ply and its constituents behave linearly elastic to fracture as is illustrated in Fig. 3; (3) the ply uniaxial strengths are associated with their respective fracture modes shown in Fig. 4; and (4) there is complete bonding at the interface of the constituents. Though the principles of solid mechanics can be used with various levels of mathematical sophistication, the mechanics of materials was used in deriving the equations summarized herein because it leads to explicit equations of simple form for each property.

Properties along the fiber direction (1-axis, Fig. 2) are conventionally called longitudinal; those transverse to the fiber direction (2-axis, Fig. 2) are called transverse; the in-plane shear is also called intralaminar shear (1-2 plane, Fig. 2). Those through the thickness (3-axis, Fig. 2) are called interlaminar properties. All ply properties are defined with respect to the ply material axes denoted by 1, 2, and 3 in Fig. 2 for description/analysis purposes. Most ply properties are denoted by a letter with suitable subscripts. The subscripts are selected to identify type of property (ply, fiber, matrix), plane, direction, and sense in the case of strengths. For example, $S_{\ell 11T}$ denotes ply longitudinal tensile strength while S_{fT} denotes fiber tensile strength. Though this notation may seem cumbersome, it is necessary to properly differentiate among the multitude of ply and constituent properties.

A variety of fibers have been used to make composites. Some of these are summarized in Table 1 with their re-

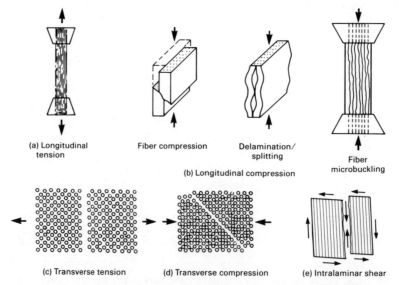

Fig. 4. In-plane fracture modes of unidirectional (ply) fiber composites

Table 1. Fiber properties(a)

Property	Symbol	Units	Boron	HMS	AS	T300	KEV	S-G	E-G
Number of fibers per end	N_f	...	1	10 000	10 000	3000	580	204	204
Fiber diameter	d_f	in.	0.0056	0.0003	0.0003	0.0003	0.00046	0.00036	0.00036
Density	ρ_f	lb/in.³	0.095	0.070	0.063	0.064	0.053	0.090	0.090
Longitudinal modulus	E_{f11}	10^6 psi	58	55.0	31.0	32.0	22	12.4	10.6
Transverse modulus	E_{f22}	10^6 psi	58	0.90	2.0	2.0	0.6	12.4	10.6
Longitudinal shear modulus	G_{f12}	10^6 psi	24.2	1.1	2.0	1.3	0.42	5.17	4.37
Transverse shear modulus	G_{f23}	10^6 psi	24.2	0.7	1.0	0.7	0.22	5.17	4.37
Longitudinal Poisson's ratio	ν_{f12}	...	0.20	0.20	0.20	0.20	0.35	0.20	0.22
Transverse Poisson's ratio	ν_{f23}	...	0.20	0.25	0.25	0.25	0.35	0.20	0.22
Heat capacity	C_f	btu/lb/°F	0.31	0.20	0.20	0.22	0.25	0.17	0.17
Longitudinal heat conductivity	K_{f11}	btu/hr/ft²/°F/in.	22	580	580	580	1.7	21	7.5
Transverse heat conductivity	K_{f22}	btu/hr/ft²/°F/in.	22	58	58	58	1.7	21	7.5
Longitudinal thermal expansion coefficient	α_{f11}	10^{-6} in./in./°F	2.8	−0.55	−0.55	0.55	−2.2	2.8	2.8
Transverse thermal expansion coefficient	α_{f22}	10^{-6} in./in./°F	2.8	5.6	5.6	5.6	30	2.8	2.8
Longitudinal tensile strength	S_{fT}	ksi	600	250	350	350	400	600	400
Longitudinal compression strength	S_{fC}	ksi	700	200	260	300	75	...	...
Shear strength	S_{fS}	ksi	100	...	...	...	...	...	...

(a) Transverse, shear, and compression properties are estimates inferred from corresponding composite properties.

Table 2. Matrix properties(a)

Name	Symbol	Units	LM	IMLS	IMHS	HM	Polyimide	PMR
Density	ρ_m	lb/in.³	0.042	0.046	0.044	0.045	0.044	0.044
Modulus	E_m	10^6 psi	0.32	0.50	0.50	0.75	0.50	0.47
Shear modulus	G_m	10^6 psi	...	...	...	...	...	...
Poisson's ratio	ν_m	...	0.43	0.41	0.35	0.35	0.35	0.36
Heat capacity	C_m	Btu/lb/°F	0.25	0.25	0.25	0.25	0.25	0.25
Heat conductivity	K_m	Btu/hr/ft²/°F/in.	1.25	1.25	1.25	1.25	1.25	1.25
Thermal expansion coefficient	α_m	10^{-6} in./in./°F	57	57	36	40	20	28
Diffusivity	D_m	10^{-10} in.²/sec	0.6	0.6	0.6	0.6	0.6	0.6
Moisture expansion coefficient	β_m	in./in./M	0.33	0.33	0.33	0.33	0.33	0.33
Tensile strength	S_{mT}	ksi	8	7	15	20	15	8
Compression strength	S_{mC}	ksi	15	21	35	50	30	16
Shear strength	S_{mS}	ksi	8	7	13	15	13	8
Tensile fracture strain	ϵ_{mT}	in./in. (%)	8.1	1.4	2.0	2.0	2.0	2.0
Compressive fracture strain	ϵ_{mC}	in./in. (%)	15	4.2	5.0	5.0	4.0	3.5
Shear fracture strain	ϵ_{mS}	in./in. (%)	10	3.2	3.5	4.0	3.5	5.0
Air heat conductivity	K_v	Btu/hr/ft²/°F/in.	0.225	0.225	0.225	0.225	0.225	0.225
Glass transition temperature (dry)	T_{GD}	°F	350	420	420	420	700	700

(a) LM = low modulus; IMLS = intermediate modulus low strength; IMHS = intermediate modulus high strength; HM = high modulus. Thermal, hygral, compression, and shear properties are estimates only; $G_m = E_m/2(1 + \nu_m)$.

spective properties needed for composite micromechanics. Similarly, some typical matrix resins are summarized in Table 2.

1.0. Geometric Relationships

Several geometric relationships are important in composite micromechanics. These range from constituent material volume ratios to interfiber spacing. A schematic of a ply that can be used to derive equations for various geometric relationships is illustrated in Fig. 5. Micromechanics equations for some geometric relationships including density (ρ) are summarized in Fig. 6. Note that the fiber diameter and the fiber volume ratio are important parameters (variables). Oftentimes the amount of fiber or resin in the composite is given in terms of weight percent. The weight percent can be used to determine the fiber or resin volume ratio from the equations. These equations are expressed in terms of weight

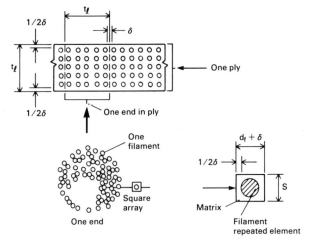

Fig. 5. Schematic of a ply

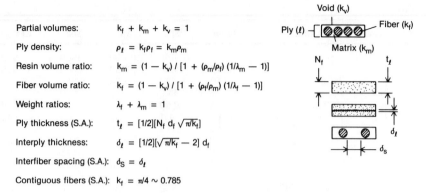

Partial volumes:	$k_f + k_m + k_v = 1$
Ply density:	$\rho_\ell = k_f \rho_f = k_m \rho_m$
Resin volume ratio:	$k_m = (1 - k_v) / [1 + (\rho_m/\rho_f)(1/\lambda_m - 1)]$
Fiber volume ratio:	$k_f = (1 - k_v) / [1 + (\rho_f/\rho_m)(1/\lambda_f - 1)]$
Weight ratios:	$\lambda_f + \lambda_m = 1$
Ply thickness (S.A.):	$t_\ell = [1/2][N_f\, d_f\, \sqrt{\pi/k_f}]$
Interply thickness:	$\delta_\ell = [1/2][\sqrt{\pi/k_f} - 2]\, d_f$
Interfiber spacing (S.A.):	$\delta_S = \delta_\ell$
Contiguous fibers (S.A.):	$k_f = \pi/4 \sim 0.785$

Fig. 6. Micromechanics: geometric relationships

ratio (λ_m or λ_f) and constituent densities (ρ_m and ρ_f), which are normally known.

The equations for the geometric relationships summarized in Fig. 6 can be used in a number of ways.

Example 1.1. The number of fibers through a ply thickness can be determined by solving the ply thickness (t_ℓ) equation for N_f. Using representative values for a graphite fiber/resin composite ($t_\ell = 0.005$ in., $d_f = 0.0003$ in., and $k_f = 0.57$), N_f equals 14 fibers through the ply, which is relatively large considering the small ply thickness. The interfiber spacing (δ_ℓ) is determined from its respective equation (Fig. 6). The interply (matrix layer) thickness can be determined from the interfiber spacing (δ_ℓ) equation.

Example 1.2. The interply thickness of a graphite fiber/resin composite is needed. This composite has 40% resin by weight and 2% voids. The fiber volume ratio and the fiber diameter are needed to determine the interply thickness. Assume the fiber density to be 0.063 lb/in.^{3} and the resin density to be 0.044 lb/in.^{3} In addition, the fiber diameter (d_f) is 0.0003 in. and the resin weight ratio (λ_m) is 0.4 (and $\lambda_f = 0.6 : \lambda_m + \lambda_f = 1.0$). Using these values, the fiber volume ratio is 0.5 and the interply thickness is 0.000076 in. This thickness is about 25% of the fiber diameter and about 1.5% of the ply thickness (0.005 in.). Obviously, the interply thickness is very thin. Use of the equations in Fig. 6 with other examples is instructive and provides insight/appreciation of the interrelationships among the various composite micromechanics geometric variables.

2.0. Mechanical Properties: Elastic Constants

The simple equations for relating ply elastic constants (moduli and Poisson's ratios) are summarized in Fig. 7. Note that the properties in the third direction are the same as those in the second because they were obtained by assuming that the ply is transversely isotropic in the 2–3 plane. Therefore, a total of five independent elastic constants are required ($E_{\ell 11}$, $E_{\ell 22}$, $G_{\ell 12}$, $G_{\ell 23}$, and $\nu_{\ell 12}$). In order to use the equations in Fig. 7, the fiber volume ratio (k_f), five fiber properties (E_{f11}, E_{f22}, G_{f12}, G_{f23}, and $\nu_{\ell 12}$), and two matrix properties (E_m and ν_m) are needed. The matrix shear modulus is related to E_m and ν_m by $G_m = E_m/2(1 + \nu_m)$ since the matrix is assumed to be isotropic.

The void effects are incorporated into these equations by using the fiber and matrix ratios (k_f and k_m) predicted by the appropriate equations in Fig. 6. This is illustrated by the following example:

Example 2.1. Determine the longitudinal transverse and shear moduli of a ply made from AS-graphite fiber/epoxy-resin composite with 70% fibers by weight and 1% void volume ratio. The fiber volume ratio for this case is 0.624 (k_f equation, Fig. 6). Using this value for k_f, and respective values for AS fibers from Table 1 and for matrix (IMLS) from Table 2, yields $E_{\ell 11} = 19.3 \times 10^6$ psi, $E_{\ell 22} = 1.23 \times 10^6$ psi, and $G_{\ell 12} = 0.63 \times 10^6$ psi. The corresponding values for zero voids are: $k_f = 0.630$, $E_{\ell 11} = 19.7 \times 10^6$ psi, $E_{\ell 22} = 1.24 \times 10^6$ psi, and $G_{\ell 12} = 0.64 \times 10^6$ psi. This example illustrates the interrelationships of the various equations as well as the negligible effect of 1% voids on these properties.

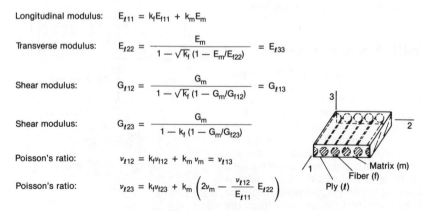

Longitudinal modulus:	$E_{\ell 11} = k_f E_{f11} + k_m E_m$
Transverse modulus:	$E_{\ell 22} = \dfrac{E_m}{1 - \sqrt{k_f}\,(1 - E_m/E_{f22})} = E_{\ell 33}$
Shear modulus:	$G_{\ell 12} = \dfrac{G_m}{1 - \sqrt{k_f}\,(1 - G_m/G_{f12})} = G_{\ell 13}$
Shear modulus:	$G_{\ell 23} = \dfrac{G_m}{1 - k_f\,(1 - G_m/G_{f23})}$
Poisson's ratio:	$\nu_{\ell 12} = k_f \nu_{f12} + k_m \nu_m = \nu_{\ell 13}$
Poisson's ratio:	$\nu_{\ell 23} = k_f \nu_{f23} + k_m \left(2\nu_m - \dfrac{\nu_{\ell 12}}{E_{\ell 11}} E_{\ell 22}\right)$

Fig. 7. Composite micromechanics: mechanical properties

Example 2.2. Determine the $G_{\ell 23}$ shear modulus, $\nu_{\ell 12}$, and $\nu_{\ell 23}$ for the same composite and "zero" voids. Again, with $k_f = 0.63$ and respective values from Table 1 for the fiber and Table 2 for the matrix, $G_{\ell 23} = 0.46 \times 10^6$ psi and $\nu_{\ell 12} = 0.278$. The $\nu_{\ell 23}$ value is determined from its respective equation in Fig. 7. Using the previously calculated value $E_{\ell 22} = 1.24 \times 10^6$ psi and $G_{\ell 23} = 0.46 \times 10^6$ psi yields $\nu_{\ell 23} = 0.348$. This value is greater than the $\nu_{\ell 12}$ value of 0.278. The examples just described illustrate the ease with which the various properties can be determined. This is more significant for $G_{\ell 23}$ and $\nu_{\ell 23}$, which are needed for any three-dimensional analysis including finite element.

3.0 Mechanical Properties: In-Plane Uniaxial Strengths

There are five in-plane ply uniaxial strengths. These are identified as: (1) longitudinal tension ($S_{\ell 11T}$); (2) longitudinal compression ($S_{\ell 11C}$); (3) transverse tension ($S_{\ell 22T}$); (4) transverse compression ($S_{\ell 22C}$); and (5) in-plane or intralaminar shear ($S_{\ell 22S}$). The fracture modes associated with each uniaxial strength are depicted schematically in Fig. 4. Note that there are three different and distinct fracture modes for longitudinal compression (Fig. 4b): (1) fiber compression (shear plane) fracture; (2) delamination transverse splitting or panel buckling; and (3) fiber microbuckling.

The composite micromechanics equations for the ply uniaxial strengths are summarized in Fig. 8 with attendant schematics. The schematics define the load direction, the fiber orientation, and the notation used in the micromechanics equations. The first five equations describe the in-plane uniaxial strengths, respectively: $S_{\ell 11T}$, $S_{\ell 11C}$, $S_{\ell 22T}$, $S_{\ell 22C}$, and $S_{\ell 12S}$. The last equation is for the void effect on the resin strength (S_m) and also provides lower-bound estimates on $S_{\ell 22T}$, $S_{\ell 22C}$, and $S_{\ell 12S}$ as will be described later.

The following are observed from the micromechanics equations for ply uniaxial strengths: (1) $S_{\ell 11T}$ depends on S_{fT}, and the fiber compression fracture mode for $S_{\ell 11C}$ depends on S_{fc}. These are the only two that are fiber-strength-dominated. (2) The delamination/splitting for $S_{\ell 11C}$ depends on matrix shear strength (through $S_{\ell 12S}$; Eq 5) and the matrix tensile strength and, therefore, is resin-strength-dominated. (3) The microbuckling fracture mode for $S_{\ell 11C}$ depends strongly on the shear modulus and mildly on the modular ratio (G_m/G_{f12}) and, therefore, is resin-stiffness-dominated. (4) The other three ($S_{\ell 22T}$, $S_{\ell 22C}$, and $S_{\ell 12S}$) depend strongly on the respective resin strengths and are, therefore, resin-strength-dominated. (5) The fiber volume ratio strongly affects $S_{\ell 11T}$ and $S_{\ell 11C}$ (fiber compressing or fiber microbuckling), which are the fiber-strength- or shear-stiffness-dominated ply strengths. (6) The fiber volume ratio mildly affects $S_{\ell 22T}$, $S_{\ell 22C}$, $S_{\ell 12S}$, and $S_{\ell 11C}$ (delamination/shear), which are the resin-strength-dominated ply uniaxial strengths. (7) The voids influence the matrix strength. Several examples below illustrate use of the uniaxial strength equations in Fig. 8.

Example 3.1. Calculate the ply tensile strength ($S_{\ell 11T}$) of a graphite fiber (AS)/intermediate modulus, high strength (IMHS) epoxy (AS/E) composite with 0.6 fiber volume ratio. From Table 1, $S_{ft} = 350$ ksi, and from equation 1 (Fig. 8), $S_{\ell 11T} = 210$ ksi, which is the same as the measured value in Table 3.

Example 3.2. Calculate the ply compression strength for an AS/E (IMHS) composite with 0.6 fiber volume ratio. All three equations (Eq 2, Fig. 8) should be used. To use the first equation, S_{fc} must be known. If it is not known, $S_{fc} \approx 0.9\, S_{fT}$ is a good approximation for graphite fiber/matrix composites (Ref 12). Using this approximation and respective values from Tables 1 and 2 in the first equation yields:

$$S_{\ell 11C} = 0.6 \times 0.9 \times 350 = 189 \text{ ksi}$$

From the second equation (note need to evaluate $S_{\ell 12S}$ from Eq 5, Example 3.5, with incomplete bonding):

$$S_{\ell 11C} = 10.0 \times 8.1 + 2.5 \times 15 = 118 \text{ ksi}$$

From the third equation:

$$S_{\ell 11C} = 0.185/[1.0 - 0.6\,(1.0 - 0.185/2.0)] = 406 \text{ ksi}$$

A conservative approach is to select the lowest value, or $S_{\ell 11C} = 118$ ksi. This is about 69% of the typical measured

1. Longitudinal tension:	$S_{\ell 11T} \approx k_f\, S_{fT}$	
2. Longitudinal compression:		
Fiber compression:	$S_{\ell 11C} \approx k_f\, S_{fC}$	
Delamination/shear:	$S_{\ell 11C} \approx 10 S_{\ell 12S} + 2.5\, S_{mT}$	
Microbuckling:	$S_{\ell 11C} \approx \dfrac{G_m}{1 - k_f\left(1 - \dfrac{G_m}{G_{f12}}\right)}$	
3. Transverse tension:	$S_{\ell 22T} \approx [1 - (\sqrt{k_f} - k_f)\,(1 - E_m/E_{f22})]\, S_{mT}$	
4. Transverse compression:	$S_{\ell 22C} \approx [1 - (\sqrt{k_f} - k_f)\,(1 - E_m/E_{f22})]\, S_{mC}$	
5. Intralaminar shear:	$S_{\ell 12S} \approx [1 - (\sqrt{k_f} - k_f)\,(1 - G_m/G_{f12})]\, S_{mS}$	
6. For voids:	$S_m \approx \{1 - [4k_v/(1 - k_f)\pi]^{1/2}\}\, S_m$	

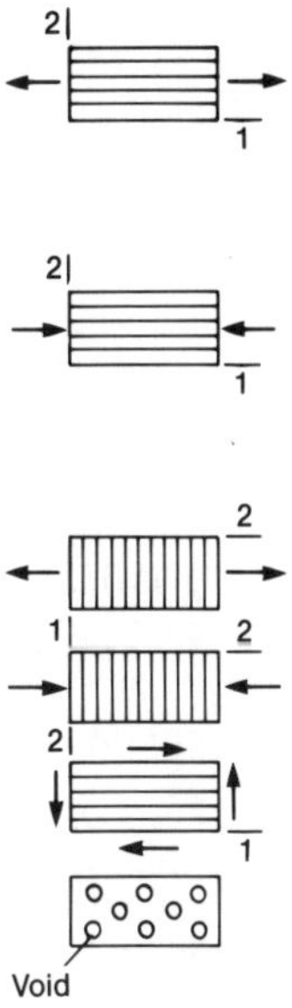

Void

All equations apply to metal matrix and some ceramic matrix composites. All apply to plastic (polymer) matrix composites. For Equation 1 and the first part of Equation 2, see "Introduction and Comments" article.

Fig. 8. Composite micromechanics: in-plane uniaxial strengths

Table 3. Typical experimental values(a) for unidirectional composite (ply) uniaxial strengths, ksi

| Fiber | Epoxy | Fiber volume ratio | Longitudinal | | Transverse | | In-plane shear | Flexural | | Inter-laminar shear |
			Tension	Compression	Tension	Compression		Longitudinal	Transverse	
Boron	505	0.50	230	360	9.1	35.0	12.0	350	40	16.0
AS	3501	0.60	210	170	7.0	36.0	9.0	230	18	14.0
HMS	934	0.60	120	90	6.7	28.6	6.5	150	7(b)	10.5
T-300	5208	0.60	210	210	6.5	36.0	9.0	260	18	14.0
KEVLAR-49	· · ·	0.60	200	40	4.0	9.4	8.7	90	6	7.0
GLASS										
S (901-S)	1002S	0.60	220	120	6.7	25.0	12.0	320	21	14.0
E	1002	0.60	160	90	4.0	20.0	12.0	165	20	14.0

(a) Data in this table were compiled from Ref 8, 9, 10, and 11. (b) Estimate.

value of 170 ksi in Table 3. The value of 189 ksi predicted by the first equation is also reasonable. This value could be used in laminates which have other than 0° plies on the outside.

Example 3.3. Calculate the ply transverse tensile strength ($S_{\ell 22T}$) for an AS/E (IMHS) composite with 0.6 fiber volume ratio. From Equation 3 and respective values from Tables 1 and 2,

$$S_{\ell 22T} = [1.0 - (\sqrt{0.6} - 0.6) \times (1.0 - 0.5/2.0)] \times 15$$
$$= 13 \text{ ksi}$$

This is about twice the measured value of 7 ksi in Table 3. One major reason for this difference is the complete bonding at the fiber/matrix interface assumed in deriving Equation 3, Fig. 8. Incomplete bonding at the interface may be approximated by assuming the presence of voids. Assuming about 5% voids by volume ($k_v = 0.05$) and using Equation 6 (Fig. 8), the reduced or degraded resin tensile strength is

$$S_{mT} = \{1.0 - [4(0.05)/(1.0 - 0.6)\pi]^{1/2}\} \times 15 = 9.0 \text{ ksi}$$

Using this reduced resin strength in Equation 3 (Fig. 8) yields $S_{\ell 22T} = 7.8$ ksi, which is a reasonable estimation compared to the measured value of 7 ksi.

The above calculations lead to the conclusion that estimation of $S_{\ell 22T}$, then, requires two steps:

1. Degradation of S_{mT} due to 5% voids by volume ($k_v = 0.05$) as predicted by Equation 6 (Fig. 8).
2. Substitution of the degraded S_{mT} in Equation 3.

Though the value of 5% voids may seem somewhat arbitrary, it is reasonable since Equation 3, Fig. 8, does not account for factors such as nonuniform fiber distribution within the ply, incomplete (partial) bonding at the fiber/matrix interface, and possible differences in the *in situ* resin matrix properties compared to the neat resin properties.

Example 3.4. Calculate the transverse compression strength ($S_{\ell 22C}$) for an AS/E (IMHS) composite with 0.6 fiber volume ratio. From Equation 4 and respective values from Tables 1 and 2,

$$S_{\ell 22C} = [1.0 - (\sqrt{0.6} - 0.6) (1.0 - 0.5/2.0)] \times 35.0$$
$$= 30.4 \text{ ksi}$$

This value is about 85% of the typical value in Table 3. It is worth noting that the interfacial bond and fiber nonuniformity are not critical in transverse compression and, therefore, do not contribute to resin compression strength degradation.

Example 3.5. Calculate the intralaminar shear strength ($S_{\ell 12S}$) for an AS/E (IMHS) composite with 0.6 fiber volume ratio. From Equation 5 and respective values from Tables 1 and 2,

$$S_{\ell 12S} = [1.0 - (\sqrt{0.6} - 0.6) (1.0 - 0.185/2.0)] \times 13$$
$$= 10.9 \text{ ksi}$$

This value is about 21% greater than the typical measured value of 9 ksi (Table 3). This value is reasonable in view of the nonuniform fiber distribution bonding condition at the interface and deviations in *in situ* properties from neat resin properties as was mentioned for $S_{\ell 22T}$ (Example 3.2). A closer estimate to the measured value may be obtained by degrading the matrix shear strength S_{mS} assuming 2% voids ($k_v = 0.02$) in Equation 6, Fig. 8. The result is $S_{\ell 12S} = 8.1$. It is worth noting that the various factors that affect the transverse tensile strength also affect the intralaminar shear strength, but not as severely.

Example 3.6. Calculate the effect of voids on the ply transverse tensile strength of an AS/E (IMHS) composite with 0.6 fiber volume ratio ($k_f = 0.6$) and 0.02 void volume ratio ($k_v = 0.02$). This is accomplished using the following three steps:

1. Voids effect on S_{mT} (Eq 6, Fig. 8):

$$S_{mT} = \{1.0 - [4(0.02)/(1.0 - 0.6)\,\pi]^{1/2}\} \times 15$$
$$= 11.2 \text{ ksi}$$

2. Incomplete interfacial bonding effects:

$$S_{mT} = \{1.0 - [4(0.05)/(1.0 - 0.6\,\pi]^{1/2}\} \times 11.2$$
$$= 6.8 \text{ ksi}$$

3. Transverse tensile strength (Eq 3, Fig. 8):

$$S_{\ell 22T} = [1.0 - (\sqrt{0.6} - 0.6)(1 - 0.5/2.0)]\,6.8$$
$$= 5.9 \text{ ksi}$$

Note that the two void ratios ($k_v = 0.02$ and $k_v = 0.05$) are not additive since Equation 6 is nonlinear.

Example 3.7. Calculate the lower bound of the ply transverse tensile strength for an AS/E (IMHS) composite with $k_f = 0.6$. For this case, we use Equation 6, Fig. 8, with $k_v = 0.6$, $k_f = 0.0$ and $S_{\ell 22T} = S_m$:

$$S_{\ell 22T} = \{1.0 - [4(0.6)/\pi]^{1/2}\} \times 15 = 1.9 \text{ ksi}$$

This value is about 27% of the typical measured value of 7 ksi and about 15% of that predicted by Equation 3, Fig. 8, without any degradation. The lower bound is overly pessi-

1. Interlaminar shear:

$$S_{\ell 13S} \approx [1 - (\sqrt{k_f} - k_f)(1 - G_m/G_{f12})]\, S_{mS}$$

$$S_{\ell 23S} \approx \left[\frac{1 - \sqrt{k_f}\,(1 - G_m/G_{f23})}{1 - k_f\,(1 - G_m/G_{f23})} \right] S_{mS}$$

2. Short-beam shear:

$$S_{\ell 13SB} \approx 1.5\, S_{\ell 13S}$$

$$S_{\ell 23SB} \approx 1.5\, S_{\ell 23S}$$

3. Flexural:

$$S_{\ell 11F} \approx \frac{3\, k_f\, S_{fT}}{1 + \dfrac{S_{fT}}{S_{fC}}}$$

$$S_{\ell 22F} \approx \frac{3\,[1 - (\sqrt{k_f} - k_f)(1 - E_m/E_{f22})]\, S_{mT}}{1 + \dfrac{S_{mT}}{S_{mC}}}$$

4. For voids:

$$S_m \approx \{1 - [4k_v/(1 - k_f)\,\pi]^{1/2}\}\, S_m$$

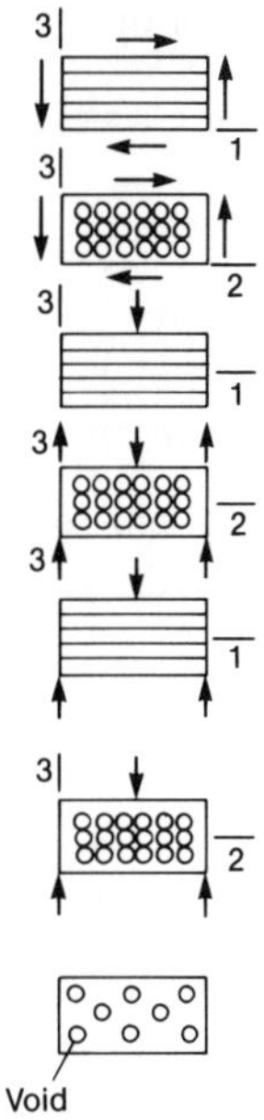

Void

All equations apply to plastic (polymer) matrix composites. All but Equation 4 apply to metal matrix and some ceramic matrix composites.

Fig. 9. Composite micromechanics: through-the-thickness uniaxial strengths

mistic for acceptable composites and should be used only in composites with no interfacial bonding. Lower-bound estimates on ply transverse compression ($S_{\ell 22C}$) and ply intralaminar shear strength ($S_{\ell 12S}$) are obtained by following the same procedure.

4.0. Mechanical Properties: Through-the-Thickness Uniaxial Strengths

There are six through-the-thickness uniaxial strengths. These are identified as: (1) longitudinal interlaminar shear (parallel to the fiber direction) ($S_{\ell 13S}$); (2) transverse interlaminar shear (transverse to the fiber direction) ($S_{\ell 23S}$); (3) longitudinal short-beam shear (parallel to the fiber direction) ($S_{\ell 13SB}$); (4) transverse short-beam shear (transverse to the fiber direction) ($S_{\ell 23SB}$); (5) longitudinal flexural (bending) ($S_{\ell 11F}$); and (6) transverse flexural ($S_{\ell 22F}$). The composite micromechanics equations for these uniaxial strengths are summarized in Fig. 9 with attendant schematics. The first six equations describe the six through-the-thickness uniaxial strengths, respectively: $S_{\ell 13S}$, $S_{\ell 23S}$, $S_{\ell 13SB}$, $S_{\ell 23SB}$, $S_{\ell 11F}$, and $S_{\ell 22F}$. The last equation describes the void effects on the resin strength and can also be used as a lower bound on ply strengths dominated by the resin, as was mentioned previously.

The following are observed from the composite micromechanics equations in Fig. 9: (1) $S_{\ell 13S}$ is the same as $S_{\ell 12S}$ in Fig. 8; (2) $S_{\ell 23S}$ depends strongly on the resin shear strength (S_{mS}) and mildly on k_f and G_m/G_{f23}; (3) the short-beam shear strengths $S_{\ell 13SB}$ and $S_{\ell 23SB}$ are 1.5 times their respective interlaminar shear strengths ($S_{\ell 13S}$ and $S_{\ell 23S}$); (4) the longitudinal flexural strength ($S_{\ell 11F}$) is fiber-dominated and, thus, depends strongly on k_f, s_{fT}, and S_{fC}; (5) the transverse flexural strength ($S_{\ell 22F}$) is matrix-strength-dominated and, thus, depends strongly on S_{mT} and S_{mC} but depends mildly on k_f and E_m/E_{f22}; (6) the voids degrade matrix strength depending nonlinearly on both k_v and k_f. Several examples below illustrate use of the equations in Fig. 9.

Example 4.1. Calculate the longitudinal interlaminar shear strength ($S_{\ell 13S}$) for an AS/E (IMHS) composite with 0.6 fiber volume ratio. This is the same as $S_{\ell 12S}$. However, we go through the steps again for completeness. Using Equation 4, the first of Equation 1 (Fig. 9), and respective property values from Tables 1 and 2:

1. Incomplete bond simulation ($k_v = 0.02$):

$$\begin{aligned} S_{mS} &= \{1.0 - [4(0.02)/(1.0 - 0.6)\pi]^{1/2}\} \times 13 \\ &= 9.7 \text{ ksi} \end{aligned}$$

2. Longitudinal interlaminar shear strength:

$$\begin{aligned} S_{\ell 13S} &= [1.0 - (\sqrt{0.6} - 0.6)(1 - 0.185/2.0)] \times 9.7 \\ &= 8.1 \text{ ksi, which is the same as } S_{\ell 12S}, \text{ as expected.} \end{aligned}$$

Example 4.2. Calculate the transverse interlaminar shear strength ($S_{\ell 23S}$) for an AS/E (IMHS) composite with 0.6 fiber volume ratio. Using S_{mS} from Example 4.1, Fig. 9, and respective properties from Tables 1 and 2 in the second part of Equation 1, Fig. 9, yields:

$$S_{\ell 23S} = \left[\frac{1.0 - \sqrt{0.6}\,(1.0 - 0.185/1.0)}{1.0 - 0.6\,(1.0 - 0.185/1.0)} \right] \times 9.7 = 7.0 \text{ ksi}$$

Note that $S_{\ell 23S} = 0.85\, S_{\ell 13S}$, indicating that the ply is weaker in transverse interlaminar shear than in longitudinal interlaminar shear.

Example 4.3. Calculate (1) the longitudinal short-beam shear strength ($S_{\ell 13SB}$) for the composite in Example 4.1, and (2) the transverse short-beam shear strength ($S_{\ell 23SB}$) in Example 4.2:

1. Using $S_{\ell 13S} = 8.1$ in the first part of Equation 2, Fig. 9:

$$S_{\ell 13SB} = 1.5 \times 8.1 = 12.1 \text{ ksi}$$

This value is in reasonably good agreement, underestimating by 14%, with the typical measured value of 14 ksi in Table 3.

2. Using $S_{\ell 23S} = 7.0$ in the second part of Equation 2, Fig. 9:

$$S_{\ell 23SB} = 1.5 \times 7.0 = 10.5 \text{ ksi}$$

Measured values for this strength are not available for comparison.

Example 4.4. Calculate the ply longitudinal flexural strength ($S_{\ell 11F}$) for an AS/E (IMHS) composite with 0.6 fiber volume ratio. Using $k_f = 0.6$, $S_{fC} = 0.9 \, S_{fT}$, and $S_{fT} = 350$ ksi (Table 1) in the first part of Equation 3, Fig. 9:

$$S_{\ell 11F} = \frac{3(0.6)(350)}{1.0 + 1.0/0.9} = 298 \text{ ksi}$$

This value overestimates the typical measured data of 230 ksi by about 30%. A lower estimate for this strength is obtained by using $S_{\ell 11T} = 210$ ksi (from Example 3.1) and $S_{\ell 11C} = 118$ ksi (from Example 3.2) in the following equation:

$$S_{\ell 11F} = \frac{3 S_{\ell 11T}}{1 + S_{\ell 11T}/S_{\ell 11C}} = \frac{3 \times 210}{1.0 + 210/118} = 227 \text{ ksi}$$

which estimates the measured value of 230 ksi almost exactly. This is definitely a very good estimate considering the simplicity of the equations and the uncertainties associated with longitudinal compression failure (Ref 12). It also illustrates, in part, that flexural failure is probably a complex combination of tension, compression, and intralaminar shear failures.

Example 4.5. Calculate the ply transverse flexural strength of an AS/E (IMHS) composite with 0.6 fiber volume ratio. Using $k_f = 0.6$ and respective property values from Tables 1 and 2 in the second part of Equation 3, Fig. 9:

$$S_{\ell 22F} = \left\{ \frac{3[1.0 - (\sqrt{0.6} - 0.6)(1.0 - 0.5/2.0)]}{1.0 + 15/35} \right\} 15$$
$$= 27.4 \text{ ksi}$$

This overestimates the typical value of 18 ksi (Table 3) by about 52%. As was the case for longitudinal flexural strength, a lower estimate can be obtained by substituting $S_{\ell 22T} = 7.8$ ksi (from Example 3.3 with partial interfacial bond) and $S_{\ell 22C} = 30.4$ ksi (Example 3.4) in the following equation:

$$S_{\ell 22F} = \frac{3 \times S_{\ell 22T}}{1 + S_{\ell 22T}/S_{\ell 22C}} = \frac{3 \times 7.8}{1.0 + 7.8/30.4} = 18.6 \text{ ksi}$$

which is almost equal to the typical measured values of 18 ksi (Table 3).

Example 4.6. Calculate the effect of 3% voids on the longitudinal flexural strength for an AS/E (IMHS) composite with 0.6 fiber volume ratio. The first part of Equation 3, Fig. 9, shows no void effect. However, the lower-estimate equation in Example 4.4 indicated that the compression strength predicted by the second part of Equation 3 should be used. This is calculated using the following steps:

1. S_{mS} degraded for voids (Eq 4, Fig. 9):

$$S_{mS} = \{1.0 - [4(0.03)/(1.0 - 0.6)\pi]^{1/2}\} \times 13$$
$$= 9.0 \text{ ksi}$$

2. Intralaminar shear strength $S_{\ell 12S}$ (Eq 5, Fig. 8) with S_{mS} from step 1:

$$S_{\ell 12S} = [1 - (\sqrt{0.6} - 0.6)(1.0 - 0.185/2.0)] \, 9.0$$
$$= 7.6 \text{ ksi}$$

3. S_{mT} degraded for voids, with void degradation ratio same as in step 1 for S_{mS}:

$$S_{mT} = \{1.0 - [4(0.03)/(1.0 - 0.6)\,\pi]^{1/2}\} \times 15$$
$$= 10.4 \text{ ksi}$$

4. Longitudinal compression $S_{\ell 11C}$ (second part of Eq 2, Fig. 8):

$$S_{\ell 11C} = 10.0 \times 7.6 + 1.5 \times 10.4 = 102 \text{ ksi}$$

5. Longitudinal flexural strength (lower-estimate equation, Example 4.4):

$$S_{\ell 11F} = 3 \times 210/(1.0 + 210/102) = 206 \text{ ksi}$$

which is about 9% less than the 227 ksi value calculated without voids in Example 4.4. Two points are worth noting: (1) step 4 results in a void degradation of about 13% in $S_{\ell 11C}$ and (2) the void degradation is more severe for the longitudinal compression strength than for the longitudinal flexural strength (13% versus 9%, respectively).

The above calculations show that the composite micromechanics equations in Fig. 9 and the alternates in Examples 4.4 and 4.5 can be used to obtain reasonable estimates for through-the-thickness uniaxial strengths. The calculations also show that the equations can be used to interpret measured data. In either case, these equations should be used judiciously.

5.0. Uniaxial Fracture "Toughness"

Fracture toughness is a measure of a material's ability to resist defects such as holes, slits, and notches. Fracture toughness is described by fracture toughness parameters associated with distinct fracture modes (Ref 13). Three fracture modes are generally considered: opening mode (Mode I), in-plane shear (Mode II), and out-of-plane shear (Mode III). In the case of unidirectional composites and assuming full thickness penetration defects, there are three major in-plane fracture toughness parameters herein defined as: (1) longitudinal fracture toughness ($\mathscr{S}_{\ell 11T}$); (2) transverse fracture toughness ($\mathscr{S}_{\ell 22T}$); and (3) in-plane (intralaminar) shear fracture toughness ($\mathscr{S}_{\ell 12S}$). These three are parallel to the uniaxial in-plane strengths $S_{\ell 11T}$, $S_{\ell 22T}$, and $S_{\ell 12S}$, respectively. These fracture toughness parameters are used herein to denote far-field stress required to produce additional damage in the composite. It is not clear whether far-field shear stress will produce Mode II fracture in unidirectional composites or some component of Mode I (opening) fracture. In view of this we consider only Mode I, opening fracture modes $\mathscr{S}_{\ell 11T}$ and $\mathscr{S}_{\ell 22T}$.

The equations describing the longitudinal and transverse fracture toughness parameters ($\mathscr{S}_{\ell 11T}$ and $\mathscr{S}_{\ell 22T}$) are given in Fig. 10 (Ref 14) with attendant schematics. Two sets of equations are given. In the first set (Eq 1 and 2), $\mathscr{S}_{\ell 11T}$ and $\mathscr{S}_{\ell 22T}$ are expressed in terms of ply properties, while in the second set (Eq 3 and 4), they are expressed in terms of constituent properties. It can be seen in Eq 3 and 4 that: (1) $\mathscr{S}_{\ell 11T}$ depends linearly on S_{fT} and depends in a complex way on K_f, (E_{f11}/E_m), (E_m/E_{f22}), (G_m/G_{f12}), and ν_m; and (2) $\mathscr{S}_{\ell 22T}$ depends linearly on S_{mT} and in a complex way on the other constituent material properties. Examining only key parameters k_f, E_{f11}, E_m, S_{fT}, and S_{mT}, $\mathscr{S}_{\ell 11T}$ increases with increasing k_f, E_{f11}, and E_m and with decreasing E_{f11}, while $\mathscr{S}_{\ell 22T}$ increases with increasing E_{f11} and S_{mT} and with decreasing k_f and E_m. Equations 3 and 4, Fig. 10, are cumbersome to use. It is easier to

Using ply properties:

1. $\mathscr{S}_{\ell 11T} = \dfrac{S_{\ell 11T}}{1 + \left[2\left(\dfrac{E_{\ell 11}}{E_{\ell 22}} - \nu_{\ell 12}\right) + \dfrac{E_{\ell 11}}{G_{\ell 12}}\right]^{1/2}}$

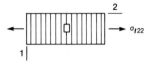

2. $\mathscr{S}_{\ell 22T} = \dfrac{S_{\ell 22T}}{1 + \left(\dfrac{E_{\ell 22}}{E_{\ell 11}}\right)^{1/2}\left[2(1 - \nu_{\ell 12}) + \dfrac{E_{\ell 11}}{G_{\ell 12}}\right]^{1/2}}$

Using constituent properties:

3. $\mathscr{S}_{\ell 11T} \approx \dfrac{k_f\, S_{fT}}{1 + (2k_f E_{f11}/E_m)^{1/2}\left\{2(1 - \sqrt{k_f}) + \sqrt{k_f}\left(\dfrac{E_m}{E_{f22}} + \dfrac{G_m}{G_{f12}}\right) + \nu_m\left[1 - \sqrt{k_f}\left(1 - \dfrac{G_m}{G_{f12}}\right)\right]\right\}^{1/2}}$

4. $\mathscr{S}_{\ell 22T} = \dfrac{(k_f E_{f11})^{1/2}\,[1 - (\sqrt{k_f} - k_f)(1 - E_m/E_{f22})]\,[1 - \sqrt{k_f}(1 - E_m/E_{f22})]^{1/2}\,S_{mT}}{\{k_f E_{f11}[1 - \sqrt{k_f}(1 - E_m/E_{f22})]\}^{1/2} + \langle E_m\{2[1 - k_f(\nu_{f12} + \nu_m) - \nu_m] + k_f[1 - \sqrt{k_f}(1 - G_m/G_{f12})](E_{f11}/G_m)\}\rangle^{1/2}}$

5. $\mathscr{S}_{\ell 22} \approx \dfrac{[1 - \sqrt{k_f} - k_f)(1 - E_m/E_{f22})]\,[1 - \sqrt{k_f}(1 - E_m/E_{f22})]^{1/2}\,S_{mT}}{[1 - \sqrt{k_f}(1 - E_m/E_{f22})]^{1/2} + [2(1 + \nu_m)]^{1/2}\,[1 - \sqrt{k_f}(1 - G_m/G_{f12})]^{1/2}}$

All equations apply to plastic (polymer) matrix composites and to metal matrix composites. Some may apply to some ceramic matrix composites.

Fig. 10. Composite micromechanics: uniaxial fracture "toughness"

use Equations 1 and 2, Fig. 10, in conjunction with the ply mechanical properties equations summarized in Fig. 7. Two examples below illustrate use of these equations and interpretation of the results.

Example 5.1. Calculate the longitudinal fracture toughness of an AS/E (IMHS) composite with 0.6 fiber volume ratio. This is accomplished using the following steps together with equations from Fig. 7 and 10, and respective constituent properties from Tables 1 and 2.

1. Calculate $S_{\ell 11T}$ (Eq 1, Fig. 8):

$S_{\ell 11T} = 0.6 \times 350 = 210$ ksi

2. Calculate $E_{\ell 11}$ (first equation, Fig. 7):

$E_{\ell 11} = 0.6 \times 31 + 0.4 \times 0.5 = 18.8$ mpsi

3. Calculate $E_{\ell 22}$ (second equation, Fig. 7):

$E_{\ell 22} = 0.5/[1.0 - \sqrt{0.6}\,(1.0 - 0.5/2.0)] = 1.2$ mpsi

4. Calculate $\nu_{\ell 12}$ (fifth equation, Fig. 7):

$\nu_{\ell 12} = 0.6 \times 0.20 + 0.4 \times 0.35 = 0.26$

5. Calculate $G_{\ell 12}$ (third equation, Fig. 7):

$G_{\ell 12} = 0.185/[1.0 - \sqrt{0.6}\,(1 - 0.185/2.0)] - 0.62$ mpsi

6. Calculate $\mathscr{S}_{\ell 11T}$ (Eq 1, Fig. 10):

$\mathscr{S}_{\ell 11T} = 210/\{1.0 + [2(18.8/1.2 - 0.26) + (18.8/0.62)]^{1/2}\} = 24$ ksi

This implies that the far-field (P/A-type) ply stress will be 24 ksi when the crack-like defect starts growing. This can also be interpreted as follows: The stress required to produce additional damage is reduced by a factor of about ten compared to that in a ply without defects.

Example 5.2. Calculate the transverse fracture toughness ($\mathscr{S}_{\ell 22T}$) in an AS/E (IMHS) composite with 0.6 fi-

ber volume ratio. This is calculated using appropriate equations from Fig. 7 and 10, respective properties from Tables 1 and 2, and the following steps.

1. Calculate $E_{\ell 11}$: From Example 5.1, step 2, $E_{\ell 11} = 18.8$ mpsi

2. Calculate $S_{\ell 22T}$ (Eq. 3, Fig. 8) with partial interfacial bonding:

   a. $S_{mT} = \{1.0 - [4(0.05)/(1.0 - 0.6)\pi]^{1/2}\}15$
       $= 9.0$ ksi

   b. $S_{\ell 11T} = [1.0 - (\sqrt{0.6} - 0.6)(1.0 - 0.5/2.0)] \times 9.0$
       $= 7.8$ ksi

3. Calculate $E_{\ell 22}$: From Example 5.1, step 3, $E_{\ell 22} = 1.2$ mpsi

4. Calculate $\nu_{\ell 12}$: From Example 5.1, step 4, $\nu_{\ell 12} = 0.26$

5. Calculate $G_{\ell 12}$: From Example 5.1, step 5, $G_{\ell 12} = 0.62$ mpsi

6. Calculate $\mathscr{S}_{\ell 22T}$: From Equation 2, Fig. 10,

$\mathscr{S}_{\ell 22T} = 7.8/\{1.0 + (1.2/18.8)^{1/2}[2(1.0 - 0.26) + 18.8/0.62]^{1/2}\} = 3.2$ ksi

This implies that the far-field (P/A-type) ply stress will be 3.2 ksi when additional damage in the vicinity of the defect will occur. Or, alternatively, the stress required to produce additional damage is reduced by a factor of about 2.5 compared to that in a ply without defects. It is worth noting that this relatively low value of 3.2 ksi required to produce additional damage is a major contributor to the brittle-like strength behavior transverse to the fiber direction.

6.0. In-Plane Uniaxial Impact Resistance

Uniaxial impact resistance of unidirectional composites is defined herein as an in-plane uniaxial impact energy density. It is denoted by the generic symbol $\mathscr{E}_\ell$ and is associated with a corresponding in-plane uniaxial impact stress. There

1. Longitudinal tension: $\mathscr{K}_{\ell 11T} \approx k_f \, S_{fT}^2/2E_{f11}$

2. Longitudinal compression: $\mathscr{K}_{\ell 11C} \approx k_f \, S_{fC}^2/2E_{f11}$

3. Transverse tension: $\mathscr{K}_{\ell 22T} \approx [1 - (\sqrt{k_f} - k_f) \, (1 - E_m/E_{f22})]^2$
$\times \, [1 - \sqrt{k_f} \, (1 - E_m/E_{f22})] \, S_{mT}^2/2E_m$

4. Transverse compression: $\mathscr{K}_{\ell 22C} \approx [1 - (\sqrt{k_f} - k_f) \, (1 - E_m/E_{f22})]^2$
$\times \, [1 - \sqrt{k_f} \, (1 - E_m/E_{f22})] \, S_{mC}^2/2E_m$

5. Intralaminar shear: $\mathscr{K}_{\ell 12S} \approx [1 - (\sqrt{k_f} - k_f) \, (1 - G_m/G_{f12})]^2$
$\times \, [1 - \sqrt{k_f} \, (1 - G_m/G_{f12})] \, S_{mS}^2/2G_m$

6. For voids: $S_m \approx \{1 - [4 \, k_v/(1 - k_f) \, \pi]^{1/2}\} \, S_m$

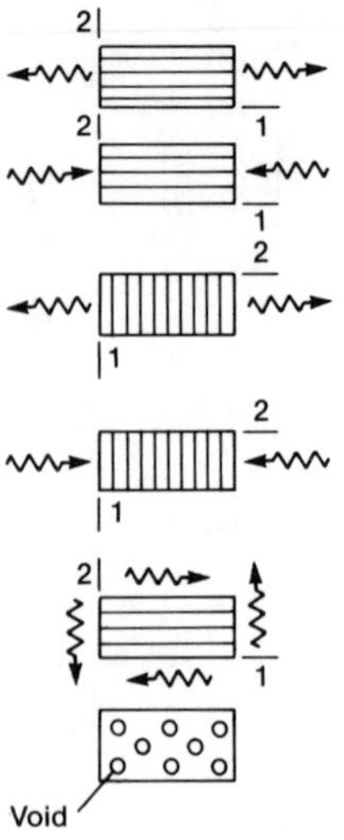

All equations apply to plastic (polymer) matrix composites. All but Equation 6 apply to metal matrix and some ceramic matrix composites.

Fig. 11. Composite micromechanics: in-plane uniaxial impact resistance (energy absorbed per unit volume)

are five impact energy densities: (1) longitudinal tension ($\mathscr{K}_{\ell 11T}$); (2) longitudinal compression ($\mathscr{K}_{\ell 11C}$); (3) transverse tension ($\mathscr{K}_{\ell 22T}$); (4) transverse compression ($\mathscr{K}_{\ell 22C}$); and (5) intralaminar shear ($\mathscr{K}_{\ell 12S}$). The composite micromechanics equations for these impact energy densities are summarized in Fig. 11 with attendant schematics. The wiggly arrows in the schematics denote dynamic stresses. These equations are derived by assuming linear stress-strain behavior to fracture (Fig. 3) under dynamic stress. The first five equations describe the five in-plane uniaxial impact energy densities; the last equation describes the void degradation effect as mentioned previously.

The following are observed from the equations in Fig. 11: (1) $\mathscr{K}_{\ell 11T}$ varies linearly with k_f, quadratically with S_{fC}, and inversely with E_{f11}; (2) $\mathscr{K}_{\ell 11C}$ also varies linearly with k_f, quadratically with S_{fC} (assuming fiber compressive fracture), and inversely with E_{f11}; (3) $\mathscr{K}_{\ell 22T}$ decreases nonlinearly with k_f, increases quadratically with S_{mT}, decreases inversely with E_m, and increases nonlinearly with increasing ratio (E_m/E_{f22}); (4) $\mathscr{K}_{\ell 22C}$ and $\mathscr{K}_{\ell 12S}$ are matrix-dominated; and (6) the matrix-dominated impact energy densities decrease nonlinearly with increasing void content. Several examples below illustrate use of the equations in Fig. 11.

Example 6.1. Calculate the longitudinal tensile impact energy density for an AS/E (IMHS) unidirectional composite with 0.6 fiber volume ratio. Using respective properties from Table 1 and $k_f = 0.6$ in Equation 1, Fig. 11,

$\mathscr{K}_{\ell 11T} = 0.6 \times 350\,000^2/2 \times 31\,000\,000 = 1185$ (lb/sq in.)/cu in.

Example 6.2. Calculate the longitudinal compressive impact energy density for an AS/E (IMHS) unidirectional composite with 0.6 fiber volume ratio. Using $S_{fC} = 0.9 S_{fT}$, $S_{fT} = 350$ ksi, $E_{f11} = 31$ mpsi, and $k_f = 0.6$ in Equation 2, Fig. 11,

$\mathscr{K}_{\ell 11C} = 0.6 \times (0.9 \times 350\,000)^2/2 \times 31\,000\,000$
$= 960$ (lb/sq in.)/cu in.

It is instructive to calculate $\mathscr{K}_{\ell 11C}$ assuming delamination/shear fracture mode (Example 3.2). For this case, $\mathscr{K}_{\ell 1C} = 370$ (lb/sq in.)/cu in., or a decrease of about 61%.

Example 6.3. Calculate the transverse tensile impact energy density ($\mathscr{K}_{\ell 22T}$) for an AS/E (IMHS) unidirectional composite with 0.6 fiber volume ratio. Using re-

spective property values from Tables 1 and 2 and $k_f = 0.6$ in Equation 3, Fig. 11, and degrading S_{mT} for incomplete interfacial bonding (Example 3.3),

$$S_{mT} = \left\{1.0 - \left(\frac{4(0.05)}{1.0 - 0.6}\right)\pi\right]^{1/2}\right\} 15 = 9 \text{ ksi}$$

$$\mathscr{K}_{\ell 22T} = \left[1.0 - (\sqrt{0.6} - 0.6)\left(1.0 - \frac{0.5}{2.0}\right)\right]^2$$
$$\times \left[1.0 - \sqrt{0.6}\left(1.0 - \frac{0.5}{2.0}\right)\right]$$
$$\times \frac{9000^2}{2} \times 500\,000$$
$$= 26 \text{ (lb/sq in.)/cu in.}$$

This value is about 2 percent of the longitudinal tensile value [1185 (lb/sq in.)/cu in., Example 6.1] and illustrates the fragile nature of unidirectional composites when subjected to transverse loads.

Example 6.4. Calculate the transverse compressive impact energy density for an AS/E (IMHS) unidirectional composite with 0.6 fiber volume ratio. Recall that incomplete bonding does not degrade the transverse compressive behavior (Example 3.4). Using respective property values from Tables 1 and 2 and $k_f = 0.6$ in Equation 4, Fig. 11,

$$\mathscr{K}_{\ell 22c} = \left[1.0 - (\sqrt{0.6} - 0.6)\left(1.0 - \frac{0.5}{2.0}\right)\right]^2$$
$$\times \left[1.0 - 0.6\left(1 - \frac{0.5}{2.0}\right)\right]$$
$$\times \frac{35\,000^2}{2} \times 500\,000$$
$$= 388 \text{ (lb/sq in.)/cu in.}$$

which is about 15 times $\mathscr{K}_{\ell 22T}$, indicating substantially "tougher" behavior in transverse compression. Also, $\mathscr{K}_{\ell 22C}$ is about the same as the $\mathscr{K}_{\ell 11C}$ [370 (lb/sq/in.)/cu. in.] calculated by assuming delamination/shear compression fracture mode (Example 6.2). This implies that longitudinal compression and transverse compression fractures probably occur simultaneously during normal impact.

Example 6.5. Calculate the intralaminar shear energy density for an AS/E (IMHS) unidirectional composite with 0.6 fiber volume ratio. Using respective property values from Tables 1 and 2 and $k_f = 0.6$ in Equation 5, Fig. 11, and degrading S_{ms} for incomplete interfacial bonding (Example 3.4),

$$\mathcal{K}_{\ell12S} = \left[1.0 - (\sqrt{0.6} - 0.6)\left(1.0 - \frac{0.185}{2.0}\right)\right]^2$$

$$\times \left[1.0 - \sqrt{0.6}\left(1.0 - \frac{0.185}{2.0}\right)\right]$$

$$\times \frac{9700^2}{2} \times 185\,000$$

$$= 53.5 \text{ (lb/sq in.)/cu in.}$$

$$S_{ms} = \left\{1.0 - \left[4\left(\frac{0.02}{1.0 - 0.6}\right)\pi\right]^{1/2}\right\} \times 13 = 9.7 \text{ ksi}$$

The effects of voids on matrix-dominated impact energy densities can be calculated by degrading the matrix strength by first using Equation 6, Fig. 11, and then substituting the degraded value into the appropriate equation. See also Example 3.6.

7.0. Through-the-Thickness Uniaxial Impact Resistance

Through-the-thickness impact resistance in unidirectional composites results from out-of-plane normal impacts. These are defined herein as impact energy densities, are denoted by the generic symbol $\mathcal{K}_\ell$, and are, respectively: (1) longitudinal interlaminar shear ($\mathcal{K}_{\ell13S}$); (2) transverse interlaminar shear ($\mathcal{K}_{\ell23S}$); (3) longitudinal flexure ($\mathcal{K}_{\ell11F}$); and (4) transverse flexure ($\mathcal{K}_{\ell22F}$). Each of these impact energy densities is associated with a dynamic stress corresponding, respectively, to: $S_{\ell13S}$, $S_{\ell23S}$, $S_{\ell11F}$, and $S_{\ell22F}$. There is also a through-the-thickness normal impact energy density. However, this impact energy density is the same as the in-plane impact energy density $\mathcal{K}_{\ell22T}$ or $\mathcal{K}_{\ell22C}$ described in section 6.0.

The composite micromechanics equations for through-the-thickness impact energy densities are summarized in Fig. 12 with attendant schematics. The following are observed from the equations in Fig. 12: (1) $\mathcal{K}_{\ell13S}$ is the same as $\mathcal{K}_{\ell12S}$; it decreases nonlinearly with increasing k_f, increases nonlinearly with increasing ratio (G_m/G_{f12}), increases quadratically

with S_{mS}, and increases inversely as G_m decreases; (2) $\mathcal{K}_{\ell23S}$ has about the same behavior as $\mathcal{K}_{\ell13S}$; (3) $\mathcal{K}_{\ell11F}$ increases linearly with k_f and quadratically with S_{fT}, and increases inversely as E_{f11} and the ratio (S_{fT}/S_{fc}) decrease; (4) $\mathcal{K}_{\ell22F}$ decreases nonlinearly with increasing k_f and with increasing ratio (E_m/E_{f22}), increases quadratically with S_{mT}, increases inversely as the square of the (S_{mT}/S_{mc}) ratio, and increases as E_m decreases. The several examples below illustrate use of the equations in Fig. 12.

Example 7.1. Calculate the longitudinal interlaminar shear impact energy density ($\mathcal{K}_{\ell13S}$) for an AS/E (IMHS) unidirectional composite with 0.6 fiber volume ratio. Since $\mathcal{K}_{\ell13S}$ is the same as $\mathcal{K}_{\ell12S}$ (the equation for $\mathcal{K}_{\ell13S}$ is identical to Eq 5, Fig. 11), from Example 6.5, $\mathcal{K}_{\ell13S} = \mathcal{K}_{\ell12S} = 53.5$ (lb/sq in.)/cu in.

Example 7.2. Calculate the transverse interlaminar shear impact energy density ($\mathcal{K}_{\ell23S}$) for an AS/E (IMHS) unidirectional composite with 0.6 fiber volume ratio. Using respective property values from Tables 1 and 2 and $k_f = 0.6$ in the equation for $\mathcal{K}_{\ell23S}$, Fig. 12, and degrading S_{mS} for incomplete interfacial bonding,

$$S_{mS} = \left\{\left[1.0 - \frac{4(0.02)}{1.0 - 0.6}\pi\right]^{1/2}\right\} 13 = 9.7 \text{ ksi}$$

$$\mathcal{K}_{\ell23S} = \left[1.0 - 0.6\left(1.0 - \frac{0.185}{1.0}\right)\right]^2 \frac{9700^2}{2}$$

$$\times 185\,000\left[1 - 0.6\left(1.0 - \frac{0.185}{1.0}\right)\right] = 67.7 \text{ (lb/sq in.)/cu in.}$$

It is interesting to note that for this example $\mathcal{K}_{\ell23S}$ is about 22% greater than $\mathcal{K}_{\ell13S}$. This increase is mainly due to G_{f23}, which is about 50% of G_{f12}. Based on the relative values for $\mathcal{K}_{\ell13S}$ and $\mathcal{K}_{\ell23S}$, interlaminar damage will occur first due to dynamic $\sigma_{\ell23}$.

Example 7.3. Calculate the longitudinal flexural impact energy density ($\mathcal{K}_{\ell11F}$) for an AS/E (IMHS) unidirectional composite with 0.6 fiber volume ratio. Using respective property values from Tables 1 and 2 and $k_f = 0.6$ in the equation for $\mathcal{K}_{\ell11F}$, Fig. 12 (assuming $S_{fc} = 0.9 \, S_{fT}$),

$$\mathcal{K}_{\ell11F} = 4.5 \times 0.6 \times \frac{350\,000^2}{31\,000\,000} \times \left(1.0 + \frac{1.0}{0.9}\right)^2$$

$$= 2394 \text{ (lb/sq in.)/cu in.}$$

1. Interlaminar shear:

$$\mathcal{K}_{\ell13S} \approx [1 - (\sqrt{k_f} - k_f)(1 - G_m/G_{f12})]^2$$
$$\times [1 - \sqrt{k_f}(1 - G_m/G_{f12})] \, S_{mS}^2/2G_m$$

$$\mathcal{K}_{\ell23S} \approx \frac{[1 - \sqrt{k_f}(1 - G_m/G_{f23})]^2 \, S_{mS}^2}{2G_m[1 - k_f(1 - G_m/G_{f23})]}$$

2. Flexural:

$$\mathcal{K}_{\ell11F} \approx \frac{4.5 \, k_f \, S_{fT}^2}{E_{f11}\left(1 + \frac{S_{fT}}{S_{fC}}\right)^2}$$

$$\mathcal{K}_{\ell22F} \approx 4.5 \, [1 - \sqrt{k_f}(1 - E_m/E_{f22})]$$
$$\times \left\{\frac{[1 - (\sqrt{k_f} - k_f)(1 - E_m/E_{f22})]}{(1 + S_{mT}/S_{mC})}\right\}^2 \left(\frac{S_{mT}^2}{E_m}\right)$$

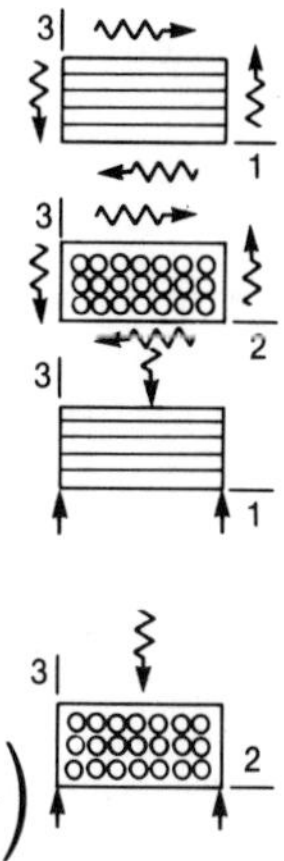

All equations apply to plastic (polymer) matrix, metal matrix, and some ceramic matrix composites.

Fig. 12. Composite micromechanics: through-the-thickness uniaxial impact resistance (energy absorbed per unit volume)

An alternative method of estimation is to use $S_{\ell 11T} = 210$ ksi from Example 3.1, $S_{\ell 11C} = 118$ ksi from Example 3.2, and $E_{\ell 11} = 18.8$ mpsi from Example 5.1 in the following equation:

$$\mathscr{E}_{\ell 11F} = 4.5 \, \frac{S^2_{\ell 11T}}{E_{\ell 11}} \left(1.0 + \frac{S_{\ell 11T}}{S_{\ell 11C}}\right)^2$$

$$\mathscr{E}_{\ell 11F} = 4.5 \times \frac{210\,000^2}{18\,800\,000} \times \left(1.0 + \frac{210}{118}\right)^2$$

$$= 1366 \text{ (lb/sq in.)/cu in.}$$

It is worth noting that the first estimate corresponds to $S_{\ell 11F}$ = 298 ksi for fiber compression fracture; the second estimate corresponds to $S_{\ell 11F}$ = 227 ksi for delamination/shear compression fracture (Example 4.4). Also, the second estimate is about 57% smaller than the first, indicating that delamination/shear is a much more severe fracture mode under impact.

Example 7.4. Calculate the transverse flexural impact energy density ($\mathscr{E}_{\ell 22F}$) for an AS/E (IMHS) unidirectional composite with 0.6 fiber volume ratio. Using respective property values from Tables 1 and 2, degraded S_{mT} for incomplete interfacial bonding (Eq 6, Fig. 11, with $k_v = 0.05$), and $k_f = 0.6$ in the equation for $\mathscr{E}_{\ell 22F}$, Fig. 12,

$$S_{mT} = \left\{1.0 - \left[4\left(\frac{0.05}{1.0 - 0.6}\right)\pi\right]^{1/2}\right\} 15 = 9.0 \text{ ksi}$$

$$\mathscr{E}_{\ell 22F} = 4.5 \left[1.0 - 0.6\left(1.0 - \frac{0.5}{2.0}\right)\right]$$

$$\times \left[\frac{1.0 - (\sqrt{0.6} - 0.6)(1.0 - 0.5/2.0)}{1 + 9/35}\right]^2 \times (9000^2/500\,000)$$

$$= 146 \text{ (lb/sq in.)/cu in.}$$

It is worth noting that this value corresponds to $S_{\ell 22F}$ = 18.6 ksi, which is the lower estimate in Example 4.5. Also, $\mathscr{E}_{\ell 22F}$ is about 10% of $\mathscr{E}_{\ell 11F}$, the lower estimate in Example 7.3.

The effect of voids on any of the through-the-thickness impact energy densities is determined by degrading S_{mT} or S_{mS} by first using Equation 6, Fig. 11, and then substituting this degraded S_m value into the applicable equation, Fig. 12. The remaining steps are identical to those in Examples 7.1 to 7.4.

8.0. Thermal Properties

The simple equations for predicting the ply thermal properties from constituent properties are summarized in Fig. 13. The thermal properties in this figure include heat capacity (C_ℓ), heat conductivity (K_ℓ), and thermal expansion coefficient (α_ℓ). All these thermal properties are expressed in terms of the respective constituent properties, the fiber volume ratio (k_f), the matrix volume ratio (k_m), the void volume ratio (k_v), and the heat conductivity of the air (K_v). The thermal expansion coefficients are also related to ply properties ($E_{\ell 11}$ and $E_{\ell 22}$).

The following examples illustrate the use of these equations.

Example 8.1. Calculate the transverse heat conductivity of S-Glass fiber/epoxy composite with 4% voids and 75% fiber by weight. The properties of the S-Glass fiber are obtained from Table 1 and those for the matrix (IMLS) from Table 2. Using densities from these tables and the appropriate equations in Fig. 6, $k_f = 0.58$ and $k_m = 0.38$. Using the appropriate equations in Fig. 13 with $K_v = 0.225$ Btu/hr/ft^2/°F/in. and $K_m = 1.130$ Btu/hr/ft^2/°F/in., $K_{\ell 22} = 3.35$ Btu/hr/ft^2/°F/in. The corresponding ply transverse conductivity without voids ($k_f = 0.605$ for this case) is $K_{\ell 22} = 3.90$ Btu/hr/ft^2/°F/in. It is interesting to note that a 4% void fraction decreased the transverse conductivity by about 14%, indicating that this conductivity is very sensitive to void content.

Example 8.2. Determine the longitudinal and transverse thermal expansion coefficients for the same S-glass/epoxy composite without voids. To determine these coefficients from the equations in Fig. 13, the ply longitudinal modulus is needed. This is calculated from the appropriate equation in Fig. 7 using properties from Table 1 for the fiber and Table 2 for the matrix, and using the previously determined k_f value of 0.605. The longitudinal modulus $E_{\ell 11} = 7.70 \times 10^6$ psi. The longitudinal thermal expansion coefficient $\alpha_{\ell 11} = 4.19 \times 10^{-6}$ in./in./°F. The transverse thermal expansion coefficient $\alpha_{\ell 22} = 19.9 \times 10^{-6}$ in./in./°F.

9.0. Hygral Properties

The simple micromechanics equations for determining ply hygral properties from constituent properties are summarized in Fig. 14. The ply hygral properties summarized in-

Heat capacity:
$$C_\ell = \frac{1}{\rho_\ell}(k_f\rho_f C_f + k_m\rho_m C_m)$$

Longitudinal conductivity:
$$K_{\ell 11} = k_f K_{f11} + k_m K_m$$

Transverse conductivity:
$$K_{\ell 22} = (1 - \sqrt{k_f})K_m + \frac{K_m\sqrt{k_f}}{1 - \sqrt{k_f}(1 - K_m/K_{f22})} = K_{\ell 33}$$

For voids:
$$K_m = (1 - \sqrt{k_v})K_m + \frac{K_m\sqrt{k_v}}{1 - \sqrt{k_v}(1 - K_m/K_v)}$$

Longitudinal thermal expansion coefficient:
$$\alpha_{\ell 11} = \frac{k_f\alpha_{f11}E_{f11} + k_m\alpha_m E_m}{E_{\ell 11}}$$

Transverse thermal expansion coefficient:
$$\alpha_{\ell 22} = \alpha_{f22}\sqrt{k_f} + (1 - \sqrt{k_f})(1 + k_f\nu_m E_{f11}/E_{\ell 11})\alpha_m$$
$$= \alpha_{\ell 33}$$

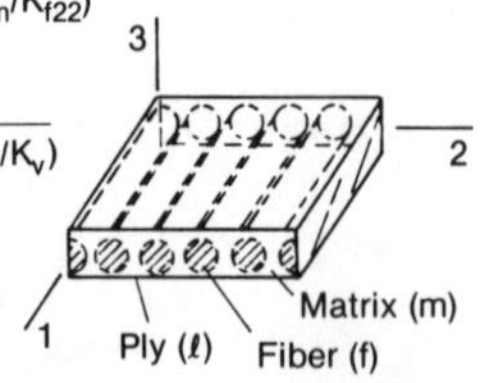

All equations apply to plastic (polymer) matrix, metal matrix, and some ceramic matrix composites.

Fig. 13. Composite micromechanics: thermal properties

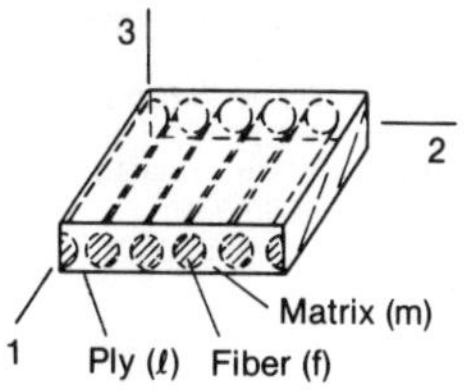

Longitudinal diffusivity:
$$D_{\ell 11} = (1 - k_f)\, D_m$$

Transverse diffusivity:
$$D_{\ell 22} = (1 - \sqrt{k_f})\, D_m = D_{\ell 33}$$

Longitudinal moisture expansion coefficient:
$$\beta_{\ell 11} = \beta_m\,(1 - k_f)\, E_m/E_{\ell 11}$$

Transverse moisture expansion coefficient:
$$\beta_{\ell 22} = \beta_m\,(1 - \sqrt{k_f})\left[1 + \frac{\sqrt{k_f}\,(1 - \sqrt{k_f})\,E_m}{\sqrt{k_f}E_{\ell 22} + (1 - \sqrt{k_f})\,E_m}\right] = \beta_{\ell 33}$$

For incompressible matrix:
$$\beta_{\ell 11} = 0$$

$$\beta_{\ell 22} = \beta_{m\rho}{}^{1/2}{}_{\rho m} = \beta_{\ell 33}$$

All equations apply to plastic (polymer) matrix composites.

Fig. 14. Composite micromechanics: hygral properties

clude: diffusivity (D_ℓ) and moisture expansion coefficients. To determine ply hygral properties summarized in Fig. 14, the respective properties of the matrix and the fiber volume ratio are needed. The longitudinal moisture expansion coefficient $(\beta_{\ell 11})$ depends also on the ply longitudinal modulus $(E_{\ell 11})$ while the transverse moisture expansion coefficient $(\beta_{\ell 22})$ depends on the ply transverse modulus $(E_{\ell 22})$. The following examples will illustrate the use of these equations.

Example 9.1. Calculate the ply longitudinal and transverse diffusivities for an AS-graphite-fiber/epoxy composite with 35% epoxy (resin) by weight and "zero" voids. Using the equation in Fig. 6, the fiber volume ratio $k_f = 0.58$. Using the matrix diffusivity from Table 2 of 6×10^{-11} in.^{2}/sec, the ply longitudinal diffusivity $D_{\ell 11} = 2.52 \times 10^{-11}$ in.^{2}/sec, and the transverse diffusivity $D_{\ell 22} = 1.43 \times 10^{-11}$ in.^{2}/sec. The ply transverse diffusivity is about 60% of the longitudinal. This implies that exposed fiber ends enhance moisture absorption/desorption.

Example 9.2. Determine the ply longitudinal moisture expansion coefficient for the composite in the previous example. First, the ply longitudinal modulus needs to be determined from the appropriate equation in Fig. 7. Using respective values for the constituents from Table 1 for the AS-graphite fiber and from Table 2 for the matrix (IMHS), and $k_f = 0.58$ (determined previously), the ply longitudinal modulus $E_{\ell 11} = 18.19 \times 10^6$ psi. The ply longitudinal moisture expansion coefficient $\beta_{\ell 11} = 0.0038$ in./in./%M (percent moisture by weight).

Example 9.3. Determine the corresponding ply transverse moisture expansion coefficient. First, the transverse ply modulus is needed. Using respective properties for the constituents and $k_f = 0.58$ in the appropriate equation in Fig. 7, the ply transverse modulus $E_{\ell 22} = 1.17 \times 10^6$ psi. Using the equation for the ply transverse moisture coefficient $(\beta_{\ell 22})$ in Fig. 14, the respective resin properties for the matrix (IMHS) from Table 2, and the above values for $E_{\ell 22} = 1.17 \times 10^6$ and $k_f = 0.58$, $\beta_{\ell 22} = 0.086$ in./in./%M (percent moisture by weight).

Example 9.4. Determine $\beta_{\ell 22}$ in the above example assuming an incompressible matrix. The ply density (ρ_ℓ) is needed to perform this calculation using the equation in Fig. 14. Using respective constituent material densities in the ply density equation in Fig. 6, $\rho_\ell = 0.056$ lb/in.^{3} and from the equation in Fig. 14, $\beta_{\ell 22} = 0.201$ in./in./%M (percent moisture by weight). Thus,

the ply moisture expansion coefficient of a composite with an incompressible matrix is about three times greater than that of one with a compressible matrix.

10.0. Moisture Absorption

The micromechanics equations for estimating moisture in the resin and composite as a function of relative humidity ratio (RHR = 1.0 for 100% relative humidity) are summarized in Fig. 15. The equations in this figure are for the moisture in the matrix (M_m) and the moisture in the ply (M_ℓ). To use these equations, the lineal moisture expansion coefficient of the matrix β_m and the saturation moisture of the matrix at 100% relative humidity are needed. If β_m is not known, it can be estimated from the equation in Fig. 15 by using the wet and dry density of the matrix. The saturation moisture M_∞ at 100% relative humidty for the particular resin is also needed. If it is not known, $M_\infty \sim 7\%$ by weight is a reasonable approximation. The following examples illustrate use of the equations in Fig. 15.

Example 10.1. Determine the matrix moisture for 70% relative humidity exposure. Using $M_\infty = 7\%$ (assuming that M_∞ is not known) and RHR = 0.7 in the matrix moisture equation (Fig. 15), $M_m = 4.9\%$ by weight.

Example 10.2. Determine the ply moisture for the previous example for an AS-graphite-fiber/epoxy matrix (IMLS) composite with 35% resin by weight and zero voids. Using the matrix volume ratio (k_m) equation in Fig. 6 and respective constituent material densities from Tables 1 and 2, $k_m = 0.42$. The corresponding ply density $\rho_\ell = 0.056$ lb/in.^{3} Using these values, $M_m = 4.9\%$, and respective values for the other variables in the ply moisture (M_ℓ) equation in Fig. 15, $M_\ell = 1.7\%$ by weight, which is about 1/3 of the moisture in the resin in the previous example.

11.0. Hygrothermal Effects (Environmental Effects)

Environmental effects refer to the effects caused by the presence of moisture and temperature in composites. The combined effects are usually called hygrothermal effects. Hygrothermal effects influence all resin-dominated properties: Uniaxial strengths, fracture toughness, and impact resistance. Hygrothermal effects are estimated using an empirical expression (Ref 15).

The equations for predicting hygrothermal degradation ef-

Moisture pickup in matrix:	$M_m \approx M_\infty \, (RHR)$
Moisture pickup in composite:	$M_\ell = M_m \, (3\beta_m k_m + k_v) \, \rho_m/\rho_\ell$
Matrix moisture expansion coefficient (lineal):	$\beta_m \sim [\rho_m/\rho_{mw}) - k_v]/3$

Notation: M_∞ = Matrix saturation moisture at 100% relative humidity and room temp. (Use $M_\infty \approx 7\%$ by weight if unknown.)

        RHR = Relative humidity ratio (100% = 1.0)

Subscripts: ℓ = Composite (ply) property

        m = Matrix

        w = Wet

        v = Void

All equations apply to plastic (polymer) matrix composites.

Fig. 15. Composite micromechanics: moisture absorption

fects in composites using micromechanics are summarized in Fig. 16. The equations in this figure are for the glass transition temperature of the wet matrix (T_{GW}), the degraded mechanical property (P_{HTM}), and the degraded thermal property (P_{HTT}). Note that two equations are given for the glass transition temperature of the wet resin (T_{GW}). One equation is in terms of the matrix moisture content (M_m), and the other is in terms of a hygrothermally degraded mechanical property (P_{HTM}). A hygrothermally degraded property can be used with the equation for P_{HTT} as well. Note also that the effects on the thermal properties are the reciprocals of those on the mechanical properties.

It is worth noting that the equations in Fig. 16 were obtained by curve-fitting experimental data. Consequently, they should be used judiciously and cross checked with available data for a specific case. The following examples illustrate use of the equations in Fig. 16.

Example 11.1. Calculate the hygrothermal effects on the ply transverse strength assuming an AS/E (IMHS) unidirectional composite, $k_f = 0.6$, T = 270 °F, T = 70 °F and 1 percent moisture by weight. Several steps are required for this calculation:

1. $T_{GD} = 420$ °F (Table 2)
2. $T_{GW} = [0.005 \, (1)^2 - 0.1 \, (1) + 1.0] \, 420 = 380$ °F
3. The hygrothermal degradation ratio (P_{HTM}/P_0) for resin-dominated properties is:

$$P_{HTM}/P_0 = [(380 - 270)/(420 - 70)]^{1/2} = 0.56$$

This means that all resin-dependent properties (E_m, G_m, and S_m) must be reduced by this ratio prior to their use in the applicable equation.

4. The reduced matrix properties (Table 2) are:

$S_{mT} = 0.56 \times 15 = 8.4$ ksi

$E_m = 0.56 \times 0.5 = 0.28$ mpsi

5. Degrade S_{mT} for partial interfacial bonding assuming 5% voids by volume:

$S_{mT} = \{1.0 - [4(0.05)/(1 - 0.6)\pi]^{1/2}\}8.4 = 5.0$ ksi

6. Using Equation 3, Fig. 8, with the respective degraded properties,

$S_{\ell 22T} = [1.0 - (\sqrt{0.6} - 0.6)(1.0 - 0.25/2.0)]5.0$
    $= 4.2$ ksi

which is a decrease of 46% compared to room-temperature dry (7.8 ksi: Example 3.3). Obviously, this is severe degradation of the hygrothermal environment assumed in the example. It is important to note that the ratio of environmentally degraded to room-temperature dry (4.2/7.8) is 0.54, which is very close to the 0.56 predicted in step 3 above. This indicates that the hygrothermal degradation ratio can be applied to either (1) resin or (2) resin-dominated composite properties equally well (Ref 7). It is recommended that the glass transition temperature of the composite be used for the

Glass transition temperature of wet resin:	$T_{GW} = (0.005M_m^2 - 0.10M_m + 1.0) \, T_{GD}$
Effects on mechanical properties:	$\dfrac{P_{HTM}}{P_0} = \left[\dfrac{T_{GW} - T}{T_{GD} - T_0}\right]^{1/2}$
Effects on thermal properties:	$\dfrac{P_{HTT}}{P_0} = \left[\dfrac{T_{GD} - T_0}{T_{GW} - T}\right]^{1/2}$
Glass transition temperature of wet resin:	$T_{GW} = T + (T_{GD} - T_0) \, (P_{HTM}/P_0)^2$

Temperature (T) any consistent units

Moisture (M) weight percent (M $\leqslant$ 10%)

Subscripts: G = Transition; D = Dry; W = Wet; 0 = Reference
    HTM = Hygrothermal mechanical; HTT = Hygrothermal thermal

All equations apply to plastic (polymer) matrix composites.

Fig. 16. Governing equations: micromechanics–hygrothermal effects

second case. The glass transition temperature of the composite is about 50 °F greater than that of the resin.

Example 11.2. Calculate the transverse flexural strength for the same composite and environmental conditions as in Example 11.1. Again, several steps are required.

1. From example 11.1, step 3,

   $P_{HTM}/P_0 = 0.56$

   from which follows: $S_{mC} = 0.56 \times 35 = 19.6$ ksi and $E_m = 0.28$ mpsi

2. The transverse compression stress is

   $S_{\ell22c} = [1.0 - (\sqrt{0.6} - 0.6)(1.0 - 0.28/2.0)]19.6$
   $= 16.7$ ksi

3. Using $S_{\ell22C} = 16.7$ ksi and $S_{\ell22T} = 4.2$ ksi (Example 11.1, step 6) in the lower-estimate equation (Example 4.5),

   $S_{\ell22F} = 3 \times 4.2/(1.0 + 4.2/16.7) = 10.1$ ksi

which is about 54% of 18.6 ksi, the room-temperature dry value calculated in Example 4.5. This calculation also illsutrates that the hygrothermal degradation can be applied to a resin-dominated composite property.

Example 11.3. Calculate the transverse fracture toughness. ($\mathscr{S}_{\ell22T}$) for the composite and hygrothermal environment in Example 11.1. The procedure for this calculation is the same as that in Example 5.2.

1. Using $P_{HTM}/P_0 = 0.56$ in the equations for $E_{\ell22}$ and $G_{\ell12}$, Fig. 7,

   $E_{\ell22} = 0.28/[1.0 - \sqrt{0.6}\,(1.0 - 0.28/2.0)]$
   $= 0.839$ mpsi

   $G_{\ell12} = 0.103/[1.0 - \sqrt{0.6}\,(1.0 - 0.103/2.0)]$
   $= 0.39$ mpsi

2. $E_{\ell11}$ and $\nu_{\ell12}$ remain practically unchanged:

   $E_{\ell11} = 18.7$ and $\nu_{\ell12} = 0.26$ (Example 5.1, steps 2 and 4)

3. $S_{\ell22T} = 4.2$ ksi (Example 11.1, step 6)

4. Substituting respective values from steps 1, 2, and 3 in Equation 2, Fig. 10,

$$\mathscr{L}_{\ell22T} = \frac{4.2}{1.0} + \left(\frac{0.84}{18.7}\right)^{1/2}$$

$$\times \left[2(1.0 - 0.26) + \frac{18.7}{0.39}\right]^{1/2} = 1.7 \text{ ksi}$$

which is about 53% of the value calculated in Example 5.2. Even in this complex expression, the environment degrades the composite resin-dominated property in about the same ratio as the resin property.

Example 11.4. Calculate the transverse impact energy density ($\mathscr{K}_{\ell22T}$) for the composite and environmental conditions in Example 11.1. The procedure for this calculation is the same as that in Example 6.3.

1. The degraded properties needed for Equation 3, Fig. 11, are $E_m = 0.28$ mpsi (Example 11.1, step 4) and $S_{mT} = 5.0$ ksi (Example 11.1, step 5).

2. Using respective values in Equation 3, Fig. 11,

   $\mathscr{K}_{\ell22T} = [1.0 - (\sqrt{0.6} - 0.6)(1.0 - 0.28/2.0)]^2$
   $\times [1.0 - \sqrt{0.6}\,(1.0 - 0.28/2.0)] \times 5000^2/2 \times$
   $280\,000 = 10.8$ (lb/sq in.)/cu in.

which is about 42% of 26 (lb/sq in.)/cu in., the room-

temperature dry value in Example 6.3. This ratio corresponds to a decrease about equal to the hygrothermal degradation ratio raised to the 3/2 power, or $(0.56)^{3/2}$, indicating again that resin-dominated composite properties degrade in the same ratio as the resin properties when subjected to hygrothermal environments.

Example 11.5. Determine the ply transverse modulus of an AS-graphite-fiber/epoxy matrix (IMLS) composite with fiber volume ratio $k_f = 0.60$ and "zero" voids in a hygrothermal environment of 80% relative humidity and 250 °F.

The moisture in the matrix $M_m = 5.6\%$ (equation in Fig. 15 and $M_\infty = 7\%$). The ply density $\rho_\ell = 0.056$ lb/in.³ (equation in Fig. 6, and Tables 1 and 2). The ply moisture $M_\ell = 1.82\%$ (equation in Fig. 15). The glass transition temperature of the wet matrix in the ply is $T_{GW} = 353.3$ °F (equation in Fig. 16 and T_{GD} from Table 2). The ratio of the degraded mechanical *in situ* matrix property (P_{HTM}/P_0) $= 0.543$ ($T_{GW} = 353.3$ °F, $T_{GD} = 420$ °F, $T_0 = 70$ °F and $T = 250$ °F). Note that the corresponding matrix thermal property ratio (P_{HTT}/P_0) $= 1.842$, the reciprocal of 0.543. Using the degraded ratio 0.543, the corresponding degraded matrix modulus $E_m = 0.272 \times 10^6$ psi (degradation ratio times IMLS matrix modulus, Table 2). Using this value for E_m, $k_f = 0.6$, and $E_{f22} = 2.0 \times 10^6$ psi in the appropriate equation in Fig. 16, the degraded ply transverse modulus $E_{\ell22} = 0.822 \times 10^6$ psi. This is a substantial reduction (about 31%) compared to the undegraded $E_{\ell22} = 1.19 \times 10^6$ psi, indicating that matrix-dominated ply properties are very sensitive to hygrothermal environments.

Example 11.6. Determine the ply transverse thermal expansion coefficient for the composite and environmental conditions of the previous example, using the equation for $\alpha_{\ell22}$ in Fig. 13.

First, the changes in α_m are determined from the equation in Fig. 16. Using the thermal property degradation ratio $P_{HTT}/P_0 = 1.842$ (determined previously) and the reference $\alpha_m = 57 \times 10^{-6}$ in./in./°F from Table 2, the degraded value of $\alpha_m = (P_{HTT}/P_0)\,\alpha_m$ (reference) or $\alpha_m = 1.84 \times (57 \times 10^{-6})$ in./in./°F $= 105 \times 10^{-6}$ in./in./°F. Next, the degraded ply longitudinal modulus is determined from the equation in Fig. 7 with $E_{f11} = 31 \times 10^6$ psi, $E_m = 0.272 \times 10^6$ psi, and $k_f = 0.6$, or $E_{\ell11} = 18.71 \times 10^6$ psi, in comparison with 18.8×10^6 psi for the dry conditions (insignificant change, as would be expected). The matrix Poisson's ratio (ν_m) is not degraded by the hygrothermal environment (Ref 7). Using the values just determined and respective constituent material properties from Tables 1 and 2 in the equation for $\alpha_{\ell22}$ in Fig. 13, $\alpha_{\ell22} = 37.7 \times 10^{-6}$ in./in./°F. This is a significant increase (about 68%) compared to $\alpha_{\ell22} = 22.4 \times 10^{-6}$ in./in./°F for the dry room-temperature condition. The numerical values from these examples show that fiber-dominated properties are not sensitive to hygrothermal environments but that the matrix-dependent properties are very sensitive.

Example 11.7. Determine the matrix glass transition temperature and moisture in a composite made from S-Glass-fiber/IMHS-epoxy with a ply transverse modulus of $E_{\ell22} = 1.6 \times 10^6$ psi at 200 °F and with a fiber volume ratio $k_f = 0.65$. The corresponding ply transverse modulus $E_{\ell22} = 2.2 \times 10^6$ psi at room temperature. The mechanical property degradation ratio (P_{HTM}/P_0) $= E_{\ell22}$ (wet)/$E_{\ell22}$ (dry) $= (1.6/2.2) = 0.72$. Using this value together with T = 200 °F, T_{GD} = 420 °F, T_0 = 70 °F, and the appropriate equation in Fig. 16, T_{GW} = 385 °F, which is the glass transition temperature of the wet resin. Using the ratio T_{GW}/T_{GD} = 385 °F/420 °F = 0.917 in the first equation in Fig. 16 yields M_ℓ = 0.87% by weight. The corresponding moisture in the resin can be determined only by solving the M_ℓ equation for M_m in Fig. 15. The desired result is M_m = 4.23% by weight. The relative humidity corresponding to this moisture, assuming M_∞ = 7%, is 60%. The interesting conclusion from the previous example is that considerable information about the composite behavior may be obtained with relatively little measured or known data.

General Discussion and Summary

The several examples presented illustrate the usefulness and advantage of having a unifed set of micromechanics equations, summarized in Fig. 6 to 16, for strength, fracture toughness, impact resistance, and thermal, hygrothermal, and hygral properties. The examples also illustrate how the various strengths and other mechanical properties are interrelated. In addition, they provide detailed and quantitative insight into the micromechanic strength behavior of composites. Furthermore, the various equations can be selectively used to conduct parametric studies as well as sensitivity analyses to assess acceptable ranges of various constituent materials and environmental factors.

Limited comparisons were provided between predicted values and available measured data for some of the numerical examples. It is important to note that the primary purpose of this report is to describe a unified set of simple, working equations and illustrate its versatility with a variety of numerical examples. These examples demonstrate computational effectiveness and illustrate interrelationships of various strengths and other properties at the micromechanistic level. It is highly recommended that the reader use this unified set of micromechanics equations to predict various properties of interest to him and compare them with measured data or with known values. This provides a direct approach to assess the application and limitations of these equations as well as guidelines on how to modify them.

Another important aspect of having this unified set of micromechanics strength equations is that they can be used to plan and guide experimental programs for maximum benefit with minimum testing. These micromechanics equations can be advantageous in a number of other ways. Many of these other ways become "self evident" after some familiarity has been obtained.

The two tables summarizing constituent material properties illustrate the amount of data needed for effective use of a unified set of micromechanics equations. The data in these tables were compiled from many sources, and many values are estimates which were inferred from predicted results and curve fits. The data are included for three main reasons: (1) to illustrate that the micromechanics equations need numerous properties; (2) to bring attention to the fact that many of these properties have not been measured and, hopefully, to stimulate enough interest to develop experimental methods to measure them; and (3) to provide indicative ranges of properties of both fibers and matrices. It cannot be overemphasized that the data should be considered dynamic in the sense that they should be continuously modified if better values are known or become available.

Lastly, the unified set of micromechanics equations described herein, in conjunction with classical laminate theories and combined stress failure criteria, can be used to calculate laminate strength based on first ply failure. In addition, in conjunction with laminate theories (Ref 15), they can be used to generate all the ply hygrothermomechanical properties needed to perform thermal and structural analyses of plastic matrix/fiber composite structures.

References

1. C. C. Chamis, "Micromechanics Strength Theories," *Fracture and Fatigue*, L. J. Broutman, Ed., Academic Press, 1974, p 94–148
2. C. C. Chamis and G. P. Sendeckyj, Critique on Theories Predicting Thermoelastic Properties of Fibrous Composites, *J. Composite Mater.*, Vol 2, No. 3, 1968, p 332–358
3. G. P. Sendeckyj, Ed., *Mechanics of Composite Materials*, Vol 1; L. J. Broutman and R. H. Krock, Eds. *Composite Materials*, Vol 2, Academic Press, New York, 1974
4. B. D. Agarwal and L. J. Broutman, *Analysis and Performance of Fiber Composites,* John Wiley and Sons, New York, 1980
5. R. Delasi and J. B. Whiteside, "Effect of Moisture on Epoxy Resins and Composites," *Advanced Composite Materials–Environmental Effects,* ASTM STP-658, J. R. Vinson, Ed., American Society for Testing and Materials, 1978, p 2–20
6. C. C. Chamis, "Designing with Fiber-Reinforced Plastics (Planar Random Composites)," NASA TM-82812, National Aeronautics and Space Administration, Washington, DC, 1982
7. C. C. Chamis, R. F. Lark, and J. H. Sinclair, "Integrated Theory for Predicting the Hygrothermomechanical Response of Advanced Composite Structural Components," *Advanced Composite Materials–Environmental Effects,* ASTM STP-658, J. R. Vinson, Ed., American Society for Testing and Materials, 1978, p 160–192
8. W. J. Renton, Ed., *Hybrid and Select Metal Matrix Composites: A State-of-The-Art Review,* American Institute of Aeronautics and Astronautics, 1977, p 15–16
9. C. C. Chamis and J. H. Sinclair, "Prediction of Composite Hygral Behavior Made Simple," Proceedings of the Thirty-Seventh Annual Conference of the Society of Plastics Industry (SPI), Reinforced Plastics/Composites Institute, 1982
10. L. A. Friedrich and J. L. Preston, Jr., "Impact Resistance of Fiber Composite Blades Used in Aircraft Turbine Engines," PWA-4727, Pratt & Whitney Aircraft, East Hartford, CT, NASA CR-134502, May 1973
11. C. C. Chamis, R. F. Lark, and J. H. Sinclair, "Mechanical Property Characterization of Intraply Hybrid Composites," *Test Methods and Design Allowables for Fibrous Composites,* ASTM STP-734, C. C. Chamis, Ed., American Society for Testing and Materials, 1981, p 261–280
12. J. H. Sinclair and C. C. Chamis, "Compression Behavior of Unidirectional Fibrous Composite," NASA TM 82833, 1982
13. P. C. Paris and G. C. Sih, "Stress Analysis of Cracks," *Fracture Toughness Testing and Its Applications,* ASTM STP-381, American Society for Testing and Materials, 1965, p 30–83
14. C. C. Chamis and G. T. Smith, "Resin Selection Criteria for Tough Composite Structures," NASA TM 83449, 1983
15. C. C. Chamis, "Computerized Multilevel Analysis for Multilayered Fiber Composites," *Comput. Struct.,* Vol 3, 1973, p 467–482

SECTION 4
Case Histories in Design and Applications of Composite Structures

Composites are used as alternatives or substitutes for conventional materials because of their high strength-to-weight ratio relative to other materials, their stiffness, which improves fabrication and formulating techniques, cost effectiveness, increased availability of raw materials, an admirable applications track record, and the engineerability of composite systems, which can be designed to meet specific criteria. Although composite material systems are not the answer to every materials design problem, the base of applications for which composites are used continues to broaden.

The case histories presented in this section deal with applied composites technology. Each case history is an abstract of a previously published study. The complete source of the original is provided to assist the reader in obtaining more information. Although not all reported case studies were successful in outcome, most were. Any negative conclusions drawn from product tests are as instructive as successful applications. An attempt was made to present case studies from existing literature, which covers a variety, but not necessarily the complete gamut, of applications and material combinations. Despite this attempt, epoxy reinforced with glass or graphite fibers and applications in the aerospace, automotive, and leisure products industries dominate.

The case studies presented are based on work done as far back as 1979, and further updates on some of the cases presented may have since been published. No attempt was made to include these updates, but recent articles were chosen for inclusion wherever possible.

The cases were chosen to form a capsulized base of the state-of-the-art knowledge in applied composites. The intent is to instruct by summarizing existing literature and to stimulate ideas for future applications through the experiences of those who authored the original papers and articles.

Recreational Applications

Golf Shaft Developed from Glass- and Graphite-Reinforced Plastic

The development of engineering criteria necessary to produce a suitable composite golf shaft begins with a stress analysis on No. 1 woods of various makes, tested by a number of good, consistent players. The use of lacquer stress coats and strain gage measurements provided the following benefits:

1 It is nearly impossible to statically reproduce the dynamic stresses in a golf shaft as it is swung, but some static deflections can be related to dynamic performance.
2 Each club has a unique vibrational frequency, which is related to variables such as shaft and head weight, shaft stiffness, and length.
3 Club head speed is inversely related to the moment of inertia of the club.
4 Because the club head path achieves its smallest radius just prior to contact with the ball, it catches up and leads the shaft into contact with the ball, rather than lagging the shaft.

5 Shaft torsional stiffness and strength during impact play little or no part in the flight of the ball, assuming that minimum strength requirements are met.
6 The flexural pattern of the shaft is most effective if it is even across the entire length of the shaft.
7 The lighter the shaft, the better.

Based on the above conclusions, a light shaft with the correct torsional and flexural properties should produce the best results in terms of distance traveled by the ball. A computer model was developed to aid in the selection of composite reinforcing fibers, matrix materials, geometry, and fiber/matrix ratios. The model significantly shortened computation time in the acquisition of test data, even though it took 2 months to develop. Having defined all design criteria, equipment was purchased to build prototypes as well as production runs.

The result is the Shakespeare TORFIL line, available in different designs varied by shaft weight and flexural pattern. These clubs utilize Union Carbide Thornel 400 graphite yarn and Owens-Corning glass fibers. The shafts include a 90° fiber layer for hoop strength, filament winding layers for torsional and flexural properties, and 0° layers for flexural strength and stiffness properties.

Prototype testing showed it was possible to drive a ball further and/or the same distance, but with more accuracy than with conventional clubs.

Abstracted from "The Development of a Glass and Graphite Reinforced Plastic Golf Shaft," by Frank W. Thomas, Shakespeare Co., Central Engineering, Columbia, South Carolina, in *Composite Materials in Engineering Design,* Proceedings of the 6th St. Louis Symposium, 11–12 May 1972, sponsored by Monsanto Co. and Washington University Association, Bryan R. Noton, Ed., American Society for Metals, p 277–282, 1973.

Graphite Fiber-Reinforced Golf Club Heads

For functional and generic reasons, golf clubs traditionally have been classified into two groups — woods and irons. Although the functional distinction still exists, the generic distinction no longer applies, because wood clubs are now made from steel and reinforced plastics, as well as from other materials. Irons also are made from alternative materials. Nevertheless, the design requirements and performance characteristics of each type of club have not changed.

Wood materials of preference for woods have been persimmon, followed by less costly and more plentiful laminated maple. An investment cast metal head was the first metal head to gain acceptance, because it had improved playability over wooden heads. More recently, graphite fiber-reinforced woods have been developed, and they compare favorably to both wooden and metallic club heads in several ways. A quality graphite/epoxy golf head can be produced by either injection molding or compression molding techniques.

Injection-molded golf club heads use only thermoplastic resins, usually ABS or polycarbonate, selected for their structural properties. These are reinforced by chopped graphite fibers measuring 3.2 mm (1/8 in.) or less and limited to 40 wt% (30 vol%) of the club head to allow adequate material flow in the molding process.

Two different compression molding techniques are used: high-pressure compression molding and low-pressure match metal molding. In high-pressure molding, long or continuous fiber-reinforced plastic is inserted into a heated mold cavity. The two mold halves are closed under pressure, and the heat/pressure combination causes the material to soften and become compressed to the configuration of the mold. In long or continuous fiber-reinforced plastics, mechanical properties remain high, but the polymer (almost always a thermoset resin) is not reprocessable. Finishing costs are also lower than with injection-molded clubs, because there are no gate marks to remove.

Low-pressure match metal molding is a more recent modified form of compression molding. In this technique, a pre-

form, usually of continuous fibers in a thermoset resin, is placed in the mold cavity. Heat is applied after the mold is closed, and as resin softening begins, pressure is applied from within the preform, forcing the material outward to conform to the mold walls. Internal pressure usually is applied by using an intumescent foam at the core of the preform, which expands upon heating. A pressurized internal bladder sometimes is used. More consistent parts are produced with the bladder method, because greater internal pressure can be produced. In general, high-pressure injection molding is the method of choice for quality clubs.

For golf club heads, the reinforcing fiber most commonly used to provide strength and stiffness is graphite. A variety of matrix materials may be used. These are summarized in Table 1. Graphite/epoxy is the best product in terms of mechanical properties, but it is also the most expensive.

Materials selection is important, but must be made within the context of the end product. Golfers and manufacturers well know that things such as the "feel" of a club are important and include such things as weighting, balance, sting after impact, and other factors. Furthermore, in a composite club, the selling point is not the extra distance the ball may travel, but increased accuracy over the same distance. If longer distance occurs, it is an additional asset. Through various means composite clubs can be made to have the feel of conventional clubs. This is done by means of added weights, balancing, and sizing and shaping of the head. The theory of the moment of inertia of the club on mishit balls and why some clubs are more forgiving than others is also discussed.

Abstracted from "A New Look in Golf Graphite Fiber Reinforced Heads," by Paul A. Roy, Aldila, Inc., in *Technology Vectors,* Vol 29, 29th National SAMPE Symposium and Exhibition, Society for the Advancement of Material and Process Engineering, p 1592–1608, 1984.

Composite Material Use in Competition Bobsleds

The United States has not done well in international bobsledding competition, due in large measure to the superior equipment designs of foreign competitors. Merlin Technologies has applied its aerospace composites experience to the task of designing bobsleds capable of superior performance. Historically, evolution of the bobsled has not been rapid, and changes in dominance of the sport have been primarily due to innovative equipment design.

Conventional bobsleds consist of a steel chassis and a fiberglass body. Italian manufacturers have dominated production for most of the last 30 years. Their designs dictate that the center of gravity of a loaded sled lies behind the

Table 1. Properties of common composite golf club head materials

Property	Injection molded		Compression molded	
	Graphite/ABS	Graphite/polycarbonate	Graphite/epoxy	Graphite/epoxy
Fiber length, mm (in.)	3.2 (0.125)	3.2 (0.125)	50.8 (2.0)	Continuous
Fiber, wt %	40	40	41	63
Fiber, vol %	27	28	33	60
Specific gravity, g/cm³ (lb/in.³)	1.24 (0.045)	1.36 (0.049)	1.42 (0.051)	1.56 (0.056)
Impact strength (notched), J/cm (ft·lb/in.)	0.5 (1.0)	0.8 (1.5)	9.6 (18)	. . .
Flexural strength, MPa (psi)	117 (17,000)	158 (23,000)	469 (68,000)	1,462 (212,000)
Flexural modulus, kPa (psi × 10⁶)	19.3 (2.8)	20.0 (2.9)	37.9 (5.5)	117.2 (17.0)
Rockwell hardness	110(R)	119(R)	105(M)	105(M)
Cost, $/kg ($/lb)	26 (12)	26 (12)	33 (15)	55 (25)

center of aerodynamic pressure, resulting in poor handling characteristics.

As the original article was being written, carbon/epoxy bobsleds were being manufactured for use in the 1984 Winter Olympics. The sleds have a slimmer body, and this aerodynamic design is partially the result of wind tunnel tests. Carbon/epoxy was selected to reduce weight and to redistribute the weight to maximize sled performance. The composite material was used primarily in the two-piece body construction of the sleds. Additional weight savings could have resulted by the use of carbon/epoxy components in the chassis, but time constraints were prohibitive.

A typical fiberglass sled body weighs about 27 kg (60 lb), which is distributed fairly evenly along its length. A carbon/graphite body, in contrast, weighs less than 4.5 kg (10 lb), and a roll bar was added for safety. Materials used in the construction included Celion 3000 carbon fiber prepregged with a Fiberite 7714A resin system. The material was hand layed-up in fiberglass molds that were fabricated from plaster models. The severe reduction in the weight of the sled allowed weight to be added as a ballast in the nose of the sled, thus improving the relationship between the centers of gravity and aerodynamic pressure of the sled.

The revolutionary changes in body design dictated changes in the chassis and steering mechanism. The chassis became narrower; outrigger bumpers were added to the design, and the steering mechanism (and its weight) was moved farther forward, as was the point of articulation between front and back sled runners. Because of the narrow body design, the steering mechanism was reengineered and employs a longitudinally mounted steering arm operated via four pulleys.

At the time this paper was written, no prototypes had yet been manufactured from which performance conclusions could be drawn.

Abstracted from ''Carbon/Epoxy Bobsleds to Challenge the Europeans at the 1984 Winter Olympics,'' by Robert D. Torczyner, John C. Presta, and Donald R. Sidwell, Merlin Technologies, Inc., Campbell, California, and Boris Said, Olympic Bobsled Racer, in *Technology Vectors*, Vol 29, 29th National SAMPE Symposium and Exhibition, Society for the Advancement of Material and Process Engineering, p 1313–1319, 1984.

Advanced Composites for Archery Equipment

Recent developments in materials selection for archery equipment stems from the desire for more arrow speed, which flattens the trajectory of the arrow and increases accuracy. Archery bows and arrows have gone through evolutionary changes in a quest for more speed, and three major bow designs (Fig. 1) have resulted. The recurve design was first and was made of high-strength unidirectional E-glass skins laminated to hardwood veneer cores. Along with this innovation came the fiberglass arrow shaft. The next step was the compound bow, which used round wheel eccentrics to compound the limb action into a shorter power stroke and gave the archer a significant increase in arrow speed. Aluminum arrows became popular with this bow, because they were stiffer than fiberglass and could be shot lighter out of a compound bow. The energy cam bow is similar to the compound bow, but the geometry is different, and cams rather than wheels are used.

Because the stresses on a cam bow at loading can exceed the maximum shear loading that maple core laminates can endure, solid-fiber-reinforced plastic (FRP) limbs are used almost industry-wide. Other materials were used for cam bows and were found to be unacceptable before E-glass/epoxy limbs

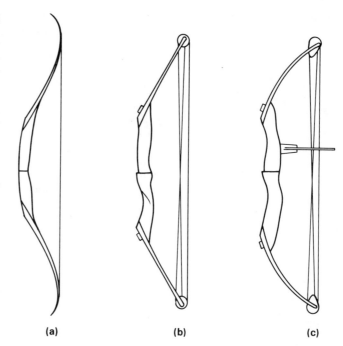

Fig. 1. Archery bow designs: (a) recurve; (b) compound; (c) cam.

were manufactured. These outlasted previous materials in fatigue cycle tests by a factor of 5 without showing signs of weakening.

Use of graphite composites in cam bows has met with only limited success. Relative to the additional cost, increases in arrow speed and bow performance have been poor. Design modifications may be possible, but the high cost would still be prohibitive unless gains in bow efficiency are sufficient to increase arrow speed by at least 6 m/s (20 ft/s).

There are three different types of composite arrow shafts on the market. For the competition archer, 100% graphite composite and aluminum-graphite composite shafts are produced. For the field archer and hunting enthusiasts, glass/graphite hybrid designs have been developed. The all-graphite composite arrow is very light and fast, but it has not replaced the aluminum target arrow, because it is not as durable. Glass/graphite hybrid arrows have been well received thus far, due to superior performance and better durability and straightenability relative to competitive materials. Use of advanced composites in archery has not become widespread, but continued achievement of performance goals will increase their use.

Abstracted from "Development of Advanced Composites for Archery Equipment," by D. Michael Gordon, Gordon Plastics Inc., Vista, California, in *Technology Vectors*, Vol 29, 29th National SAMPE Symposium and Exhibition, Society for the Advancement of Material and Process Engineering, p 1320–1325, 1984.

Use of Composite Materials in Canoes

Canoeing has changed from a means of transportation to a sport and a way of maintaining cardiovascular fitness. For this reason, many facets of canoe sport have evolved, including recreational canoeing, tripping, whitewater, competition, and solo freestyle.

Each style of canoeing makes specific demands on the craft. For this reason, a large number of canoe designs are available. Selection of materials is an important aspect of canoe design, and common materials in use include ABS Royalex, fiberglass, and Kevlar. Composed of three materials — a vinyl outer skin, a foam core, and an ABS substrate — ABS Royalex is vacuum formed into a canoe hull from sheet material. This method and material produce tough canoes that are reasonable in cost and appropriate for whitewater use.

In fiberglass craft, fiberglass fabric saturated with polyester resin is built up by layers in a mold, forming the canoe hull. Although fiberglass canoes are the cheapest on the market, they are nevertheless a durable craft suitable for general recreational use. Kevlar is sometimes used instead of fiberglass and is combined with a vinyl ester resin to make the strongest possible laminate that is stiff, tough, and lightweight.

Even more weight savings can be achieved by using a sandwich laminate of epoxy prepreg Kevlar fabric and Nomex honeycomb, which results in one of the lightest-weight canoes possible with no sacrifice in stiffness and strength. These are used in demanding competitive circumstances.

Each technological advance in canoe design leads to an increase in cost, and price is directly related to increased performance. Also, the higher the materials technology, the less the canoe weighs. ABS Royalex and fiberglass make good canoes of excellent value, but Kevlar is the material of preference for the more demanding design criteria characteristic of the high end of the market.

Abstracted from "Modern Canoe Technology," by James A. Henry, Mad River Canoe, Inc., Waitsfield, Vermont, in *Technology Vectors*, Vol 29, 29th National SAMPE Symposium and Exhibition, Society for the Advancement of Material and Process Engineering, p 1307–1312, 1984.

Design Analysis of Composites in Determining Snow Ski Characteristics

Ski design requirements depend on performance level (racing, recreational, or novice) and the increasing skill of the skiing population. Design criteria are even more critical when applied to high-performance skis. Current ski production methods include laminated sandwich construction, wet wrap/torsion box construction, or reaction injection molding. Injection-molded skis are produced for the low end of the market, while the wet wrap/torsion box skis are at the high end. This case study focuses on laminated skis.

Laminated ski construction methods offer the designer considerable flexibility, because the only constraint is the side geometry for a given mold design. The thickness of the ski, materials used, and other parameters are all variable. Many different materials are used in skis, and important factors governing their selection include fatigue and impact resistance, tensile and compressive strength, stiffness, elongation, and minimal temperature effects on mechanical properties.

ABS is used on top surfaces and sidewalls as a decorative layer in most ski manufacturing. The plastic top edge component is a deterrent to top edge cutting, which is caused by the bottom steel edge crossing the top of the ski. Edging materials used include glass-filled ABS, polyurethane, or polyarylether. In addition to cut resistance, these materials possess the same qualities as the top and edge surfaces. Top, side, and edge materials contribute only minimally to structural and physical characteristics of the ski.

Most laminated ski structures depend partly on fiberglass laminates for strength and stiffness. A pultruded unidirectional laminate has a tensile modulus around 44,127 MPa (6,400 ksi) and a 1241-MPa (180-ksi) tensile break strength at 70% glass loading. Graphite and Kevlar have limited use in skis, because of some negative characteristics including

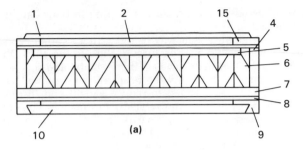

(a)

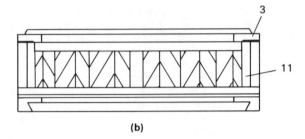

(b)

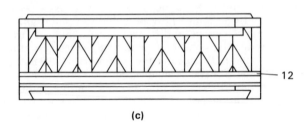

(c)

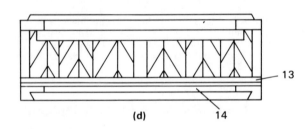

(d)

1   ABS top surface
2   1.3-mm (0.051-in.) unidirectional glass
3   7075-T6 aluminum top edge
4   0.5-mm (0.020-in.) woven glass
5   2.0-mm (0.080-in.) random fiber-reinforced plastic
6   Aspen/birch wood core
7   1.1-mm (0.043-in.) unidirectional glass
8   0.3-mm (0.012-in.) rubber foil
9   50 R$_C$ steel edge
10 Polyethylene running surface
11 Aluminum ribbed wood core
12 0.5-mm (0.020-in.) bias-ply glass
13 0.4-mm (0.016-in.) woven glass
14 0.5-mm (0.020-in.) 7075-T6 aluminum
15 Plastic top edge, polyarylether

Fig. 2. Ski cross sections and typical materials: (a) Olin Mark IV Comp.; (b) Olin 930; (c) Olin Mark VI SL; (d) Olin Mark VI GS.

negative thermal expansion coefficients.

Random glass-fiber laminates are used for ski torsional stability and for binding reinforcement plates. The glass content for these applications is between 45 and 50 vol% with an epoxy or polyester matrix.

A very important element in ski design is the material used in the ski core. Ski cores experience high bending and compressive stresses, must meet certain fatigue requirements, and must withstand significant impacts requiring core material resiliency. Materials used in ski cores include wood, rigid polyurethane foam, and honeycomb materials of aluminum, fiberglass, or aramid fiber. Rigid foam polyurethane is normally reinforced with fiberglass, phenolic, or aluminum to add strength and stiffness. Wood offers the advantages of high strength-to-weight ratio, maintenance of physical properties

with temperature change, easy processing, and low cost. Honeycomb materials have the highest cost.

The article from which this case history is abstracted evaluates the torsional and vibrational characteristics of ski designs through a set of design equations and the testing of fabricated samples. Following these steps, a prototype product was produced based on data gathered, and certain desired torsional and vibrational characteristics were produced (Fig. 2).

Abstracted from "Expanded Design Analysis of the Use of Composites in Determining Snow Ski Characteristics," by Edward D. Pilpel, Olin Ski Co., Middletown, Connecticut, in *Materials Overview for 1982*, Vol 27, 27th National SAMPE Symposium and Exhibition, Society for the Advancement of Material and Process Engineering, p 616–627, 1982.

Automotive Applications

Composite Automotive Rear Floor Pans

In 1981, Ford Motor Co. research engineers began a program to develop a structural composite design methodology. A requisite of this program was to design for an Escort model a rear floor pan with approximate dimensions of 120 cm (48 in.) in width by 150 cm (60 in.) in length with a spare tire well about 30 cm (12 in.) in depth.

A critical floor pan section was selected and molded before the entire floor pan was designed so that the preliminary fabrication and design experience would be beneficial. This section was compression molded. Continuous glass reinforcement was used to conform to highly curved areas without extensive cutting and fitting. Testing of this initial section approximated finite element model results, which predicted that the composite would not be as stiff as a steel part, but that it would be twice as strong.

Following the completion of test section work, the full floor pan was fabricated using glass/vinyl ester sheet (SMC) and directionally reinforced sheet (XMC) molding compounds. All attachments were fastened with adhesives and were stock Escort components. The final configuration consolidated ten steel components into one composite molding with a weight saving of 15%. This weight reduction was not as much as anticipated because of the inclusion of production attachment hardware.

The rear floor was prototyped using three different materials: a vinyl-ester-based R-65 SMC and XMC, a polyester-based R-65 SMC and XMC material, and a glass-reinforced polypropylene sheet material. The first two materials were molded using compression-molding techniques at 143 °C (290 °F). The third material was thermoplastically stamped using a stamping press and a zinc alloy tool at 93 °C (200 °F).

The composite floor pans tested were installed in running and mock-up vehicles. The installation procedure was the same for both types and involved three major steps: trimming the composite floor pan and cutting away the steel floor pan; coating all the attachment hardware with a primer paint per adhesive manufacturer instructions and locating the attachments with adhesive on the component, followed by an adhesive cure; and mating the surface of the vehicle with the profile of the floor pan, applying a urethane adhesive, and trimming the finished installation.

A variety of material property tests, strain gage tests under simulated loading conditions, static bending and torsional loading tests, durability tests, and noise tests were conducted on the prototype installations. In all tests, the prototype specimens performed as predicted by modeling techniques.

This program demonstrated the feasibility of molding a large structural part using selective continuous reinforcements. Static and dynamic testing verified the structural integrity of the part and short-term adhesive durability was also proved. Adhesive durability for longer runs is still an issue.

Abstracted from "A Composite Rear Floor Pan," by N. G. Chavka and C. F. Johnson, Ford Motor Co., Proceedings of the 40th Annual Conference, Reinforced Plastics/Composites Institute, 28 Jan–1 Feb 1985, Session 14-D, The Society of the Plastics Industry, Inc., p 1–6.

Composite Elliptic Springs for Automotive Suspensions

Fiber-reinforced plastic composites have higher energy storage capacity than steel, but poor resistance to shear stresses. Use of composites in coil springs would therefore require that the coil be of larger diameter. This would eliminate some, if not all, of the weight advantage of a composite material relative to its steel counterpart. Also, proper fiber orientation is difficult to maintain in a coil spring as it is being wound into shape. Because the application of composites in coil springs has been largely unsuccessful, an elliptic spring configuration was designed and developed.

The composite elliptic spring, designed to replace steel coil springs used in current automobiles, consists of a number of hollow elliptic elements joined together to obtain necessary characteristics. The configuration is illustrated in Fig. 3. Materials evaluated in this program were unidirectional and quasi-isotropic E-glass fiber-reinforced epoxy. The unidirectional material has considerably higher tensile and compressive strength and tensile modulus than the quasi-isotropic material.

Fourteen elliptic spring elements were manufactured by winding fiber-reinforced epoxy tapes to various thicknesses over a collapsible mandrel. After curing for 14 h at 149 °C (300 °F), the spring components were removed from the mandrel and tested in static compression with a crosshead speed of 13 mm/min (0.5 in./min). Due to a lower modulus, quasi-isotropic springs were found to have a lower spring rate than unidirectional springs. Springs behaved similarly

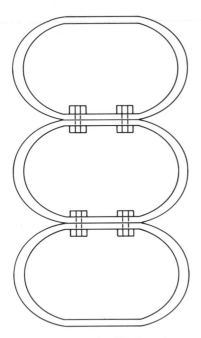

Fig. 3. Composite elliptic spring.

regardless of whether load was being applied or removed.

After testing individual spring elements, three unidirectional springs were stacked in series after the elliptic elements were slightly flattened at their tops and bottoms. The spring elements were mechanically fastened to one another. The actual behavior of the spring elements in series was predictable in the design stage through an equation based on the spring constants of each individual element and the number of elements used.

Work to date indicates that FRP springs have considerable potential as a substitute for steel coil springs. Advantages of the composite design include a weight saving of almost 50%, easier repairability, and the potential elimination of shock absorbers due to the high damping characteristics inherent in fiber-reinforced plastics.

Abstracted from ''Design and Development of Composite Elliptic Springs for Automotive Suspensions,'' by P. K. Mallick, University of Michigan–Dearborn, Dearborn, Michigan, Proceedings of the 40th Annual Conference, Reinforced Plastics/Composites Institute, 28 Jan– 1 Feb 1985, Session 14-C, The Society of the Plastics Industry, Inc., p 1–5.

Fiber-Reinforced Plastic Engine Components

Parts that oscillate or rotate and that are subject to a variety of high loads and stresses are good engine component candidates for fiber-reinforced plastic designs. Reduction in mass of parts such as connecting rods, piston pins, and pistons can ultimately result in engine vibration and noise emission improvements. Furthermore, weight reduction in some key engine components can permit secondary weight reductions in other components. In engine components where mass reduction is critical, carbon-fiber-reinforced plastics are used. In applications where the high stiffness/low weight combination is not as critical, glass-fiber-reinforced plastics may be used because of their lower cost.

The development of mass production processing technology for fiber, matrix, and part production is the key to the future of fiber-reinforced plastics in the automotive industry. Single-purpose automated machinery dedicated to a certain family of composite parts will be essential for high-speed processing. Composite components are less energy intensive to produce than their metallic counterparts, even exclusive of lower machining and finishing costs, which brightens the outlook for fiber-reinforced plastics in automotive applications.

In a research program sponsored by the West German Ministry for Research and Technology, connecting rods and piston pins for a gasoline engine were designed and manufactured using carbon-fiber-reinforced plastic materials. The goal was to reduce oscillating masses by 50%. These parts were designed to be used with an aluminum piston developed for the application, which was 30% lighter than a production-grade piston.

The harsh gasoline engine environment, with high oil temperatures, additives, oxidation, and degradation by-products, required careful study to determine which polymer matrix to use. A number of high-temperature epoxy, polyimide, and polyester resins were tested for their suitability. Preliminary tests studied viscosity as a function of temperature, torsion vibration tests, analysis of weight losses, and various tests for determining glass transition temperature, degree of polymerization, and thermal operating limits. Eleven resins passed these tests and were subjected to further screening tests, some of which are still underway, to determine their ultimate suitability for the application.

Resins were selected based on their mechanical properties, but prototype fabrication of four different designs was done using epoxy resin because of its ease of processing for fabrication. In testing, adverse interactions between carbon-fiber-reinforced plastic laminate parts, metal parts, and oil with additives and combustion residues were experienced.

After designing and testing of prototypes, a second generation of compression rods was designed as a flat bar (instead of the original ovaloid shape) press-formed from 70 layers of prepreg laminate. The resulting bar was 8 mm (0.3 in.) thick and was designed for fatigue life under engine loads.

Redesigned wound filament loops under proper pretensioning prevented separation of the elements during operation, as experienced with initial, hybrid component designs. With this design part, weight was reduced by 60%. Another second generation design simplifies the tensioning loop assembly concept. This concept reduces weight by 70%, but does not lend itself to mass production techniques. The design may be appropriate to race car applications, however. These designs performed very well in static and dynamic testing. After 10 million cycles at loads less than ultimate, the residual tensile strength was greater than the ultimate tensile strength of the virgin parts. This was probably due to relaxation of the filaments within the loop. Installation of these parts in a geometrically adapted engine is in progress, and motored and fired tests are scheduled to be run.

Preliminary tests using carbon-fiber-reinforced plastic piston pins indicated that the necessary surface quality and abrasion resistance properties could not be met without the use of a metallic housing. This restricted the choice of laminate layers, because the metal housing had to have the same coefficient of thermal expansion as the pins. Originally, the pins were hollow, but solid pins were found to have lower localized pressures and were preferable. Another pin design had an asterisk-shaped cross section. The piston pin designs resulted in weight reductions of up to 70%. Static and dynamic testing showed that a metal piston pin cannot be replaced by a carbon-fiber-reinforced plastic design.

Abstracted from "Fiber Reinforced Plastics Move Inside Engine," by Hans-Dieter Beckmann and Hermann Oetting, Volkswagenwerk AG, Wolfsburg, West Germany, in *Automotive Engineering,* p 34–41, May 1985. Based on SAE Paper 850520, "Fiber Reinforced Plastics for Light Weight Engine Parts."

Reinforced Plastic Bumpers for Passenger Cars

In 1976, the Ford Motor Co. started investigating the feasibility of using reinforced plastic bumper systems as an alternative to steel or aluminum. These studies were aimed at determining design and manufacturing parameters of a typical mid-size vehicle. This case study is an outgrowth of that effort, only the emphasis is on a particular car model, specifically the 1979 Mustang.

This particular model was chosen because it was within the weight range for which the original bumper had been designed, a number of representative vehicles were available for support testing, and the bumper would be hidden behind a urethane fascia, which obviates the need for a highly polished surface on the composite component. The reinforced plastic bumper was designed to be a bolt-for-bolt replacement of the existing aluminum reinforcement detailed in Fig. 4.

Initial studies showed that selection of materials and adhesives was critical to the performance of the assembly. Three adhesive systems (epoxy, modified acrylic, and polyurethane) and 14 material systems, including polyester, vinyl ester, and epoxy resins, were evaluated. The system that performed best under testing was one in which all parts were molded from a 20% calcium-carbonate-filled, 50% glass-vinyl-ester compound bonded with a two-component urethane adhesive.

The design optimization of the bumper system focused on weight reduction, and because of time limitations, only three approaches to weight reduction were considered. The 1979 Mustang went into production using both steel and aluminum bumpers. The advent of the reinforced plastic design would have most likely displaced the steel system, but there was not sufficient time before new model introduction to develop full confidence in the composite bumper. The aluminum design was used, and the composite system was redirected toward future models.

From a design and performance standpoint, a reinforced plastic bumper system is feasible. Styling and packaging limitations can be met by composite bumpers, and the composite bumper can be both cost and weight competitive with other lightweight materials.

Abstracted from "High-Strength Reinforced Plastic Bumpers for Passenger Car Applications," by N. A. Hull and D. H. Bergstrom, Ford Motor Co., in *Efficient Materials and Coatings Applications for Improved Design and Corrosion Resistance,* Proceedings of the American Society for Metals Highway and Off-Highway Vehicles, 13–15 Nov 1979, Activity Session, Chicago, Illinois, Materials & Processing Congress, p 3–13, 1981.

Composite Automotive Driveshafts

At the 1979 Daytona 24-h race, a modified Mazda RX-7 sustained a transmission tailshaft failure caused by a whipping driveshaft. The abnormally high rotational speeds of driveshafts, especially in racing vehicles, magnifies small discrepancies in balance, alignment, or straightness. Composite driveshafts were used to resolve these problems. Because of the lower weight of a composite tube, the same amount of misalignment is not as critical to the balance of a composite driveshaft.

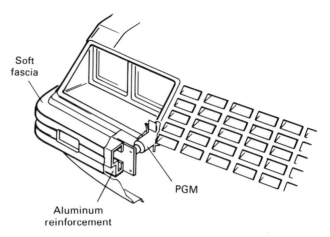

Fig. 4. 1979 Ford Mustang front end construction.

Merlin Technologies and Celanese Corporation developed carbon/glass fiber composite driveshafts that, in addition to weight savings, offer reduced complexity, warranty savings, lower maintenance, cost savings, and noise and vibration reduction relative to their metal counterparts.

Carbon-fiber composites are well suited to the driveshaft application because of their high specific stiffness and because certain drivetrain requirements can be met by varying fiber direction and quantity. Engineering criteria important to the design include matching the shaft whirl speed to various power train frequencies and forcing functions, adequate resistance to torsional buckling, and the attachment of metal sleeves.

The production method chosen for fabricating the driveshafts was a filament winding technique, in which a resin-wetted fiber is wrapped around a mandrel under tension to ensure a tight wrap. A special winder, oscillating mandrel, and dual-fiber (carbon-glass) delivery system have been developed for the purpose of manufacturing composite tubular goods. The wet filament winding technique and hybrid glass/carbon design have proven effective in road tests. Since early 1982, composite-driveshaft Mazdas have logged nearly 20,000 miles in competition at speeds averaging 148 km/h (92 mph).

Abstracted from "Composite Driveshafts — Dream or Reality," by D. R. Sidwell, M. Fisk, and D. Oeser, Merlin Technologies, Inc., Campbell, California, in *New Composite Materials and Technology,* the American Institute of Chemical Engineers, p 8–11, 1982.

Composite Leaf Springs for Heavy Trucks

Composite materials make it possible to manufacture lighter components without any reduction in strength and/or stiffness. This study explores the use of composites for truck leaf springs, as illustrated in Fig. 5. The material used must have equal properties in tension and compression, because the magnitude of these two types of stresses will be equivalent in application. Because fiberglass and aramid fiber systems have better tensile than compressive properties, their use was somewhat disadvantageous. High load-bearing properties would favor high-strength graphite/epoxy systems, but this material is expensive, and cost is also a consideration. The chosen material, after reviewing all the relevant criteria and material properties, was fiberglass/epoxy.

Design of the component must take into account the three-point loading pattern. Interchangeability of the composite design with the steel part is a design requirement. The loading pattern and required interchangeability dictate that the

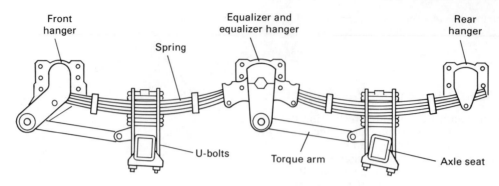

Fig. 5. Truck rear axle suspension.

thickness of the loaded span be of parabolic profile, except at the ends where constant thickness is necessary. Other design criteria for the spring include: a rated 1-g load of 3,402 kg (7,500 lb); a 2-g vertical, 0.75-g transverse, and 0.8-g longitudinal design loads; an assembly capable of some longitudinal twist; a fatigue life of 100,000 cycles under vertical load; and the capability to withstand accelerating and braking torques. A fiberglass/epoxy design using a steel main leaf was selected for the application.

Two fabrication techniques were considered — a compression-molding process and a resin injection-molding technique. The compression-molding process involved the use of fiberglass sheet molding compound, which is cut into patterns and fit into a matched metal die. Using pressures of 3,450–10,340 kPa (500–1,500 psi) and temperatures from 425 to 475 K, the part cures in 10 to 20 min and is then ready for finishing operations.

Fiber orientation distortion under pressure and the ability to achieve uniform fiber and matrix distribution throughout the part are disadvantages in using this method. The injection-molding process involves the placement of dry reinforcement in a matched mold cavity and injecting liquid resin under pressure and vacuum conditions to infiltrate the reinforcement. The part is cured by heating the mold. For the volumes required, both techniques are adequate, but the compression method was selected because the injection technique is in the preliminary stages of development.

Various mechanical tests were conducted on the finished parts, and design requirements for the component can be met using composites while achieving a minimum of 40% weight reduction over steel springs. Spring damping characteristics need to be fleet tested for final evaluation.

Abstracted from "Composite Leaf Springs in Heavy Truck Applications," by R. L. Daugherty, Exxon Enterprises, Greer, South Carolina, in *Composite Materials: Mechanics, Mechanical Properties, and Fabrication*, Proceedings of Japan-U.S. Conference, 1981, Tokyo, Japan, The Japan Society for Composite Materials, p 529–538, 1981.

Composite Truck Frame Rails

A composite heavy truck frame of graphite and Kevlar fibers has been developed by the Convair Division of General Dynamics and has been tested for 30,000 km (18,640 mi.) on a GMC truck without any problems. The composite frame rails weigh 62% less than steel, but have the same strength and stiffness.

The truck frame rails were C-channels made only of composite material. Because of the size of the autoclave required for curing, the rails were limited in length to 5.8 m (19 ft). Epoxy resin content of the finished product was about 32

wt%, and fiber consumption was roughly two parts graphite per one part Kevlar.

Fabrication of the prepreg laminate beams required removal of all solvents, water, and trapped air before curing; proper resin bleeding to obtain uniform resin distribution through the thickness; and proper application of pressure for adequate ply consolidation. The orientation and thickness of each varied, but 62 plies were used on the flange and 38 plies on the web. After final layup and curing in an autoclave, drilling operations were performed. Nondestructive tests on the rails and beams provided favorable results.

The rails were assembled into a GM truck. Inadvertently, one of the rails was damaged and suffered delamination. Epoxy was forced into the voids, and the flange was clamped between two steel plates for 20 h. En route to the proving grounds, delamination occurred in the repaired areas. The repaired area was removed and rebuilt with glass cloth and epoxy resin. One year (30,000 km, or 18,640 mi.) after the frame was installed, there was no evidence of structural damage. Bolt holes maintained their integrity, and there was no significant creep of the resin matrix.

Abstracted from "Composite Truck Frame Rails," by Gerald L. May, GM Truck and Coach Division, and Curtis Tanner, Convair Division of General Dynamics, in *Automotive Engineering*, p 77–79, Nov 1979.

Sheet Molding Compound Automotive Deck Lid

In response to a design competition sponsored by the Society of Automotive Engineers, a designer at Owens-Corning Fiberglas developed a deck lid that provided equivalent performance to a steel unit and satisfied class A surface requirements. Additionally, the designs (two design approaches were used) had to satisfy assorted strength, stiffness, dent resistance, and dimensional criteria. The one-piece design consisted of an integral skin and rib construction, while the two-piece design consisted of an outer skin with bonded inner hat section reinforcement. Molding constraints and surface finish requirements were of paramount importance.

Laminate formulations used for the deck lid were a sheet molding compound low-profile polyester system with the following approximate weight ratios: glass fiber (28%), resin (29%), and filler (43%). The two-piece design was made with 51-mm (2-in.) glass fiber randomly oriented for the outer skin and 25- or 51-mm (1- or 2-in.) randomly oriented glass-fiber reinforcing members. The skin surface of the one-piece design used 51-mm (2-in.) randomly oriented glass fibers to bridge the ribs and 13-mm (0.5-in.) randomly oriented fibers to fill the ribs.

The deck lid was designed to match the contours of the steel part. Two designs were evaluated for the two-piece construction. Both had undersupports around the perimeter of the piece, but center supports were of an X configuration in one and a Y configuration in the other. The X design was better for torsional stability and broad skin support against deflection. The Y design offered considerable material and weight savings, but was not as structurally sound or dent resistant as the X design.

The one-piece design was similar to the X design in the two-piece approach. Dual ribbing was adopted for the perimeter and X diagonal reinforcements. Additional single ribs were used for latch structure and panel centerline reinforcement.

The attachment of hinges to the sheet molding compound deck lids crushed the laminate as the bolts were torqued down. Oversized washers helped somewhat, but laminate bosses on either side of the holes alleviated this problem, as well as served a spacing and stiffening function. The same considerations apply to the hood latch.

Abstracted from "SMC Deck Lid: A Design Study," by Eugene Gray, Jr., Owens-Corning Fiberglas, in *Automotive Engineering*, p 65–66, Oct 1981.

Fiber-Reinforced Plastic Rear Axle Design

Properties of reinforced plastics are such that they can be considered for load-bearing parts. However, because their characteristics are different from those of metals, radical component design changes are necessary. An oriented-fiber-reinforced plastic rear axle design (see Fig. 6) resulted in a 57% weight saving at an estimated cost increase of about $1.20 per kilogram (2.2 lb) of weight saved. The durability of the prototype was short of expectations, but further work is expected to improve its performance.

Fig. 6. Volkswagen Auto-2000 fiber-reinforced plastic rear axle.

Design goals for the fiber-reinforced plastic axle structure were as follows: stiffness; strength; packaging, or fitting the vehicle with no major modifications; manufacturing with mass production technology within 5 years; minimize mass with respect to cost; and keep the cost increase per kilogram (2.2 lb) of weight saved to $1.00. Upon studying the fiber-reinforced plastic/steel axle trade-offs, designers drew several important conclusions. First, the axle would have to be of a radically different design concept to improve its stiffness, or it would have to use oriented fibers, or both. Second, because oriented fibers would almost certainly have to be used, manufacturability would be a major design consideration so that the product could be mass produced. Third, joint placement and attachment would be a major concern. Fourth, the project team would have to design the geometry, material, and manufacturing system with very little applied experience as background. Because of a stringent timetable, the use of finite-element analytical techniques was necessary and had

to take the place of trial-and-error experience.

The design team had considered three major design approaches and for various reasons, including the avoidance of adhesive joints in critical areas and manufacturing ease, settled on a design consisting of two identical halves bonded in the center along the horizontal midplane (see Fig. 7). With the help of finite-element techniques, the iterative process of design refinement was accomplished, resulting in changes of the originally proposed flange design, center section, and the addition of more reinforcing ribs. Another area of design concern was the joint between the axle and wheel carrier

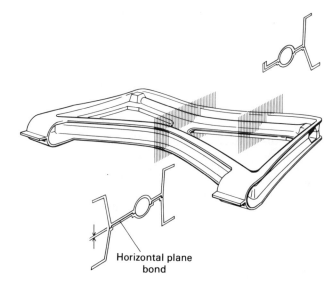

Horizontal plane bond

Fig. 7. Detail of fiber-reinforced plastic Volkswagen rear axle design.

system. The carrier was designed as a metal (aluminum) part and bonded over a large area to the composite axle. This design was found to be sound.

The material selected for the axle was 60% glass/polyester, but retrospective evaluations indicate that vinyl ester may have been a better material. Using higher-performance materials to reduce weight even further is possible, but the cost effectiveness of further weight reduction becomes prohibitive.

Prototype manufacture utilized careful charge placement via premolded oriented-fiber inserts. A single set of tooling was used to produce the top and bottom halves of the axle assembly. The prototype was tested for durability using the standard durability test schedule and fatigue life was found to be below expectations, consistent with predictions. Shortcomings were quantified for the next iterative design.

Abstracted from "Rear Axle Designed in Oriented FRP," by Jack M. Thompson, Structural Dynamics Research Corporation, and A. Bauer and D. Brodowsky, Volkswagenwerk AG, in *Automotive Engineering*, p 71–78, Aug 1982.

Composite Tank Trucks

Composite tank trucks used for transporting chemicals have been used in Europe for 20 years, but their development and use in the United States has been stalled by various codes and regulations. The advantages of composite tank trucks are superior corrosion resistance; more effective insulating properties; and reduced wear, maintenance, and repair costs. Their good track record in overseas use prompted the U.S. Department of Transportation (DOT) to grant an exemption to

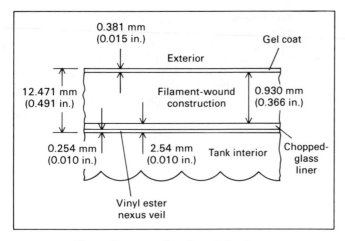

Fig. 8. Structure of tank truck laminate.

build a prototype vehicle in which safety and environmental concerns associated with the transport of hazardous chemicals were given top design priority. The experience gained with this prototype resulted in DOT approval for composite tankers on U.S. roads as of September, 1984.

Many of the design requirements for composite tankers under various load conditions were the same as for metal tankers that carry toxic chemicals. However, stress and buckling analyses were not applicable, because design experience for tanker trucks was nonexistent. Engineers had to draw from their experience in composite structural components.

A laminar construction was developed through a series of finite-element models, shown in Fig. 8. The interior layer is vinyl ester, surrounded by a chopped-glass liner, a filament-wound layer (15 plies) consisting of polyester and E-glass, and a gel-coated exterior to provide protection from ultraviolet radiation.

Stress analysis of the structure was performed with the SAP IV computer program and buckling analysis with the BOSOR IV program. For safety reasons, the composite tanker trucks were rated at stress levels below their actual tolerable limits.

Abstracted from "Composite Tanker Trucks Have Made the Grade," by Joseph M. Plecnik, North Carolina State University, Robert Short, Composite Engineering, Westminster, California, and John Plecnik, Belmont Abbey College, Belmont, North Carolina, in *Plastics Engineering*, p 63–65, March 1985.

Designing for Minimum Weight in Double-Layer Automotive Panels: Sheet Molding Compound Versus Steel

Use of composite materials promises to reduce automotive structural weight to help meet fuel economy standards, but changing automotive design constraints require a general design approach that allows for variable design criteria in designing composite automotive components. A typical automotive double-layer panel, utilizing a smooth outer skin and a more complex inner layer that has stamped or pressed sections used for stiffening, is used as an example in the design approach (Fig. 9).

The main structural requirement for this part was stiffness. The four different stiffness parameters considered were torsional, bending, edge bending, and oil-canning stiffness. In metal designs, oil-canning resistance requirements are satisfied by imposing minimum gage requirements, but this is unacceptable for sheet molding compound (SMC) designs. Oil-canning resistance is designed into sheet molding compound components by molding thin ribs into the thin outer sheet to stiffen unsupported areas.

The sheet molding compound selected for this component

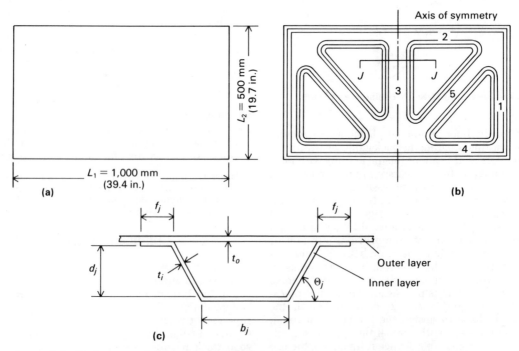

Fig. 9. Stamped double-layer panel: (a) outer layer; (b) inner layer; (c) cross section J-J (with outer layer attached).

consisted of a polyester resin reinforced by chopped-glass fibers about 25 mm (1 in.) long. Design constraints included maximum allowable panel depth, minimum allowable inner/outer layer thickness, and minimum allowable stiffness. The allowable panel depth was allowed to vary continuously in the search for a least-weight design. In a steel part, the thicknesses of the outer and inner layers are the same, but in a sheet molding compound part, the thickness of the outer layer is greater than the inner layer (about 1.5:1 ratio), reflecting current manufacturing practice designed to improve outer surface finish. Stiffness evaluation on least-weight designs considered two criteria: baseline stiffness determined by testing a production panel and a half-stiffness allowable. Changes in any of the variables in the design procedure necessitated repetitive calculating, for which an in-house program and optimization algorithm (CONMIN) was used.

Steel and sheet molding compound panels were designed for comparison. For steel, reducing the stiffness constraint led to small weight savings and a shallower maximum usable depth. Reduced stiffness requirements resulted in a more substantial weight saving for the sheet molding compound panel.

Thin-gage steel and sheet molding compound panels were also designed. Lowering the minimum allowable thickness led to an increase in the maximum usable panel depth and a decrease in optimal weight for both materials. Optimal panel designs for both materials, subject to the half-stiffness criteria and thin-gage constraints, showed that sheet molding compound designs were much more sensitive to changes in maximum allowable depth than steel designs.

All analyses assumed perfect bonding between inner and outer layers. Imperfect bonding decreases structural stiffness and compensating for this adds a weight penalty to the part.

The best sheet molding compound design does not scale with respect to material properties relative to the best steel design, and there is a maximum usable part depth (different for steel and sheet molding compound), beyond which the optimal weight remains constant. When stiffness requirements are relaxed, sheet molding compound designs benefit considerably more from weight savings than steel designs, and bonding between inner and outer layers is more critical in high-efficiency structures.

Abstracted from "Minimum Weight Design in Doublelayer Panels: Sheet Molding Compound versus Steel," by D. C. Chang and M. R. Barone, General Motors Research Laboratories, Warren, Michigan, in *ASTM Special Technical Publication 674*, Composite Materials: Testing and Design (Fifth Conference), 20–22 March 1978, New Orleans, Louisiana, sponsored by American Society for Testing and Materials, p 59–70, 1979.

Designing Continuous-Fiber-Reinforced Automotive Components

Many of the fiber-reinforced plastics used in automotive applications are chopped-fiber-reinforced materials used in semistructural applications. However, continuous-fiber components, which display excellent fatigue strength, high specific strength, and high modulus, are ideal for load-bearing members and where stiffness is required. Weight reduction and resultant fuel efficiency also are realized. By using the most successful design procedures established for aerospace applications, Ford Motor Co. built an experimental vehicle in 1979 using graphite-fiber-reinforced plastic (GrFRP) components.

The design cycle for composite components consists of five phases: laminate design, stress analysis, fabrication methods, cost analysis, and failure analysis. The latter three factors are common to conventional components as well, but the laminate design and stress analysis steps require a unique approach for composite application.

The laminate design step requires the establishment of layer parameters. These include type and orientation of fibers, matrix material selection, and form of layers (cloth or tape). Laminate parameters, or the stacking sequence of layers, are also established.

Stress analysis for composites differs from that for conventional materials in that lamination theory, or the analysis of stress on individual layers, is used. Stress tests are performed on a composite model, and each layer is examined for failure under load. Through iterative changes in stacking sequence and fiber orientation, a part meeting design criteria can be produced.

In the design of a graphite-fiber-reinforced plastic wheel rim, two basic criteria are considered. Specific load-carrying capabilities are evaluated by rim roll fatigue, rotary fatigue, and impact tests. The part must also be adequately strong in all radial directions, due to the nature of loading. Dominant load criteria were those required to pass impact tests. After computing the number of layers required for satisfactory performance, 0°, 90°, and ±45° cloth orientations were made to ensure axisymmetric strength in radial directions. The wheel was fabricated using conventional aerospace layup techniques and successfully passed all the tests. Each wheel represented a 45% weight savings over its steel counterpart. This was considered a difficult achievement, considering the nature of the part and design limitations imposed by severe loading and safety requirements.

To evaluate a stiffness-critical component, a graphite-fiber-reinforced plastic front end structure was designed. Component parts of this system were divided into two groups: those requiring bending moment stiffness and those requiring torsional stiffness. Local fastening areas were analyzed separately. Given the directional properties of the graphite-fiber-reinforced plastic components, they could match either the bending moment or torsional stiffness of steel parts, but not both. Regardless of which property was matched, the composite component only had to have 67% of the steel part stiffness to match the same deflection criteria. Stress analyses were done on both steel and composite front end assemblies, and the composite front end was expected to act as a structure of increased stiffness relative to the steel assembly. Conventional hand layup techniques were used to manufacture the composite front end, which was mounted on the experimental vehicle.

Aerospace fabricating techniques were found to be adaptable to automotive applications. Aside from achieving significant weight savings, anisotropic materials were successfully used in the design of complex geometric components. Providing that constraining factors such as cost and mass manufacturing techniques can be overcome, the potential widespread use of continuous-fiber reinforcement is feasible.

Abstracted from "Designing Automotive Components with Continuous Fiber Composites," by H. T. Kulkarni and P. Beardmore, Ford Motor Co. Research Staff, Dearborn, Michigan, in *NBS Special Publication 563*, Proceedings of the 29th Meeting of the Mechanical Failures Prevention Group, 23–25 May 1979, held at the National Bureau of Standards, Gaithersburg, Maryland, p 135–143, issued October 1979.

Aerospace Applications

Polyimide Matrix Composites in Supersonic Tactical Missiles

The development of polymers that can withstand high temperatures and still retain their specific strength and stiffness offers a significant benefit to tactical missiles, if processing and manufacturing methods can be developed. Most significant among the high-temperature resins are polyimides (PI), polybenzimidazole, and polyphenylquinoxaline (PPQ). Many different fibers may be used with these matrix materials, including graphite, glass, quartz, and Kevlar.

Tactical missiles include air-to-surface missiles, surface-to-surface missiles, and air-to-air missiles. This case study focuses on air-to-surface missile applications, with typical mission parameters of a speed of Mach 3.0, altitude of 10,670 m (35,000 ft), flight time of 3 min, a stagnation temperature of 427 °C (800 °F), and a structural temperature from 260 to 427 °C (500 to 800 °F). The air-to-surface missile has the configuration shown in Fig. 10. With the use of composites, a weight saving of from 50 to 67% can be achieved on components such as fins, inlet fairings, and nose cones. This case study focuses on inlet fairings for feasibility and demonstration purposes.

Four flightworthy graphite/polyimide fairings were fabricated using a modified autoclave cure, in which imidization and curing were accomplished in one autoclave run. Special tooling was developed because of extreme autoclave conditions (316 °C and 1380 kPa, or 600 °F and 200 psi), and several fabrication methods were investigated to yield a smooth aerodynamic external surface. The best surface finish was provided by the use of an effective mold release on the smooth tooling surface. Due to autoclave conditions, conventional bagging materials were not suitable so a polyimide film bag sealed with a special adhesive was used.

Graphite/polyimide prepreg unidirectional tape was used with a thickness of 0.127 mm (0.005 in.) per ply. The acceptable resin content of this material was 40%. Prepregs with resin contents of 33 and 44% were tried, but did not yield acceptable parts. Each composite fairing was inspected by X-ray and ultrasonic techniques and each showed minor resin-rich areas near the corners and some small foreign object inclusion, but they were deemed acceptable for flight testing.

Tests required to qualify the fairings included material, element static, and wind tunnel tests. Destructive and nondestructive quality evaluations were also conducted on fabricated parts.

Advantages of polyimide matrix components included ease of fabrication, lack of volatiles during processing, superior thermal endurance, superior specific strength and stiffness when reinforced with graphite fibers, and weight reductions. In missile applications, weight savings effectively improve missile performance.

The program behind this case study has shown that polyimide composites offer significant benefit to missile performance and can be put into widescale use. The parts can be produced in an efficient one-step cure cycle and can provide significant weight reductions. Wind tunnel tests have demonstrated the ability of the graphite/polyimide material to survive in a supersonic tactical missile environment.

Abstracted from "The Application of Polyimide Matrix Composites to Supersonic Tactical Missiles," by Charles H. Standard, Robert C. Van Siclen, and Bill G. Pinkerton, Vought Corporation, in *Materials and Processes — Continuing Innovations*, Vol 28, 28th National SAMPE Symposium and Exhibition, 12–14 April 1983, Anaheim, California, Society for the Advancement of Material and Process Engineering, p 1095–1106, 1983.

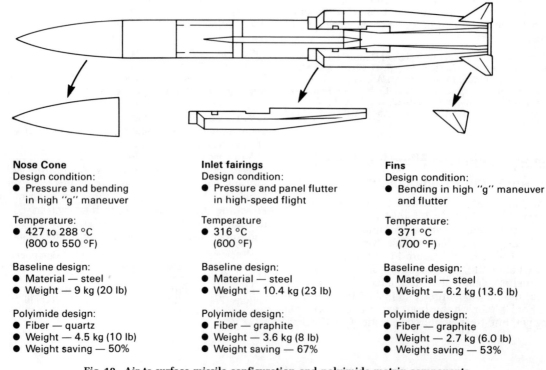

Nose Cone
Design condition:
● Pressure and bending in high "g" maneuver

Temperature:
● 427 to 288 °C (800 to 550 °F)

Baseline design:
● Material — steel
● Weight — 9 kg (20 lb)

Polyimide design:
● Fiber — quartz
● Weight — 4.5 kg (10 lb)
● Weight saving — 50%

Inlet fairings
Design condition:
● Pressure and panel flutter in high-speed flight

Temperature
● 316 °C (600 °F)

Baseline design:
● Material — steel
● Weight — 10.4 kg (23 lb)

Polyimide design:
● Fiber — graphite
● Weight — 3.6 kg (8 lb)
● Weight saving — 67%

Fins
Design condition:
● Bending in high "g" maneuver and flutter

Temperature:
● 371 °C (700 °F)

Baseline design:
● Material — steel
● Weight — 6.2 kg (13.6 lb)

Polyimide design:
● Fiber — graphite
● Weight — 2.7 kg (6.0 lb)
● Weight saving — 53%

Fig. 10. Air-to-surface missile configuration and polyimide matrix components.

Composite Satellite Antenna Reflectors

Since 1958, RCA has been involved in the design of satellite systems and hardware. Many communications satellites have been launched in the past two decades, and the antenna reflectors used on these spacecraft are dimensionally precise and environmentally stable. These antenna systems fall into three categories: frequency-reuse, dual-shell reflectors; dual-polarized, single-shell reflectors; and solid, single-shell reflectors. They may be fixed or deployable in space.

The fixed, frequency-reuse, dual-shell reflector generally consists of two parabolic sandwich shells. These are located one in front of the other and attached by precisely contoured sandwich ribs. The two shells are polarized orthogonally. The arrangement allows for compact packing of the reflectors within the allowable launch vehicle volume. The sandwich shells are fabricated from Kevlar 49 fabric-epoxy skins sandwiching a honeycomb core. Top and bottom skins are of two-ply construction with the radiofrequency-reflecting grids bonded to the parabolic surface. The two reflectors are bonded to each other by a rib subassembly, also of sandwich construction. The entire reflector assembly is supported by fixed legs of graphite/epoxy composites with titanium mounting brackets bonded to the bottom of the legs.

Fixed dual-polarized, single-shell reflectors also consist of Kevlar 49/epoxy honeycomb core sandwich shells, Kevlar 49/epoxy sandwich ribs, and graphite/epoxy fixed support legs. Dual polarization on a single shell is achieved by bonding two different polarization grids on different areas of the reflecting surface. One polarization is for the transmit and one is for the receive beams. This type of reflector is used when a single-feed assembly is employed in both receive and transmit modes.

Solid, single-shell reflectors usually consist of an aluminum honeycomb core sandwiched by graphite/epoxy skins. Modified designs may employ a hybrid of graphite/epoxy and Kevlar/epoxy for the skin and Kevlar/epoxy honeycomb for the core material. Single- and dual-shell configurations (both deployable) are illustrated in Fig. 11.

RCA has established reflector fabrication techniques resulting in extremely precise dimensional tolerances. Mechanical and thermal design concepts have helped improve the dimensional stability of the hardware. Additionally, the use of composite materials results in minimum-weight designs for the reflectors.

Developed composite antenna designs offer precise dimensional tolerances and environmental stability, resulting in better radiofrequency performances. These improvements have been achieved with minimum-weight structures.

Abstracted from "Advanced Composite Antenna Reflectors for Communications Satellites," by R. N. Gounder, C. F. Shu, and B. D. Jacobs, RCA Astro-Electronics, Princeton, New Jersey, in *Materials and Processes — Continuing Innovations*, Vol 28, 28th National SAMPE Symposium and Exhibition, 12–14 April 1983, Anaheim, California, Society for the Advancement of Material and Process Engineering, p 678–686, 1983.

Composite Hardware on the INTELSAT V Spacecraft

The INTELSAT V spacecraft is a satellite designed for geosynchronous orbit, providing 12,000 communication circuits for the 100-member nations of the INTELSAT organization. The spacecraft uses composite solar arrays, a composite antenna support module, and a composite rectangular box main

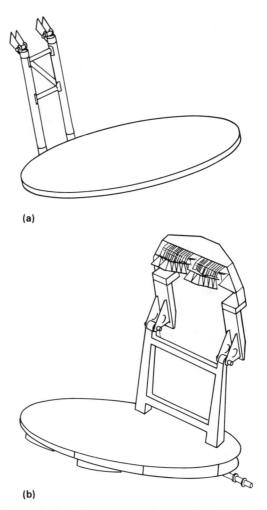

(a)

(b)

Fig. 11. Deployable antenna reflectors: (a) solid, single-shell reflector; (b) dual-shell reflector.

body containing communications support equipment. The intent of this case study is to highlight specific areas and components on the satellite in which unique materials were used.

The spacecraft contains 11 antennas, three of which are deployed after orbit is achieved. These consist of four parabolic reflectors and five conical horns, two of which include two antenna functions built into one horn assembly. Antenna reflectors were manufactured primarily from graphite/epoxy and Kevlar/epoxy materials. The larger reflectors utilized a Kevlar honeycomb core with two-ply graphite skins on the outside and Kevlar 49 bonded to the inner surface. The smaller reflectors employed graphite inner and outer skins sandwiching an aluminum honeycomb core. In all cases, reflector surfaces were bare graphite painted with thermal control paint. The horn antennas were honeycomb core sandwiches bonded with graphite skins.

Two antenna feed arrays are on the craft, each utilizing 88 feed elements. These are fabricated from graphite/epoxy fabric with a copper coating on the inside for desired electrical performance. The feed elements are attached to a feed network array made of copper-clad graphite/epoxy skins surrounding an aluminum honeycomb structure. These feed arrays are, in turn, bolted to another aluminum honeycomb structure clad with graphite/epoxy skins, which is a part of the antenna support structure.

Graphite/epoxy was used on the antenna tower in place of aluminum to provide a 30% reduction in weight, and because a number of odd shapes or bends were required in the structural elements of the tower. Also, the graphite/epoxy waveguide had a coefficient of thermal expansion close to that of the antenna support module, thus minimizing expansion problems. The graphite/epoxy tower has proven successful in that weight is considerably reduced, dimensional stability is greater than for aluminum, and performance is electrically equivalent to conventional waveguide materials.

The antenna support structure, used to support antennas, antenna feed arrays, and waveguides, is made of graphite/epoxy tubular members of various diameters and wall thicknesses bonded together to form a truss. Five mounting platforms, which are a part of the structure, are constructed of graphite face skins over an aluminum honeycomb core. Major design criteria in the selection of these materials were the low coefficient of thermal expansion and stiffness necessary for structural frequency requirements.

The development of graphite/epoxy multiplexers for the craft was a difficult task. There are over 1,100 separate graphite/epoxy multiplexer parts per spacecraft. Requirements of the part dictated the use of graphite/epoxy materials, as well as pure graphite instead of Invar, which is six times heavier.

The solar array consists of two wings, each with three hinged panels and a yoke structure. Construction of these panels consisted of graphite/epoxy members framing an aluminum honeycomb core, to which two graphite fabric/epoxy face skins are bonded. The yoke structure is also of graphite/epoxy.

The use of composite material components achieves requirements of low thermal expansion, stiffness, strength, lighter weight, and thermal stability. Over 4,000 separate parts of the spacecraft, accounting for 25% of the spacecraft weight, are of composite materials. Without them, the effective payload of the satellite would have been effectively reduced.

Abstracted from "Advanced Composites Hardware Utilized on the INTELSAT V Spacecraft," by H. L. Hillesland, Ford Aerospace and Communications Corporation, Palo Alto, California, in *The 1980's — Payoff Decade for Advanced Materials*, Vol 25, 25th National SAMPE Symposium and Exhibition, 6–8 May 1980, San Diego, California, Society for the Advancement of Material and Process Engineering, p 202–211, 1980.

Composite Pressure Vessels in Transportation Applications

Generally, composite pressure vessels are light metal liners overwrapped with continuous high-strength fibers embedded in a resin matrix. The transfer of this technology from the aerospace industry to the automotive industry required the development of liner materials for long-term storage, composite wrappings to carry most of the structural load, and environmentally resistant resin systems. Resulting pressure vessels were 20 to 50% lighter than metal vessels.

In an attempt to commercialize government-sponsored technological development, the Johnson Spacecraft Center awarded contracts to the private sector to develop composite pressure vessel products. Among these are a lightweight commercial fireman's breathing system, composite cylinders for escape slide inflation aboard commercial aircraft, and other filament-wound vessels since developed for commercial use.

Composite pressure vessels are usually cylinders or spheres. To comply with Department of Transportation (DOT) regulations, these vessels must have a design burst pressure equal to three times the operating pressure. The simplest form of composite pressure vessel is the "hoop only," or selectively reinforced cylinder. A fully wrapped cylinder has helically and circumferentially wrapped fibers for added reinforcement in directions other than radial. A fully wrapped vessel provides a higher-performance component of lighter weight and with higher structural margins. A spherical vessel consists of an internal metal liner overwrapped with many discrete winding patterns between the poles and the equator.

The metal serves as a winding mandrel during fabrication and in addition prevents leakage, provides a boss for the valve, and shares pressure load. The liner usually is made of seamless aluminum, but may also be of steel, stainless steel, Inconel, titanium, or another material. E-glass, S-glass, or Kevlar fibers usually are used for overwrapping. Epoxy resin usually is the matrix material of choice.

Prior to 1976, filament-wound pressure vessels had not been proven in commercial applications, and there were no regulations covering their use. The DOT had to grant exemptions for their use in interstate commerce applications. Extensive qualification tests are necessary for each type of pressure vessel to determine its equivalence to steel vessels. These include temperature and pressure cycle, burst, gunfire, bonfire, flaw growth resistance, and impact tests. Nondestructive tests also are performed on a random sampling of like containers. Tailor-made pressure vessels are tested according to customer specifications. These tests have included dragging the vessels over pavement and up curbs at moderate speeds and dropping weighted test specimens off buildings. The parts are shipped to the customer after quality specifications are met.

Transportation uses for composite pressure vessels include spacecraft, missile, aircraft and helicopter, hydrofoil ship, and ground service applications. Their advantages include long life, freedom from rust, low maintenance, good cyclic fatigue and static loading properties, compatibility with a variety of fluids, nonshatterability under ballistic impact, good resistance to a wide variety of environmental conditions, and interchangeability with conventional metallic systems.

Abstracted from "Filament Wound Composite Pressure Vessels in Transportation Applications," by Edgar E. Morris, Structural Composites Industries Inc., Pomona, California, in *Materials and Processes — Continuing Innovations*, Vol 28, 28th National SAMPE Symposium and Exhibition, 12–14 April 1983, Anaheim, California, Society for the Advancement of Material and Process Engineering, p 905–909, 1983.

Composite Applications in RCA Satellites

RCA has been involved in the design and development of satellite systems for many applications. Structures for these satellites have difficult strength, weight, alignment, and stiffness requirements, as well as constraints on size and weight imposed by the launch vehicle. Advanced fiber-reinforced composites with customized stiffness characteristics offer great potential to meet the above requirements.

The strength and stiffness of advanced fiber composites is superior to conventional metallic materials. The superiority of composites over conventional materials is even more striking when compared on an equivalent weight basis. Comparison of material properties makes it evident that composites can provide properties equal to metals at much less weight. Unidirectional high-modulus graphite/epoxy, for example, is five times stiffer and over three times stronger than an equivalent weight of 7075-T6 aluminum.

Because the goal of satellite design is to provide the maximum number of payloads at minimum weight, advanced fiber composites have been applied to satellite structures wherever possible. One such application is a Kevlar/epoxy polarized, parabolic antenna reflector of sandwich construction. Aside from good strength and stiffness characteristics, this material was selected because it can be designed with a zero thermal expansion coefficient and because its low-loss dielectric characteristics are desirable.

In communications satellites, graphite/epoxy designs have been developed for lightweight feed towers, waveguides, and feed horns. These structures are coated with metallic films to protect the composite from moisture penetration and to provide radiofrequency conductive interior surfaces. Graphite/epoxy designs also have been used for multiplex microwave filters on communications satellites and on precision mounting platforms used in weather satellites to precisely align certain instruments and keep them aligned in service.

Modeling and stress analysis techniques are employed in the design of composite satellite structures. Computer programs and finite-element modeling are used extensively in the design and analysis of laminate structures and fiber orientations.

Prepregs in the form of tapes or woven fabrics are used in the layup of many parts. Using a template, prepregs can be cut to shape and stacked upon each other at the desired fiber orientation. Simultaneous application of heat and pressure is then used to form the composite. Different methods of forming and curing may be used, depending on the materials and the dimensional characteristics of the component. Filament winding of structures is also a common method of fabricating tubes, cylinders, and pressure vessels. In this method, dry fibers are passed through a resin bath and positioned over a mandrel in a helical pattern, which is then cured in an oven.

In the future, satellite systems will require increased payloads, weight and size constraints, shuttle launch compatibility, cost efficiencies, and the ability to operate reliably under severe conditions. Many of these requirements have already been satisfied by the application of advanced fiber composite components.

Abstracted from "Advanced Composites Applications in RCA Satellites," by Raj N. Gounder, RCA Astro-Electronics, Princeton, New Jersey, in *Material and Process Applications: Land, Sea, Air, Space*, Vol 26, 26th National SAMPE Symposium, 28–30 April 1981, Los Angeles, California, Society for the Advancement of Material and Process Engineering, p 216–223, 1981.

Composite Inertial Upper Stages for Space Vehicles

This case study explores the feasibility of using advanced composite structures to improve the performance of the Department of Defense two-stage space vehicle and airborne support equipment (ASE). The two-stage vehicle will be borne into space via the Space Shuttle, which will place the 14,515-kg (32,000-lb) two-stage vehicle into orbit. This vehicle can then place a 2,270-kg (5,000-lb) spacecraft into geosynchronous orbit, or in a three-stage design, send a 2,040-kg (4,500-lb) vehicle to other planets.

Figure 12 illustrates the preliminary Stage I composite design, in which the need for three intermediate rings and 6,000 fasteners was eliminated. The Stage II design is shown in Fig. 13 and uses fabric and tape laminates for the payload ring, eight longerons, an equipment deck strut and radial supports, and a motor support ring. To reduce the Stage II part count, an integrally stiffened cocured concept and single-piece honeycomb construction were selected. Figure 14 illustrates the baseline aluminum construction for the airborne support equipment frame with composite cap plates designed to achieve bending stiffness.

The composite components are all made of graphite/epoxy materials. A variety of manufacturing processes were used and included elastomeric molding cocured, solid laminate and cocure lay-up, and honeycomb sandwich techniques. All of these techniques utilize net resin prepregs to eliminate the need for prebleeding and excessive resin flow during cure. The various subsystems are fastened to each other mechanically.

The integration of composite components into the space

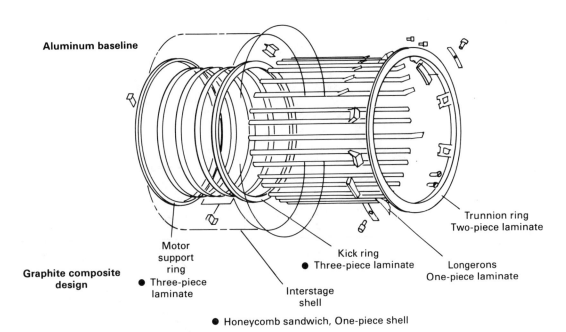

Aluminum baseline

Graphite composite design

Motor support ring
● Three-piece laminate

Interstage shell

Kick ring
● Three-piece laminate

Trunnion ring
Two-piece laminate

Longerons
One-piece laminate

● Honeycomb sandwich, One-piece shell

Fig. 12. Stage I space vehicle design.

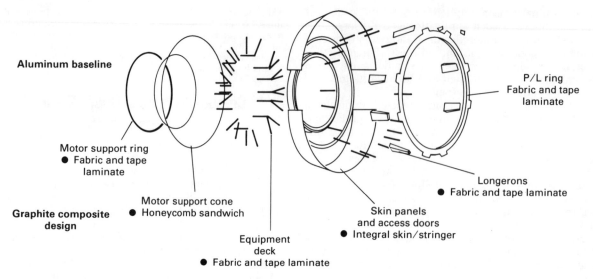

Aluminum baseline

Motor support ring
● Fabric and tape laminate

Graphite composite design

Motor support cone
● Honeycomb sandwich

Equipment deck
● Fabric and tape laminate

P/L ring
Fabric and tape laminate

Longerons
● Fabric and tape laminate

Skin panels and access doors
● Integral skin/stringer

Fig. 13. Stage II space vehicle design.

vehicle reduces the weight of the Stage I structure by 46%, the Stage II structure by 28%, and the cap plates of the airborne support equipment frame by 17%. The projected cost to produce 65 units of the composite frame is 1.2% less than for a metallic frame.

Two full-scale elements of the assembly, a Stage II longeron and an intermediate shear panel, were fabricated and tested. Test results verify the critical aspects of design for the parts.

The results indicate that composite designs for inertial upper stage (IUS) vehicles can reduce weight considerably, reduce cost slightly compared to metallic construction, and show that composite structures are feasible using existing and near-term state-of-the-art technology.

Abstracted from "Inertial Upper Stage (IUS) Advanced Composite Structure," by James E. Bell, Boeing Aerospace Co., Seattle, Washington, and Theodore J. Muha, Air Force Flight Dynamics Laboratory, Wright-Patterson Air Force Base, Dayton, Ohio, in *Fibrous Composites in Structural Design,* Proceedings of the Fourth Conference on Fibrous Composites in Structural Design, 14–17 Nov 1978, San Diego, California, Plenum Press, p 181–193, 1980.

Tungsten-Fiber-Reinforced Superalloy Turbine Blades

Aircraft gas turbine engine performance and durability can be improved by the development of higher-strength turbine blade materials. Because nickel and cobalt, the current blade

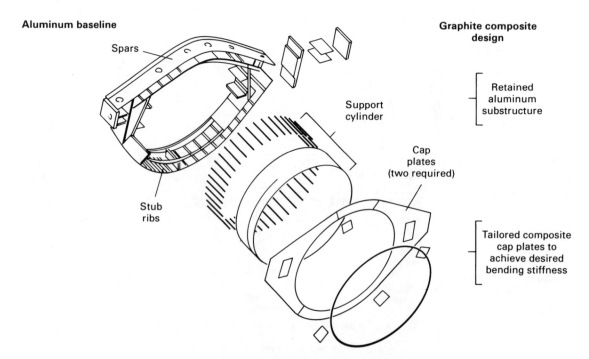

Aluminum baseline

Spars

Stub ribs

Support cylinder

Cap plates (two required)

Graphite composite design

Retained aluminum substructure

Tailored composite cap plates to achieve desired bending stiffness

Fig. 14. Airborne support equipment design.

materials, are already being pushed close to their physical limits on some turbine applications, an attempt to study tungsten-fiber-reinforced superalloy (TFRS) turbine blades was undertaken.

The primary general requirements for any turbine blade material are adequate creep-rupture strength, oxidation resistance, thermal and mechanical fatigue resistance, and resistance to impact damage. Any tungsten-fiber-reinforced superalloy candidate material must have a combination of these properties, which in turn are a function of fiber properties and matrix properties in combination.

Fiber property requirements include:

1 Higher creep-rupture and mechanical fatigue strength than conventional materials
2 Adequate toughness at operating temperature to resist foreign object impact damage
3 Low cost

Property requirements for the matrix materials include:

1 High oxidation and hot corrosion resistance
2 Compatibility with the fiber material at high temperatures
3 High mechanical and thermal fatigue resistance at elevated temperatures
4 Low density to offset the high density of tungsten fibers
5 High toughness and ductility at low temperatures, because at low temperatures the matrix supplies impact damage resistance and protects the fibers
6 Adequate shear creep strength to allow fiber angle plying for strength in various directions

After a review of existing information on tungsten-fiber-reinforced superalloy materials, a thoriated tungsten-fiber-reinforced FeCrAlY matrix composite (W/FeCrAlY) was selected to best meet material requirements. Tungsten fibers provide strength and high-temperature impact resistance to the turbine blade. The tungsten fibers, strengthened by dispersion of thoria or hafnium carbide, are adequately strong for turbine blades. The strongest such fibers are very high in

cost, so lamp filament wire was used, based on availability and low cost. Many composite properties are controlled by the choice of the matrix material, but no single material is superior in all required respects. The FeCrAlY material offers the best compromise for the combination of required matrix material properties. Both fiber and matrix properties determine the density, elastic constants, impact resistance, and thermal conductivity of the composite component.

The specific composite under consideration, through a series of tests, was determined to possess adequate properties for turbine blade use. The next step was therefore to explore whether complex blades could be fabricated effectively from the material. The technique with the most promise (for a hollow blade) is the diffusion bonding of monolayer composite plies. The approach can supply adequate fiber alignment and distribution and limits fiber-matrix interdiffusion during manufacture. The technique involves the sandwiching of composite monotape plies between layers of FeCrAlY. The technique has been used to make aluminum/boron blades, as illustrated in Fig. 15. The process yields a part close to finished dimensions, with only root machining and touch-up grinding required.

The material properties of the composite and the reasonable anticipated fabrication costs prompted NASA to produce an actual blade design for fabrication. A JT9D first-stage convection-cooled blade was selected for the first attempt. This particular blade was selected because it is a relatively low-stress, high-temperature blade and permits efficient use of the blade strength versus temperature properties. It has a long airfoil and minimal twist, making it easier to design and fabricate. Finally, this blade presently is used in many aircraft engines so that successful fabrication may have near-term value. The blade had not been fabricated at the time this article was written.

The W/FeCrAlY composite is a promising composite blade material with an excellent combination of properties. Although much work is still needed on fabrication and design, this material is the correct choice for the transition from laboratory data to a testable prototype. Furthermore, W/Fe-

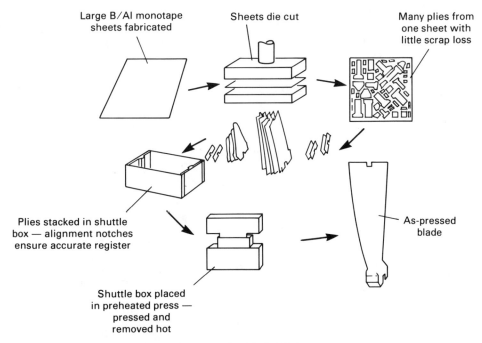

Fig. 15. Boron-aluminum fan blade fabrication sequence.

CrAlY is a first-generation tungsten-fiber-reinforced super-alloy, and stronger fibers do exist that can result in stronger blades in the future. Their expense is presently high, but it can be reduced when manufactured in larger quantities.

Abstracted from "Tungsten Fiber Reinforced FeCrAlY — A First Generation Composite Turbine Blade Material," by D. W. Petrasek, E. A. Winsa, L. J. Westfall, and R. A. Signorelli, NASA Lewis Research Center, Cleveland, Ohio, in *Advanced Fibers and Composites for Elevated Temperatures,* Proceedings of a symposium sponsored by The Metallurgical Society of AIME and American Society for Metals Joint Composite Materials Committee, 108th AIME Annual Meeting, 20–21 Feb 1979, New Orleans, Louisiana, American Institute of Mining, Metallurgical and Petroleum Engineers, p 136–155, 1980.

Automated Production of Composite Airplane Components

Higher material costs of composite aircraft components can be compensated for only by savings in the manufacturing process. Production cost savings are possible only if the proper selection of materials and their efficient and complete utilization is made, the component is designed with cost savings in mind, and a high degree of mechanization or automation is possible in the production process. A program to realize cost-effective production rates was started in 1979 for manufacturing a carbon-fiber-reinforced plastic (CFRP) fin box for the Airbus A300/A310. A module production method was selected, and although the cost of materials compared with that for an aluminum fin box increased 35%, total manufacturing costs were lowered by 65 to 85%.

Shell assemblies, spars, and ribs of the fin box were stiffened by a "modular" grid, so called because the grid is comprised of similar units (Fig. 16). Each module is an aluminum core around which the prepreg carbon-fiber-reinforced plastic is wrapped automatically. Each module is arranged into a grid and mechanically pressed together longitudinally and transversely to a specified size. The skin is also laid up mechanically with reinforcements in the spar and rib areas. After a few additional operations, the complete shell assembly is cured to create a component with a high degree of integrity. The aluminum cores are each made of three sections, which contract relative to the carbon-fiber-reinforced plastic material in cooling, making them easier to remove.

To determine the production method qualifications for series production, a number of tests were run in a laboratory environment. Several different machining and handling units (including robots) were developed, built, and/or tested. Among these were:

1 Prepreg bandage cutters requiring a high degree of precision
2 Prepreg bandage handlers consisting of a vacuum drum controlled by a robot
3 Robot-controlled vacuum drums used to wrap the prepreg around the aluminum core
4 Pneumatic pressure arms that fold the prepreg bandage over itself once and compact it against the aluminum core
5 Compacting chamber for each wrapped core
6 Vacuum suction tools that combine with the robot to transfer each core from the compacting chamber to the module grid
7 Laying heads that work in conjunction with the robot to layup the stringer strips

These devices were integrated into an automated procedure for manufacturing the box fins. The procedure, although complex, has proven successful in the serial production of the component for which it was designed and indicates the possibility of the total exchange from aircraft construction, based on metal processing techniques.

Abstracted from "Principal Tests for the Automated Production of the Airbus Fin Assembly with Fiber Composite Materials," by Branko Sarh, Messerschmidt Bulkow Blohm, Hamburg, West Germany, in *Technology Vectors,* Vol 29, 29th National SAMPE Symposium and Exhibition, Society for the Advancement of Material and Process Engineering, p 1477–1488, 1984.

Composite Designs for Spherical Pressure Vessels

Fiber-metal composite spherical pressure vessels provide improved operating performance, less weight, and improved safety features in aerospace and military applications. This case study considers an improved spherical vessel design made of an Inconel 718 liner completely overwrapped with Kevlar 49 fibers, which is used for helium storage on a space launch vehicle (Fig. 17).

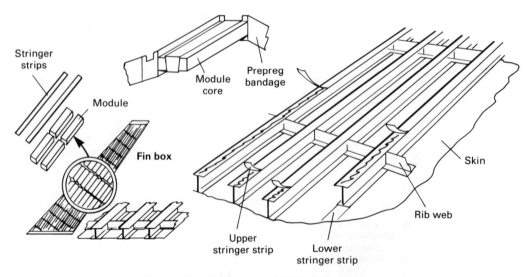

Fig. 16. Modular design and fabrication technique.

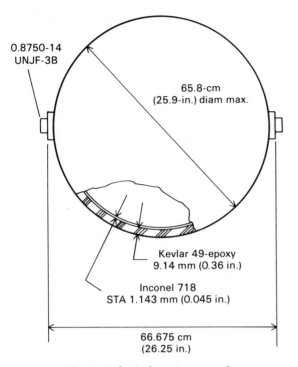

0.8750-14
UNJF-3B

65.8-cm
(25.9-in.) diam max.

Kevlar 49-epoxy
9.14 mm (0.36 in.)

Inconel 718
STA 1.143 mm (0.045 in.)

66.675 cm
(26.25 in.)

Fig. 17. Spherical pressure vessel.

A composite pressure vessel should be designed for maximum operating performance at minimum weight, while maintaining safety design features. In optimizing design parameters, the following factors must be considered: load and strain compatibility between the two types of material, wrap buckling strength of the metal shell, effects of prestress into the plastic region of the metal shell, thermal properties of component materials, and effects of cyclic and sustained loading on vessel life. Additionally, special design attention must be paid to the required burst factor of safety, adequate metal shell compressive strength so that adhesive bonding is not required to stop buckling, and other factors.

Inconel 718 was chosen for the liner material because it had the best combination of properties, producibility, low cost, and short lead time. Kevlar 49 reinforcing fibers were chosen because of their superior strength/weight ratio, good static loading and fatigue resistance, and excellent processability for the overwrap application.

The Inconel 718 shell was formed by electron-beam welding of two hemispherical pieces. The hemispherical pieces were actually subassemblies and included polar bosses that were also electron-beam welded to openings machined into the hemispheres. The Inconel liner was then overwrapped with Kevlar 49 fibers that were impregnated with epoxy resin in an asymmetrical multi-angular pattern. The vessel was then cured at 121 °C (250 °F) and hydrostatically pressure sized.

The finished vessel passed the performance and acceptance tests performed on the vessel, except for the burst pressure test where the mode of failure was determined. Failure occurred at one "end" of the sphere, initiating in the composite material. Overall, this sphere had one of the highest efficiencies of any composite pressure vessel demonstrated to date.

Abstracted from "A High-Performance Composite Fiber/Metal Sphere," by Robert E. Landes, Structural Composites Industries, Inc., Pomona, California, in *Technology Vectors*, Vol 29, 29th National SAMPE Symposium and Exhibition, Society for the Advancement of Material and Process Engineering, p 92–98, 1984.

Use of Advanced Composites in Aerobatic Aircraft

Aircraft flown in aerobatic competitions require unusually high strength combined with high thrust-to-weight ratios. Additionally, minimum pitch and yaw axes at mass moment of inertia are desirable, coupled with pilot skill and endurance, and are judged during flight routines.

Current aircraft used in these events, such as the American-designed Pitts and Christen Eagles biplanes, strive for maximum available thrust for a minimum of empty weight to complete upward vertical maneuvers while retaining sufficient energy to exit and continue gracefully. Nevertheless, many aerobatic planes do not have sufficient strength-to-weight ratios to complete maneuvers such as outside snap rolls gracefully in competition. A new aircraft, the Laser 200, attempts to minimize performance limitations by incorporating an optimized tubular steel fuselage and advanced composite wings, horizontal and vertical tail, landing gear legs, and secondary structure and fairings.

Advanced design optimizations were used in constructing the tubular truss fuselage, and major weight reductions were achieved, despite the nonuse of composites in this area. In addition to weight reduction and the satisfaction of structural requirements, the fuselage design also included provisions for more pilot comfort, increased pilot field of view, and incorporated the engine mount directly to the framework.

The wings were designed to include honeycomb sandwich wing cover panels, foam sandwich box beam main and rear spars, and foam sandwich ribs. The entire wing was constructed of graphite/epoxy structural material using Fiberite HMF 133 and 176 fabrics and Hy-E 1034 unidirectional tape. With help received in determining airloads from NASA/Langley and pilot experience, critical ultimate design conditions were established, and a finite-element model of the wing was created to validate the design (Fig. 18). Design optimization procedures were aided by use of the NASTRAN computer program.

The wing, a symmetrical airfoil, required only a minimum number of tools to fabricate. Full-sized CAD/CAM-generated vellums were adhesively bonded to make various templates. The wing lay-up was a complex process due to the nature of the part and its internal components, but a 60 to 62% fiber volume was maintained, and weight variance for each group of components was less than 1% of the anticipated.

Horizontal and vertical tail surfaces were of full-depth foam sandwich designs. The foam was laminated from sheets and carved to shape using a router on a trammel arrangement. The entire assembly was overwrapped with cloth and tape and cocured.

In most aerobatic craft, landing gear legs are made of spring steel or possibly aluminum. Using composite materials, a landing gear assembly was fabricated and bolted to the fuselage, which is reinforced in that area. Composite gear legs (including fuselage reinforcement) create a gear weight of 4.1 kg (9 lb), comparing favorably with steel (20.4 kg, or 45 lb) and aluminum (12.25 kg, or 27 lb).

Advanced composite materials can improve the performance of aerobatic craft, as evidenced by the design in this

Fig. 18. Wing cross section.

study. The weight saving benefit effectively increased the thrust available for maneuvers.

Abstracted from "Advanced Composite Materials in Aerospace Aircraft," by Hans D. Neubert, H. D. Neubert & Associates, Anaheim, California, and Leo E. Loudenslager, Aerobatic Champion Pilot, in *Materials Overview for 1982*, Vol 27, 27th National SAMPE Symposium and Exhibition, Society for the Advancement of Material and Process Engineering, p 995–1001, 1982.

Composite Panel Stitching as a Fastening Method

Engineers at the Manufacturing Research Division of Lockheed California Company have experimented with sewing graphite/epoxy stiffeners to graphite/epoxy panels. They found that stitching uncured stiffeners to uncured panels with Kevlar 29 fiber produces a strong bond after curing that requires no additional fasteners. The method is still experimental and in need of refinement and testing. Application of this method would be particularly useful for fastening stiffeners under the skin of a wing surface.

Early methods of joining composite panels consisted of first curing both components, then joining them via adhesives or mechanical fasteners. For high-stress applications, mechanical fasteners are required, and it is difficult and expensive to drill holes in cured composites. Consequently, stitching was proposed, because it could be used on uncured components, and the whole assembly could then be cured simultaneously.

Three decisions had to be made at the outset: type of sewing machine, type of stitch, and type of thread material to be used. A sewing machine used in the manufacture of shoes was selected, and the needle easily penetrated the uncured, tacky material to yield a good stitch. Two stitching methods were considered (see Fig. 19). The first was a lockstitch requiring the use of two threads, fed one each from the top and bottom. The second, preferred method is the chainstitch, which utilizes only one thread fed from the bottom. The fiber selected had to withstand the abrasive effects of being pushed through the graphite and both manufacturing and operating environments. Thread materials considered along with the

selected Kevlar 29 material included nylon, polyester, and fiberglass.

Other variables included the number of stitching rows required, how far from the edge stitches could be made, and the maximum thickness of panels and stiffeners that could be handled. Destructive testing showed that, relative to an unstitched panel and stiffener combination, one row of stitching yielded a 31% improvement in pull-off strength, two rows yielded a 72% improvement, and mechanical fasteners yielded a 107% improvement. Further tests may show that a sufficient number of stitching rows can exceed results of mechanical fasteners.

Abstracted from "Future Composite Aircraft Structures may be Sewn Together," by Daniel J. Holt, Automotive Engineering Magazine, in *Automotive Engineering*, p 46–49, July 1982.

Design and Fabrication of the de Havilland DHC-7 Nose Avionics Compartment Using Aramid Composites

The advent of commercially available aramid fiber prepregs and their substitutability for fiberglass led to a design program for a nonmetallic nose avionics compartment for the de Havilland DHC-7. Using Kevlar prepregs with a Nomex honeycomb core, a 30% weight saving over glass skin/balsa core construction could be realized. Aramid fabrics also could be used with existing tooling and lay-up procedures.

A material specification was prepared covering aramid cloth requirements and included specifics such as resin matrix, cured part properties, tensile and flexural strengths, and flammability. Materials from a number of companies were evaluated, and a 17-mesh, plain-weave Kevlar 49 fabric was chosen, based on production requirements, impact resistance, and cost. For a given thickness of Kevlar 49, highest impact resistance was obtained by minimizing the number of plies and maximizing the weave crimper.

Design criteria for the nose avionics compartment include the bearing of aerodynamic and static loads, stability under a wide variety of environmental conditions, and protection from lightning strikes. The sandwich construction must meet a bending moment of 6.36 J/cm (143 in.·lb/in.) of width, and a shear value of 9.5 kg/cm (53 lb/in.) of width, as determined by engineering calculations. Lightning protection is necessary to protect the avionics equipment, as well as to ensure the structural integrity of the compartment.

Several structural configurations were evaluated. The final construction selected included a 120-mesh aluminum outer screen to provide lightning protection at the lowest weight, an outer skin of two-ply Kevlar (style 281), a Nomex honeycomb core, and an inner skin of two-ply 281 Kevlar. The material had a nominal thickness of 9.1 mm (0.36 in.) and weight of 2.34 kg/m² (0.48 lb/ft²). Aircraft certification requires testing of materials in flat sheets and in a full-sized nose compartment. Preliminary results of mechanical, temperature, and weathering tests were favorable.

Design details for the shell structure call for local stiffening at the aft face and the access door area. The aft end of the nose tapers from a sandwich cross section (described above) to a laminate roughly 2.5 mm (0.1 in.) thick where an aluminum joint strap (nose to fuselage) is riveted to the compartment. The access door surround reinforcement is the same as the aft stiffener, but is 19 mm (0.75 in.) thick. Where door hinge support brackets are located, epoxy and aluminum plate inserts are used between the outer skin and core materials. The six avionics support shelves are located three on each

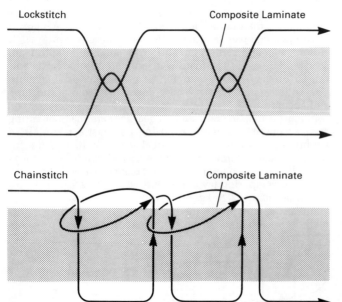

Lockstitch   Composite Laminate

Chainstitch   Composite Laminate

Fig. 19. Stitching methods.

side and supported and fastened at the forward and aft sections of the nose. The shelves are a sandwich construction of aluminum skins bonded to a balsa core. Installation of other interior components is done by reinforcing the sandwich core, potting compounds, or by tapering the sandwich construction to a solid laminate. The exterior finish includes one coat of chromated epoxy primer and two of polyurethane enamel. A PVF film is used to line the compartment interior and to provide a moisture barrier.

Manufacturing development consisted first of producing feasibility panels to study the basic shell sandwich structure for aluminum screen lightning protection, the shelf support Nomex core stiffening ring, and the access door frame structure. A full-scale compartment was also constructed to serve as a test unit and to verify manufacturing and assembly procedures and production tooling. Detailed manufacturing procedures such as lay-up, curing, and trimming and assembly details were also established. Tooling used for the compartment was made of fiberglass-reinforced plastic. Plastic tools require shorter fabrication flow time, weigh 25% less than equivalent metal tools, are easier to repair, and have a thermal expansion coefficient more closely matching the part. Tooling was manufactured from a dimensionally stable master model, from which female layup mandrels and other tools were constructed using various shape transfer techniques.

The actual manufacture of the avionics nose compartment used procedures similar to those used for structural fiberglass components. An autoclave curing process was used to form the aluminum mesh/Kevlar/Nomex shapes. Conventional drilling and routing equipment was used for trimming and cut-out operations. Adhesive bonding and mechanical fastening techniques were used for interior compartment completion. Shapes were formed using the lay-up mandrels coated with a release agent, upon which Nomex, Kevlar, and aluminum mesh layers were laid. Metal parts were cleaned with solvent and located within the lay-up by positioning templates. The access door was fabricated in the same way. Final trimming and assembly used mechanical fastening techniques and required tungsten carbide bits for drilling and routing.

Full-scale testing of the avionics compartment consisted of lightning and structural performance procedures. Exposure to laboratory-induced lightning flashes resulted in some minor compartment modifications until the shielding afforded by the design was acceptable. Mechanical tests were centered around the nose/fuselage joint under vertically loaded conditions and a production run nose loaded under the same conditions. Structural integrity of the nose was maintained during the tests, but one of the interior shelves failed under 200% ultimate loading.

Abstracted from "The Design and Fabrication of the de Havilland DHC-7 Nose Avionics Compartment Using Aramid Composites," by Leonard K. John and Lyle L. Bryson, Proceedings of the 1978 International Conference on Composite Materials, 16–20 April 1978, Toronto, Canada, American Institute of Mining, Metallurgical and Petroleum Engineers, Inc., p 1424–1441, 1978.

Marine Applications

Composite Hydrofoil Control Flap

Expecting significant weight savings from the use of composite components on high-performance vessels, the Naval Sea Systems Command sponsored a project to design and fabricate full-scale graphite/epoxy aft inboard control flaps for the PCH-1 Hydrofoil craft (Fig. 20). Based on the service requirements of the hydrofoil, a preliminary composite flap design was developed to provide the same service as the steel component. Criteria used for the design included critical pressure loading of 30,760 kg/m² (6,300 lb/ft²) without buckling, freedom from loss of control or hydroelastic instability to a speed of 100 knots, a service life of 15 years, and a range of environmental conditions.

The flap was designed to be completely interchangeable with the steel part with a length of 206.8 cm (81.4 in.) and a chord length of 48.8 cm (19.2 in.). The upper and lower cover panels, in response to stiffness requirements, consisted of 36 plies of graphite/epoxy fabric oriented at 45° with 10 mil of titanium bonded on their surface. The graphite plies were fastened to a titanium substructure by both adhesive and mechanical bonding techniques. The crank spar assembly was machined from titanium. Figure 21 shows the design details.

A series of titanium-clad material specimens were tested under hydrofoil environmental conditions. Half of the specimens were tested as fabricated, and half were placed in salt water for 90 days prior to testing. Both specimen types were tested statically and in fatigue. Results of these tests verified the applicability of the titanium-clad graphite/epoxy system for hydrofoil applications. In-plane strengths and fatigue characteristics did not decrease after the 90-day submersion in salt water.

To develop the techniques necessary to build the full-scale composite flap and demonstrate its ability to comply with environmental requirements, a demonstration component was built that consisted of the crank end and part of the composite flap. This part was tested in salt water using a loading spectrum developed from actual flap performance data over a 15-year life. Relative movements were detected between the substructure and covered areas shortly after cyclic loading began. Loading continued until the component failed prematurely after 12,000 cycles. The primary failure consisted of a titanium rib fracture, which was preceded by a bond failure along the nose of the flap. With the help of the NASTRAN finite-element model and analysis of the failed section, design modifications were made. The assembly was refabricated according to the revised design, and the manufacturing techniques were changed to produce a product with improved performance. The rebuilt assembly was again tested and failed after 145,000 cycles at a weld termination near the base of the crank shaft in the forward cell of the titanium closure rib.

The attachment failures experienced in the demonstrator components led to the evaluation of mechanical fastener joints. A number of fastener designs were considered, and torqued bolts were chosen for use on the flap.

A final design was developed based on the original concept. The composite covers were not changed, but the use of torqued bolts for fastening throughout the structure and changes in the spar/crank design were necessary to ensure component performance and life expectancy. To permit the installation of torqued-through bolts, threaded inserts were incorporated in the lower flap cover.

Using techniques developed in preliminary stages and design changes made as a result of testing and analysis, a full-

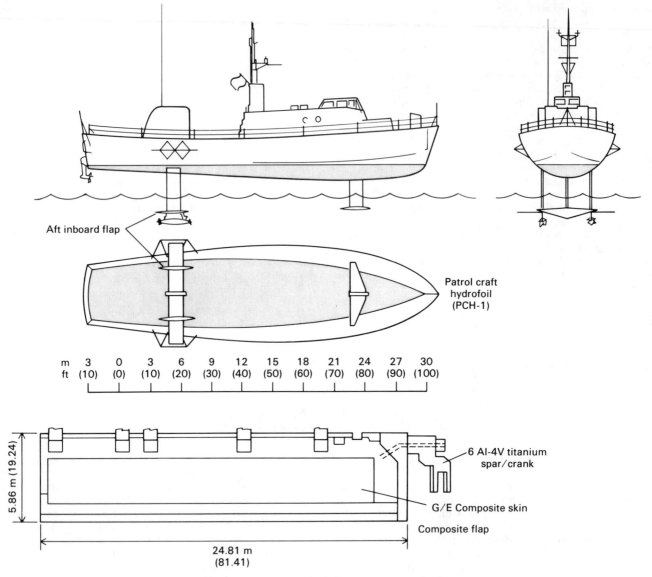

Fig. 20. Patrol craft hydrofoil (PCH-1) composite flap.

scale component was fabricated. The composite covers were laid up on steel tools, bagged, and cured in an autoclave at 177 °C (350 °F). Cured laminates were machined and holes were drilled into which threaded titanium inserts were bonded. After preparation of all metal parts, the composite covers and metal assembly were stacked for final assembly using a film adhesive at the bond interface surfaces. The entire assembly was bagged and again placed in an autoclave for adhesive cure. The emerging part appeared satisfactory, and a few completion operations were performed. During fabrication, 52 strain gages and one moisture detector were mounted on the inner surfaces of the composite covers. These were used to monitor in-service tests aboard the PCH-1 for about 6 months.

Abstracted from "Development of an Advanced Composite Hydrofoil Flap," by Samuel Oken, Boeing Aerospace Company, and Roy W. Deppa and David W. Taylor, Naval Ship Research and Development Center, in *Fibrous Composites in Structural Design*, Proceedings of the Fourth Conference on Fibrous Composites in Structural Design, 14–17 Nov 1978, San Diego, California, Plenum Press, p 659–674, 1980.

Composite Structures for Underwater Pressure Hull Applications

Underwater ordnance systems such as hulls could benefit from the use of lightweight advanced composite materials. Aside from weight savings, the inherent vibration damping characteristics of composite materials make them more resistant to extraneous vibration than metallic hulls. This is an important factor in obtaining proper operation of high-performance sonar systems. Ordnance of lighter weight improves depth capability and maneuverability and allows greater payload/endurance ratios.

Previous Navy programs focused on the development of fiberglass filament-wound honeycomb sandwich shells for the DEXTOR (Deep-Depth Experimental Torpedo) research vehicle. Following completion of this phase, a design incorporating a filament-wound, carbon fiber/organic matrix composite shell was developed. During the previous projects, no development work was done on nose shell fabrication. Because transducer noise interference is concentrated in the nose

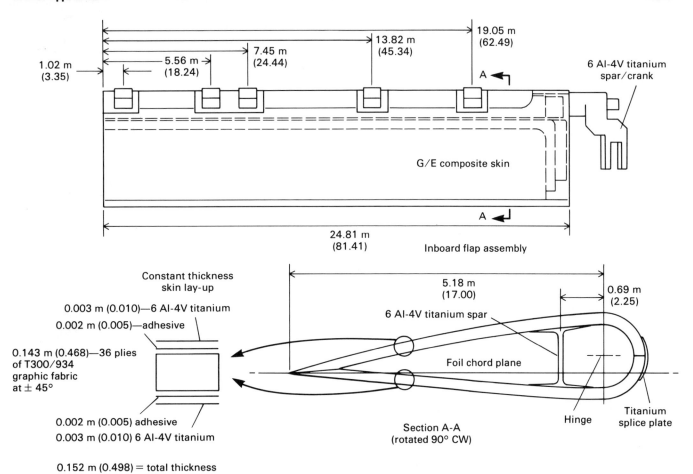

Constant thickness
skin lay-up

0.003 m (0.010)—6 Al-4V titanium

0.002 m (0.005)—adhesive

0.143 m (0.468)—36 plies
of T300/934
graphic fabric
at ± 45°

0.002 m (0.005) adhesive

0.003 m (0.010) 6 Al-4V titanium

0.152 m (0.498) = total thickness

Fig. 21. Composite flap concept.

shell area, a damping nose shell material would provide performance benefits. The objectives of the project were to design, fabricate, and test a sandwich composite sonar nose shell.

For the nose shell, a hybrid composite construction consisting of a sandwich with carbon/epoxy facings on a foam core was selected. The assembly cross section is shown in Fig. 22. Carbon/epoxy laminate facings were made with a woven prepreg fabric interplied with circumferentially wound continuous carbon fibers. The syntactic foam core was a mixture of epoxy resin with an *m*-phenylenediamine curing agent, hollow-glass microspheres, and short glass fibers. A casting resin and curative were used for the elastic foundation between the composite shell and the aluminum forward end ring, which was bonded to the composite shell. To prevent galvanic corrosion, a layer of glass fiber mat was inserted into the bond line between the aluminum rings and carbon/epoxy shell. A polyurethane coating was cast over all outside surfaces prior to in-water testing to prevent leakage. The selected design configuration was analyzed using finite-element modeling techniques.

The decision to use a foam core material instead of a honeycomb was based on the advantages that foam offers in forming complex configurations; its variability of density, which permits tailoring of the density to achieve localized mechanical component properties; its surface machinability, and its cocuring capabilities, which result in laminate facings without the dimpling effect experienced with honeycomb cores.

Fabrication of the entire laminate shell structure was done in a single mold. A mandrel machined from a graphite rod

provided a lay-up tool, curing tool, and lathe fixture for machining the part after core assembly. Advantages of using a bulk graphite mandrel include dimensional stability, closer coefficient of thermal expansion to the composite part, and low cost. The other piece of tooling was an assembly fixture used to align the trimmed composite shell. Aluminum end rings were placed axially and radially during bonding operations. The fixture was also used for final machining of the outside composite surface.

The sandwich shell was fabricated from the inside out

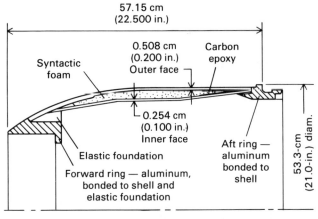

Fig. 22. Carbon-epoxy hybrid shell structure.

starting with graphite prepreg lay-up and fiber winding on the mandrel. The inner skin was precured, and the preformed foam "tiles" were bonded to the inner skin of the shell. Cracks between the tiles were grouted with an uncured foam mixture and cured in an oven. The outer surface of the foam core was then ground to the dimensions of the inner surface of the outer skin. The outer skin was then laid up, and the filament was wound in the same way as the inner skin. Extra plies were added to ensure excess thickness. Upon curing of the outer skin, the shell was stripped from the mandrel and machined to finished dimensions after the aluminum end rings were bonded to the part.

Testing then proceeded on the finished parts. A unit was fitted with pressure transducers, and preliminary noise damping tests indicated a 3 to 4 dB reduction in noise levels in the nose section compared with an aluminum nose. Another pressure test was to be conducted to determine failure loads and to verify materials, design, and fabrication procedures.

The feasibility of applying composites to underwater ordnance systems was demonstrated by this project. Early testing confirmed improved noise damping characteristics. The carbon/epoxy sandwich shell resulted in a 10.4-kg (23-lb) weight reduction over an aluminum shell. Complete compatibility with other materials was also demonstrated. Corrosion problems would also be eliminated, but materials and fabrication costs would be higher. Other areas requiring further study include thermal management, EMI protection, and moisture effects. Structural design problems and hardware attachment techniques must also be addressed.

Abstracted from "Carbon/Epoxy Composite Structures for Underwater Pressure Hull Applications," by Paul W. Harruff, McDonnell Douglas Astronautics Co., Huntington Beach, California, and Dr. Bruce E. Sandman, Naval Underwater Systems Center, Newport, Rhode Island, in *Materials and Processes — Continuing Innovations*, Vol 28, 28th National SAMPE Symposium and Exhibition, 12–14 April 1983, Anaheim, California, Society for the Advancement of Material and Process Engineering, p 40–50.

Filament Winding of Large Ship Hulls

Filament winding is a process in which prepreg fibers are wound around a mandrel. Upon curing of the resin, a strong, lightweight composite structure is formed in the shape of the mandrel. Typically, a filament winding machine has a headstock and tailstock in which the mandrel is held and revolves to receive the filament drawn from a spool and passed through a resin bath.

Filament winding has the following advantages:

1 Fibers are used in their lowest-cost form
2 Labor is minimized
3 Accuracy and repeatability are great
4 Fibers can be oriented at any angle
5 Other costs attendant to machine operation are low

Finding fiber paths that would adequately support the hull in this case study presented problems, due to its configuration. These problems included making the transition between deck levels where there was a step-down, supporting the mandrel, fibers bridging the mandrel surface due to the concave shape of the hull, and winding the changing curvature and cross-section near the bow. This case study concentrates only on the last of these problems, but all were resolved.

The first step was to wind a 1/48-scale model hull. Two such hulls were produced, with a total winding time for each hull of 6 h. The finished hull walls contained six layers of

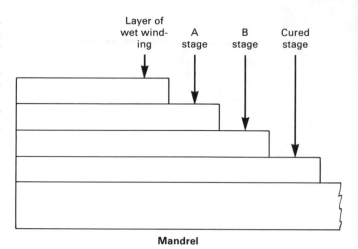

Fig. 23. Controlled cure during winding.

fibers with a total thickness of about 1.5 mm (0.060 in.) Pads cast to match the concave section of the hull were used to hold the fibers close to the mandrel. Pads were removed during each placement of fibers in the area, coated with release agent, and then clamped in place until after the winding was cured.

Analyses indicated that the full-size hull thicknesses required to provide adequate strength ranged from 0.318 to 12.06 cm (0.125 to 4.75 in.), using a combination of E-glass/epoxy and graphite/epoxy composites. Hull sections 7.62 cm (3 in.) thick were wound on a mandrel having both flat and curved sides. The winding took 9 h and consisted of about 75 layers. A progressively curing dual-resin system was employed to reduce shrinkage while maintaining a primary bond between winding layers. The progressively curing layers are illustrated in Fig. 23.

This case study confirms the feasibility of filament winding complex curved structures with thick walls required for a large ship hull. These structures can be cured without added heat. Although some conditions in winding a scale model hull would be different in winding a full-size hull, the nonrepresentative conditions do not deter from the feasibility of the process.

Abstracted from "Demonstration of the Feasibility of Filament Winding Large Ship Hulls," by J. L. McLarty, McClean Anderson Inc. Laboratory, Menomonee Falls, Wisconsin, in *Materials and Processes — Continuing Innovations*, Vol 28, 28th National SAMPE Symposium and Exhibition, 12–14 April 1983, Anaheim, California, Society for the Advancement of Material and Process Engineering, p 50–55, 1983.

Design, Fabrication, and Evaluation of an Advanced Composite Foil Tapered Box Beam

A program to evaluate the use of advanced composites for naval ship structures has been conducted at the David W. Taylor Naval Ship Research and Development Center. The hydrofoil strut/foil system (tapered box beam) was used to assess the feasibility of advanced composites for marine applications. Components of hybrid graphite/epoxy skins and HY-130 steel spars, representing a 25% weight saving over metallic strut foil systems, were fabricated for testing (see Fig. 24).

The laminate skin selected consisted of 53 plies of Thornel 300 yarn and 32 plies of GY-70 fibers, with all 85 plies comprising a skin 13 mm (0.5 in.) thick. T-300 fibers were used

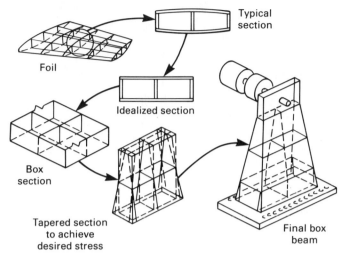

Fig. 24. Box beam configuration.

for strength and GY-70 fibers for bending and torsional stiffness.

For attaching composite skins to the internal metal substructure, three types of joints were evaluated: bonded joints using two different adhesives, bolted joints, and bonded-bolted joints (Fig. 25). Joints were fabricated by making a hybrid component panel using a detailed manufacturing and curing process and were then inspected by through-transmission, ultrasonic C-scan. Having found no defects, the panel was

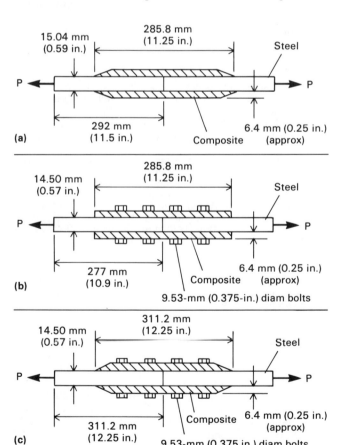

Fig. 25. Joint configurations: (a) bonded; (b) bolted; (c) bonded-bolted.

Table 2. Joint evaluation data

Joint configuration	Adhesive type	Load at failure, kN (tons)	Joint efficiency	Failure mode
Bonded	EA 9309	302.5 (34.0)	60	Adherend
	EA 951	234.0 (26.3)	46	Adherend
Bolted	None	110.3 (12.4)	21	Shear-out
Bonded-bolted ...	EA 951	175.7 (19.7)	31	Adherend, shear-out

cut into narrow strips, upon which the joint designs were tested.

Table 2 gives the results of experimental work on each type of joint. These are average values for three specimens of each joint design. Joint efficiencies were calculated by dividing specimen stress at failure by the ultimate strength of the laminate. Failures of bonded joints were all within the 45° GY-70 layers of the composite adherend and appear to have initiated at the tapered region of the test specimens and progressed inward toward the specimen center. The thickness of composite that remained bonded to the steel at failure ranged from 0.53 to 1.50 mm (0.02 to 0.06 in.). Failures of bolted joints were shear failures that started at the end bolts as they picked up more load than the center bolts. This is a failure mode not characteristic of bolted steel plates, where material yielding creates a more uniform shear stress pattern on bolts in a row. For this specific composite hybrid layup, bolted joints were found to be only one-third as efficient as bonded joints. Failures of bonded-bolted joints were similar to those of bonded joints, with failure initiating at the tapered ends of the composite and progressing to the center. Composite still adhering to the steel then became overstressed and failed in tension where the steel bars butted against each other. The joint then behaved as a bolted joint and failed by shear-out.

HY-130 I-beams were fabricated with the same composite bonded/bolted to each cap of the "I" using the previous methods of fabricating, curing, and attachment. Two reinforced I-beams were tested in static three-point flexure, and three were tested in cyclic three-point flexure. Failure in the statically flexed beams occurred in the Thornel layers, on the tensile side first, followed by the compressive side. After initial outer layer failures, the composite failed completely through on the compressive side with no further failure on the tensile side. No shear-out bolt failures were observed. Flexural fatigue tests on two specimens resulted in failures in the steel portion of the I-beam with no adhesive failure on either specimen. However, one of the specimens was statically preloaded prior to fatigue testing, because it was expected that preloading would induce beneficial residual stresses in the steel. Static preloading increased fatigue life by 120%.

A scarf joint (see Fig. 26) was then fabricated for testing as the step prior to final box beam design. Tensile failure in

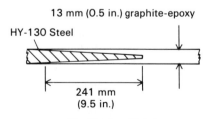

Fig. 26. Scarf joint.

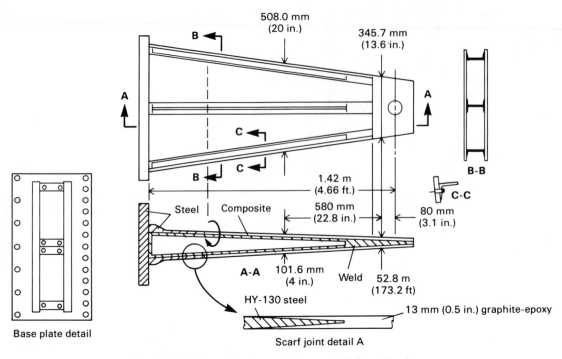

Fig. 27. Hybrid graphite/epoxy skin – steel skeleton box beam.

these specimens usually occurred in the composite adherend, and the same was true of compressive failure. The bond used was strictly adhesive. The thickness of the tip of the composite material in the joint was critical to performance. As this approached 0 (a perfect point), the compressive strength of the joint approached the compressive strength of the hybrid composite.

Within the context of previous test results, the final box beam was specified. Bolted-bonded joints were selected based on reliability considerations. Final fabrication consisted of hybrid composite skins of 58% T-300/5208 prepreg at 0° orientation and 42% GY-70/5208 prepreg at ±45° orientations. Composite skins were bonded-bolted to the steel internal structure detailed in Fig. 27. The whole assembly was nondestructively tested by various means, and no delaminations, debonds, or defects existed in the skins or scarf joint regions.

Conclusions of this study include:

1 The various joint configurations have differing efficiencies relative to each other and the hybrid material alone

2 Test results for bonded-bolted I-beam components verify this method of attaching hybrid composite skins to the substructure
3 Cyclic I-beam tests indicate the method of attachment will withstand fatigue loading conditions expected for box beams in air
4 Static preloading greatly aids fatigue life
5 Static and fatigue tests of scarf joints in air led to a 33% reduction in the design ultimate strength of the hybrid laminate
6 Nondestructive evaluation revealed no defects as a result of laminate fabrication or box beam assembly

Abstracted from "Design, Fabrication, and Nondestructive Evaluation of an Advanced Composite Foil Test Component (Tapered Box Beam)," by L. B. Greszczuk, McDonnell Douglas Astronautics Co., and W. P. Couch, David W. Taylor Naval Ship Research and Development Center, in *ASTM Special Technical Publication 674,* Composite Materials: Testing and Design (Fifth Conference), sponsored by American Society of Testing and Materials, 20–22 March 1978, New Orleans, Louisiana, p 84–100, 1979.

Medical Applications

Orthopedic Applications for Advanced Composites

Carbon fibers recently have been recognized as a biocompatible material in medicine. Graphite-reinforced resin composites have been suggested for use or have been used in applications that include repair of tendons or ligaments, soft tissue augmentation, external fixation devices, and as a surface coating for attachment of orthopedic implants.

Frequently, invasive techniques to repair bone fractures are required, using bone plates made of stainless steel and co-

balt-chromium or titanium alloys. Using canine test subjects, researchers have tried various composite systems such as bone plate materials, graphite/polymethyl methacrylate, glass/epoxy, graphite/polysulfone, and graphite/polypropylene with varying degrees of success. In Great Britain, graphite/epoxy bone plates were used on sheep tibia, and after 24 weeks, the resultant bone had 60% of its original stiffness. Favorable results have been reported in human clinical trials after 18 months of testing.

Another approach involved graphite with polyactic acid as a matrix material. As the polyactic acid biodegraded, the load bearing would slowly transfer from the stiff composite to the repaired bone itself. This technique has not yet proved successful, and the composite design will require further refinement.

Patients sometimes must undergo total joint replacement to correct a number of degenerative joint conditions. Most joint replacement implants consist of a metallic surface bearing on an ultrahigh molecular weight polyethylene surface. The metal used is usually a cobalt-chromium-molybdenum or titanium alloy. For one of the bearing surfaces in hip joints, a carbon-fiber-filled ultrahigh molecular weight polyethylene material has been used opposite the femoral component in the acetabulum part of the pelvis. Failures in some patients have been observed, but reasons for the failures have only been theorized.

In attempting to address failure problems, canine joint research has been conducted using carbon-fiber-reinforced polymers and carbon-fiber-reinforced carbon. A hip involving a carbon-fiber-reinforced carbon joint and an aluminum oxide head was cemented into place in ten subjects without an observed failure.

The application of composites to hip joints is in its early stages, but one good feature of carbon-reinforced polymers for this application is that fiber content and orientation can be varied to provide the mechanical properties desired for specific cases. Polymers still being evaluated as matrix materials for graphite systems in 1983 included polymethyl methacrylate, polysulfone, epoxy, and polyurethane. Thermoplastic materials generally showed higher mechanical properties than thermosets, with polysulfone showing the highest ultimate strength. The polysulfone composite also showed the most resistance to loss of mechanical properties under severe, accelerated environmental conditions.

Abstracted from "Applications of Advanced Composites in Orthopaedic Implants," by Kenneth R. St. John, Hexcel Corp., Dublin, California, in *Biocompatible Polymers, Metals, and Composites,* the Society of Plastics Engineers, Inc., chapter 27, p 861–871, 1983.

Composite Artificial Limbs and External Bracing Systems

Carbon-fiber-reinforced plastics (CFRP) are attractive materials for prosthetic and external bracing devices because of their weight saving potential. The reduction in weight relative to metallic devices allows the patient to use less energy to effect movement.

The first carbon-fiber-reinforced plastic clinical application occurred in 1970, when artificial limb designs were developed for use by thalidomide victims. These limbs had to be externally powered, and the requirements for the harness to which they were fit included stability, lightness, and ease of manufacture; CFRP offered the necessary combination of stiffness, strength, lightness, and ease of fabrication. The carbon-fiber-reinforced plastic was prepared using prepregging

techniques with epoxy resin, resulting in strands roughly 60 vol % loaded with fiber. The construction of the harness was designed to use techniques normally accessible in clinical practice, such as those used for the working of plaster of Paris. A plaster mold of the patient's rib cage was made, with appropriate attachments for addition of the artificial limbs. The mold for the carbon-fiber-reinforced plastic, made of polyethylene foam semicircular channels, was attached to this plaster torso. The mold itself was made part of the final harness, providing a soft surface against the patient's body. Carbon-fiber-reinforced plastic prepregs were laid in the channel of the mold and compressed by enveloping the assembly in an evacuated PVC bag and curing the resin. The resultant harness and artificial limb assembly was 20% lighter overall, and the harness part alone was 60% lighter. Problems with the method of attachment of other system components to the harness originally experienced were overcome by the use of pretaped metal inserts for support at high-stress points.

In another application, a bracing device was developed for use by spina bifida children. These are called orthoses, or devices that are worn on the body to provide external support or constraint. Traditionally, these devices were of steel construction, weighing up to 15% of the child's body weight. Because each orthosis is a custom item and because plaster of Paris molds are not used for lower limb orthoses, a method of forming the carbon-fiber-reinforced plastic shape and retaining the shape during resin cure was required. The selected method of doing this was to lay carbon-fiber-reinforced plastic prepregs in the channels on both sides of an aluminum I-beam, which was then wound with two layers of porous heat shrink tape to consolidate the composite. A film adhesive was used in the I-beam channels to bond the carbon-fiber-reinforced plastic to the metal, followed by curing at 150 °C (302 °F).

Cured struts are formed into the necessary contours by slow and gradual mechanical bending procedures. Conventional orthoses use riveted joints to connect side struts to corset sections and thigh and calf bands that allow removal of individual struts for repair or replacement. This characteristic was somewhat more difficult to build into a carbon-fiber-reinforced plastic orthosis, but special mechanical fastening techniques were developed using threaded metal inserts in the composite and steel interface connections that clamp onto the carbon-fiber-reinforced plastic composite hybrid strut.

The completed hybrid composite orthosis weighed 40% less than a conventional device, and in the one clinical test case, the wearer showed faster walking times, presumably due to the lighter device. The device was worn daily until it failed after 18 days due to delamination at the metal-composite interface as a result of abusive torsional loading.

Abstracted from "Artificial Limbs and External Bracing Systems," by R. L. Nelham, East Sussex Area Health Authority, Lewes, Sussex, United Kingdom, in *Fibre Composite Hybrid Materials,* Applied Science Publishers Ltd., United Kingdom, p 261–283, 1981.

Electrical/Electronic Applications

Composite Battery Electrodes

A large number of commercial power needs are not ideally met because of the limited capacities (such as volumetric and gravimetric energy densities) of batteries. These and other deficiencies can be improved through the use of composite electrodes, which are lightweight battery electrodes using a

substructure of graphite mat fibers and nickel with active electrochemical material deposits. Because these electrodes can be made thinner than conventional ones, either the volumetric or gravimetric energy density of the system can be increased to suit the application. In applications such as deep diving submersibles and electric vehicles, the use of composite electrodes can result in a reduction of overall battery weight of 30 to 40%, increased energy per unit of weight or volume, increased service life and vehicle range, and minimized use of expensive raw materials such as silver.

Graphite mat fibers were chosen to form the substrate of the composite because of: (1) their large surface area, which can be coated with the metal; (2) nonreactivity; (3) ability to withstand high sintering temperatures of the metal coating without melting or other phase change; (4) electrical conductivity; and (5) low cost.

The electrode is formed by electrochemically plating electroless nickel onto the graphite mat. One mat is placed on either side of a nickel grid to form a sandwich, which is placed between two stainless steel plates that were previously coated with magnesium oxide as a parting agent. The entire assembly is compacted and sintered in a hydrogen atmosphere. The composite precursor is electrochemically impregnated and converted to form the active nickel hydroxide electrode.

After several electrochemical performance tests, the composite electrode behaved as would a sintered nickel hydroxide electrode, but delivered a 47% increase in usable power. Because mass production techniques can be used to manufacture the composite electrode, its cost will be equivalent to, if not less than, conventional electrodes. An important, but unresolved, technical issue is the effect of graphite fiber on the electrochemical performance of the electrode during 300 to 1000 charge/discharge cycles.

Abstracted from "Electrochemical Measurements on Lightweight Composite Nickel-Graphite Battery Electrodes," by R. A. Sutula and C. R. Crowe, Naval Surface Weapons Center, Silver Spring, Maryland, in *Materials to Supply the Energy Demand,* Proceedings of an International Conference, 11–16 May 1980, Harrison, British Columbia, Canada, American Society for Metals, p 835–848, 1981.

Kevlar Fiber-Reinforced Low-Expansion Printed Circuit Boards

Integrated circuits (IC) are enclosed in various packages to protect the circuitry from damage during manufacture and use. Higher-density electronic circuitry has resulted in square ceramic packages with leads on all four sides, known as ceramic chip carriers (CCC). Ceramic chip carriers can be leaded or leadless. Leadless carriers have solder pads in place of pins, which permit direct bonding to contacts on printed circuit boards (PCB) and eliminate bent leads caused by misalignment, shipping, handling, and assembly. Because the solder joint thickness between the leadless ceramic chip carrier and the printed circuit board is small, it is affected more by mechanical and thermal stresses. Thus, laminates are needed that can be tailored for necessary expansion characteristics and that can fulfill other requirements for reliable printed circuit boards.

Standard glass-fabric-reinforced epoxy boards have been shown to cause premature thermal fatigue failure at the solder joints, but Kevlar aramid fibers have the necessary characteristics. Composites fabricated from Kevlar 29/epoxy and Kevlar 49/epoxy prepregged undirectional tapes show negative coefficients of thermal expansion. This is important in circuit board applications, because either fiber can be effective in lowering expansion of the composite circuit board material.

Fabrics and not undirectional tapes are used to reinforce leadless ceramic chip carriers in practice. Controlling the Kevlar fiber volume to between 30 and 40 vol % results in a composite with an expansion value to match the leadless ceramic chip carrier. The choice of resin is secondary. Fiber volume loading may need to be higher in multilayer printed circuit boards to compensate for the expansion of copper circuitry. Out-of-plane expansion coefficients, however, are greater for Kevlar-reinforced than for glass-reinforced boards.

Kevlar 49 and 29 fabrics can be tailored to produce in-plane expansion characteristics to match leadless ceramic chip carriers; thermal expansion coefficients of both Kevlar fibers are negative parallel to the filament direction, and positive in radial filament directions. Longitudinal fiber expansion dominates in-plane thermal expansion, whereas out-of-plane expansion is affected by resin characteristics.

Abstracted from "Low Expansion Printed Circuit Boards Reinforced by Kevlar Aramid Fibers," by Edward W. Tokarsky, E. I. du Pont de Nemours & Co., Wilmington, Delaware, in *Materials and Processes — Continuing Innovations,* Vol 28, 28th National SAMPE Symposium and Exhibition, 12–14 April 1983, Anaheim, California, Society for the Advancement of Material and Process Engineering, p 1251–1256, 1983.

Structural Elements

Design of Leak Failure Mode for Composite Overwrapped Metal Tanks

Filament-wound tankage frequently requires the use of a metal liner as a permeation barrier to meet requirements for low allowable leakage rate and exposure to extreme temperatures. A desirable feature of such tankage, in addition to light weight and better damage tolerance, is benign failure without catastrophic rupture. Designing this property into tankage is accomplished mainly by ensuring that the minimum cyclic fatigue life of the outer composite shell is roughly an order of magnitude greater than that of the metal liner.

Once composite and liner fatigue lives are known, the designer needs to be sure that a fatigue crack will not propagate through the wall to critical size and that the composite shell strength is sufficient to withstand pressure and shock loading in the event of sudden or complete liner failure. Consequently, a proof test is the final operation in the manufacturing process to demonstrate structural integrity.

The proof test imparts a permanent compressive stress in the metal liner, permitting more efficient structural utilization of the metal than in a monolithic tank (the example under consideration is spherical). Because the proof test plastically deforms the metal liner, and it is common to use aluminum alloy liners, which plastically deform at the maximum expected operating pressure, an empirical design approach is used.

Equations used in the empirical design approach take into account behavior of part-through cracks near ultimate tensile strength, calculations of critical flaw size, determination of the period of natural vibration of the composite shell, calculation of crack tip velocity, and assumptions concerning crack propagation distance estimation. A design flow diagram is used for this procedure. Two actual numerical design calculations are made, using the theoretically outlined procedure for tanks made of two different composite materials,

S-glass/epoxy and Kevlar 49/epoxy. Calculations illustrate one acceptable and one unacceptable set of design parameters. The design method has worked well in designing metal-lined composite tanks that have satisfied design and performance criteria.

Abstracted from "Design Assurance of a Leak Failure Mode for Composite Overwrapped Metal Tankage," by W. W. Schmidt, Brunswick Corp./Defense Division, Lincoln, Nebraska, in *NBS Special Publication 563,* Proceedings of the 29th Meeting of the Mechanical Failures Prevention Group, held at the National Bureau of Standards, 23–25 May 1979, Gaithersburg, Maryland, p 198–207, issued October 1979.

Graphite/Epoxy Cylinders

For many years, wet filament winding processes have been used extensively to fabricate rocket motor cases, gas storage bottles, high-pressure pipes, and other large cylinders. Because the stresses in such applications are primarily tensile, the demands placed on wet filament winding are not that great. When compression and bending are the primary stresses the void content of the composite must be held near 0%, which requires the use of high winding tension to flush excess resin and voids during the lay-up procedure. Because of the friability of graphite, only low winding tension can be used, necessitating alternate lay-up methods.

For test purposes, cylinders with a 10-cm (4-in.) inside diameter and 28 cm (11 in.) long were used for characterization and feasibility studies. The filament winding process used in-line resin impregnation, circumferential winding, helical winding, and hand lay-up techniques to achieve the desired fiber orientation. Resulting cylinders had fiber volumes of 50 to 56%, resin contents of 34 to 40%, and void contents of 0.1 to 1.9%. Cylinders were machined into rings 25 mm (1 in.) wide for testing. Test results appeared favorable, and consistent results were obtained for hoop, tensile, and compressive stresses.

The next step was to try winding cylinders in which fiber orientations were changed. For the winding of a 234-cm- (92-in.-) diam cylinder, a lightweight steel mandrel was fabricated and an existing filament-winding machine was modified to add a hydraulic drive system and a helical winding capability. The winding of this cylinder took 32 h, and the plies consisted of 0-, 90-, and 45-degree fiber orientations. The zero-degree plies were laid-up by hand. The graphite fibers were impregnated in a dip tank prior to being wound onto the mandrel at 27 to 32 kg (60 to 70 lb) tension (for 90-degree windings). The helical plies were wound with 14 to 18 kg (30 to 40 lb) of tension. The lay-up was then cured at 121 °C (250 °F) for 2 h and at 177 °C (350 °F) for 4 h.

Specimens from this cylinder were tested to determine the feasibility of wet filament winding for large cylinders. These tests included physical and mechanical properties, C-scan ultrasonic tests, photomicrography, and scanning electron microscopy. The results of these tests indicate that the properties of the wet filament winding part are comparable to autoclave-cured parts. A low value obtained for hoop compressive strength of the 234-cm (92-in.) cylinder is believed to be the result of specimen configuration rather than inherent materials properties.

Conclusions drawn from this case study indicate that the wet filament winding process can produce hardware with properties equivalent to autoclave-cured parts; that lightweight steel mandrels are suitable for filament-wound skin fabrication, and that wet filament winding offers cost advantages over autoclave curing techniques.

Abstracted from "Filament Winding of Large Graphite/Epoxy Cylinders," by H. J. Kruger, Jr., R. C. Curley, and V. L. Freeman, McDonnell Douglas Astronautics Co., Huntington Beach, California, in *The 1980's — Payoff Decade for Advanced Materials,* Vol 25, 25th National SAMPE Symposium and Exhibition, San Diego, California, Society for the Advancement of Material and Process Engineering, p 233–242, 1980.

Carbon-Fiber-Reinforced Plastic Structural Elements

Carbon-fiber-reinforced plastic (CFRP) materials generally are superior to metals and other materials in specific strength and modulus, but only in the direction of the reinforcing fibers. They also are weak at cuts and near machined holes. As a result, two new structural shapes were designed, in which forces flow in the fiber direction. One is a cylindrical grid structure using continuous yarns with bolt holes cast in the piece during fabrication. The other design is a pipe with threads cast into it.

The cylindrical fiber-reinforced grid structure was formed by continuous winding of carbon yarn wetted with epoxy resin onto a mandrel. The mandrel was collapsible and contained pivot pins, around which the yarn could be wrapped to form a pattern, as in Fig. 28. The wrapping pattern was repeated layer upon layer, and after forming, the mandrel was removed, leaving a cylindrical composite grid structure. This trial demonstrated that filament-wound grid structures could be formed precisely.

Stress analysis was done on the grid structure using the NASTRAN computer program. The coincidence of calculated and measured stresses indicated the resultant structure had uniform properties. Also, the use of the NASTRAN program as a useful design tool was verified.

Using the same reinforcing grid structure, threaded pipe sections were tried next. The grid wrapping pattern was the same, except that graphite fibers wetted with resin were wrapped circumferentially at the threads. After curing, the pipe sections were finished by machining. These structures were fabricated for a satellite structural application. Tensile, compressive, and bending stresses were measured on prototype samples, and the results indicated the viability of the reinforcing method and the finished products. However, further refinement on both structural member designs is needed.

Abstracted from "New CFRP Structural Elements," by T. Hosomura and Takashi Kawashima, Nissan Motor Company, Tokyo, Japan, and D. Mori, Institute of Space and Aeronautical Science, Tokyo, Japan, in *Composite Materials: Mechanics, Mechanical Properties, and Fabrication,* Proceedings of Japan-U.S. Conference, 1981, Tokyo, Japan, Japan Society for Composite Materials, p 447–452, 1981.

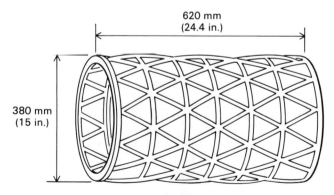

Fig. 28. Cylindrical grid.

Composite Box Beams for Mobile Military Bridging

The U.S. Army is pursuing a program of improved mobile bridging using advanced materials technologies. One concept features a vehicle-mounted three-section folding bridge positioned across an obstruction by means of a launch beam. This beam is cantilevered during deployment, functioning as a simple supported beam upon reaching the far side of the obstruction.

Beam design requirements are summarized in Table 3. Several materials and cross-sectional beam configurations were considered. A simple hollow box cross section constructed from graphite/epoxy composite was selected. The highest flexural rigidity consistent with weight limitations and manufacturing considerations occurs when the maximum possible amount of longitudinal fibers is placed in the flanges. Analysis of the full-scale beam anticipated acceptable beam performance with adequate margins of safety under expected loading conditions.

A half-scale beam was fabricated with a winding mandrel and a mold. Beam cross-sectional thicknesses and plying sequences were built up by alternately wrapping the mandrel with ±45° layers of prepreg tape and laying 0° continuous prepreg tape along the flanges. The assembly is then placed in a four-sided mold and cured. Figure 29 gives the cross-sectional detail of the half-scale beam. Actual beam testing had not yet begun when the paper was presented.

A graphite/epoxy box beam achieves a 50% weight saving and a 16% increase in flexural stiffness over an aluminum beam of comparable design. Commercial production feasibility of such beams was demonstrated via the half-scale beams. Full-scale designs are expected to evolve from the techniques demonstrated to this point.

Table 3. Box beam design requirements

Element	Requirement
Cross-section envelope	30.76 × 63.50 cm (12.11 × 25.00 in.)
Length	6.48 m (21 ft, 3 in.)
Maximum weight	29.5 kg/m (20 lb/ft)
Minimum bending moment capacity	6.24×10^7 J (5.52×10^6 in. · lb)
Minimum shear force capacity	9,072 kg (20,000 lb)
Flexural rigidity	Maximum, greater than comparable aluminum beam
Operating temperature	−54 to 74 °C (−65 to +165 °F)
Humidity, refrigeration hardened	98%
Fatigue life	10^4 cycles
Margin of safety	>0.33 for tension, compression, shear, and buckling

Abstracted from "Graphite Composite Box Beams for U.S. Army Mobile Bridging," by Dr. P. Ronald Evans, Hercules Inc., Cumberland, Maryland, and Dr. John M. Slepetz, Army Materials and Mechanics Research Center, Watertown, Massachusetts, in *Fibrous Composites in Structural Design,* Proceedings of the Fourth Conference on Fibrous Composites in Structural Design, 14–17 Nov 1978, San Diego, California, p 687–700, 1980.

Design and Fabrication of Composite Tubes for Compression Applications

In applications such as aircraft components, composite materials such as graphite/epoxy or Kevlar/epoxy theoretically can result in weight savings of up to 70% over conventional materials. This magnitude of weight reduction is rarely realized, however, because of design considerations such

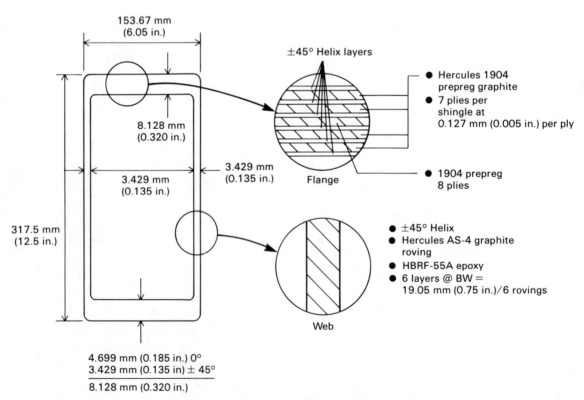

Fig. 29. Half-scale box beam cross-sectional detail.

as combined stress states, presence of stress concentrations, reinforcement required for fasteners, and other factors. In contrast, tubular structures, because loads usually are introduced only at the ends, requiring minimal reinforcement, and because most of the component operates under a uniform, uniaxial stress state, are not encumbered by many of these design limitations. Examples of applications requiring tubular components compressively stressed axially include satellite trusses, flight control push rods, and large space structure elements.

The design of such a compression column is dictated by a requirement to prevent member instability overall, or in localized areas. To achieve this design criterion, the optimum combination of tube diameter and wall thickness that will withstand required stresses and still meet minimum weight requirements must be determined. As tube diameter increases, the wall thickness required decreases, and the larger the diameter, the higher the structural efficiency. The upper limit on tube diameter is reached at the point where the tube wall becomes too thin to withstand the necessary stresses. Starting with a given design in an aluminum part, the designer can either change wall thickness or change the tube diameter. Three areas of improvement guide the designer: achieving maximum weight savings, minimizing the tube diameter without weight penalty, or a combination of reduced weight and decreased diameter.

A series of stress equations are used in conjunction with a computer program (TUBEOPT) to optimize the various parameters involved, including the graphite/epoxy laminate make-up. Under conditions given in the article, lowest weight was achieved with a 60% 0°, 40% ±45° laminate. This represents an unconstrained design situation, however, and designs that result have large diameters and very thin walls. The members can withstand applied stresses, but minimum wall thickness is often constrained for reasons of impact damage tolerance. Adding the minimum wall thickness constraint and using the TUBEOPT program to help with the stress calculations, a different tube design is achieved, and its performance is heavily linked to the axial modulus of the material used.

Another design constraint may be a limit on the maximum diameter of the tube due to clearance limitations, to minimize areas vulnerable to external ballistic threats, or to reduce drag in external strut applications. With this constraint, the material property of merit is the inverse of specific stiffness, and optimum stress is independent of load and is a function of member component geometry.

Loads are introduced into the composite tube through some means of attachment, typically machined metal end fittings that mate with the outer diameter or inner diameter of the tube. Because of their galvanic compatibility with graphite and low coefficients of thermal expansion, titanium and stainless steel are often used as end fitting materials. Thermal environments and expansion compatibilities determine whether the fitting attachment is made inside or outside the tube. Bonding or mechanical attachment to the tube is possible and mathematical rules of thumb are given to calculate the length of fitting overlap required for each type of attachment.

The wrapped mandrel method, a lower cost/higher volume process, is adequate for this type of part. Prepreg material of any orientation is roll-wrapped over a stainless steel mandrel and then covered with a shrink tape or tensioned film. When placed in a curing oven, the combination of shrink tape contraction and metal mandrel expansion results in pressures sufficiently high to produce a dense, high-quality laminate. Fiber volumes (and the resultant part modulus) are controlled easily by using net resin volume prepregs. Also, resin selection is less constrained than it would be using alternate production methods such as filament winding or pultrusion techniques.

This case study concludes that: (1) composite tubes offer a weight saving potential (of 40 to 60%) due to simple load paths, optimum geometry, and lack of fasteners; (2) composite tube design offers considerable flexibility; (3) three design situations exist — unconstrained, minimum gage, and maximum diameter; (4) the material property of merit and loading index parameters are different for all three design situations; (5) the wrapped mandrel method is ideally suited for these components; and (6) the ability to use standard prepreg systems is a significant advantage over competing fabrication methods.

Abstracted from "Design and Manufacture of Composite Tubes for Compression Applications," by David P. Maass, Advanced Composite Products, in *Materials Overview for 1982*, Vol 27, 27th National SAMPE Symposium and Exhibition, Society for the Advancement of Material and Process Engineering, p 1–10, 1982.

Composites for Cryogenic Structures

The feasibility of design and construction of laminated composite struts for superconductive energy storage magnets in cryogenic applications is the subject of this case study. Energy storage magnets are large solenoids in deep underground tunnels. The composite structures support the solenoid and transmit magnetic forces to bedrock.

Materials selected for preliminary analysis in the application included glass cloth/epoxy, glass cloth/polyester, and unidirectional glass/epoxy. The large mass of material needed for structural support and the cost of manufacturing the struts dictated the selection of low-cost materials that are easily fabricated and installed. Because glass-polyester is at maximum half as expensive as glass/epoxy, it was selected for the strut design.

The strut designs presented were for a laminated composite fastened to the vertical rock of a tunnel wall. Loading of the strut occurred via an aluminum bulkhead to the free edge of the strut. The bulkhead is a load-bearing member between the strut and the liquid helium dewar. The strut was cantilevered to the rock wall with rock bolts as fasteners and covered with superinsulation to minimize radiation heat leak.

Three preliminary designs were analyzed using finite-element models. The first two designs used slabs of glass cloth/epoxy and slabs of glass cloth/polyester. The third design was a symmetric 20-ply unidirectional glass/epoxy laminate with angular arrangements in various radial directions. The finite-element model showed that the radial stiffnesses for all three designs were the same, but that resistance to shear deformation of the variable-angle composite was considerably better than the two slab designs. Failure analysis was used in determining the minimum thicknesses required for the struts. Analysis of relative stresses in the struts and of material properties indicated that shear strength was the critical factor in determining required thicknesses. Also, high shear stresses resulted in inefficient use of composite struts. The design was not complementary to loading conditions. Alternate designs included bracing below the strut, trussed web girders, and modular jointless struts of continuously wound glass fibers or unidirectional plies.

Abstracted from "Laminated Fiberglass Composites for Cryogenic Structures in Underground Superconductive Energy Storage Magnets," by S. G. Ladkany and E. L. Stone, University of Wisconsin, Madison, Wisconsin, in *Nonmetallic Materials and Composites at Low Temperatures*, A. F. Clark, R. P. Reed, and G. Hartwig, Ed., Plenum Press, p 377–385, 1979.

Nonautomotive Machinery and Parts

Composite Bellows

The bellows considered in this case study is a critical part of a large bubble chamber. The part contains one convolution, is 800 mm (31.5 in.) in diameter, operates in a magnetic environment, and has a stroke and pressure cycle that necessitates the use of high-quality glass-reinforced plastic material and precludes the use of wet lay-up techniques for manufacturing. The part requires a high and uniform glass content and no matrix-heavy areas or surfaces.

A special mold using an aluminum outer shell and a self-releasing silicone rubber inner mold part was designed. The laminating procedure involved cutting strips of glass cloth at a 45° bias and wrapping these strips around the silicone rubber form until 12 complete layers were in place. The mold was then assembled and impregnation of the fibers with resin was effected via a vacuum technique. After 24 h of temperature and pressure mold preconditioning, the resin (Bisphenol A diglycidyl ether with an acid anhydride curing agent) was transferred directly from its degassing container to the mold cavity. The application of pressure to the mold cavity aided resin impregnation. Following molding and depressurization, the part was demolded.

Assembly and testing were then conducted on the finished part. The joints between bellows and both piston and flange were adhesively bonded using a room-temperature-curing epoxy, vacuum impregnation, and a layer of glass fiber between both adherends. The prototype bellows had been submitted to more than 700,000 cycles, and had experienced no problems at the time this paper was presented.

The method outlined above can be used to produce quality glass/epoxy composite parts to close tolerances. Nevertheless, the method is a time-consuming one and is not applicable to high-volume production parts.

Abstracted from "The Manufacture and Properties of a Glass Fabric/Epoxy Composite Bellows," by D. Evans, J. U. D. Langridge, and J. T. Morgan, Rutherford Laboratory, Chilton, Didcot, Oxon, England, in *Nonmetallic Materials and Composites at Low Temperatures*, A. F. Clark, R. P. Reed, and G. Hartwig, Ed., Plenum Press, p 365–375, 1979.

Composite Material Flywheels

Composite flywheel energy storage systems and their potential application to urban vehicles were investigated in this case study. Several such projects are briefly described, tracing an evolutionary history of composite flywheel design as contracted by various agencies to AiResearch Manufacturing Company of California (ARMC).

The major requirement of a high-energy-density rotor is the effective use of a high strength-to-weight ratio composite material. A feature of the ARMC flywheel design is the composite multiring rim, which operates mostly in uniaxial tension and limits centrifugal stresses in the radial direction to less than the cross-fiber allowable strength of the composite rim. Compressive radial rim forces, caused by the rim being sprung upon an oversized aluminum hub, attach the rim to the hub.

This fundamental design was applied to a program for the U.S. Army Mobility Equipment Research and Development Command. The system consisted of an aluminum hub, to which a rim comprised of 12 rings was attached. The inner ring was fiberglass/epoxy material, and the outer rings were of Kevlar/epoxy composite.

The Department of Energy Near-Term Electric Vehicle (NTEV) program presented an opportunity to apply a composite flywheel in a vehicular environment. An S-glass and Kevlar multiring rim was used. The unit was tested in 1,000 cycles between 13,000 rpm and a top speed of 25,000 rpm using vehicular bearings and seals. The flywheel was then run in an NTEV housing with air introduced into the flywheel chamber. Failure occurred by friction heating, and the rim disintegration was totally contained in the lightweight steel housing.

Other contracts, with Lawrence Livermore Laboratories and with Sandia Laboratories, used the fundamental NTEV program flywheel design. However, the Sandia contract employed a graphite composite material for the hub to reduce weight.

Fabrication for the various designs consisted of rim filament winding, hub machining, and rotor assembly. The composite matrix for the rim was epoxy resin, and the multiring rim consisted of an inner ring of fiberglass/epoxy material and outer rings with successive layers of Kevlar 29 and Kevlar 49/epoxy composites. Metal hubs were machined from 7075 aluminum and were designed for equal stress throughout. Resulting products were statically and dynamically tested.

Abstracted from "Advancements in Composite Material Flywheels," by David L. Satchwell and Dennis A. Towgood, AiResearch Manufacturing Company of California, in *Fibrous Composites in Structural Design*, Proceedings of the Fourth Conference on Fibrous Composites in Structural Design, 14–17 Nov 1978, San Diego, California, p 701–709, 1980.

Sewing Machines with Reinforced Plastic Components

The typical sewing machine is made of C-frame construction, and until the 1950's, it was produced by machining heavy iron castings. As portability and aesthetics became more important to consumers, the first aluminum diecast structures were introduced, and more sophisticated features and controls were also added. Escalating machining, energy, and raw materials costs in the 1970's prompted Singer to evaluate alternatives to diecast aluminum. An injection-molded reinforced polyester held the most promise for the application. This material offered excellent dimensional control due to its low shrinkage rate, was hard and rigid compared with similarly priced plastics, had no significant loss of properties under operating environment conditions, and had good insulation properties necessary for consumer safety requirements.

Design requirements for the sewing machine were centered around the most critical tolerance between the hook point and the eye of the needle. Additional requirements included a UL flammability rating of V-0, falling ball impact resistance of 6.8 J (5 ft·lb), temperature resistance from −40 to 66 °C (−40 to 150 °F), and the ability to withstand multiple insertions of self-threading screws.

The first step in the design approach was to construct a finite-element-analysis model for an aluminum structure and to compare the predictions against actual measurements, thus confirming the technique. Results from these tests led to appropriate placement of support ribs, definition of wall thicknesses in certain areas, and increasing certain radii to minimize stress concentrations. Fatigue tests and thermal stress and dimensional analyses were also performed in the design stage.

A number of candidate reinforced plastic materials were evaluated with tests of tensile and flexural strength, room-temperature fatigue, creep behavior, room-temperature impact strength, thermal expansion coefficient, thermal conductivity, and paintability. Paintability was a major consideration, because the product is viewed critically from a short distance by the consumer and because fabric moving across painted surfaces is surprisingly abrasive. A glass-reinforced polyester material was selected.

The last step was to finalize the mold designs, which are critical because glass fiber orientations affect component properties and are in turn affected by the ability of the reinforced plastic to flow through the mold. A number of mold adjustments and modifications were made to accommodate the fabrication of quality parts.

The compliance of the arm/bed structure was ascertained using a photoelastic method, a brittle lacquer technique, and holographic interferometry. Mixing and molding procedures were also thoroughly tested, and optimum procedures were developed.

The result of this work is the Singer Duratec sewing machine, presently in production. The product demonstrates the feasibility of molding precision-reinforced plastic structural materials without the need for secondary machining operations. The project also showed that finite-element models and test samples can expedite the early stages of a design process and that reinforced plastic components are cost-effective for this application.

Abstracted from "Singer's All-Plastic Sewing Machine," by Robert L. Sedlatschek and Frederick R. Wiehl, Singer International Engineering, Fairfield, New Jersey, and John L. Rutherford, Singer/Kearfott Division, Little Falls, New Jersey, in *Plastics Design Forum*, p 25–30, March/April 1985.

SECTION 5
Property Data: Reinforcements

5.1 Fiber and Whisker Reinforcements

5.1.1. General Fiber Reinforcement Data and Specifications

Classification of Some Fibers

Category	Material
Metal	Beryllium
	Molybdenum
	Steel
	Tungsten
Glass	Vitreous silica
	E glass
	S glass
Carbonaceous	PAN[a], high strength
(Material refers to starting process)	PAN[a], high modulus
	Pitch
	Rayon, very high modulus
	Rayon, high modulus
Polymer	Aramid
	Olefin
	Nylon
	Rayon
Inorganic	Alumina (monocrystal)
	Alumina (polycrystal)
	Alumina (whisker)
	Alumina silicates
	Asbestos
	Boron (tungsten core)
	Boron nitride
	Silicon carbide (carbon core)
	Silicon carbide (polycrystal)
	Silicon carbide (whisker)
	Silicon nitride (whisker)
	Zirconia (polycrystal)

[a]PAN is polyacrylonitrile.

Characterization of Reinforcements

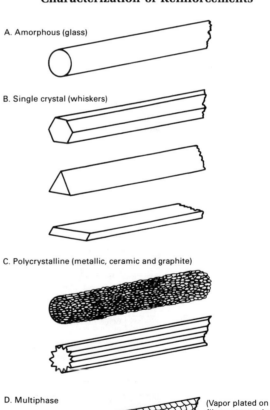

A. Amorphous (glass)

B. Single crystal (whiskers)

C. Polycrystalline (metallic, ceramic and graphite)

D. Multiphase

(Vapor plated on filamentary substrate)

Core sheath

ASTM and Military Standards and Specifications for Fibrous Reinforcements

ASTM Standards/Specifications of General Interest

ASTM D76, "Tensile Testing Machines for Textile Materials."
ASTM D123, "Definition of Terms Relating to Textile Materials."
ASTM D1117, "Testing Nonwoven Fabrics."
ASTM D4158, "Abrasion Resistance of Textile Fabrics."
ASTM D1682, "Breaking Load and Elongation of Textile Fabrics."
ASTM D1776, "Conditioning Textiles and Textile Products for Testing."
ASTM D1777, "Measuring Thickness of Textile Materials."
ASTM D2654, "Test for Amount of Moisture Content and Moisture Regain in Textile Materials."

ASTM Standards for Fibrous Glass

ASTM D578, "Specification and Testing of Glass Yarns."
ASTM D579, "Specification for and Testing Glass Fabrics."
ASTM D580, "Testing and Tolerances for Woven Glass Tapes."
ASTM D2150, "Woven Roving Glass Fabric for Polyester-Glass Laminates."
ASTM D2343, "Test for Tensile Properties of Glass Fiber Strands, Yarns, and Rovings Used in Reinforced Plastics."
ASTM D2408, "Woven Glass Fabric Cleaned and After-Finished with Amino-Silane Type Finishes, for Plastic Laminates."
ASTM D2409, "Woven Glass Fabric, Cleaned and After-Finished with Vinyl-Silane Type Finishes, for Plastic Laminates."
ASTM D2410, "Woven Glass Fabric, Cleaned and After-Finished with Chrome Complexes, for Plastic Laminates."

ASTM D2587, "Test for Acetone Extraction and Ignition of Strands, Yarns, and Roving for Reinforced Plastics."
ASTM D2660, "Woven Glass Fabric, Cleaned and After-Finished with Acrylic-Silane Finishes, for Plastic Laminates."
ASTM D3098, "Woven Glass Fabric, Cleaned and After-Finished with Epoxy-Functional Silane Type Finishes for Plastic Laminates."

ASTM High Modulus Fiber Tests

ASTM D3317, "High Modulus Organic Yarn and Roving."
ASTM D3318, "Woven Cloth from High Modulus Organic Fiber."
ASTM D3544, "Guide for Reporting Test Results on High Modulus Fibers."

Military Specifications

MIL-Y-1140, "Yarn, Cord, Sleeving, Cloth and Tape — Glass."
MIL-C-9084, "Cloth, Glass, Finished for Polyester Laminates."
MIL-F-9118, "Finish, for Glass Cloth."
MIL-F-12298, "Fabric, Glass Woven."
MIL-M-15617, "Mats, Fibrous Glass, for Reinforcing Plastics."
MIL-C-19663, "Cloth, Glass, Woven Roving, for Plastic Laminates."
MIL-R-60346, "Roving, Glass, Fibrous (for Filament Winding Applications)."
MIL-M-43248, "Mats."

Effect of Forming Process on Strength of Various Fibers (Ref 11, p 22)

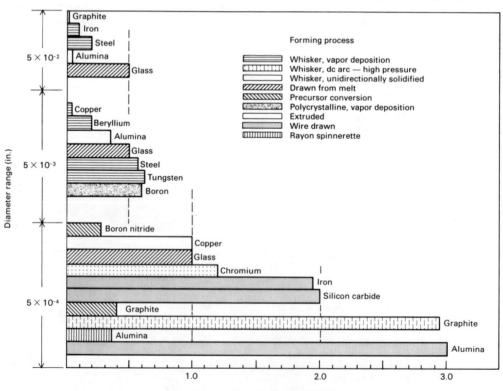

Selected Commercial Fiber Manufacturers, Product Names, and Descriptions

Product	Description	Manufacturer
Magnamite HMS	Continuous graphite fiber	Hercules
Magnamite HMU	Continuous graphite fiber	Hercules
Magnamite	Chopped graphite fiber	Hercules
Magnamite AS1	Continuous graphite fiber	Hercules
Magnamite AS2	Continuous graphite fiber	Hercules
Magnamite AS4	Continuous graphite fiber	Hercules
Magnamite AS6	Continuous graphite fiber	Hercules
Magnamite A193-P	Plain weave graphite fabric	Hercules
Magnamite A370-5H	5-harness graphite satin weave fabric	Hercules
Magnamite A370-8H	8-harness graphite satin weave fabric	Hercules
Magnamite IM6	Continuous carbon fiber	Hercules
Magnamite HTS	Graphite fiber	Hercules
HDG	High-density isotropic graphite	Fiber Materials, Inc.
Microfil 40	Ultrahigh-strength carbon fiber	Fiber Materials, Inc.
Microfil 55	High-strength/high-modulus graphite fiber	Fiber Materials, Inc.
Celion C-6S	Chopped carbon fiber – sized	Celanese
Celion G-50	High-modulus carbon fiber – sized	Celanese
Celion GY-70	High-modulus undirectional graphite fiber tape	Celanese
Celion GY-80	High-modulus PAN-based graphite fiber	Celanese
Celion 1000	High-strength carbon fiber – sized	Celanese
Celion 3000	High-strength carbon fiber – sized	Celanese
Celion 6000	High-strength carbon fiber – sized	Celanese
Celion 12000	High-strength carbon fiber – sized	Celanese
Panex~CF-30	Chopped carbon fiber – sized	Stackpole
Panex 30	Continuous carbon – filament	Stackpole
Panex 30Y/300d	Carbon fiber yarn – roving	Stackpole
Panex 30Y/800d	Carbon fiber yarn	Stackpole
Panex 30R	Carbon fiber yarn – roving	Stackpole
Panex CFP 30-05	Carbon fiber paper	Stackpole
Panex KFB	Carbon "Jersey Knit" fabric	Stackpole
Panex PWB-3	Plain weave carbon fabric	Stackpole
Panex PWB-6	Spun yarn plain weave fabric	Stackpole
Panex SWB-8	Spun yarn 8-harness satin weave carbon fiber	Stackpole
Panex WRB-14	Spun yarn 2 × 2 basketweave carbon fabric	Stackpole
Fortafil 3(C)	Chopped or continuous carbon fiber – sized	Great Lakes Carbon
Fortafil 3(O)	Chopped or continuous carbon fiber – sized	Great Lakes Carbon
Fortafil 5(O)	Chopped or continuous carbon fiber – sized	Great Lakes Carbon
Fortafil OPF(C)	Oxidized polyacrylonitrile chopped fiber	Great Lakes Carbon
Fortafil OPF(O)	Oxidized polyacrylonitrile chopped fiber	Great Lakes Carbon
Thornel T-40	PAN-carbon, continuous fiber	Union Carbide
Thornel T-50	PAN-carbon, continuous fiber	Union Carbide
Thornel 75	Carbon fiber	Union Carbide
Thornel T-300	PAN-carbon, continuous fiber	Union Carbide
Thornel 400	Carbon fiber	Union Carbide
Thornel T-500	PAN-carbon, continuous fiber	Union Carbide
Thornel 700	PAN-carbon fiber	Union Carbide
Pan 50	PAN-carbon fiber	Union Carbide
VSA-11 (VS-0032)	Mesophase-pitch fiber	Union Carbide
P-VSB-32	Pitch-carbon fiber	Union Carbide
P-VS-0053	Pitch-carbon fiber	Union Carbide
Grafil XA-S	Standard high-performance and high-strain grades – carbon/graphite fibers	Hysol Grafil Co.
Grafil IM-S	Graphite fibers	Hysol Grafil Co.
Grafil HM-S/10K	Graphite fibers	Hysol Grafil Co.
Grafil HM-S/6K	Graphite fibers	Hysol Grafil Co.
Hi-Tex	Graphite fibers	Hitco Materials Inc.
Hi-Tex HS	Graphite fibers	Hitco Materials Inc.
Nextel 312	Polycrystalline metal oxide fibers – 62% Al_2O_3, 24% SiO_2, 14% B_2O_3	3M
Nextel 440	70% Al_2O_3, 28% SiO_2, 2% B_2O_3	3M
Fiberfrax	Standard ceramic fiber – 49.5% Al_2O_3, 48.3% SiO_2, 0.85 – 1.1% Fe_2O_3, 1.00 – 1.83% TiO_2	Standard Oil Engineered Materials Co.
Fibermax	Mullite ceramic fiber – 72% Al_2O_3, 27% SiO_2, 0.02% Fe_2O_3, 0.001% TiO_2	Standard Oil Engineered Materials Co.
FP Alumina	Polycrystalline alumina	Du Pont
Saffil	Alumina fibers	Imperial Chem. Inds.

(continued)

Selected Commercial Fiber Manufacturers, Product Names, and Descriptions, continued

Product	Description	Manufacturer
APA-1	Alumina papers	Zircar
APA-2	Alumina papers	Zircar
APA-3	Alumina papers	Zircar
ALBF-1	Bulk alumina fibers	Zircar
ZYBF-2	Zirconia bulk fibers	Zircar
ZYW-15	Zirconia fabric	Zircar
ZYK-15	Zirconia fabric	Zircar
ZYW-30A	Satin weave zirconia – sized	Zircar
Nicalon	SiC Fiber	Nippon Carbon
X PV1	"Cobweb" whisker – predominately amorphous silica	S. M. Huber
SCW	SiC whisker	Tateho Chemical Industries Co. Ltd.
SNW	Si_3N_4 whisker	Tateho Chemical Industries Co. Ltd.
Ribtec-OC 330	Stainless steel fiber	Ribtec
Ribtec-310	Stainless steel fiber	Ribtec
Ribtec-OS 446	Stainless steel fiber	Ribtec
Ribtec-GR 304	Stainless steel fiber	Ribtec
Ribtec-LR 430	Stainless steel fiber	Ribtec
Ribtec-HT	Stainless steel fiber	Ribtec
Fiberglas	Glass fiber	Owens-Corning
Astroquartz	Quartz fibers – 99.5% SiO_2	J. P. Stevens & Co.
Alphaquartz	Quartz fibers	Alpha Associates, Inc.
Kevlar 29	Aramid fiber	Du Pont
Kevlar 49	High-modulus aramid fiber	Du Pont
Spectra-900	High-tenacity, high-modulus polyethylene and polypropylene fiber	Allied Fibers

5.1.2. Fiber Property Data

5.1.2.1. Refractory Fibers

Maximum Use Temperatures of Some Refractory Fibers in Oxidizing and Nonoxidizing Atmospheres

	Maximum use temperature			
	Oxidizing atmosphere		Nonoxidizing atmosphere	
Fiber type	°C	°F	°C	°F
Al_2O_3	1540	2805	1600	2910
ZrO_2	1650	3000	1650	3000
SiO_2	1060	1940	1060	1940
Al_2O_3-SiO_2	1300	2370	1300	2370
Al_2O_3-SiO_2-Cr_2O_3	1427	2600	1427	2600
Al_2O_3-SiO_2-B_2O_3	1427	2600	1427	2600
C	400	750	2500	4530
B	560	1040	1200	2190
BN	700	1290	1650	3000
SiC	1800	3270	1800	3270
Si_3N_4	1300	2370	1800	3270

Tensile Strengths of α-Al_2O_3 at Elevated Temperatures

Temperature		Tensile strength	
°C	°F	GPa	ksi
250	480	6.07	880
500	930	5.52	800
800	1470	5.17	750
1000	1830	3.93	570
1500	2730	2.07	300
1900	3450	1.03	150

Strength Retention of Nextel After Heat Treatment (3M)

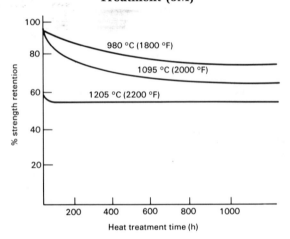

Room-Temperature Strengths of Some Ceramic Continuous Filaments After Heat Treatment (Ref 11, p 129)

Fiber[a]	Room temperature tensile strength		Young's modulus	
	GPa	10^5 psi	GPa	10^6 psi
95% Al_2O_3	1.72[b]	2.5[b]	···	···
90% SiO_2	3.93[c]	5.7[c]	80	11.6
Modified Mullite	2.21[d]	3.2[d]	172	25.0
Stabilized ZrO_2	1.72[e]	2.5[e]	···	···

[a]Fibers are circular in cross section. [b]1.72 GPa (2.5×10^5 psi) after heating for 1 h at 1095 °C (2000 °F) and tested at room temperature. [c]0.34 GPa (0.5×10^5 psi) at 1095 °C (2000 °F). [d]1.17 GPa (1.7×10^5 psi) after heating for 1 h at 1370 °C (2500 °F) and tested at room temperature. [e]0.76 GPa (1.1×10^5 psi) after heating for 1 h at 1370 °C (2500 °F) and tested at room temperature.

High-Temperature Strength of FP Fibers Vs. Other Fibers (Du Pont Co.)

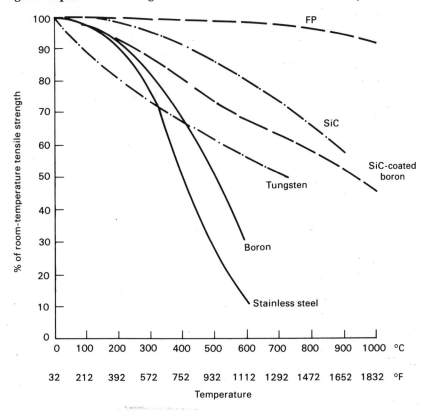

Mechanical Properties and Maximum Use Temperatures of Refractory Fibers

Fiber	Fiber diameter μm	mils	Density Mg/m³	lb/in.³	Tensile strength MPa	ksi	Tensile modulus GPa	10⁶ psi	Maximum use temperature °C	°F
Avco boron	102	4.0	2.57	0.093	3520	510	400	58	⋯	⋯
	142	5.6	2.49	0.090	3520	510	400	58	⋯	⋯
	203	8.0	2.46	0.089	3520	510	400	58	⋯	⋯
Boron (tungsten core)	51	2.0	3.38	0.122	2760	400	400–415	58–60	⋯	⋯
Boron on carbon	102	4.0	2.24	0.081	3280	475	365	53	315	600
	142	5.6	2.27	0.082	3280	475	380	55	315	600
	203	8.0	2.30	0.083	3170	460	345	50	⋯	⋯
SiC-coated boron	107	4.2	2.66	0.096	2410	350	400–415	58–60	⋯	⋯
	145	5.7	2.57	0.093	2410	350	400–415	58–60	⋯	⋯
Boron carbide	102	4.0	2.35	0.085	2690	390	425	62	315	600
Boron nitride	6.9	0.27	1.91	0.069	1380	200	90	13	1095	2000
Titanium boride (TiB₂)	⋯	⋯	4.48	0.162	105	15	510	74	2205	4000
TiC	⋯	⋯	4.90	0.177	1540	224	450	65	⋯	⋯
Zirconium oxide	⋯	⋯	4.84	0.175	2070	300	345	50	1925	3500
Nextel 312	9.9–11.9	0.39–0.47	2.71	0.098	1380–1720	200–250	150	22	1205	2200
Nextel 440	9.9–11.9	0.39–0.47	3.10	0.112	1380–2070	200–300	205–240	30–35	⋯	⋯
Fiberfrax	2–12	0.08–0.47	2.60	0.094	1030–1720	150–250	105	15	1795[a]	3260[a]
Al₂O₃ (polycrystalline)	⋯	⋯	3.15	0.114	2070	300	170	25	1650	3000
FP Al₂O₃	15.2–25.4	0.6–1.0	3.71	0.134	1380[b]	200[b]	345	50	2045[a]	3710[a]
SiO₂-coated alumina	⋯	⋯	3.71	0.134	1900[b]	275[b]	380	55	⋯	⋯
Al₂O₃ monocrystal (sapphire)	⋯	⋯	3.96	0.143	2550	370	470	68	2040[a]	3700[a]
Nicalon	⋯	⋯	⋯	⋯	>2410	>350	180	26	1095	2000
Avco SiC	142	5.6	3.04	0.110	3450	500	425	62	⋯	⋯
SiC	102	4.0	3.46	0.125	2280	330	450	65	⋯	⋯
SiC (carbon core)	142	5.6	3.29	0.119	3790[b]	550[b]	345	50	⋯	⋯
SiC (tungsten core)	142	5.6	3.29	0.119	3790[b]	550[b]	415	60	600	1110
Al₂O₃·Cr₂O₃ monocrystal (ruby)	⋯	⋯	3.99	0.144	3450–4140	500–600	470	68	2040[a]	3700[a]
Borsic (SiC/B/W)	107–145	4.2–5.7	2.77	0.100	2930	425	415	60	2300[a]	4170[a]
Borsic/C	107–145	4.2–5.7	2.30	0.083	3170	460	350–365	51–53	⋯	⋯
Saffil alumina, RF grade	⋯	⋯	3.29	0.119	2000	290	295	43	1600	2910

[a]Melting or softening point. [b]Ultimate tensile strength.

Strengths of Sapphire Rods for Different Surface Conditions (Ref 8, p 331)

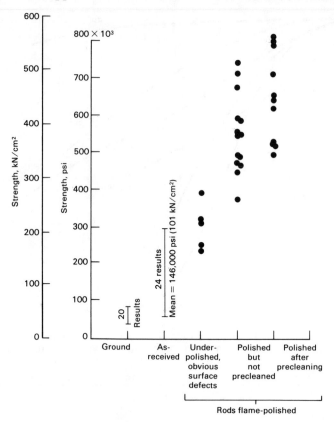

5.1.2.2. Carbon/Graphite Fibers

Classification of Carbon-Base Fibers (Ref 70, p 34)

Classification	Carbon content, %	Maximum processing temperature °C	°F	X-ray diffraction crystal structure	Crystallite orientation	Treatment	Approximate modulus of elasticity GPa	10⁶ psi	Approximate tensile strength MPa	ksi
Carbon	>80	<1000	<1830	Crystallites too small to be detected	"Amorphous"	Carbonization	34	5	690	100
Graphite	>99	>2500	>4530	Crystallites large enough to be detected	Similar to precursor "random"	Graphitization	97	14	1035	150
Structural carbon or structural graphite ..	>99	>2500	>4530	Crystallite number and size greater than in graphite fiber	Preferred orientation of graphite crystallites in a carbon matrix ("turbostratic")	Combination thermomechanical treatments	>172	>25	>1240	>180

Classification of Carbon Fibers by Modulus and Strength (Ref 31, p 248)

Fiber type	State-of-the-art Modulus GPa	Modulus 10⁶ psi	Strength GPa	Strength ksi	Newer Fibers Modulus GPa	Modulus 10⁶ psi	Strength GPa	Strength ksi
High strength	228	33	3.45	500	241	35	4.14	600
					255	37	5.17	750
Intermediate modulus					283	41	4.83	700
High modulus	379	55	2.41	350				
Very high modulus	517	75	2.07	300				
Ultrahigh modulus					690	100	2.24	325
					827	120	2.41	350

Fabrication of Rayon-Precursor Carbon-Base Fibers (Ref 70, p 40)

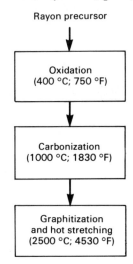

Fabrication of PAN-Precursor Carbon-Base Fibers (Ref 70, p 40)

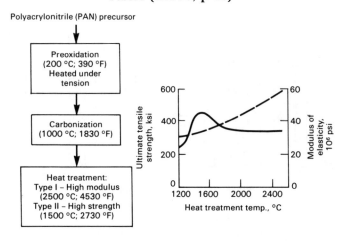

Carbon Fiber Manufacturing Process (Ref 31, p 255)

The carbon fiber manufacturing process is outlined below. The precursor polymers may be *cellulose (rayon), polyacrylonitrile (PAN),* or *pitch. Today, rayon is used only for very-low-modulus fibers.* The precursor polymer is converted into a fiber by an extrusion process which may be either wet spinning, dry spinning, or melt spinning. The originally coarse fibers are stretched or drawn into finer fibers. The next step is to stabilize the fiber by oxidation in air at a temperature of approximately 400 °C (750 °F). The fiber is next carbonized to drive off most of the volatiles, leaving only the carbon behind. The carbonization temperature is between 1000 and 2000 °C (1830 and 3630 °F). It is carried out in an inert atmosphere. *A final heat treatment, to achieve higher modulus or higher crystallinity, is performed at temperatures above 2000 °C (3630 °F) and sometimes as high as 3000 °C (5430 °F).*

Precursor polymer	Cellulose (rayon) Polyacrylonitrile Pitch (mesophase)
Fiberize	Wet-, dry-, melt-spin Draw
Stabilize	Oxidize Heat to 400 °C (750 °F)
Carbonize	To 1500 ± 500 °C (2730 ± 900 °F) in inert atmosphere
Heat treat	To 2500 ± 500 °C (4530 ± 900 °F)

The Graphite-Layer Plane (Ref 31, p 249)

All carbon fibers are based on the graphite-layer structure, which is sometimes a very imperfect structure. The perfect graphite layer consists of a continuous "chicken-wire" network of carbon atoms. The layer planes are stacked together in a parallel arrangement. However, many carbon fibers, particularly the high-strength varieties, have very small layer sizes and imperfect structures with missing carbon atom sites at which cross-linking can take place between the adjacent layers.

Perfect

Imperfect

Crystal Structure of Graphite

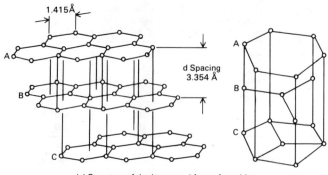

(a) Structure of the hexagonal form of graphite

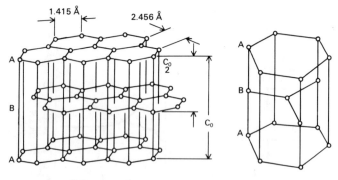

(b) Structure of the rhombohedral form of graphite

The Graphite Crystal (Ref 31, p 250)

The figure below represents the graphite crystal edge-on, showing a parallel stacking of the graphite layers. For the perfect graphite crystal, the Young's modulus is 1020 GPa (148×10^6 psi) in the layer plane direction. If the layers are imperfect and hence distorted and cross-linked, the modulus is lower. The shear modulus, on the other hand, is extremely low: only 4.1 GPa (0.6×10^6 psi) or even less. If it is an imperfect structure with cross-linking, one can expect a higher shear modulus. To estimate tensile strengths of graphite structures, one must take into account these same differences in structure: very high tensile strengths are realized for pure tension parallel to the basal plane (as high as 20.7 GPa, or 3×10^6 psi, has been measured in graphite whiskers), but if appreciable shear stress is introduced, the strength is much lower, being limited by the weak bonding between graphite layers.

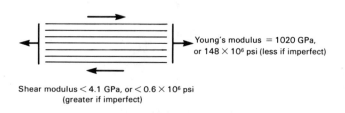

Young's modulus = 1020 GPa, or 148×10^6 psi (less if imperfect)

Shear modulus < 4.1 GPa, or < 0.6×10^6 psi (greater if imperfect)

Determination of Carbon-Fiber Properties by Structure (Ref 31, p 251)

The main structural features of carbon fibers are indicated below. Orientation can be high or low depending upon whether the layers are straight and parallel to the fiber axis. Another important structural parameter is the crystallinity. If the linear dimensions of the perfect crystalline regions are large, the structure is said to possess a high degree of crystallinity; such regions tend to behave similarly to the perfect graphite crystal. Finally, defect content is very important, particularly with regard to fiber strength.

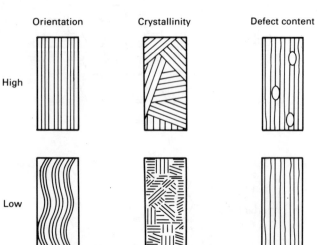

Effects of Orientation, Crystallinity, and Defects on Carbon-Fiber Properties (Ref 31, p 252–254)

As Orientation Is Improved

These properties increase:
Longitudinal tensile strength
Longitudinal tensile modulus
Thermal conductivity
Electrical conductivity
Longitudinal negative CTE

These properties decrease:
Transverse strength
Transverse moduli

As Crystallinity Is Improved

These properties increase:
Thermal conductivity
Electrical conductivity
Longitudinal negative CTE
Oxidation resistance

These properties decrease:
Longitudinal tensile strength
Longitudinal compressive strength
Transverse strength and moduli
Longitudinal shear modulus

As Defects Are Removed

These properties increase:
Tensile strength
Thermal conductivity
Electrical conductivity
Oxidation resistance

Mechanical Properties of Selected Carbon/Graphite Commercial Fibers

Fiber	Fiber diameter		Density		Tensile strength		Tensile modulus		Ultimate elongation,
	μm	mils	Mg/m³	lb/in.³	MPa	ksi	GPa	10⁶ psi	%
Magnamite HMS	8.00	0.315	1.83	0.066	2210	320	345	50	0.58
Magnamite HMU	8.00	0.315	1.85	0.067	2760	400	380	55	0.70
Magnamite chopped fiber	8.00	0.315	1.77	0.064	2480	360	205	30	1.2
Magnamite AS1	8.00	0.315	1.80	0.065	3100	450	230	33	1.32
Magnamite AS2	8.00	0.315	1.80	0.065	2760	400	230	33	1.3[a]
Magnamite AS4	8.00	0.315	1.80	0.065	3590	520	235	34	1.53
Magnamite AS6	···	···	1.82	0.0657	4140	600	243	35.3	1.65[a]
Magnamite IM6	···	···	1.74	0.0627	4380	635	279	40.4	1.50
Microfil 55	4.32	0.170	1.77	0.064	3620	525	380	55	1.00
Microfil 40	4.32	0.170	1.69	0.061	4480	650	275	40	1.65
Celion C-6S	7.11	0.280	1.77	0.064	3790	550	231	33.5	1.64
Celion G-50	6.60	0.260	1.77	0.064	2480	360	360	52	0.7
Celion GY-70	8.38	0.330	1.91–1.97	0.069–0.071	1520	220	485	70	0.38
Celion GY-80	···	···	1.91–1.97	0.069–0.071	1520	220	550	80	···
Celion 1000	7.11	0.280	1.77	0.064	3240	470	235	34	1.4
Celion 3000	7.11	0.280	1.77	0.064	3790	550	231	33.5	1.64
Panex ¹/₄CF-30	7.92	0.312	1.74	0.063	2410	350	205	30	1.2
Panex 30	7.92	0.312	1.74	0.063	2590	375	220	32	1.3
Fortafil 3(C)	7.37	0.290	1.77	0.064	3100	450	230	33	1.4
Fortafil 3(O)	5.33–13.97	0.210–0.550	1.77	0.064	2760	400	230	33	1.2
Fortafil 5(O)	4.32–11.94	1.170–0.470	1.77	0.064	3100	450	345	50	0.9
Grafil XA-S (standard)	···	···	1.79	0.0646	3100	450	234	34.0	1.31
Grafil XA-S (high performance)	···	···	1.79	0.0646	3450	500	234	34.0	1.45
Grafil XA-S (high strain) ..	···	···	1.79	0.0648	3860	560	234	34.0	1.65
Grafil IM-S	···	···	1.76	0.0635	3100	450	290	42.0	1.07
Grafil HM-S/10/K	···	···	1.85	0.067	2480	360	345	50.0	0.73
Grafil HM-S/16/K	···	···	1.85	0.067	2760	400	372	54.0	0.74
Hi-Tex	···	···	1.80	0.065	3100–3240	450–470	228	33.0	···
Hi-Tex HS	···	···	1.80	0.065	3620–3690	525–535	234	34.0	···
Thornel P-25W 4K	10.92	0.430	1.91	0.069	1380	200	160	23	0.90
Thornel T-40 12K	5.94	0.234	1.80	0.065	5650	820	275–290	40–42	2.0
Thornel T-50 3K	6.45	0.254	1.80	0.065	2410	350	395	57	0.70
Thornel 75	5.56	0.219	1.83	0.066	2620	380	545	79	···
Thornel T-300 6K	6.93	0.273	1.77	0.064	3240	470	231	33.5	···
Thornel 400	···	···	1.77	0.064	3100	450	205	30	···
Thornel T-500	6.93	0.273	1.80	0.065	3860	560	241	35.0	1.5
Thornel T-600	···	···	1.80	0.065	4140	600	241	35.0	1.7
Thornel T-700	···	···	1.80	0.065	4480	650	250	36	1.8
Modmor I	7.75	0.305	1.97	0.071	1380	200	380	55	···
Modmor II	8.03	0.316	1.74	0.063	2410	350	240	35	···

[a]Minimum elongation.

Control of Carbon-Fiber Structure (Ref 31, p 256)

1. Orientation is improved by fiber drawing or by restraining the fiber so that it can't shrink during the heat treatments. The precursor fiber structure helps to determine the degree of orientation in the final carbon fiber; starting with an oriented fiber structure, one tends to improve orientation still further by heat treatment.
2. Crystallinity is largely determined ahead of time by the precursor chemistry. It is also strongly affected by the heat treatment — that is, by the final processing temperature.
3. Defect content is controlled by the purity of the raw materials and by the mechanics of fiber handling.

Structural parameter:	Controlled by:
Orientation	1. Fiber drawing
	2. Precursor fiber structure and heat treatment
Crystallinity	1. Precursor chemistry
	2. Heat treatment
Defect content	1. Precursor purity
	2. Process handling, etc.

Stress-Strain Behavior of Carbon Fibers (Ref 94, p 372)

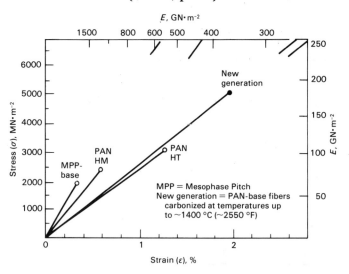

Comparison of Specific Strengths and Specific Moduli of Carbon-Base Monofilaments With Those of Other Filaments and Conventional Materials (Ref 70, p 45)

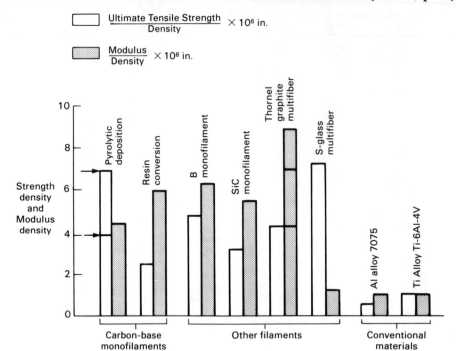

Weight Loss of Carbon Fibers After 24 h at 375 °C (705 °F) (Celanese)

Fiber variant	Weight loss, %
Celion GY-70	0.12
Magnamite HMS	0.16
Celion 3000 with epoxy finish	1.3
Celion 6000 with epoxy finish	1.9
Celion 6000 without finish	0.5
Celion 6000 with polyimide finish	1.7
Thornel 300	14.9
Magnamite AS-1	36.5

Weight Loss of Carbon Fibers After Exposure in Air at 315 °C (600 °F) (Celanese)

Fiber variant	Weight loss, %, for exposure time, h, of:					
	24	50	100	200	500	1000
Celion 6000 with epoxy finish	0.9	1.0	1.2	1.3	3.3	...
Celion 3000 with epoxy finish	1.4	1.7	2.0	2.8	5.1	7.3
Celion 3000 without finish	0.3	0.4	0.9	1.5	3.3	7.1
Thornel 300	1.0	1.4	3.0	6.5	22.8	49.7

Longitudinal CTE of Selected Commercial Carbon/Graphite Fibers

Fiber	Thermal expansion, %, from 25 to 1500 °C (77 to 2730 °F)
GSCY2-5	0.57
WYB 85½	0.40
CX5	0.46
Thornel 25	0.15
Modmor II	0.21
Thornel 50	0.08 (1.77)[a]
Fortafil 5-Y	0.17
HMG 50	0.12
Modmor I	0.14
Thornel 75	0.08

[a]Expansion transverse to fiber axis.

Electrical Resistivity of Selected Commercial Carbon/Graphite Fibers

Fiber	Electrical resistivity, $10^{-6}\ \Omega \cdot cm$	Fiber	Electrical resistivity, $10^{-6}\ \Omega \cdot cm$
Magnamite IM6	1404	Fortafil 5(O)	950
Magnamite AS6	1826	Fortafil 3(C)	1670
Celion GY-70SE	650	Fortafil 3(O)	1820
Celion G-50	1000	Thornel T-50	950
Celion C-6S	1500	Thornel T-300	1800
Celion 1000	1500	Thornel T-500	1800
Celion 3000	1500	Thornel P-25W	1300
Celion 6000	1500	Thornel P-55S	750
Celion 12000	1500	Thornel P-75S	500
		Thornel P-100	250

Axial Thermal Conductivity and Thermal Expansion of Some Commercial Carbon/Graphite Fibers

Fiber	Axial thermal conductivity			Axial thermal expansion	
	W/cm·°C	W/m·°C	Btu/ft-lb·°F	10^{-6}/°C	10^{-8}/°F
Fortafil 3(O)	0.20	20	11.6	−0.11	−6.11
Fortafil 3	0.20	20	11.6	−0.11	−6.11
Fortafil 5	1.44	144	83.2	−0.5	−27.8
Thornel 300, WYP 90-1/0	···	0.2051	0.1186	···	···
Thornel 300, WYP 30-1/0	···	0.2051	0.1186	···	···

Variation of Carbon-Fiber Tensile Strength With Young's Modulus (Ref 93, p 297)

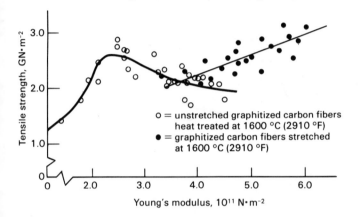

Fiber Tensile Strength as a Function of $(E/C)^{1/2}$ (Ref 93, p 297)

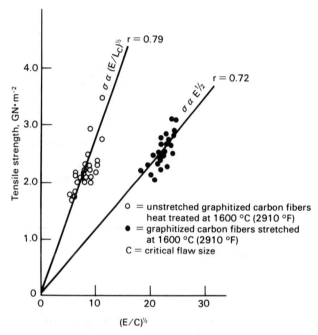

Thermal Conductivity of Several Carbon/Graphite Fibers

Fiber	Thermal conductivity	
	W/m·K	cal/cm·s·°C
VYB 70$^1/_2$	25	0.06
WYB 85$^1/_2$	38	0.09
Thornel 50	121	0.29
Thornel 75	155	0.37

Electrical Conductivity of Selected Carbon/Graphite Fibers

Fiber	Electrical conductivity	
	$\Omega \cdot cm^{-1}$	$\Omega \cdot m^{-1} \times 10^{-4}$
Fortafil 3(O)	570	5.7
Fortafil 3	570	5.7
Fortafil 5	1050	10.5
Celion GY-70	1538	15.38
Celion 6000	667	6.67
Celion 3000	667	6.67
Celion 1000	667	6.67

Effect of Temperature on Thermal Properties of Graphite (Great Lakes Carbon)

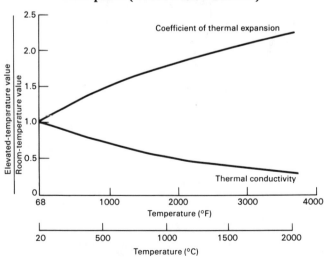

Effect of Final Heat Treatment Temperature, f_{HT}, on Mechanical Properties of Carbon Fibers: (a) literature data; (b) results of Muller (1984) and Frohs (1984) (Ref 94, p 369)

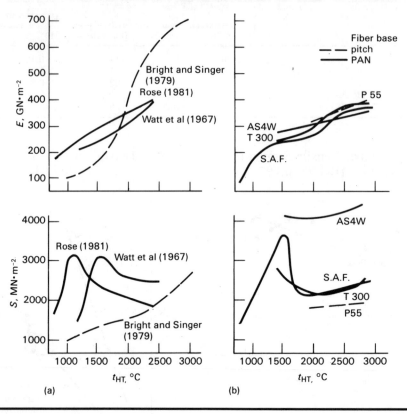

(a)   (b)

5.1.2.3. Silicon Carbide Fibers

Schematic Representations of (a) the CVD SiC Fiber Cross Section and (b and c) the Silicon Content in Two Types of Fiber Coatings (Ref 68, p 12)

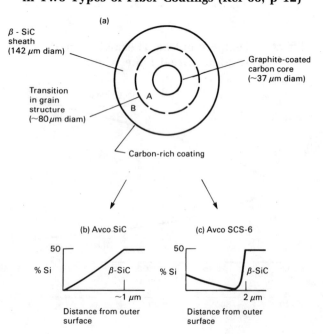

Typical Properties of Nicalon SiC Fiber (Ref 68, p 10)

Production method	Si- and C-containing polymer spun, cured, and pyrolyzed
Diameter	10 to 20 μm, or 0.39 to 0.79 mil (500 per yarn)
Modulus	180 GPa, or 26.1 × 10⁶ psi (420 GPa, or 60.9 × 10⁶ psi, for β-SiC)

Strength at 20 °C (68 °F):
  As-produced .... 2 GPa (0.29 × 10⁶ psi)
  After treatment at 1400 °C
  or 2550 °F (argon) .... <1 GPa (<0.15 × 10⁶ psi)
Strength at 1400 °C or 2550 °F
  (oxygen) .... <0.5 GPa (<0.07 × 10⁶ psi)
Creep strain at 1300 °C (2370 °F),
  0.6 GPa (0.09 × 10⁶ psi), 20 h 4.5%

Ultimate Tensile Strengths of Tungsten-Coated and Uncoated SiC Filaments at Various Temperatures (Ref 6, p 165)

Temperature		UTS[a] (coated[b])		UTS (uncoated)		
°C	°F	GPa	ksi	GPa	ksi	σ/σ_0
R.T.	R.T.	1.835	266	2.828	410	0.649
800	1470	1.648	239	2.359	342	0.699
1000	1830	1.531	222	2.083	302	0.735
1200	2190	1.414	205	1.883	273	0.751
1400	2550	1.083	157	1.359	197	0.797
1600	2910	0.855	124	1.062	154	0.805

[a]Mean ultimate tensile strength. [b]Coating thickness, 12.7 μm (0.0005 in.).

Stress-Rupture Strengths (100-Hour) of SiC/C and SiC/W Filaments (Ref 6, p 168)

Filament	1093 °C (2000 °F)		1204 °C (2200 °F)		1316 °C (2400 °F)	
	GPa (ksi)	$S/\rho \times 10^6$ in.	GPa (ksi)	$S/\rho \times 10^6$ in.	GPa (ksi)	$S/\rho \times 10^6$ in.
SiC/C[a]	1.93 (280)	2.3	1.034 (150)	1.4	0.69 (100)	0.96
SiC/W[b]	1.07 (155)	1.3	0.863 (126)	1.1	0.34 (50)	0.43

[a]C is used as a substrate. [b]W is used as a substrate.

Strength Retention for SiC Fibers Exposed for 15 Min in Argon or Nitrogen Environment (Ref 68, p 13)

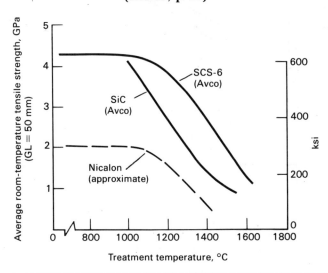

Strength at Temperature for SiC Fibers Exposed for 15 Min in Oxygen (Ref 68, p 13)

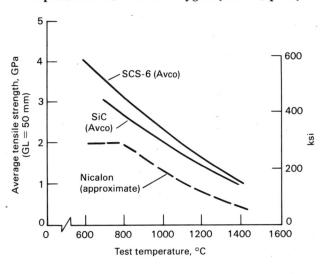

5.1.2.4. Boron Fibers

Strength Properties of Large-Diameter Boron Fibers Before and After Chemical Polishing (Ref 68, p 9)

Diameter		As-produced[a]			Slight polish[b]		
		Average strength[c]		COV[d],	Average strength[c]		COV[d],
μm	mils	GPa	ksi	%	GPa	ksi	%
203	8	4.0	580	7	4.4	640	3
280	11	3.6	520	12	4.2	610	4
406	16	2.1	300	14	4.6	660	4

[a]Fracture sources: surface and core. [b]Fracture source: core only. [c]Gauge length, 25 mm (1 in.). [d]Coefficient of variation = standard deviation/average value.

Strength Properties of Improved Large-Diameter Boron Fibers (Ref 68, p 10)

Diameter		Treatment	Strength		COV[b],	Relative fracture energy
μm	mils		Average[a]		%	
			GPa	ksi		
142	5.6	As-produced	3,8	550	10	1.0
406	16	As-produced	2.1	300	14	0.3
382	15	Chemical polish	4.6	660	4	1.4
382	15	Oxygen + polish	5.7	820	4	2.2

[a]Gauge length, 25 mm (1 in.). [b]Coefficient of variation = standard deviation/average value.

Comparison of Relative Strength-to-Weight Ratios for Boron Filaments and Other Fibers (Avco)

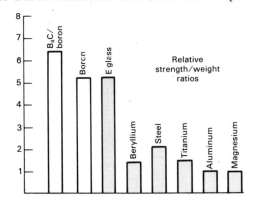

Comparison of Relative Stiffness-to-Weight Ratios for Boron Filaments and Other Fibers (Avco)

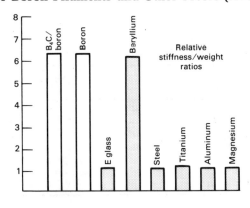

Tensile Strength of Boron Filaments (Avco)

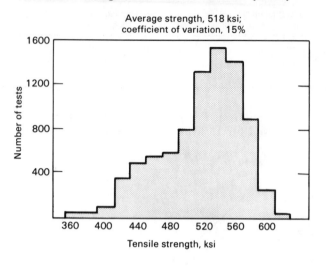

Effects of Oxidation-Induced Contraction (Plus Surface Polishing) on Strength of Large-Diameter Boron Fibers (Ref 68, p 12)

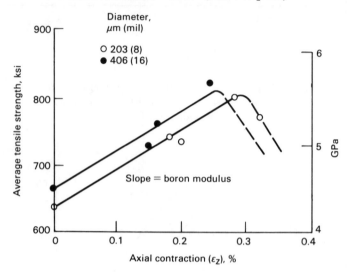

5.1.2.5. Glass and Quartz Fibers

Glass Fiber Terminology (Hexcel-Trevarno)

A number of years ago a system of **letters** and **numbers** was developed to identify the different glass yarns. Letters designate the basic strand by chemical *composition, type* and *diameter.*

Composition

"E" designates "electrical," the most widely used all-purpose glass yarn. "S" designates "high strength," and "C" designates chemical resistant," two special-purpose glass yarns. These yarns vary in their chemical makeups.

Letters also designate whether the yarn is continuous filament ("C") or staple ("S").

Continuous filaments of glass are drawn or pulled from a batch of molten glass through a precise multi-hole bushing and combined into strands. The number of filaments in a strand may range from 51 to 1224 depending on the bushing size and the requirements of the strand. Continuous filament yarns, which are twisted and plied together from several strands, are stronger than any other form of glass yarn and are the most widely used.

Diameter

In addition, letters designate the average diameter of the individual filaments in a strand, as follows:

Continuous filament designations	Nominal filament diameter, in.
C	0.000175
D	0.000225
DE	0.000250
E	0.000275
G	0.000375
H	0.000425
K	0.000525

An example of filament and yarn identifying terms:

ECG 150 2/3

E–All-purpose "electrical" glass
C–Continuous filament yarn
G–Average filament diameter of 0.000375 in.

The Numbers

The first number in the example, 150, represents *1/100 of the total approximate bare glass yardage in a single pound of strand.* This is known as the strand count. Therefore, by multiplying this number by 100 (150 × 100), the resultant figure (15 000) is the number of yards in one pound of strand.

The next set of numbers in the example, 2/3, reveals *the number of basic strands in the yarn.* The first digit of the series, "2" in this instance, expresses the number of strands twisted together in making the yarn.

*The second digit, "3," separated by the diagonal, expresses the number of twisted strands which have been plied together. To find the total number of strands used in the product, multiply the digits on either side of the diagonal by each other. (0 is multipled as 1.) In our example, there are six strands. Two basic strands have been twisted together and three of these twisted strands have been plied to form the yarn.

To compute the approximate number of yards per pound of plied product, divide the strand count multiplied by 100 by the total number of strands:

$$ECG\text{-}150\ 2/3 = \frac{150 \times 100}{2 \times 3}$$

$$\frac{15\ 000\ yd}{6\ strands} = 2500\ yd/lb\ of\ plied\ yarn$$

TEX System of Yarn Identification (Ref 64, p 30, 31)

In Europe, glass fibers are classified by "weight per unit length." In this system, called "TEX," the unit of length is 1000 m. As an example of the TEX system, a basic strand 1000 m long composed of approximately 200 filaments, each having a 9-micron diameter, weighs 33 g. Therefore, the strand count is 33 g per 100 m, or simply 33. Filament diameters are listed in "microns" (a number) in the TEX system as opposed to the letter designation used in the "yardage" system. Table 1 provides a comparison of filament designations between the TEX and YARDAGE systems.

Table 1. Average Standard Filament Diameter Designations

Metric designation, microns[a]	U.S. designation (letter)	Nominal fiber diameter, in.
3.5	B	0.00015
4.5	C	0.00018
5	D	0.00021
6	DE	0.000250
7	E	0.00029
9	G	0.00036
10	H	0.00043
13	K	0.00051

[a]1 micron = ~0.00004 in.

Another comparison shows not only filament diameters but also corresponding strand counts, for a 200-filament strand.

It can be seen in Table 2 that as the filament diameter increases so does the strand count in TEX. The opposite is true, however, in the system of U.S. yardage. This inverse relationship can result in considerable confusion if it is not kept carefully in mind.

Table 2. Basic Strand Fiber Diameters and Count Designations[a]

Filament diameter designation Metric (microns)	U.S.	Strand count in TEX, g/100 m	U.S. strand count	Strand yards per pound
3.5	B	5	B -900	90 000
4.5	C	9	C -600	60 000
5	D	11	D -450	45 000
6	DE	17	DE-300	30 000
7	E	22	E -225	22 500
9	G	33	G -150	15 000
10	H	39	H -125	12 500
13	K	66	K - 75	7 500

[a]For 200-filament basic strand.

Yarn Yield and Twist. A comparison between U.S. and TEX yarn systems, specified by strand count, construction, and twist, is shown in Table 3.

Yardage Conversions

To obtain the basic fiber yardage per pound of other commonly used fibers, the following conversion formulas should be used.

TEX System. Yards per pound is obtained by dividing 496 053 by the TEX strand count.
Example: TEX strand count 33:

$$\frac{496\ 053}{33} = 15\ 000 \text{ yd/lb}$$

Grams per 1000 meters (TEX) is obtained by dividing 496 053 by yards per pound.
Example: 7500 yards per pound:

$$\frac{496\ 053}{7500} = 66 \text{ g/km}$$

Denier System. Yards per pound is obtained by dividing the number 4 464 500 by the denier size.
Example: a 300 denier yarn:

$$\frac{4\ 464\ 500}{300} = 14\ 882 \text{ yd/lb}$$

Cotton System. Yardage per pound is obtained by multiplying the cotton yarn number by 840.
Example: No. 16 cotton yarn:

$$16 \times 840 = 13\ 440 \text{ yd/lb}$$

Table 3. Specification for Yarn Yield, Construction, and Twist

System	Single yarns					Plied yarns					
TEX	EC	9	33	40	Z	EC	9	33	x2	80	S
U.S.	EC	G	150	1/0	1.0	Z	EC	G	150	1/2	2.0	S
Note	(1,2)	(3)	(4)	(5)		(6)	(7)	(1,2)	(3)	(4)	(5)	(6) (7)

(1) Glass Type — "E" signifies that electrical grade glass is used.

(2) Fiber Type — "C" signifies that the yarn is made from continuous fiber. If staple fiber were used, "S" would be used to signify that type.

(3) Filament Diameter — The 9 and G are the corresponding TEX and U.S. designations for 9-μm filament diameters.

(4) Yarn Count — the 33 (TEX) and 150 (YARDAGE–h.y.p.p.) designate the yarn count as indicated previously.

(5) Yarn Constructions — Since the two systems differ somewhat, they will be explained in some detail.

Single Yarn
— TEX System:
The convention in this system is to designate a single yarn as understood when no multiplier is indicated immediately after the count in TEX. The multiplier × 1 is not indicated and is understood by convention.
— U.S. System:
The 1/0 designates a single yarn uncombined or plied with any others. The first digit signifies one strand. The second (the zero after the slash) indicates that the strand is unplied.

Plied Yarns
— TEX System:
The designation × 2 signifies that the yarn is plied. Two single yarns twisted together form the yarn construction.
— U.S. System:
The construction designation 1/2 signifies one strand plied with another yarn of the same type. The yarn in Example 1 would be constructed of two single yarns twisted together.

(6) Magnitude of Twist — The TEX and U.S. twist equivalences are 40 and 1.0, respectively, for the single-yarn example in TEX, the 40 signifying forty turns per meter (TPM), which is the same as one turn per inch (TPI) or 1.0 in the U.S. system.

In the case of the single and plied yarns, the 80 and the 2.0 again indicate the magnitude of twist in turns per meter and turns per inch, respectively.

(7) The twist is designated by the letters Z or S, and these letters have the normal significance for either a left-handed or right-handed twist, respectively.

Commercial Forms of Glass-Fiber Reinforcements (Ref 4)

Nominal form	General description	Process	Nominal glass content of typical laminates, %	Typical applications
Rovings	Continuous strands of glass fibers	Filament winding, continuous panel, preforming (matched-die molding), spray-up, pultrusion	25–80	Pipe, automobile bodies, rod stock, rocket-motor cases, ordnance
Chopped strands	Strands cut to lengths of 0.125–2 in. (3.2–50.8 mm)	Premix molding, wet slurry preforming	15–40	Electrical and appliance parts, ordnance components
Reinforcing mats	Continuous or chopped strands in random matting	Matched-die molding, hand lay-up, centrifugal casting	20–45	Translucent sheets, truck and auto body panels
Surfacing and overlaying mats	Nonreinforcing random mat	Matched-die molding, hand lay-up, filament winding	5–15	Where smooth surfaces are required (automobile bodies, some housings)
Yarns	Twisted strands	Weaving, filament winding	60–80	Aircraft, marine, electrical laminates
Woven fabrics	Woven cloths from glass fiber yarns	Hand lay-up, vacuum bag, autoclave, high-pressure laminating	45–65	Aircraft structures, marine, ordnance hardware, electrical flat sheet and tubing
Woven roving	Woven glass fiber strands (coarser and heavier than fabrics)	Hand lay-up	40–70	Marine, large containers
Nonwoven fabrics	Unidirectional and parallel rovings in sheet form	Hand lay-up, filament winding	60–80	Aircraft structures

Comparative Property Data for Glass Fibers

Property	S-2 Glass[a]	E glass	S glass	C glass	D glass
Physical Properties[b]					
Specific gravity, g/cm³	2.49	2.60	2.48	2.49	...
Density, lb/in.³	0.090	0.094	...	...	...
Mechanical Properties[b]					
Virgin tensile strength at 70 °F, ksi	665	500	665	480	...
Modulus of elasticity:					
At 72 °F (without heat compaction), 10⁶ psi	12.6	10.5	12.4	10.0	...
At 72 °F (after heat compaction), 10⁶ psi	13.5	12.4	...	...	...
At 1000 °F (after heat compaction), 10⁶ psi	12.9	11.8	...	...	...
Elongation at 72 °F, %	5.4	4.8	5.7	4.8	...
Thermal Properties[c]					
Coefficient of expansion, in./in.·°F × 10⁶	3.1	2.8	3.1	4.0	1.7
Specific heat at 75 °F, Btu/lb·°F	0.176	0.192	0.176	0.212	0.175
Softening point, °F	1 778	1 155	1 778	1 380	1 420
Strain point, °F	1 400	1 140	1 400	...	890
Annealing point, °F	1 490	1 215	1 490	...	970
Electrical Properties[c]					
Dielectric constant at 72 °F:					
At 1 MHz	5.34	6.33	...	...	...
At 10 kHz	5.21	6.13	...	...	...
Power factor (loss tangent) at 72 °F:					
At 1 MHz	0.002	0.001	...	...	...
At 10 kHz	0.0068	0.0039	...	...	...
Acoustical Properties[b]					
Velocity of sound (calculated), ft/s	19 200	17 500	...	...	...
Velocity of sound (measured), ft/s	...	18 000	...	...	...
Optical Property[c]					
Index of refraction	1.523	1.547	1.523	1.541	...

[a]S-2 Glass®, Owens-Corning Fiberglas. [b]Properties determined on glass fibers. [c]Properties determined on bulk glass.

Glass Fiber Compositions (Wt %)

Component	Grade of glass			
	A (high alkali)	C (chemical)	S (high strength)	E (electrical)
Silicon oxide	72.0	64.6	64.2	54.3
Aluminum oxide........	0.6	4.1	24.8	15.2
Ferrous oxide...........	···	···	0.21	···
Calcium oxide	10.0	13.2	0.01	17.2
Magnesium oxide	2.5	3.3	10.27	4.7
Sodium oxide	14.2	7.7	0.27	0.6
Potassium oxide	···	1.7	···	···
Boron oxide	···	4.7	0.01	8.0
Barium oxide	···	0.9	0.2	···
Miscellaneous	0.7	···	···	···

Chemical Compositions of High-Silica and Quartz Fibers (J. P. Stevens)

Component	Content[a]	
	High silica (high purity)	Astroquartz
SiO_2 (exclusive of binders)	99.23	99.95
Aluminum	900	100
Antimony	···	0.5
Boron	205	10
Cadmium.......................	···	0.5
Calcium	5	23
Chromium	150	···
Copper	2	1
Iron...........................	16	3
Lithium	1	1
Magnesium	6	2
Manganese	1	2
Phosphorus.....................	···	3
Potassium	3	5
Sodium	4	9
Titanium	2800	12

[a]SiO_2 in percent; all other components (elements) in parts per million.

Strength of Silica Fibers Vs. Temperature (Rolls-Royce, Ltd.)

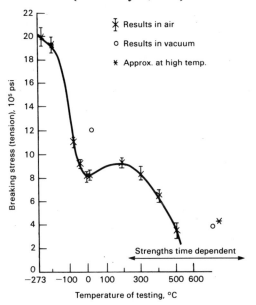

Melt Viscosity Vs. Temperature for Quartz and High-Silica Fibers (J. P. Stevens)

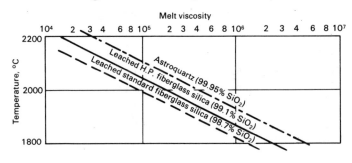

Properties of Glass Fibers by Composition Compared With E Glass (Ref 11, p 292, 293)

Component or property	E glass	Composition, %					Range by weight
		I	II	III	IV	V	
SiO_2..................................		45.5	45.5	50.75	47.7	42.0	40–55
B_2O_3		19.0	19.0	17.2	14.2	18.2	10–20
Li_2O.................................		0.5	0.5	0.45	0.3	0.5	0.1–1
MgO		17.5	10.6	6.3	15.6	12.7	5–20
Al_2O_3...............................		17.5	24.4	25.3	22.2	19.5	19–29
BeO		···	···	···	···	6.6	0–8
Fe_2O_3		···	···	···	···	0.5	···
Fiber density:							
Mg/m^3..................	2.54	2.33	2.05	2.00	2.20	2.20	<2.33
$lb/in.^3$	0.0918	0.0842	0.0741	0.0723	0.0795	0.0795	<0.0842
Tensile strength:							
MPa	3450	4050	3970	4060	4360	5030	3970–5030
ksi	500	587	576	589	633	730	576–730
Al_2O_3/MgO.................	···	1.0	2.3	4.0	1.43	1.54	1:1–5:1
Modulus:							
GPa.....................	74.5	75.2	62.7	62.1	73.1	82.0	···
10^6 psi	10.8	10.9	9.1	9.0	10.6	11.9	···

Properties of Some Refractory Glass Fibers (Ref 56, p 126, 127)

Fiber composition, wt %		Average length		Average diameter		High		Low		Median		Fusion temperature		Thermal expansion, calculated, × 10⁻⁶	
		mm	in.	μm	mil	MPa	ksi	MPa	ksi	MPa	ksi	°C	°F	m/m·K	in./in.·°F
SiO₂	36.0	75–380	3–15	25.4	1.0	2310	335	400	58	1125	163	1475	2685	4.28	2.38
Al₂O₃	48.0														
MgO	16.0														
SiO₂	50.0	50–255	2–10	16.0	0.63	2495	362	1220	177	1875	272	1450	2640	5.20	2.89
Al₂O₃	22.5					3825ᵇ	555ᵇ	980ᵇ	142ᵇ	2040ᵇ	296ᵇ				
MgO	7.5														
ZrO₂	20.0														
SiO₂	60.0	25–150	1–6	25.4	1.0	2525	366	1105	160	1770	257	1605	2920	4.28	2.38
Al₂O₃	7.5														
MgO	2.5														
ZrO₂	30.0														
SiO₂	45.0	25–50	1–2	…	…	…	…	…	…	…	…	1680	3055	…	…
ZrO₂	45.0														
PbO	10.0														
SiO₂	40.0	25–50	1–2	…	…	…	…	…	…	…	…	1680	3055	…	…
ZrO₂	40.0														
ZnO	20.0														
SiO₂	35.0	50–100	2–4	24.9	0.98	3510	509	910	132	1800	261	1810	3290	4.01	2.23
Al₂O₃	32.5														
ZnO	32.5														
SiO₂	25.0	50–100	2–4	…	…	…	…	…	…	…	…	1620	2950	…	…
Al₂O₃	62.5														
MgO	12.5														
SiO₂	50.0	50–205	2–8	20.1	0.79	4625	671	177.5	25.75	1735	252	1580	2875	4.99	2.77
Al₂O₃	27.0														
MgO	3.0														
ZrO₂	20.0														
SiO₂	17.9	25–50	1–2	13.5	0.53	7345	1065	713.6	103.5	1995	289	>1830	>3325	5.04	2.80
Al₂O₃	35.7														
ZnO	35.7														
Sb₂O₃	10.7														
SiO₂	18.5	13–50	½–2	…	…	…	…	…	…	…	…	1830	3325	…	…
Al₂O₃	37.0														
ZnO	37.0														
Sb₂O₃	7.5														
SiO₂	17.9	13–25	½–1	…	…	…	…	…	…	…	…	1750	3180	…	…
Al₂O₃	35.7														
ZnO	35.7														
CeO₂	10.7														
SiO₂	18.5	25–50	1–2	…	…	…	…	…	…	…	…	1750	3180	…	…
Al₂O₃	37.0														
ZnO	37.0														
Y₂O₃	7.5														
SiO₂	60.0	25–75	1–3	25.4	1.0	2785	404	600	87	1480	215	1575	2865	4.18	2.32
Al₂O₃	10.0														
ZrO₂	30.0														
SiO₂	58.2	25–75	1–3	23.1	0.91	2345	340	490	71	1350	196	1575	2865	4.48	2.49
Al₂O₃	7.3														
MgO	2.4														
ZrO₂	29.2														
Sb₂O₃	2.9														
SiO₂	45.0	25–75	1–3	…	…	…	…	…	…	…	…	1790	3255	…	…
Al₂O₃	40.0														
P₂O₅	15.0														
Al₂O₃	57.0	25–50	1–2	20.6	0.81	3420	496	1020	148	1930	280	1530	2785	15.93	8.85
CaF₂	43.0														

ᵃFor 10 specimens. ᵇWith anhydrous A-1100 finish.

Mechanical Properties and Maximum Service Temperatures of Some Glass and Quartz Fibers

Fiber	Fiber diameter μm	mils	Density Mg/m³	lb/in.³	Tensile strength MPa	ksi	Tensile modulus GPa	10⁶ psi	Maximum service temperature °C	°F
E glass:										
Monofilament	0.4–12.7	0.015–0.5	2.55	0.092	3450	500	72.4	10.5	550	1020
Strand	0.4–12.7	0.015–0.5	2.55	0.092	2760	400	72.4	10.5	550	1020
Hollow	0.4–12.7	0.015–0.5	2.55	0.092	3450	500	72.4	10.5	550	1020
D glass	...	...	2.16	0.078	2410	350	51.7	7.5	770[a]	1420[a]
S glass:										
Monofilament	9.7	0.38	2.49	0.090	4590	665	85.5	12.4	650	1200
Strand	9.7	0.38	2.49	0.090	3790	550	85.5	12.4	650	1200
SiO₂	35.6	1.4	2.19	0.079	5860	850	72.4	10.5	1660	3020
Aluminum-coated silica	50.8	2.0	2.46	0.089	3720	540	...	...	...	...
Quartz[b]	8.9	0.35	...	...	3450	500	69.0	10.0	...	...

[a]Melting or softening point. [b]Astroquartz II.

Properties of Quartz[a] Fibers (J. P. Stevens)

Ultimate tensile strength, MPa (ksi) 895 (130)
Tensile modulus, GPa (10⁶ psi) 68.9 (10)
Ultimate tensile elongation at 22 °C (72°F), % 0.17
Index of refraction at 32 °C (90 °F) 1.4585
Coefficient of thermal expansion,
10⁻⁶ cm/cm·°C (10⁻⁶ in./in.·°F) 0.54 (0.3)
Specific heat at 24 °C (75 °F), J/kg·K (cal/g·°C) 963 (0.23)
Softening point, °C (°F) 1670 (3038)
Dielectric constant at 22 °C (72 °F) and 1 MHz 3.7
Power factor (tangent loss) at 22 °C (72 °F) and 1 MHz..... 0.0002

[a]Astroquartz.

Tensile Strengths of Quartz Vs. Glass Fiber Rovings (J. P. Stevens)

Product	Tensile strength MPa	ksi
Astroquartz II 9779	3660	530.5
Astroquartz 9288	2560	371.0
S 901 fiberglas	3840[a]	557.0[a]
S-2 fiberglas 463	3585[a]	519.9[a]
E fiberglas 462	1870[a]	271.5[a]
E fiberglas 456	1915[a]	277.5[a]

[a]Data from Owens-Corning Fiberglas Corp. Publication No. 5-ASP 10139-A.

5.1.2.6. High-Temperature Alloys, Superalloy and Refractory Metal Alloy Wire and Fibers

Typical Properties of 7-μm-Diam Stainless Steel Fibers and of Sized and Chopped Stainless Steel Fiber Bundles (Ref 60, p 31)

Fibers[a]

Alloy composition, %:
Chromium .... 18 to 20
Nickel .... 8 to 12
Manganese 2
Silicon ....≤1
Carbon ....≤0.08
Iron .... Rem
Ultimate tensile strength, 10⁵ psi 2.10
Tensile strength at yield, 10⁵ psi 1.80
Tensile elongation, % ....1.2 to 1.5
Rockwell hardness 38
Electrical resistivity, μΩ·cm 75
Density, g/cm³ 7.9

Fiber Bundles

Fiber diameter, μm 7.26
Number of fibers per bundle 1159
Bundle size, 10⁻² in. .... 1.0 by 1.8 by 25
Sizing content, wt % .... 6 to 8
Bulk density[b], lb/ft³ 55

[a]Type 304 Brunsmet stainless steel fibers, made by Brunswick Corp. [b]For a 6.35-mm-long bundle.

Tensile Strengths of Some Superalloy Filaments (Ref 8, p 331)

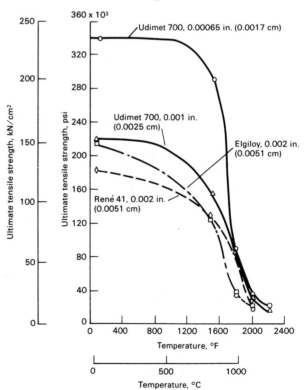

Mechanical Properties and Melting Points of Stainless Steel and Other Metallic Fibers

Fiber	Fiber diameter μm	mils	Density Mg/m³	lb/in.³	Tensile strength MPa	ksi	Tensile modulus GPa	10⁶ psi	Melting point °C	°F
AFC-77	150–1270	6–50	7.75	0.280	3640–4135	528–600	207	30.0	1370	2500
René 41	25–50	1–2	8.25	0.298	2000–2345	290–340	220	31.9	1370	2500
Udimet 700	255	10	7.92	0.286	1515–2330	220–338	221	32.0	1404	2559
Ribtec-HT[a]	205–510	8–20	···	···	57	8.3	82.7	12.0	1480/1530	2700/2790
Ribtec-LR 430[a]	205–510	8–20	···	···	47	6.8	82.7	12.0	1480/1530	2700/2790
Ribtec-GR 304[a]	205–510	8–20	···	···	124	18	124	18.0	1400/1455	2550/2650
Ribtec-OS 446[a]	205–510	8–20	···	···	52	7.6	96.5	14.0	1425/1510	2600/2750
Ribtec-310[a]	205–510	8–20	···	···	151	22	124	18.0	1400/1455	2550/2650
Ribtec-OC 330[a]	205–510	8–20	···	···	193	28	134	19.5	1345/1425	2450/2600
Steel (wire)	25	1	···	···	3445	500	207	30.0	···	···
Aluminum	···	···	2.68	0.097	620	90	73.1	10.6	···	···

[a]Modulus of elasticity computed at 315 °C (600 °F); tensile strength computed at 870 °C (1600 °F).

Compositions of Superalloys That Have Been Drawn Into Filaments or Wires (Ref 8, p 271)

Alloy	Mn	Si	Cr	Ni	Co	Mo	Ti	Al	Fe	W
A-286	1.35	0.5	15	26.0	···	1.25	2.0	···	Rem	···
Elgiloy	2.00	···	20	15.0	Rem	7.0	···	···	16	···
Hastelloy C	···	···	16	58.0	···	15.5	···	···	6.0	3.5
Chromel R	···	···	20	74.0	···	···	···	3.0	3.0	···
Inconel 702	···	···	15.6	79.5	···	···	0.70	3.4	0.35	···
Hastelloy B	1.0	1.0	1.0	Rem	2.5	28	···	···	6.0	···
M-252	0.5	0.5	19	Rem	10	10	2.5	1.0	2.0	···
René 41	···	···	19	Rem	11	10	3.1	1.5	3.0	···
Udimet 500	···	···	19	Rem	19	4	3.0	2.9	4.0	···
Udimet 700	···	···	15	Rem	19	5	3.5	4.5	5.0	···
Waspaloy	···	···	19	Rem	15	4	3.0	1.3	1.0	···

Tensile Properties of Refractory Metal Wires (Ref 91)

Wire material	Wire diameter cm	in.	Test temperature °C	°F	Ultimate tensile strength MPa	ksi	Elongation in 2.5 cm or 1 in., %	Reduction in area, %
W-Hf-C (in-process annealed)	0.038	0.015	Room		2700	392	5.4	21.1
			1093	2000	1430	207	···	67.8
			1204	2200	1390	201	···	70.9
W-Hf-C (hard drawn)	0.038	0.015	Room		2250	326	2.8	1.9
			1093	2000	1740	253	···	44.2
			1204	2200	1540	224	···	46.9
W-Re-Hf-C (hard drawn)	0.038	0.015	Room		3160	458	4.8	27.5
			1093	2000	2160	314	···	24.7
			1204	2200	1940	281	···	37.6
ASTAR 811C	0.051	0.020	Room		1700	247	6.9	51.0
			1093	2000	744	108	···	80.8
			1204	2200	490	71	···	89.8
	0.038	0.015	Room		1740	253	5.3	42.9
			1093	2000	779	113	···	66.4
			1204	2200	550	80	···	66.9
B-88	0.051	0.020	Room		1480	215	4.8	26.5
			1093	2000	530	77	···	87.4
			1204	2200	350	50	···	97.9
	0.038	0.015	Room		1620	235	7.7	54.8
			1093	2000	490	71	···	94.5
			1204	2200	310	45	···	95.7
W-2ThO₂	0.038	0.015	Room		2650	384	5.5	14.2
			1093	2000	1190	173	···	50.2
			1204	2200	1030	150	···	51.0

Mechanical Properties and Melting Points of Refractory Metallic Fibers/Wires (Ref 8, p 265)

Type of filament or wire	Nominal composition	Melting point °C	Melting point °F	Density, ρ g/cm³	Density, ρ lb/in.³	Tensile strength, σ kN/cm²	Tensile strength, σ ksi	Specific strength, σ/ρ 10⁶ cm	Specific strength, σ/ρ 10⁶ in.	Modulus of elasticity, E MN/cm²	Modulus of elasticity, E 10⁶ psi	Specific modulus, E/ρ 10⁶ cm	Specific modulus, E/ρ 10⁶ in.	Typical cross section (diameter) 10⁻³ cm	Typical cross section (diameter) 10⁻³ in.
Cr	Cr	1865	3390	7.2	0.26	159	230	2.2	0.89	29	42	411	162	25	10
Nb-Su16	Nb-11W-3Mo-2Hf-0.08C	~2590	~4700	9.27	0.335	69-89	129-142	0.98-1.1	0.39-0.42	12.1-13.4	17.6-19.5	~159-177	~63-70	25-89	10-35
Nb-Su31	Nb-17W-3 5Hf-0.12C	~2590	~4700	9.46	0.342	105-150	150-218	1.1-1.6	0.44-0.64	12.2-13.5	17.7-19.6	~160-178	~63-70	61-102	24-40
Nb, FS85	Nb-28Ta-10W-1Zr-0.005C	2590	4695	10.6	0.383	151	219	1.5	0.57	14	20	132	52	13	5
Nb, AS30	Nb-20W-1Zr	~2590	~4700	9.6	0.347	176	255	1.9	0.74	~20	~20	~147	~58	13	5
Nb, B88	Nb-28W-2Hf-0.06C	~2590	~4700	10.3	0.372	162	235	1.6	0.63	~14	~20	~137	~54	38	15
Mo	Mo	2610	4730	10.2	0.369	221	320	2.2	0.87	36	52	358	141	15	6
Mo + 0.5 Ti	Mo-0.5Ti	2610	4730	10.1	0.367	179	260	1.8	0.71	32	46	318	125	13	5
Mo, TZM	Mo-0.5Ti-C.08Zr-0.015C	~2610	~4730	10.2	0.369	197	285	2.0	0.77	32	46	318	125	20-25	8-10
Mo, TZC	Mo-1.25Ti-0.3Zr-0.15C	~2610	~4730	10.1	0.367	227	329	2.3	0.89	32	46	318	125	13	5
Ta, ASTAR 811C	Ta-8W-1Re-0.9Hf-0.03C	~2990	~5400	16.9	0.610	170	247	1.0	0.41	20	29	122	48	38-51	15-20
W	W	3410	6170	19.3	0.697	169-327	239-474	0.9-1.7	0.34-0.68	41	59	216	85	5.1-127	2-50
W, 218CS	W	3410	6170	19.2	0.695	239-266	346-386	1.3-1.4	0.5-0.56	41	59	216	85	20-38	8-15
W + 1 ThO₂ (NF)	W-1ThO₂	~3410	~6170	19.1	0.691	225-231	327-335	1.2-1.2	0.47-0.48	41	59	216	85	20-51	8-20
W + 2 ThO₂	W-2ThO₂	~3410	~6170	18.9	0.683	265-275	384-399	1.3-1.5	0.56-0.58	41	59	218	86	20-51	8-20
W + 3 Re (3D)	W-3Re	~3410	~6170	19.4	0.70	279	404	1.5	0.58	41	59	213	84	7.6-20	3-8
W + 5 Re	W-5Re	~3410	~6170	19.4	0.70	169-265	245-384	0.9-1.4	0.35-0.55	41	59	213	84	25-127	10-50
W + Hf + C	W-0.03Hf-0.036C	~3410	~6170	19.4	0.70	225-270	326-342	1.2-1.4	0.47-0.56	41	59	213	84	38	15
W + Re + Hf + C	W-4Re-0.38Hf-0.02C	~3410	~6170	19.4	0.70	316	458	1.7	0.65	41	59	213	84	38	15

Retention of Room-Temperature Tensile Strength for Wires at Two Test Temperatures (Ref 91)

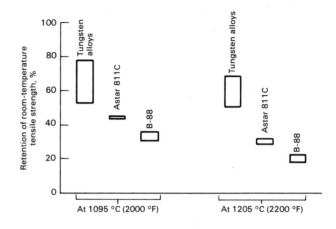

Tensile Strength Vs. Test Temperature for Tungsten Fibers (Ref 10, p 159)

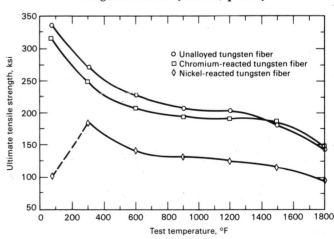

Finest Diameters Achieved by Drawing Superalloys Through Diamond Dies (Ref 8, p 271)

Alloy	Finest diameter achieved cm	in.	Degree of die wear
A-286	0.0030	0.0012	Excessive
Elgiloy	0.0013	0.0005	Slight
Hastelloy C	0.0013	0.0005	Excessive
Chromel R	0.0013	0.0005	Slight
Inconel 702	0.0018–0.0023	0.0007–0.0009	(a)
Hastelloy B	0.0015	0.0006	Slight
M-252	0.0025	0.0010	Excessive
René 41	0.0015	0.0006	Excessive
Udimet 500	0.0025	0.0010	Excessive
Udimet 700	0.0017	0.00065	Excessive
TD nickel	0.0025	0.0010	(a)
"Ductile" chromium	(b)	(b)	(b)

[a]Die wear could not be established due to wire breakage in dies. [b]Could not be drawn.

Stress-Rupture Properties of Tungsten-Base Wire Materials (Ref 91)

Wire material(a)	Test temperature °C	°F	Stress MPa	ksi	Rupture time, h	Reduction in area, %
W-Hf-C (thermally annealed during drawing)	1095	2000	1300	189	4.4	44.2
			1290	187	10.3	58.4
			1230	178	21.1	23.2
			1210	175	19.1	35.0
			1150	167	61.5	44.5
			1110	161	108.5	18.0
	1205	2200	918	133	28.3	15.3
			841	122	42.9	21.9
			765	111	104.3	11.5
			689	100	188.4	28.5
W-Hf-C (hard drawn)	1095	2000	1310	190	17.7	44.2
			1240	180	139.4	37.0
			1230	178	88.6	22.6
			1210	175	262.0	57.3
			1100	160	(b)	⋯
	1205	2200	1170	170	6.0	29.4
			1140	165	4.3	30.6
			1100	160	11.1	20.2
			1040	150	22.5	24.9
			965	140	18.6	20.2
			896	130	37.4	16.6
			827	120	63.0	22.6
			793	115	74.3	17.8
			758	110	334.1	50.2
			689	100	329.6	65.8
W-Re-Hf-C (hard drawn)	1095	2000	1590	230	15.6	15.7
			1520	220	36.2	19.5
			1480	215	42.8	34.0
			1450	210	72.1	27.1
			1380	200	442.6	34.7
			1310	190	104.2	16.7
			1240	180	522.3	37.3
	1205	2200	1140	165	14.4	32.0
			1040	150	18.4	43.2
			965	140	39.7	23.0
			896	130	49.8	32.8
			862	125	365.5	33.7
			827	120	345.5	44.3
			793	115	342.2	43.2

(a) Wire diameter, 0.038 cm (0.015 in.). (b) Test stopped at 233.4 h.

100-Hour Rupture Strengths and Specific 100-Hour Rupture Strengths of Refractory Metal Wires at 1095 and 1205 °C (2000 and 2200 °F) (Ref 91)

Wire material	Approximate wire density		Wire diameter		100-hour rupture strength		100-hour rupture strength-to-density	
	gm/cm³	lb/in.³	cm	in.	MPa	ksi	m	in.
Test Temperature, 1095 °C (2000 °F)								
W-Hf-C (hard drawn)	19.37	0.7	0.038	0.015	1240	180	6450	257 000
W-Re-Hf-C (hard drawn)	19.37	0.7	0.038	0.015	1410	205	7440	293 000
W-Hf-C (in-process annealed)	19.37	0.7	0.038	0.015	1110	161	5840	230 000
Astar 811C	16.9	0.61	0.051	0.020	580	84	3500	138 000
			0.038	0.015	580	84	3500	138 000
B-88	10.3	0.373	0.051	0.020	330	48	3280	129 000
			0.038	0.015	300	44	3000	118 000
W-2-ThO₂	18.91	0.68	0.038	0.015	660	96	3560	141 000
Test Temperature, 1205 °C (2200 °F)								
W-Hf-C (hard drawn)	19.37	0.7	0.038	0.015	827	120	4340	171 000
W-Re-Hf-C (hard drawn)	19.37	0.7	0.038	0.015	910	132	4800	189 000
W-Hf-C (in-process annealed)	19.37	0.7	0.038	0.015	765	111	4040	159 000
Astar 811C	16.9	0.61	0.051	0.020	260	38	1600	62 000
			0.038	0.015	355	51.5	2100	84 000
B-88	10.3	0.373	0.051	0.020	190	28	1900	75 000
			0.038	0.015	200	29	2000	78 000
W-2-ThO₂	18.91	0.68	0.038	0.015	480	69	2570	101 000

Summary of Wire Microstructure Observations as Functions of Exposure Time and Temperature (Ref 91)

Wire material	As drawn	Exposed at 1095 °C (2000 °F)		Exposed at 1205 °C (2200 °F)	
		Short time (<50 h)	Long time (>80 h)	Short time (<50 h)	Long time (>80 h)
W-Hf-C (hard drawn)	Heavily worked elongated grains; HfC particles, 0.015 to 0.040 μm in size	...	Grain-width increase and subgrain formation	Grain-width increase	Grain-width increase
W-Re-Hf-C (hard drawn)	Heavily worked elongated grains; small particles, 0.010 to 0.120 μm in size	Structure similar to as-drawn condition	Grain-width increase and some subgrain formation	Grain-width increase and subgrain formation	Grain-width increase
W-Hf-C (annealed during drawing)	Heavily worked elongated grains; small particles of HfC, 0.030 to 0.100 μm in size	Structure similar to as-drawn condition	Grain-width increase	Grain-width increase and subgrain formation	Formation of long wide grains
Astar 811C	Pronounced particle alinement; particles range in size from 0.015 to 1.0 μm	Structure similar to as-drawn condition	Particle coarsening; higher particle content than for short-time exposure	Particle coarsening	Particle coarsening and higher particle content than for short-time exposure; apparent loss of fibrous structure
B-88	Heavily worked elongated grains; particles range in size from 0.030 to 2 μm	Particle coarsening	Higher particle content than for short-time exposure; particle coarsening	Particle-content increase and subgrain formation	Particle coarsening and subgrain formation

Tensile Strengths of Tungsten Wires (Ref 76, p 8)

Wire material	Wire diameter cm	in.	Tensile strength at test temperature of: 21 °C (70 °F) MPa	ksi	1095 °C (2000 °F) MPa	ksi	1205 °C (2200 °F) MPa	ksi
W-2ThO$_2$-5Re	0.051	0.020	2137	310	1276	185	1014	147
W-2ThO$_2$	0.025	0.010	2751	399	1014	147	910	132
	0.038	0.015	2648	384	1193	173	1034	150
218 CS tungsten	0.038	0.015	2386	346	765	111	648	94
W-1ThO$_2$	0.051	0.020	2310	335	800	116	738	107

Stress-Rupture Properties of Tungsten Wires (Ref 76, p 9, 10)

Wire material	Wire diameter cm	in.	Test temperature °C	°F	Stress MPa	ksi	Life, h	Reduction in area, %
W-5Re-2ThO$_2$	0.051	0.020	1095	2000	690	100	18.4	28.6
					621	90	41.2	26.9
					621	90	41.2	31.9
					586	85	85.3	48.1
					552	80	144.3	47.4
			1205	2200	552	80	4.6	33.6
					552	80	5.2	27.7
					414	60	17.1	30.3
					345	50	55.3	...
					276	40	174.1	...
W-2ThO$_2$	0.025	0.010	1095	2000	827	120	14.0	...
					758	110	24.0	29.4
					738	107	36.5	11.6
					724	105	61.0	13.5
					690	100	(a)	...
			1205	2200	621	90	13.4	...
					552	80	34.1	15.4
					517	75	66.4	13.5
					483	70	118.7	29.4
	0.038	0.015	1095	2000	690	100	6.5	45.0
					669	97	43.3	26.2
					655	95	228.1	42.2
					641	93	218.5	38.2
			1205	2200	552	80	17.8	15.4
					517	75	26.1	14.2
					503	73	51.7	...
					483	70	89.6	29.3
					483	70	120.6	49.1
					448	65	116.0	20.4
					414	60	146.6	17.8
218 CS tungsten	0.038	0.015	1095	2000	379	55	86.8	2.8
					414	60	44.2	8.0
					483	70	9.5	45.2
					552	80	4.0	83.5
			1205	2200	276	40	105.9	5.5
					296	43	36.5	9.2
					310	45	29.3	10.4
W-1ThO$_2$	0.051	0.020	1095	2000	621	90	5.1	65.7
					586	85	7.7	56.5
					572	83	23.4	56.5
					552	80	133.5	68.7
					517	75	(b)	...
			1205	2200	310	45	279.1	7.0
					345	50	79.3	13.5
					379	55	33.9	20.0
					379	55	12.2	32.7
					414	60	19.5	36.0
					414	60	21.3	28.6

(a) Test stopped after 328.9 h. (b) Test stopped after 282.5 h.

Stress-Rupture Properties of Tantalum- and Niobium-Base Wire Materials (Ref 91)

Wire material	Wire diameter cm	Wire diameter in.	Test temperature °C	Test temperature °F	Stress MPa	Stress ksi	Rupture time, h	Reduction in area, %
Astar 811C (Ta-base) 0.051	0.051	0.020	1095	2000	690	100	7.3	17.5
					620	90	68.5	8.7
					590	85	43.0	7.0
					520	75	338.2	3.8
	0.038	0.015	1095	2000	620	90	14.6	29.8
					590	85	94.6	4.2
					570	82	19.1	19.3
					570	82	162.8	6.6
					550	80	338.2	7.4
					550	80	(a)	...
	0.051	0.020	1205	2200	350	50	10.8	8.8
					310	45	28.5	9.7
					280	40	78.3	8.3
					240	35	166.7	7.3
	0.038	0.015	1205	2200	520	75	10.2	15.3
					480	70	14.7	10.3
					410	60	45.4	5.2
B-88 (Nb-base) 0.051	0.051	0.020	1095	2000	380	55	20.1	6.5
					350	50	62.7	2.8
					310	45	391.9	<1.0
					380	55	4.1	32.8
					370	53	37.1	15.8
	0.038	0.015	1095	2000	350	50	101.3	18.6
					310	45	102.1	16.6
					350	50	44.8	22.1
					310	45	55.4	23.3
	0.051	0.020	1205	2200	280	40	199.3	20.7
					280	40	2.6	34.7
					240	35	14.4	39.9
					210	30	78.5	20.9
	0.038	0.015	1205	2200	240	35	25.4	32.0
					210	30	86.8	28.6
					170	25	224.1	26.1

(a) Test stopped at 348.9 h.

5.1.2.7. Polymer Fibers

Specific Tensile Strength and Specific Tensile Modulus of Aramid Fibers (Kevlar 29, Kevlar 49) Compared With Other Reinforcing Fibers (Du Pont Co.)

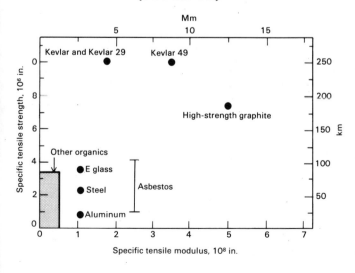

Mechanical Properties and Maximum Use Temperature of Kevlar 49 (Du Pont Co.)

Tensile strength:
  At room temperature (16 months) ....No loss in strength
  At 50 °C (120 °F) in air (2 months) ....No loss in strength
  At 100 °C (210 °F) in air, MPa (ksi) 3170 (460)
  At 200 °C (390 °F) in air, MPa (ksi) 2720 (395)
Tensile modulus:
  At room temperature (16 months) ....No loss in modulus
  At 50 °C (120 °F) in air (2 months) ....No loss in modulus
  At 100 °C (210 °F) in air, GPa (10^6 psi)...............114 (16.5)
  At 200 °C (390 °F) in air, GPa (10^6 psi)...............110 (16.0)
Long-term use temperature in air, °C (°F) 160 (320)
Decomposition temperature, °C (°F).....................500 (930)

Specific Heat of Kevlar 49 at Various Temperatures (Du Pont Co.)

Temperature °C	°F	Specific heat J/kg·°C	Specific heat Btu/lb·°F
0	32	1220	0.292
50	122	1600	0.383
100	212	1990	0.476
150	302	2370	0.565
200	392	2620	0.626
250	482	2740	0.654
300	572	2840	0.679

Mechanical Properties of Aramid, Polyamide, Polyester, and Nylon Fibers

Fiber	Density		Tensile strength		Tensile modulus		Ultimate elongation, %
	Mg/m³	lb/in.³	MPa	ksi	GPa	10⁶ psi	
Aramid-Kevlar 29	1.44	0.052	3620	525	83	12	4.4
Aramid-Kevlar 49	1.44	0.052	3620	525	124	18	2.9
Polyamide	1.13	0.041	830	120	2.8	0.4	...
Polyester-Dacron Type 68	1.38	0.050	1120	162	4.1	0.6	14.5
Nylon–Du Pont 728[a]	1.13	0.041	990	143	5.5	0.8	18.3
Spectra-900	0.97	0.035	2590	375	117	17	...

[a]Unimpregnated twisted yarn test–ASTM D2256.

Effect of Tension-Tension Fatigue on Aramid (Kevlar 29) Fibers (Du Pont Co.)

Cycled between (% of ultimate tensile strength)		No. of cycles	Break load after cycling		Decrease in tensile strength due to fatigue
High	Low		N	lb	
Control		...	552	124	...
74	45	1000	578	130	None
52	29	1000	610	137	None
31	8	1000	587	132	None
10	0	13 × 10⁶	525	118	5%

1500 denier (1670 dtex) 2-ply yarn of Kevlar 29 was tested using air-actuated 4-D cord clamps on an Instron test machine, at 254 mm (10 in.) original gage length, 10% per minute elongation, 55% R.H., and 22 °C (72 °F).

Coating Materials Used Successfully With Kevlar 29 Aramid Fiber (Du Pont)

Coating	Typical end uses
Neoprene synthetic rubber	Inflatable boats
Hypalon synthetic rubber	Pond liners, tarpaulins
Nitrile rubber	Pressure diaphragms
Nordel hydrocarbon rubber	Heat-resistant conveyor belts
Buna-N	Hoses
Urethane polymers	Inflatable structures
Silicon and fluorosilicon	Belting
Polyvinyl chloride	Air-supported structures
Teflon (TFE, FEP) fluorocarbon resin	Nonstick belts
Polyvinyl alcohol	Specialty uses
Laminations: Tedlar polyvinyl fluoride Mylar polyester	Lighter-than-air craft

Effect of Temperature on Tensile Strength of Aramid (Kevlar 29) Fiber (Du Pont Co.)

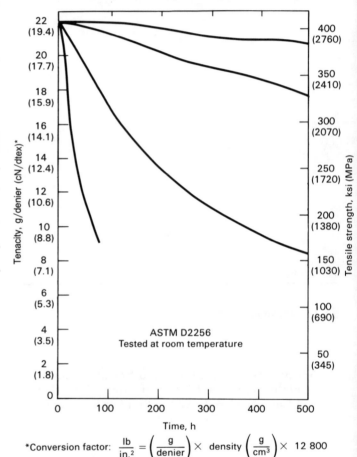

ASTM D2256
Tested at room temperature

*Conversion factor: $\dfrac{lb}{in.^2} = \left(\dfrac{g}{denier}\right) \times density \left(\dfrac{g}{cm^3}\right) \times 12\,800$

Chemical Resistance of Kevlar (Ref 34)

Chemical	Concentration, %	Temperature		Time, h	Strength loss, %	
		°C	°F		Kevlar 29	Kevlar 49
Hydrochloric acid	37	21	70	100	72	63
Hydrochloric acid	37	21	70	1000	88	81
Hydrofluoric acid	10	21	70	100	10	6
Nitric acid	1	21	70	100	16	5
Nitric acid	10	21	70	100	79	77
Sulfuric acid	10	21	70	100	9	12
Sulfuric acid	10	21	70	1000	59	31
Sodium hydroxide	10	21	70	1000	74	53
Ammonium hydroxide	28	21	70	1000	9	7
Acetone	100	21	70	1000	3	1

(continued)

Chemical Resistance of Kevlar, continued (Ref 34)

Chemical	Concentration, %	Temperature °C	Temperature °F	Time, h	Strength loss, % Kevlar 29	Strength loss, % Kevlar 49
Dimethyl formamide	100	21	70	1000	0	0
Methyl ethyl ketone	100	21	70	24	···	0
Trichloroethylene	100	21	70	24	···	1.5
Trichloroethylene	100	88	190	387	7	···
Ethyl alcohol	100	21	70	1000	1	0
Jet fuel (JP-4)	100	21	70	300	0	45
Jet fuel (JP-4)	100	200	390	100	4	···
Brake fluid	100	21	70	312	2	···
Brake fluid	100	113	235	100	33	···
Transformer oil (Texaco #55)	100	60	140	500	4.6	0
Kerosene	100	60	140	500	9.9	0
Freon 11 (Du Pont)	100	60	140	500	0	2.7
Freon 22 (Du Pont)	100	60	140	500	0	3.6
Tap water	100	100	212	100	0	2
Seawater (Ocean City, NJ)	100	···	···	1 year	1.5	1.5
Water at 10 000 psi	100	21	70	720	0	···
Water, superheated	100	138	280	40	9.3	···
Steam, saturated	100	150	300	48	28	···

Source: Du Pont.

Stress-Strain Curves at Various Temperatures for Spectra-900 (J. P. Stevens & Co.)

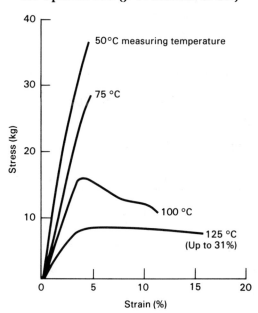

Stress-Strain Behavior of Aramid (Kevlar 29) Yarns Vs. Other Yarns (Du Pont Co.)

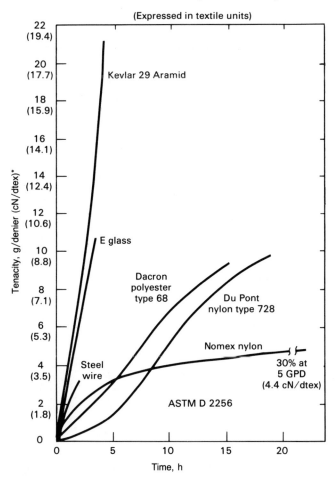

*Conversion factor: $\dfrac{lb}{in.^2} = \left(\dfrac{g}{denier}\right) \times density\left(\dfrac{g}{cm^3}\right) \times 12\,800$

Ultimate Elongation at Various Temperatures for Spectra-900 (J. P. Stevens & Co.)

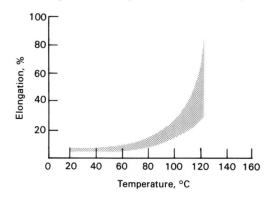

Tensile Properties of Spectra-900 at (a) Various Temperatures and (b) at Room Temperature After Annealing (J. P. Stevens & Co.)

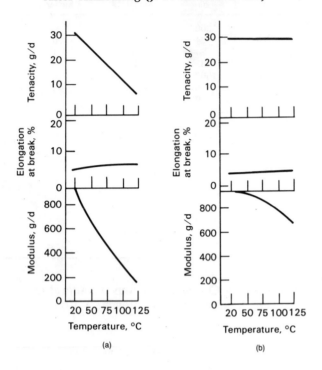

(a)

(b)

Shrinkage at Various Temperatures for Spectra-900 (J. P. Stevens & Co.)

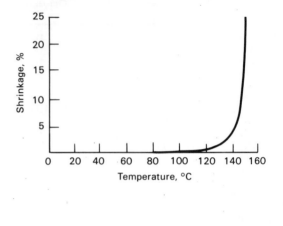

5.1.2.8. Comparative Tables of Selected Fibers and Whiskers

5.1.2.8a. Mechanical Properties

Comparative Mechanical-Property Data for Various Fibers (Ref 40, p 149)

Fiber	Melting or softening point, °F	Density (ρ), lb/in.³	Tensile strength (σ), 10³ psi	Specific strength (σ/ρ), 10⁶ psi	Young's modulus (E), 10⁶ psi	Specific modulus (E/ρ), 10⁶ psi	Typical cross section, μm[a]
Glass							
Type E....................	1290	0.092	500	5.4	10.5	114	10
Type S....................	1540	0.090	650	7.2	12.6	140	10
Type 4H-1	1650	0.096	730	7.6	14.5	151	···
SiO_2	3020	0.079	850	10.8	10.5	133	35
Polycrystalline							
Al_2O_3	3700	0.114	300	2.6	25	219	···
ZrO_2	4800	0.175	300	1.7	50	286	···
BN	5400	0.069	200	2.9	13	188	7
Graphite	6600	0.051–0.072	180–375	3.5–5.8	25–60	490–830	7
Multiphase[b]							
B	4170	0.095	400	4.2	55	578	100
B_4C	4440	0.085	330	3.9	70	824	···
SiC	4870	0.125–0.127	350–600	2.8–4.8	60–67	480–525	100–150
SiC on B	4170	≈0.1	400	≈4.0	55	≈550	108
TiB_2	5400	0.162	15	0.09	74	456	···
Metal							
W	6150	0.697	580	0.80	59	85	13->500
Mo	4750	0.369	320	0.90	52	141	25->500
René 41.................	2460	0.298	290	0.98	24	81	25
Steel...................	2550	0.280	600	2.1	29	103	13–up
Be	2340	0.066	185	2.8	35	530	127

[a]1 μm = 0.000039 in. [b]On 0.005-in.-diam tungsten-core wire.

Approximate Material Properties of Potential High-Strength Fibers (Ref 74, p 134, 135)

Material	Form	Elastic modulus E, 10^6 psi	Ultimate tensile strength, 10^3 psi	Approximate thermal expansion coefficient α, $10^{-6}/°C$	Melting point, °C	ρ (theoretical), g/cm³ (lb/in.³)	E/ρ, 10^6 in. (max)
Graphite	Whisker	145	3000	4	(3700)	2.10 (0.076)	1910
WC	Bulk	74–102	(50)	5	2600–2900	15.70 (0.568)	180
TiB₂	Bulk	54–77	19	9	2940 ± 50	4.5 ± 0.10 (0.163)	470
ZrC	Bulk	59–60 (69)	28	9	3400	6.56 (0.237)	250 (291)
Al₂O₃	Whisker	60	160–2600	6–9	2000 ± 25	3.98 (0.144)	440
BeO	Whisker	60 (114)	2000–2800	8–13	2550	3.01 (0.109)	550 (1050)
TiC	Bulk	45–60 (64)	68	9–10.2	2940–3250	4.92 (0.178)	250 (360)
SiC	Whisker	65(a)	450(b) (100–1650)	5.9	2700	3.22 (0.116)	565
B₄C	Whisker	65	965 (200–900)	4.5	2450	2.51 (0.091)	715
ZrB₂	Bulk	64(c) (50)	29	7.5	3060	6.10 (0.220)	290 (227)
W	Wire	59 (50)	400 (320–570)	4.6	3410	19.30 (0.698)	85
Be	Wire	42	220	13–18	1285	1.85 (0.067)	630
B	Fiber	64–66 (73)	350 (470–500) (1000)	···	2200	2.35 (0.083)B 2.66 (0.094)F	795 (775)
Steel	Wire	29 [21.5(d)]	500 (100–500)	11.76	1537	7.87 (0.284)	102 (102)
SiO₂	Fiber	8 (E-10.5; 994-12.2)	205–212(100–900)	4(e)	815 (softens)	2.54 (0.092)	87 (114–133)
C	Filament	6	180	···	3700	1.86 (0.067)	90
Bamboo	Fiber	4.9	50	···	···	(0.5) (0.018)	272

(a) Values reported from <40 to 123 × 10^6 psi. (b) Maximum UTS of SiC reported at 3 × 10^6 psi and 4.3% strain. (c) 91.5% dense, hot pressed. (d) Stainless steel. (e) Fused silica, 0.55 × $10^{-6}/°C$; 96% silica (vycor), 0.8 × $10^{-6}/°C$; borosilicate, 4.2 × $10^{-6}/°C$; aluminosilicate, 4.2 × $10^{-6}/°C$; Type E glass, 4.2 × $10^{-6}/°C$.

Some Fibers Used in Metal-Matrix Composites (Ref 86)

Fiber[a]	Fiber diameter μm	mils	Tensile strength MPa	ksi	Modulus GPa	10^6 psi	Use limit °C	°F	Price/lb (1985)
Boron (C)	102–103	4.0–8.0	3450	500	400	58	540	1000	$262
Carbon/graphite-PAN (C)	7.0	0.28	2410–4830	350–700	230–395	33–57	>1650[b]	>3000[b]	$17–$450
Carbon/graphite-pitch (C)	5.1–12.7	0.20–0.50	2070	300	380–690	55–100	>1650[b]	>3000[b]	$26–$1250
SiC monofilament (C)	140	5.51	4140	600	425	62	930	1700	$800
SiC (W)	6	0.24	3340	485	485–825	70–120	930	1700	$95
FP alumina (C)	20	0.79	1380	200	380	55	>1650	>3000	$200
Fiberfrax (DC)	2–5	0.08–0.20	1030	150	105	15	1150	2100	$1
Fibermax (DC)	3–6	0.12–0.24	830	120	150	22	1760	3200	$16.65

[a]C = continuous; W = whisker; DC = discontinuous. [b]Oxidation begins at lower temperatures.

Some Fibers Used in Plastic (Ref 86)

Fiber[a]	Fiber diameter μm	mil	Density Mg/m³	lb/in.³	Tensile strength MPa	ksi	Modulus GPa	10^6 psi	Use limit °C	°F	Price/lb (1985)
E glass (C)	3–20	0.12–0.79	2.49	0.000	3450	500	72.4	10.5	425	800	$0.80–$1.20
S glass (C)	10–20	0.39–0.79	2.49	0.090	4590	665	86.9	12.6	425	800	$4
Kevlar (C)	12	0.47	1.44	0.052	2760	400	125	18	425	800	$16
Carbon/graphite-PAN (C)	7.0	0.28	1.72–1.80	0.062–0.065	2410–4830	350–700	230–395	33–57	>1650[b]	>3000[b]	$17–$450
Carbon/graphite-pitch (C)	5.1–12.7	0.20–0.50	1.99–2.16	0.072–0.078	2070	300	380–690	55–100	>1650[b]	>3000[b]	$26–$1250
Processed mineral fiber (DC)	4–6	0.16–0.24	2.68	0.097	830	120	105	15	760	1400	$0.30–$0.50
Fiberfrax (DC)	2–5	0.08–0.20	2.60	0.094	1030	150	105	15	1150	2100	$1
Fibermax (DC)	3–6	0.12–0.24	2.99	0.108	860	125	150	22	1760	3200	$16.65

[a]C = continuous; DC = discontinuous. [b]Oxidation begins at lower temperatures.

Short-Time Elevated-Temperature Tensile Strengths of Various Fibers and Whiskers (Ref 95)

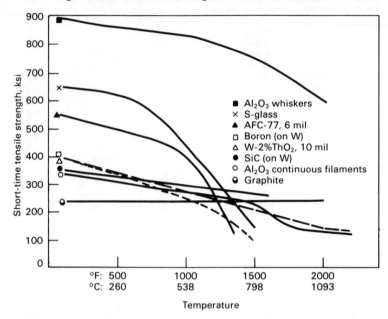

Mechanical Properties of Carbon, SiC, and Carbon-Boron Alloy Fibers (Ref 73)

Fiber	Nominal diameter μm	mils	Elastic modulus GPa	10⁶ psi	Gage length cm	in.	Mean UTS MPa	ksi	Standard deviation of UTS MPa	ksi	Mean strain to failure, cm/cm	Coefficient of variation
Carbon fiber substrate (as-received)	33	1.30	41.4	6.0	2.54	1.0	1280	186	283	41	0.03100	0.220
					10.16	4.0	848	123	124	18	0.02050	0.146
					25.4	10.0	786	114	221	32	0.01900	0.281
Carbon fiber substrate (etched)	27	1.05	37.2	5.4	2.54	1.0	1170	170	131	19	0.03148	0.112
					10.16	4.0	903	131	207	30	0.02426	0.229
					25.4	10.0	634	92	276	40	0.01704	0.435
SiC/C-core	142	5.60	406	58.9	2.54	1.0	4220	612	862	125	0.01039	0.204
					10.16	4.0	3360	488	924	134	0.00828	0.275
					25.4	10.0	2690	390	862	125	0.00662	0.321
SiC/C-core[a]	102	4.00	379[b]	55.0	0.635	0.25	5060	734	283	41	0.01335	0.056
					1.78	0.5	4900	711	338	49	0.01293	0.069
					2.54	1.0	4790	696	296	43	0.01266	0.062
					5.08	2.0	4340	630	607	88	0.01146	0.140
					7.62	3.0	3780	548	593	86	0.00997	0.157
					25.4	10.0	2850	413	669	97	0.00751	0.235
					50.8	20.0	2590	375	662	96	0.00682	0.256
					7.62	3.0	3920	569	648	94	0.01035	0.165
					25.4	10.0	2960	430	648	94	0.00782	0.219
SiC/W-core	142	5.60	434	63.0	2.54	1.0	3470	504	172	25	0.00801	0.050
					25.4	10.0	2900	420	441	64	0.00667	0.152
C-B alloys, group A[c]	86.4	3.40	177	25.7	0.635	0.25	2400	348	407	59	0.01354	0.170
					2.54	1.0	1590	231	558	81	0.00899	0.351
					10.16	4.0	1100	160	83	12	0.00623	0.075
					25.4	10.0	841	122	138	20	0.00475	0.164
C-B alloys, group B[c]	86.4	3.40	129	18.7	0.635	0.25	2320	337	345	50	0.01802	0.148
					2.54	1.0	1630	237	462	67	0.01267	0.283
					10.16	4.0	1230	179	310	45	0.00957	0.251
					25.4	10.0	1080	157	179	26	0.00840	0.166
C-B alloys, group C[c]	140	5.52	199	28.9	2.54	1.0	2780	403	621	90	0.01394	0.223
					25.4	10.0	1890	274	476	69	0.00948	0.252
C-B alloys, group D[c]	140	5.52	210	30.4	2.54	1.0	3740	543	724	105	0.01879	0.193
					25.4	10.0	3280	475	165	24	0.01644	0.051

[a]Tensile data from D. M. Kotchick, R. C. Hink, and R. E. Tressler, Gauge Length and Surface Damage Effects on the Strength Distributions of Silicon Carbide and Sapphire Filaments, *J. Compos. Mater.*, Vol 9, 1975, p 327–336. [b]Calculated from rule-of-mixtures. [c]See table at bottom right on p 5-33 for characteristics of groups.

Strengths of Several "High-Temperature" Fibers at Elevated Temperatures (Ref 8, p 330)

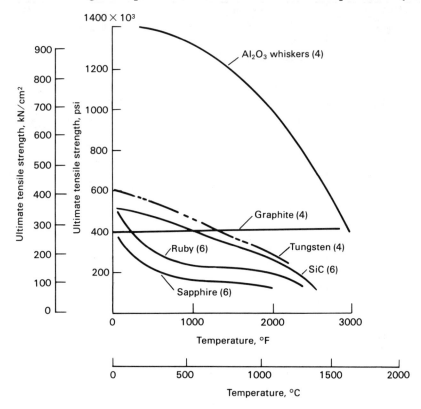

Some Filament Properties at Room and Elevated Temperatures (Ref 8, p 214)

Property	Type of filament				
	Borsic	SiC	Mo (TZM)	Sapphire	Be
Diameter:					
μm	107–145	102	102	254	127–1525
mils	4.2–5.7	4	4	10	5–60
Density:					
Mg/m³	2.768	3.460	10.186	3.958	1.855
lb/in.³	0.100	0.125	0.368	0.143	0.067
Modulus, GPa:					
Room temperature	415	450	295	470	295
205 °C (400 °F)	···	450	···	···	···
315 °C (600 °F)	···	450	···	···	···
425 °C (800 °F)	345	450	···	···	290
540 °C (1000 °F)	275	450	···	460	255
Modulus, 10^6 psi:					
Room temperature	60	65	43	68	43
205 °C (400 °F)	···	65	···	···	···
315 °C (600 °F)	···	65	···	···	···
425 °C (800 °F)	50	65	···	···	42
540 °C (1000 °F)	40	65	···	67	37
Tensile strength, MPa:					
Room temperature	2930	2275	2550	2550	690–965
205 °C (400 °F)	2860	2070	···	2035	550
315 °C (600 °F)	2655	2040	···	1895	345
425 °C (800 °F)	1795	2025	···	1795	240
540 °C (1000 °F)	1550	2015	···	1795	170
Tensile strength, ksi:					
Room temperature	425	330	370	370	100–140
205 °C (400 °F)	415	300	···	295	80
315 °C (600 °F)	385	296	···	275	50
425 °C (800 °F)	260	294	···	260	35
540 °C (1000 °F)	225	292	···	260	25
Coefficient of thermal expansion, room temperature to 540 °C (1000 °F):					
10^{-6}/°C	5.22	3.24	5.94	7.92	15.3
10^{-6}/°F	2.90	1.80	3.30	4.40	8.50

Textile Fibers Comparison Table (Ref 65)

Fiber type		Glass		Polyester			Nylon			
		Fiberglas		Dacron (Du Pont)			Nylon 6, 6			Nomex (Du Pont)
		E Glass	S Glass	Regular Tenacity	High Tenacity	Staple and Tow	Regular Tenacity	High Tenacity	Staple and Tow	Staple, Tow, and Filament
		Single filament		Filament			Monofilament Filament	Filament		
Breaking tenacity (gpd.)	Std.	15.3	19.9	2.8 to 5.6	6.0 to 9.5 / 6.0 to 9.5	2.2 to 6.0 / 2.2 to 6.0	2.3 to 6.0	5.9 to 9.5	3.5 to 7.2 / 3.0 to 6.1	4.0 to 5.3
	Wet	15.3	19.9	2.8 to 5.6		2.0 to 5.5	2.0 to 5.2	5.1 to 8.0	3.7 to 5.9	3.04 to 4.1
	Std. Loop			2.5 to 5.2		2.0 to 5.5	2.0 to 5.1	5.0 to 7.6	3.7 to 5.9	4.0 to 5.0
	Std. Knot						2.0 to 5.1	5.0 to 7.6		
Tensile strength (psi.)		450- to 550,000	650- to 700,000	50- to 99,000	106- to 168,000	39- to 106,000	40- to 106,000	86- to 134,000		90,000
Breaking elongation (%)	Std.	4.8	5.7	19 to 34	12 to 16	12 to 55	25 to 65	15 to 28	16 to 66	22 to 32
	Wet	4.8	5.7	19 to 34	12 to 16	12 to 55	30 to 70	18 to 32	18 to 68	20 to 30
Elastic recovery (%)		100	100	97 at 2% / 80 at 8%	100 at 1%	100 at 1%	100 at 5% / 98 to 100 at 10%	100 at 4% / 96 at 5% / 95 at 10%		
Average stiffness (gpd.)		320	380	12 to 27	46 to 82	12 to 17	5 to 24	21 to 58	10 to 45	
Average toughness		0.37	0.53	0.40 to 0.80	0.50 to 0.70	0.20 to 1.10	0.50 to 1.00	0.74 to 0.84	0.58 to 0.84	0.85
Specific gravity		2.54	2.48	1.38			1.13 to 1.14			1.38
Water Absorbency, %	70 F., 65% r.h.	None	None	0.4 or 0.8 (depends on type)			4.0 to 4.5			6.5
	70 F., 95% r.h.	Up to 0.3	Up to 0.3				6.1 to 8.0			12.5
Effect of Heat		Will not burn. Retains 95% tensile at 650 °F. Softens at 1,350 °F.	Will not burn. Retains 80% tensile at 650 °F. Softens at 1,560 °F.	Sticks at 445 °F. Melts at 482 °F.			Sticks at 445 °F. Melts at 480° to 500 °F. Yellows slightly at 300 °F, when held for 5 hrs.			Does not melt. Decomposes at 700 °F.
Effect of Acids and Alkalis		Resists most acids and alkalis.		Good resistance to most mineral acids. Dissolves with partial decomposition in concentrated solutions of sulfuric acids. Good resistance to weak alkalis. Moderate resistance to strong alkalis at room temperatures. Disintegrates in strong alkalis at boil.			Unaffected by most mineral acids, except hot mineral acids. Dissolves with partial decomposition in concentrated solutions of hydrochloric, sulfuric, and nitric acids. Soluble in formic acid. Substantially inert in alkalis.			Unaffected by most acids, except some strength loss after long exposure to hydrochloric, nitric, and sulfuric. Generally good resistance to alkalis
Effect of Bleaches and Solvents		Unaffected		Excellent resistance to bleaches and other oxidizing agents. Generally insoluble except in some phenolic compounds.			Can be bleached in most bleaching solutions. Generally insoluble in most organic solvents. Soluble in some phenolic compounds.			Unaffected by most bleaches and solvents except for slight strength loss from exposure to sodium chlorite.
Dyes Used		Resin-bonded pigment systems. Vat, acid, or chrome dyes will tint.		Disperse, developed, and cationic (for some types), with carrier or at high temperature.			Disperse, acid, and premetalized are usually preferred, but most other classes are also used.			Industrial yarn is nondyeable. Staple is dyeable with cationic dyes.
Resistance to Mildew, Aging, Sunlight, Abrasion		Excellent resistance to sunlight and aging. Not attacked by mildew. (Binder may be affected by mildew.)		Not weakened by mildew. Excellent resistance to aging and abrasion. Prolonged exposure to sunlight causes some strength loss but no discoloration.			Excellent resistance to mildew, aging, and abrasion. Prolonged exposure to sunlight causes some deterioration.			Excellent resistance to mildew and aging. A 50% strength loss after 60 weeks' exposure to sunlight. Good abrasion resistance
Typical Stress-Strain Curves (Load in gpd)										

Specific Tensile Strengths and Specific Moduli of Kevlar, High-Strength Graphite, Glass, and Polyester Fibers (Du Pont Co).

Property	Kevlar 29	Kevlar 49	High-strength graphite	E glass	S glass	Dacron polyester type 68
Specific tensile strength:						
10^7 cm	2.5	2.5	1.8	0.9	1.6	0.8
10^6 in.	10.1	10.1	7.1	3.8	6.4	3.3
Specific tensile modulus:						
10^8 cm	5.7	9.0	12.6	2.7	3.5	1.0
10^8 in.	2.3	3.5	5.1	1.1	1.4	0.40

Strength Vs. Modulus for Tungsten and Various Ceramics in Bulk, Fiber, and Whisker Forms (Ref 8, p 359)

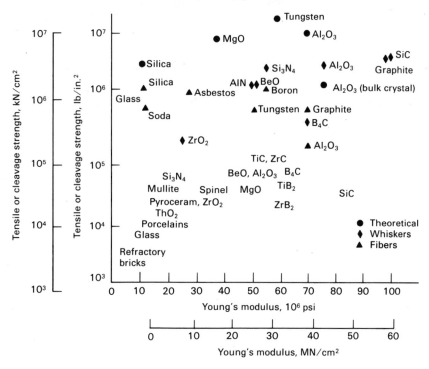

Comparative Mechanical Properties of Kevlar, Glass, and Graphite Fibers (Du Pont Co.)

Property	Kevlar 29	E glass	Graphite
Tensile strength[a]:			
MPa	3620	2415	2760
ksi	525	350	400
Tensile modulus[a]:			
GPa	82.7	68.9	220
10^6 psi	12	10	32
Density:			
g/cm^3	1.44	2.52	1.74
lb/in.^{3}	0.052	0.091	0.063
Brittleness	Tough	Brittle	Brittle
Abrasiveness	No	Yes	Yes

[a]Resin-impregnated strand test.

Specific Strength Vs. Specific Modulus for Reinforcing Fibers (Allied Fibers)

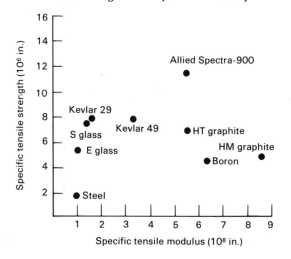

Room-Temperature Specific Strength and Specific Stiffness of Several Fibers (Ref 11, p 25)

Specific values in this figure were determined by dividing strength or modulus by density, expressed in lb/in.^{3} or kg/M^3.

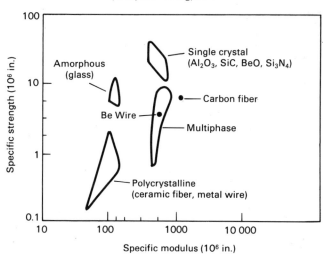

Characteristics of Carbon-Boron Alloy Fiber Groups (Ref 73)

Characteristic	Group			
	A	B	C	D
Carbon interlayer	No	No	Yes	Yes
Number of CVD reaction chambers	4	3	3	3
Boron content[a], wt %	43–46	35–39	46–48	48–51

[a]Unpublished research, D. L. McDanels, NASA Lewis Research Center.

5.1.2.8b. Thermal Properties

Thermal Properties of Selected Fibers

Fiber	Diameter, μm	Heat capacity, kJ/(kg·K)[a]	Thermal conductivity, W/(m·K)[b]	Coefficient of thermal expansion, 10^{-6}/°C
Graphite				
PAN HM	7	0.7	1003	−1.1
PAN HTS (T300)	8	0.7	1003	−1.1
Rayon (T50)	6	0.7	1003	−1.1
Thornel 75 (T75)	5	0.7	1003	−1.1
Pitch (type P)	5–10	0.7	1003	−1.1
Pitch UHM	11	0.7	1003	−1.1
Boron on tungsten	102–203	1.3	38	5.0
Borsic	102–203	1.3	38	5.0
Boron on carbon	102–203	1.3	38	5.0
Silicon carbide				
on tungsten	102–203	1.2	16	4.3
Silicon carbide				
on carbon	102	1.2	16	4.3
Beryllium	127	1.9	150	11.5
Alumina (FP)	20	...	...	8.3
S glass	9	0.7	13	5.0
E glass	9	0.7	13	5.0
Molybdenum	127	0.3	145	4.9
Steel	127	0.5	29	13.3
Tantalum	508	0.2	55	6.5
Tungsten	381	0.1	168	4.5
Whisker ceramic				
Al_2O_3	10–25	0.6	24	7.7
Metallic (Fe)	127	0.5	29	13.3

[a]To convert kJ/(kg·K) to Btu/(lb·°F), divide by 4.184. [b]To convert W/(m·K) to Btu·ft/(ft²·h·°F), divide by 1.729.

Specific Heats of Some Fibers and Materials

Fiber or material	Specific heat J/kg·°C	Specific heat Btu/lb·°F
Type A graphite fiber	710	0.17
HM graphite fiber	710	0.17
VHM graphite fiber	710	0.17
Aramid (type 49)	1380	0.33
S glass	755	0.18
Boron	840	0.20
Aluminum	960	0.23
Epoxy resin	1885	0.45

Longitudinal Coefficients of Thermal Expansion of Glass, Kevlar, and Carbon/Graphite Fibers

Fiber	Longitudinal coefficient of thermal expansion 10^{-6} m/m·K	10^{-6} in./in.·°F
E glass	5.0	2.8
S glass	5.6	3.1
Kevlar 29	−3.6	−2.0
Kevlar 49	−6.3	−3.5
High-strength graphite	−0.4	−0.2
High-modulus graphite	−0.5	−0.3
Ultrahigh-modulus graphite	−1.1[a]	−0.6[a]
Ceramic (AB-312)	2.9	1.6
Magnamite AS	−0.4	−0.2
Magnamite HTS	−0.4	−0.2
Magnamite HMS	−0.5	−0.3

[a]Estimated.

Coefficients of Thermal Expansion of Various Monofilaments and Yarns (Ref 18)

Fiber	Form[a]	Diameter μm	Diameter mils	Coefficient of thermal expansion, 10^{-6}/°C
Boron (W core)	M	100	4.0	4.7
Silicon carbide				
(C core)	M	140	5.6	4.4
Silica	M	50	2.0	0.3
Tungsten alloy	M	>250	>10	2.7
Steel	M	>50	>2	11
Carbon	Y	180–255	7–10	−0.36 to −1.8
Alumina type				
Du Pont FP	Y	510	20	5.7
3M Nextel-312	Y	255	10	2.6[b]
Silicon carbide	Y	255–380	10–15	3.1

[a]M = monofilament; Y = yarn. [b]2.8 at 150–240 °C; 3.3 at 250–350 °C; 3.0 at 360–435 °C.

5.1.3. Whisker Reinforcements

Melting Points and Specific Strengths and Moduli for Various Whiskers (Ref 40, p 149)

Fiber	Melting or softening point °C	°F	Specific strength (σ/ρ) MPa	10^6 psi	Specific modulus (E/ρ) MPa	10^6 psi	Typical cross section, μm^a
Ceramic							
Al_2O_3	2040	3700	28.95–168.9	4.2–24.5	2895–7239 (2992)[b]	420–1050 (434)[b]	0.5–10
BeO	2570	4660	126.9	18.4	3344	485	10–30
B_4C	2450	4440	151.0	21.9	5302	769	...
SiC	2690	4870	118.6–357.1	17.2–51.8	4137–8963	600–1300 (608)[b]	0.5–10
Si_3N_4	1900	3450	42.0–120.0	6.1–17.4	2399–3296	348–478	1–10
Graphite	3650	6600	326.8	47.4	11,722	1700	...
Metal							
Cr	1885	3430	34.5	5.0	924	134	...
Cu	1085	1980	9.0	1.3	386	56	...
Fe	1540	2800	46.2	6.7	703	102	...
Ni	1455	2650	11.7	1.7	662	96	...

[a] 1 μm = 0.000039 in. [b] Safe design value.

Strength as a Function of Size for Carefully Selected Alumina Whiskers (Ref 11, p 54)

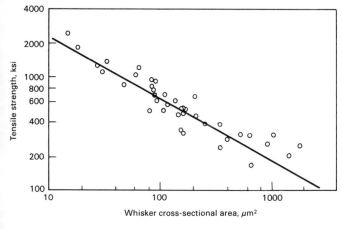

Strength as a Function of Size for Metal-Coated Al_2O_3 Whiskers Extracted From an Aluminum Matrix (Ref 11, p 43)

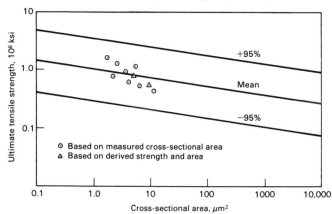

Mechanical Properties and Maximum Service Temperatures of Whiskers

	Whisker diameter μm	mils	Density kg/m³	lb/in.³	Tensile strength MPa	ksi	Tensile modulus GPa	10^6 lb/in.²	Maximum service temperature °C	°F
Al_2O_3	3–10	0.11–0.39	3958	0.143	20 685	3000	427	62	1650	3000
BeO...........	10–30	0.39–1.17	2851	0.103	13 100	1900	345	50	1925	3500
B_4C	...	...	2519	0.091	13 790	2000	483	70	1095	2000
SiC	1–3	0.03–0.11	3211	0.116	20 685	3000	483	70	600	1110
Si_3N_4	...	...	3183	0.115	13 790	2000	379	55	600	1110
Quartz	...	...	...	...	4 135	600	76	11	...	...
Sapphire	...	...	...	...	11 720	1700	510	74	...	...
Cr	...	...	7197	0.260	8 895	1290	241	35	540	1000
Cu	...	...	8913	0.322	3 275	475	124	18	260	500
Fe	...	...	7833	0.283	13 100	1900	200	29	540	1000
Ni	...	...	8968	0.324	3 860	560	200	29	540	1000
Graphite......	...	...	1661	0.060	19 615	2845	703	102	(a)	(a)
Au	...	...	...	...	1 655	240	76	11	...	...
Zirconia......	...	...	...	...	4 135	600	427	62	...	...
Silicon	...	...	...	...	3 790	550	158	23	...	...

(a) 600 °F in air; 5000 °F in inert atmosphere.

Methods Used to Fabricate Whisker Composites (Ref 77, p 328, 329)

Method used to combine constituents	Matrix	Whisker[a] Composition	Alignment	Coating	Consolidation process[b]	Forming or shaping process	Remarks
Deposition (molecular)							
Chemical vapor deposition	Ni or NiCr	Al$_2$O$_3$, SiC	R	C	H.P. or L.P.H.P.	Hot roll	SiC whiskers reacted chemically with matrix
Electro codeposit	Ni	Al$_2$O$_3$	R	C	H.P.		Problem of voids and low whisker content
	Ni	Al$_2$O$_3$, SiC	R	C, U			Problem of voids and low whisker content
	Ni	Al$_2$O$_3$	A	C, U	As deposited or hot forged		Align whiskers via flow in electroplating bath
Electroplate	Ni	Al$_2$O$_3$	A	C	H.P.		Problem of consolidation and whisker breakage
	Ni	Al$_2$O$_3$, SiC	A, R	C	Cold and hot roll, H.P.		SiC whiskers reacted with matrix, also whisker breakage
Electroform	Ni	Al$_2$O$_3$	A	C			Excellent properties, limited to very small specimens
Liquid (matrix)							
Alloy (eutectic)	Al, Nb	NiAl$_2$, Nb$_2$C	A	U	Unidirectional solidification	Extrude, roll	Whisker content fixed by alloy composition
Infiltration	Resin	Al$_2$O$_3$	A	C	Polymerize matrix		Achieved high-strength composites
	Ag	Al$_2$O$_3$	A	C	Solidification of matrix		Achieved high-strength composites
	Cu, Ni alloys	Al$_2$O$_3$	A	C	Solidification of matrix		Achieved varying degrees of success; coating stability a problem
	Al	B$_4$C	A	C	Solidification of matrix		Achieved varying degrees of success; coating stability a problem
	Al alloys	Al$_2$O$_3$	A	C	Solidification of matrix		Achieved varying degrees of success; coating stability a problem
Melting of powdered matrix	Fe, NiCr	Al$_2$O$_3$	R	C, U	L.P.H.P.		Problems of interfacial reactions and whisker segregation
Casting of melt and whiskers	Co alloys	Al$_2$O$_3$, SiC	R	C, U	Solidification of melt		Whiskers concentrated at grain boundaries, low whisker content
Spin, extrude, draw slurry of whisker, matrix powder, and carrier solution	Ag, Fe, Ni	Si$_3$N$_4$	A	U	Dry, burn off organic carrier, H.P.		Excellent whisker alignment and packing
	Al	SiC	A	U	Dry, burn off organic carrier, H.P.		Excellent whisker alignment and packing
	Al alloy	SiC	A	U	Dry, burn off organic carrier, L.P.H.P.		Excellent whisker alignment, fabrication of complex shapes
	Cu, Al, Mg	SiC	A	U	Dry, burn off, sinter, H.P.		Excellent whisker alignment
	Al alloy	SiC	A	U	Dry, burn off, H.P.		Excellent whisker alignment, problem of carbon burn-off
Filter slurry or settle out whiskers and matrix	Ag alloy	Si$_3$N$_4$	R	U	L.P.H.P.		Problem of matrix porosity and wetting
	Cu, Al alloy	Al$_2$O$_3$, SiC	R	U	H.P.	Hot roll	Little whisker alignment, much whisker breakage
	NiCr, Al alloy	SiC	A	C, U	L.P.H.P.	Clad and roll	Ni-coated whiskers were aligned magnetically during settling in slurry
	NiCr; Al alloy	Al$_2$O$_3$, SiC	A, R	C, U	L.P.H.P.	Clad and roll	Ni-coated whiskers were aligned magnetically during settling in slurry
	Mg alloy	Al$_2$O$_3$, SiC	A, R	C, U	L.P.H.P.	Hot extrusion	Successfully rolled composites with aligned whiskers

(continued)

Methods Used to Fabricate Whisker Composites, continued (Ref 77, p 328, 329)

Method used to combine constituents	Matrix	Whisker[a] Composition	Alignment	Coating	Consolidation process[b]	Forming or shaping process	Remarks
Impregnation into whisker strands and tapes	Resin	SiC	A	U	H.P. and squeeze out excess resin		Achieved excellent alignment, little whisker breakage
	Resin	Al_2O_3	A	U	H.P. and squeeze out excess resin		Achieved excellent alignment, little whisker breakage
	Resin	Si_3N_4	A	U	H.P. and squeeze out excess resin		Achieved excellent alignment, little whisker breakage
Solid state							
Powder matrix and whiskers	Ni	Al_2O_3	R	C	H.P.		Many whiskers broken
	Al	Si_3N_4	A	U	Extrude		Extended long rods, but many whiskers broken
	Ni, Ti	Al_2O_3, SiC	R	U	High-energy-rate forming		Chemical reaction minimized, but whiskers still broken
Deposit whiskers on Al foil	Al	SiC	A	U	Diffusion bond		Achieved excellent alignment of whiskers

[a]R = random whisker alignment. A = whiskers unidirectionally aligned. C = coated whiskers. U = uncoated whiskers. [b]H.P. = hot-pressing. L.P.H.P. = liquid-phase hot-pressing.

Differences Between Bulk and Whisker Strengths of Various Materials (Ref 56, p 78)

Material	Tensile strength GPa Bulk	Whisker	10^6 psi Bulk	Whisker
Iron	0.028	13.1	0.004	1.9
Copper	0.0014	2.8	0.0002	0.4
Silicon	0.034	3.79	0.005	0.55
Graphite	0.28	20.7	0.04	3.0
Boron carbide	0.1551	6.653	0.0225	0.965
Alumina	0.55	42.7	0.08	6.2
Silicon carbide	0.21	20.7[a]	0.03	3.0[a]

[a]Bend test (all other values are tensile data).

Strength as a Function of Size for SiC Whiskers (Ref 77, p 169)

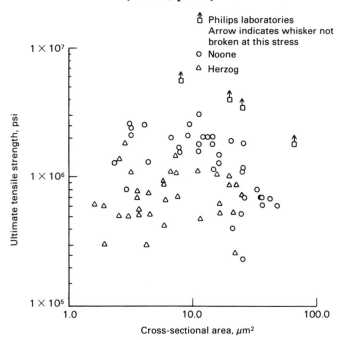

Growth Conditions for Silicon and Boron Whiskers (Ref 77, p 33)

Specimen	H_2	CO_2	N_2	$SiCl_4$	v(cm/s)	P_T(mm Hg)	Temperature, °C
SiO_2	0.80	0.20	···	0.02	1700	10	1450
Si_3N_4	0.60	···	0.2	0.02	2100	8	1600
	H_2	BCl_3	Cl_{2xs}				
B	1.40	0.02	0.02		7000	4	1700
B	0.75	0.03	0.03		9000	1	1500
B	0.75	0.03	0.04		5000	2	1600

Properties of Sapphire (Al₂O₃) and SiC Whiskers (Ref 56, p 156, 157)

Material	Whisker diameter		Aspect ratio, L/D	Tensile strength		Elastic modulus[a]		Density		Melting point	
	μm	mils		GPa	10⁶ psi	GPa	10⁶ psi	g/cm³	lb/in.³	°C	°F
Sapphire (Al₂O₃) Whiskers											
Loose needles	1–10	0.04–0.4	10–200	4.1–14	0.6–2.0	550–1035	80–150	3.97	0.143	2080	3780
Loose needles	1–30	0.04–1.2	10–200	1.4–4.1	0.2–0.6	415–690	60–100	3.97	0.143	2080	3780
Needles in powder form	1–3	0.04–0.1	10–200	14–24	2.0–3.5	690–2070	100–300	3.97	0.143	2080	3780
Cluster ball[b]	5–15	0.2–0.59	10–100	...	...	...	...	3.97	0.143	~2080	~3780
Mat form[c]	1–3	0.04–0.1	500–5000	14–24	2.0–3.5	690–2070	100–300	...	...	~2080	~3780
Mat form[c]	1–10	0.04–0.4	150–200	4.1–24	0.6–3.5	550–1035	80–150	...	...	~2080	~3780
Mat form.............	1–30	0.04–1.2	100–2000	1.4–24	0.2–3.5	415–690	60–100	...	...	~2080	~3780
β-Sic Whiskers											
Wool	1–10	0.04–0.4	1000–10 000	1.4–21	0.2–3.0	>690	>100	3.2	0.12	~2430[d]	~4400[d]
Mixed Whiskers											
50% SiC + 50% SiC/ Al₂O₃ bulk crystals..	1–5	0.04–0.2	...	...	...	...	...	3.2	0.12	...	...
Al₂O₃ major + AIN minor (submicron rods)..............	~0.2–1.0	~0.008–0.039	...	...	...	...	...	...	...	~2080	~3780
Al₂O₃ major + AIN minor (cluster ball– polycrystalline).....	5–15	0.2–0.59	...	...	...	...	...	...	...	~2080	~3780
AIN major + Al₂O₃ minor.............	3–30	0.1–1.2	100–200	3.4	0.5	345	50	3.6	0.13	~2200	~4000

[a]Estimated. [b]A 3D cluster of fine whiskers; also contains a minor amount of AIN whiskers. [c]3/16 × 4 × 15 in. (4.8 × 100 × 380 mm). [d]Sublimes.

Conditions for Al₂O₃ Whisker Growth by Hydrogen Reduction (Ref 77, p 26)

Factor studied	Temperature		Hydrogen flow rate		Hydrogen dew point		Growth appearance	Remarks
	°C	°F	m³/h	cfgh[a]	°C	°F		
Temperature	1299	2370	0.022	0.8	−45	−50	Fuzz	Growth produced directly above charge inside boat
	1399	2550	0.022	0.8	−45	−50	Fuzz	Growth produced directly above charge inside boat
	1499	2730	0.022	0.8	−45	−50	Wool and coarse	Coarse growth produced directly above charge inside boat; wool nucleated on furnace tube wall
	1599	2910	0.022	0.8	−45	−50	Wool and coarse	Coarse growth directly above charge inside boat
Dew point	1499	2730	0.022	0.8	+24	+75	Fine powder	Growth produced on charge inside boat
	1499	2730	0.022	0.8	−4	+25	Fine powder	Growth produced on charge inside boat
	1499	2730	0.022	0.8	−19	−2	Very coarse	Growth produced over charge inside boat
	1499	2730	0.022	0.8	−34	−29	Coarse and wool	Coarse growth produced over charge inside boat; wool nucleated on furnace tube wall
	1499	2730	0.022	0.8	−45	−50	Coarse and wool	Coarse growth produced over charge inside boat; wool nucleated on furnace tube in much greater quantities than at −34 °C (−29 °F) dew point
	1499	2730	0.022	0.8	−59	−75	Fuzz	Growth produced directly over charge inside boat
Flow rate	1499	2730	0.009	0.33	−45	−50	Coarse and wool	Coarse growth produced over charge inside boat; wool nucleated on furnace tube wall in large quantities
	1499	2730	0.021	0.75	−45	−50	Wool	Wool nucleated on furnace tube wall in large quantities
	1499	2730	0.042	1.50	−45	−50	Coarse and wool	Coarse growth produced over charge inside boat in large quantities; the wool nucleated on furnace tube surface in medium quantities
	1499	2730	0.084	3.00	−45	−50	Very coarse	Very coarse growth produced over charge inside boat

[a]Cubic feet of gas per hour.

Elastic Modulus Vs. Cross-Sectional Area for SiC Whiskers (Ref 77, p 172)

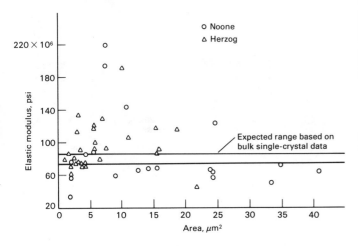

Properties of SiC (SCW[a]) Whiskers (Tateho Chemical Industries, Ltd.)

Crystalline phase	β-type
Diameter, μm	0.05–1.5
Length, μm	20–200
Aspect ratio	20–200
Density, g/cm³	3.18
Chemical composition, %:	
Magnesium	0.2 max
Calcium	0.5 max
Aluminum	0.2 max
Iron	0.1 max

[a]Tateho Silicon Carbide Whisker.

Size Categories for α-Al₂O₃ Whiskers (Ref 12, p 653)

Category	Cross-sectional area, μm²	Diameter, μm[a]	Predicted strength GPa	Predicted strength 10⁶ psi
Fine	3–60	1–9	5.5–21	0.8–3
Medium	60–400	9–23	2.1–5.5	0.3–0.8
Coarse	400–3500	23–67	0.7–2.1	0.1–0.3
X-Coarse	>3500	>67	<0.7	<0.1

[a]1 μm = 0.000039 in.

Properties of Si₃N₄ (SNW[a]) Whiskers (Tateho Chemical Industries, Ltd.)

Crystalline phase	α-type
Diameter, μm	0.1–1.6
Length, μm	20–200
Aspect ratio	20–200
Density, g/cm³	3.18
Chemical composition, %:	
Magnesium	0.2 max
Calcium	0.5 max
Aluminum	0.6 max
Iron	0.1 max

[a]Tateho Silicon Nitride Whisker.

5.2. Other Reinforcements, Additives, and Extenders

5.2.1. Textile, Paper, and Felt Reinforcements

5.2.1.1. Fabric Weaves, Composite Fabrication Combinations, and Military Specifications

Some Military Specifications for Fabric Reinforcements

MIL-Y-1140	Yarn, Code, Sleeving, Cloth, and Tape–Glass
MIL-C-9084C	Cloth, Glass, Finished, for Polyester Resin Laminates
MIL-C-20079	Fabrics for Thermal Insulation Components, i.e., Navy Board Facing
L-P-383	Plastic Material, Polyester Resin, Glass Fiber Base, Low Pressure Laminated
MIL-A-23054 (Ships)	Acoustical Absorptive Board, Fibrous Glass, Perforated Fibrous Glass Cloth, Faced

Kevlar Fabric Styles (Hexcel-Trevarno)

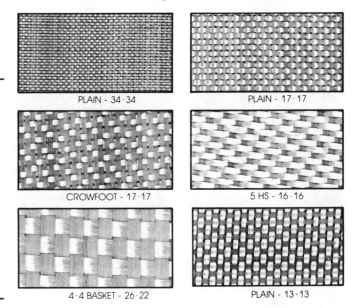

Fabric Weaves and Construction (Hexcel-Trevarno)

Both conventional and specialty perfected looms can be used to interlace warp (lengthwise) yarns and fill (crosswise) yarns. A fabric construction of 16 × 14 means 16 ends per inch in the warp direction and 14 pick yarns per inch in the fill direction.

The most commonly used weaves for industrial purposes are plain, leno, crowfoot satin, and twill.

PLAIN
In the plain weave, yarns are interlaced over and under each other in an alternating fashion. The plain weave gives the most stability and firmness to the fabric and the least yarn slippage.

LENO
When relatively few yarns are specified, the leno weave is used. In the leno weave, two or more warp threads cross over each other and interlace with one or more fill threads. It maintains uniformity of yarns and minimizes distortion.

CROWFOOT SATIN (Four Harness Satin)
In the crowfoot satin weave, a single warp yarn weaves over three and under one fill yarn. Fabrics woven to a crowfoot satin weave are pliable and conform easily to contoured surfaces.

TWILL
The twill weave, which is characterized by a diagonal rib called a twill line, provides for more yarns per unit than a plain weave.

Possible Fabric/Tape/Mat Composites (Union Carbide Corp.)

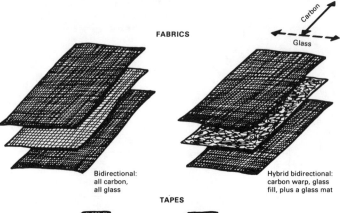

FABRICS

Bidirectional: all carbon, all glass

Hybrid bidirectional: carbon warp, glass fill, plus a glass mat

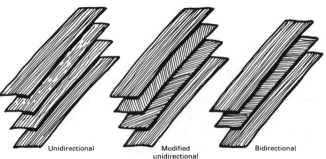

TAPES

Unidirectional

Modified unidirectional

Bidirectional

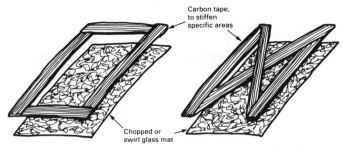

SELECTIVE HYBRIDIZATION

Carbon tape, to stiffen specific areas

Chopped or swirl glass mat

Graphite and Ceramic Fabric Styles (Hexcel-Trevarno)

Graphite

PLAIN - 10.5 · 10.5

PLAIN - 12.5 · 12.5

8 HS - 24 · 24

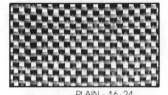
PLAIN - 16 · 24

Ceramic

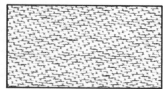
5 HS - 48 · 47

5.2.1.2. Fabric Reinforcement Property Data (Includes Papers and Felts)

Some Commercial Fabric Specifications and Uses (Fiberite)

Fiber	Weaver	Style	Weight, g/m²	Cured ply, mils	Comments
Graphite/Kevlar 49	Fiberite	8HS	340	13	50% graphite/50% Kevlar hybrid fabric
Graphite	Fiberite	8HS	367	13	Standard on missiles and commercial aircraft
Graphite	Fiberite	Plain	191	7	Standard on spacecraft and Space Shuttle
Graphite	Fiberite	Plain	194	7	Standard on commercial aircraft
Graphite	Fiberite	Plain	128	5	Standard on spacecraft, Uses 1K tow
Graphite	Fiberite	5HS	285	10	European standard
Graphite	Fiberite	5HS	375	13	Uses 1.5% strain 6K
Graphite	Fiberite	5HS	145	6	Uses pitch 75, 1K
Graphite/S-2 glass	Fiberite	Plain	199	7	All graphite warp uni-fabric (95% weight graphite)
Graphite/S-2 glass	Fiberite	Plain	250	8	50% graphite/50% S-2 glass hybrid fabric
Graphite	Fiberite	Basket	652	25	Uses 12K graphite
Kevlar 49	Clark-Schwebel	Cr. Satin	176	10	Standard for structure
Kevlar 49	Clark-Schwebel	Plain	59	4	Standard for structure
E-Glass	Clark-Schwebel	Plain	20	1	For scrim on uni-tape
E-Glass	Clark-Schwebel	Plain	35	1.5	For scrim on uni-tape
E-Glass	Clark-Schwebel	Plain	50	2	For scrim on uni-tape
E-Glass	Clark-Schwebel	Cr. Satin	110	5	For corrosion barrier; interiors
E-Glass	Clark-Schwebel	Cr. Satin	307	9	Uni-fabric, 90% weight in warp
E-Glass	Clark-Schwebel	12HS	1,380	45	For tooling
Nextel	3M	5HS	407	26	Standard for firewalls
Nicalon SiC	Nippon Carbon	5HS	400	27	Standard weave

Construction Properties of Glass Fabrics[a] (Hexcel-Trevarno)

Style, bidirectional	Standard width[b], in.	Average yards per roll	Weave	Yarn	Construction Warp × Fill	Thickness, mils	Weight, oz/yd²	Tensile strength, lb/in. Warp	Fill
108	38	500	Plain	900	60 × 47	2	1.43	70	40
112	38	500	Plain	450	40 × 39	3	2.09	82	80
116	38, 50	500	Plain	450	60 × 58	4	3.16	123	120
120	38	250	Crowfoot	450	60 × 58	4	3.40	125	120
128	38, 50	250	Plain	225	42 × 32	7	6.00	250	200
181[c]	38	125	Satin	225	57 × 54	8.5	8.90	340	330
182	38	125	Satin	225	60 × 56	13	12.40	440	400
183	38	125	Satin	225	54 × 48	18	16.75	650	620
184	38	75	Satin	225	42 × 36	27	25.90	950	800
332	49	125	Crowfoot	37	48 × 32	15	13.00	600	400
1527	44	125	Plain	150	17 × 17	15	12.90	535	485
1528	38, 50	250	Plain	150	42 × 32	7	6.00	250	200
1543	38[d]	200	Crowfoot	Mixed	49 × 30	8.6	9.40	660	70
1557	38[d]	250	Crowfoot	Mixed	57 × 30	6	5.39	370	60
1581	38, 72	125	Satin	150	57 × 54	8	9.00	350	325
1582	38	125	Satin	150	60 × 56	13	13.10	440	400
1583	38	125	Satin	150	54 × 48	18	16.42	650	620
1584	38	75	Satin	150	44 × 35	24.6	24.16	950	800
7781[c]	38, 50	125	Satin	75	57 × 54	9	8.95	350	340

[a]Values are given as a guide only and are not to be considered as specifications. All styles except 332 comply with MIL-Y-1140. [b]Others available. [c]7781 is a commercial version of mil spec 181 style fabric. [d]Unidirectional.

Thermal Conductivity of Kevlar Felts and Fabrics Vs. Other Materials (Du Pont Co.)

Fabric	Weight g/m²	Weight oz/yd²	Thickness mm	Thickness mils	Lag time, s	Thermal conductivity, cal/cm²·s·°C	Temperature rise in 25 s °C	Temperature rise in 25 s °F
Kevlar 29 (1 ply)	333	9.8	0.76	30	0	0.324	60	108
Kevlar 29 (3 ply)	998	29.4	2.16	85	3	0.162	30	54
Kevlar 29 (felt)	917	27.0	2.67	105	1.5	0.084	16	28
Fiberglass (1 ply)	285	8.4	0.30	12	0	0.600	111	200
Fiberglass (8 ply)	2282	67.2	2.16	85	5.1	0.105	19	35
Asbestos	1386	40.8	2.29	90	2.5	0.168	·31	55

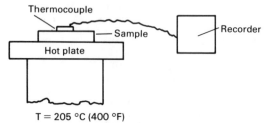

Schematic of Test Apparatus for Determining Lag Time

T = 205 °C (400 °F)

Lag time is time between placing sample in contact with hot plate and any perceptible recorder readout.

Some Typical Fabric Dimensions (Hexcel-Trevarno)

Construction	Weave	Thickness, mils	Width, in.	Weight, oz/yd²
Graphite				
12.5 × 12.5	Plain	7.2	42	5.7
24.0 × 24.0	Satin	13.5	42	10.9
10.5 × 10.5	Plain	6.0	42	5.5
16.0 × 24.0	Plain	6.1	38	4.7
Kevlar				
34 × 34	Plain	4.5	38	1.8
50 × 50	Satin	11	38	5.0
22 × 22	Plain	4.5	38	2.2
17 × 17	Plain	10	38, 50	5.0
17 × 17	Crowfoot	10	38, 50	5.0
13 × 13	Plain	10	50	5.0
16 × 16	Satin	13	50	9.0
28 × 28	Basket	20	50	10.5
26 × 22	Basket	26	44, 50, 60	13.5
17 × 30	Plain	7	38	3.1
S Glass				
24 × 22	Plain	5.5	38	3.7
18 × 18	Plain	9	38	5.8
48 × 30	Crowfoot	9	38	8.8
57 × 30	Crowfoot	5.5	38	5.4
57 × 54	Satin	8.5	38	8.9
Ceramic				
48 × 47	Satin	9.0	38	7.5

Comparative Textile Fiber Properties (Ref 90)

	Spectra 900	Spectra 1000	Aramid LM	Aramid HM	Carbon HT	Carbon HM
Denier/Number of filaments	1200/118	650/120	1500/1000	1500/1000	1730/3000	1630/3000
Tenacity, g/d	30	35	22	22	20	14
Elongation, %	3.5	2.7	3.6	2.8	1.2	0.6
Tensile modulus, g/d	1400	2000	488	976	1500	2400
Shrinkage at boil, %	<1	<1				
Specific gravity	0.97	0.97	1.44	1.44	1.73	1.81
Melting point, °C	147	147				
Filament size, μm	38	27	12	12	7.0	6.5

Carbon/Graphite Cloth (Union Carbide Corp.)

General Description

Carbon and graphite cloths, consisting entirely of flexible filaments, are produced by pyrolysis of rayon cloth at high temperatures to yield products with a high degree of purity. The cloth flexibility results from the very small carbon and graphite filament diameter of 8.9 μm (0.00035 in.). Single filaments from carbon or graphite cloths have tensile strengths of 345 to 690 MPa (50 to 100 ksi) and Young's Modulus of about 28 GPa (4×10^6 psi). The average breaking strength of the cloth is 48 to 165 MPa (7 to 24 ksi) depending on grade.

Graphite cloths and most carbon cloths are fair conductors of electricity. At room temperature, the volume resistivity of graphite cloth is about 40 times that of "Nichrome" wire. Graphite cloth resistance decreases with temperature, reaching half the room-temperature value at 1315 °C (2400 °F).

Reactivity

Graphite cloth is stable in vacuum and inert atmospheres up to temperatures approaching the sublimation temperature 3650 °C (6600 °F). Constituents of neutral or reducing atmospheres may react with graphite cloth at high temperatures. For example, pure hydrogen begins to react appreciably with graphite at about 540 °C (1000 °F). Graphite begins to oxidize in steam at about 730 °C (1350 °F) and in carbon dioxide at about 900 °C (1650 °F).

Applications

Carbon cloth/graphite cloth reinforced plastics are used in a variety of applications requiring thermal stability, high-temperature strength, good ablation characteristics, and insulation capability with an accompanying low density (on the order of 1.40 g/cm^3, or 0.05 lb/in.^{3}). Since graphite is somewhat more resistant than carbon and physically stable at elevated temperatures, graphite reinforcements are used in rocket-nozzle throats and ablation chambers. Carbon cloths are used when greater strength and lower conductivity are required such as in re-entry vehicles, rocket-nozzle entrance sections and exit cones, and critical insulation areas.

Some Carbon/Graphite Fabric Constructions and Dimensions (Hexcel-Trevarno)

Designation	Fiber	Weave	Yarn (tow) count Warp Fill	Fabric weight g/m²	oz/yd²	Thickness μm	mils
Bidirectional Fabrics							
F 1T 093 ....1K, T300		5-Harness Satin	17.8 × 17.8	93	2.74	24	3.5
F 3T 168 ....3K, T300		Plain	8 × 8	122	3.6	31	4.5
F 3T 272 ....3K, T300		4-Harness Satin	12 × 12	186	5.5	41	6.0
F 3T 282 ....3K, T300		Plain	12.5 × 12.5	193	5.7	50	7.2
F 3C 282 ....3K, Celion		Plain	12.5 × 12.5	224	6.6	55	8.0
F 3T 584 ....3K, T300		8-Harness Satin	24 × 24	370	10.9	93	13.5
F 3C 584 ....3K, Celion		8-Harness Satin	24 × 24	407	12.0	101	14.7
F 6C 510 ....6K, Celion		5-Harness Satin	10 × 10	338	9.98	83	12.0
F 6C 676 ....6K, Celion		Crowfoot Satin	15 × 2(16)	516	15.22	128	18.6
F 12C 688 ....12K, Celion		Plain	8 × 8	532	15.7	128	18.6
F 5T 699 ....15K, Celion		Basket	9.25 × 9.25	678	20.0	207	30.0
Unidirectional Fabrics							
F 1T 712 ....Warp, 1K, T300 Fill, 225-1/0 glass		4-Harness Satin	46 × 10	136	4.02	36	5.2
F 3T 496 ....3K, T300		Plain	35 × 6	315	9.3	80	11.6
F 3C 716 ....Warp, 3K Fill, 150-1/0 glass		Plain	16 × 24	159	4.70	42	6.07
F 3T 782 ....Warp, 3K, T-300 Fill, 900-¹/₂ glass		Plain	40 × 8	353	10.4	90	13.0

Tear and Tensile Strengths of Kevlar 49 Fabrics (Du Pont Co.)

Fabric designation	Weave	Trapezoidal tear strength lb	N	Tensile strength lb/in.	kN/m	Tongue tear strength lb	N
120 ....Plain		22/22	98/98	250/250	44/44	60/60	267/267
181 ....8 Harness Satin		56/56	249/249	700/700	123/123	110/110	489/489
143 ....Crowfoot		15/70	67/311	1300/125	228/22	...	...
243 ....Crowfoot		20/100	89/445	1500/300	263/53	...	...
281 ....Plain		43/43	191/191	650/650	114/114	105/105	467/467
285 ....Crowfoot		40/40	178/178	650/650	114/114	...	...
328 ....Plain		65/65	289/289	700/700	123/123	120/120	534/534

Strength[a] in warp/fill direction

[a]ASTM test methods: D1910—weight; D1117—tensile strength; D2261—tongue tear strength; D2263—trapezoidal tear strength.

Typical Properties of Thornel Carbon and Graphite Cloth (Union Carbide Corp.)

Property	Carbon		Graphite
Construction			
Weave	5 Harness Satin	8 Harness Satin	Plain
Width, cm (in.)	109–114 (43–45)	107–114 (42–45)	107–113 (42–44.5)
Weight, g/m² (oz/yd²)	254 (7.5)	271 (8.0)	244 (7.2)
Gage, mm (in.)	0.457 (0.018)	0.508 (0.020)	0.559 (0.022)
Count, yarns/cm (yarns/in.):			
Warp	15.7 (40)	20.9 (53)	11.4 (29)
Fill	15.0 (38)	20.5 (52)	8.3 (21)
Filaments/yarn bundle	980	720	1470
Filament diameter, mm (in.)	0.0089 (0.00035)	0.0089 (0.00035)	0.0089 (0.00035)
Physical			
Tensile strength, N/m (lb/in.):			
Warp	6130 (35)	7880 (45)	10 500 (60)
Fill	5250 (30)	8760 (50)	4380 (25)
Surface area, m²/g (Kr)	1	1	1
Emissivity	0.9	0.9	0.9
Density, g/cm² (oz/in.²):			
(He)	1.65 (0.375)	1.75 (0.398)	1.42 (0.323)
(Hg)	1.47 (0.334)	1.44 (0.328)	1.42 (0.323)
Chemical			
Carbon assay, %	99	97	99.9
Ash, %	0.5	0.75	0.01
pH	9.5	8.6	7.9
Electrical			
Resistivity at 20 °C (70 °F), $\Omega \cdot cm^2$ ($\Omega \cdot in.^2$):			
Warp	0.068 (0.44)	0.078 (0.50)	0.060 (0.39)
Fill	0.071 (0.46)	0.087 (0.56)	0.082 (0.53)

Typical Properties of Carbon and Graphite Felts (Fiber Materials, Inc.)

Property	CH grade[a]	GH grade[b]
Weight per unit area per 25.4 mm (1 in.) of thickness, kg/m² (lb/yd²)	2.4 (4.4)	2.3 (4.1)
Carbon content, wt %	94 min	99.7 min
Ash content, wt %	0.8	0.2
Thermal conductivity at 50 °C (120 °F), W/m·K (Btu·in./h·ft²·°F)	0.055 (0.37)	0.069 (0.48)
Specific heat, J/kg·K (Btu/lb·°F), at:		
20 °C (70 °F)	700 (0.17)	700 (0.17)
1000 °C (1830 °F)	1900 (0.45)	1900 (0.45)
2000 °C (3630 °F)	2100 (0.50)	2100 (0.50)
Electrical resistivity at 20 °C (70 °F), $10^{-3}\ \Omega \cdot cm$ ($10^{-3}\ \Omega \cdot in.$):		
Longitudinal	220–330 (90–130)	180–280 (70–110)
Transverse	330–350 (130–140)	190–210 (75–80)
Tensile strength at 20 °C (70 °F), MPa (psi):		
Longitudinal	0.5 (70)	0.09 (13)
Transverse	0.35 (40)	0.06 (8)
Splitting resistance at 20 °C (70 °F), N/m (lb/in.):		
Longitudinal	140 (0.8)	50 (0.3)
Transverse	120 (0.7)	25 (0.1)
Linear shrinkage, %	1	0

[a]CH carbon grade–heat treated at 1250 °C (2280 °F) with a minimum carbon content of 94%. [b]GH graphite grade–heat treated at 2300 °C (4170 °F) with a minimum carbon content of 99.7%.

Carbon Fiber Paper Reinforcement (Panex CFP 30-05) Properties (Stackpole Fibers Co.)

Fiber tensile strength, MPa (ksi)	2415 (350)
Fiber tensile modulus, GPa (10^6 psi)	205 (30)
Fiber density, g/cm² (oz/in.²)	1.73 (0.39)
Paper areal weight, g/m² (oz/yd²)	14 (0.413)
Electrical resistivity, $\Omega \cdot m$ ($\Omega \cdot in.$)	17.78 (0.07)

Thermal Conductivity of CH Grade Felt (130 kg/m³ or 8.2 lb/ft³)

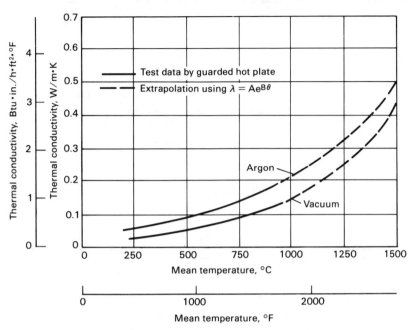

Thermal Conductivity of GH Grade Felt (100 kg/m³ or 6.2 lb/ft³)

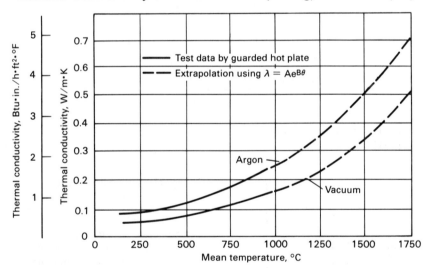

The thermal conductivity of CH and GH grade felts to elevated temperatures in both argon and vacuum environments is presented in the graphs above as a function of mean temperature. The curves are based on measurements to 1700 °F (850 °C) using a guarded hot plate apparatus (ASTM C177) with extrapolation to high temperature using a well-known exponential relationship for thermal conductivity of fibrous materials, i.e.:

$$\lambda = Ae^{B\theta}$$

The respective values of A and B are given in the table below.

Respective Values of A and B in the Exponential Relationship $\lambda = Ae^{B\theta}$

| | W/m·K unit | | | | Btu·in./h·ft²·°F unit | | | |
| | CH grade[a] | | GH grade[b] | | CH grade[a] | | GH grade[b] | |
Parameter	Argon	Vacuum	Argon	Vacuum	Argon	Vacuum	Argon	Vacuum
A	0.041	0.017	0.068	0.041	0.275	0.112	0.460	0.276
B	1.7×10^{-3}	2.2×10^{-3}	1.3×10^{-3}	1.4×10^{-3}	9.6×10^{-4}	1.2×10^{-3}	7.5×10^{-4}	8×10^{-4}
θ	°C	°C	°C	°C	°F	°F	°F	°F

[a]CH carbon grade–heat treated at 1250 °C (2280 °F) with a minimum carbon content of 94%. [b]GH graphite grade–heat treated at 2300 °C (4170 °F) with a minimum carbon content of 99.7%.

Abrasion Resistance of Refractory Textiles (3M)

Textile		Weight g/m²	oz/yd²	Thickness mm	in.	Stoll flex and abrasion resistance[a], cycles to failure With sizing	Without sizing
Leached silica	Harness satin	1220	36	1.37	0.054	80	70
Nextel 312	Harness satin	862	25	0.99	0.039	1580	400
Fused silica	Harness satin	285	8.4	0.35	0.014	25	Too brittle

[a]The samples were heated at 800 °C (1470 °F) for $1/2$ h to remove sizing or finish. The standard wear bar on the Stoll tester was replaced with a $1/4$-in. hex bar for the tests on samples without sizing. The tension applied to the sample was 0.23 kg ($1/2$ lb) for all tests.

Typical Properties of Nextel Commercial Ceramic Woven Fabrics (3M)

Style	Weight[a], g/m² (oz/yd²)	Available widths, m (in.)	Thickness[b], mm (in.)	Thread count[c] per inch, cm (in.) Warp	Fill	Yarn type[d] Warp	Fill
A-62	1000 (29.5)	0.10, 0.31, 0.76 (4, 12, 30)	1.6 (0.062)	16 (40)	8 (20)	2/2	2/2
5H-40	854 (25.2)	0.91 (36)	0.99 (0.039)	13 (32)	8 (20)	1/4	2/2
5H-26	407 (12)	0.97 (38)	0.66 (0.026)	11 (29)	10 (26)	1/2	1/2
B-14	288 (8.5)	0.97 (38)	0.36 (0.014)	8 (20)	7 (17)	1/2	1/2

Style	Air permeability m³ m²/min	ft³ ft²/min	Weave	Breaking strength, kg/cm (lb/in.) With sizing Warp	Fill	Without sizing Warp	Fill
A-62	36	118	Double layer	98 (550)	54 (300)	32 (180)	20 (110)
5H-40	11.6	38	5 harness satin	89 (500)	54 (300)	41 (230)	28 (155)
5H-26	19.8	65	5 harness satin	49 (275)	51 (285)	22 (125)	21 (120)
B-14	73	240	Plain	36 (200)	36 (200)	21 (120)	21 (120)

[a]±10%. [b]±20%. [c]±2 end and 2 picks per inch. [d] 900 denier yarn.

Temperature Limits for Refractory Textiles (3M)

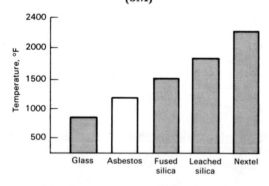

Breaking Strength of Flexweave 1000[a] (Standard Oil Engineered Materials)

Style	Weight g/m²	oz/yd²	Breaking strength (grab method) Warp N	lbf	Fill N	lbf
FW-24	622	24	2364	531.3	2047	460.1
FW-30	778	30	1983	445.6	1454	326.8
FW-35	907	35	1871	420.4	1320	296.6
FW-40	1037	40	2879	647.0	1998	449.0

[a]All construction is plain weave; 40- and 60-in. widths.

Properties of Tantalum Oxide (Ta_2O_5) Cloths and Felts (Zircar Products, Inc.)

Tantalum oxide in textile fiber form is an electrical insulator and has exceptional corrosion resistance to hot acids, particularly sulfuric and phosphoric acids.

Electrolyte absorption capacity, wicking rate, and bubble pressure have been optimized as electrode separators to satisfy requirements for acid fuel cell and battery applications. High-purity electronic-grade tantalum salts are used in their manufacture.

Typical properties of ZIRCAR tantalum oxide in three textile forms are listed in the table below.

Properties	Textile form Tricot knit	Satin weave	Needled felt
Thickness, in.	0.010	0.030	0.080
Weight, g/ft²	23	85	70
Breaking strength, lb/inch of width	1.0	3.0	1.0
Bubble pressure, psi (H_2O at 70 °F)	0.6	1–2	0.5
Wicking rate, min/5 in. (H_2O at 70 °F)	3.0	3.1	2.5

Properties of Zirconia Felt (Zircar Products, Inc.)

Property	ZYF50[a]	ZYF100[a]
Weight, kg/m² (lb/ft²)	0.34 (0.07)	0.65 (0.13)
Thickness, mm (in.)	1.3 (0.050)	2.5 (0.100)
Bulk density, g/cm³ (lb/ft³)	0.24 (15)	0.24 (15)
Porosity, % voids	96	96
Breaking strength, g/cm (lb/in.) width	107 (0.6)	286 (1.6)
Flexibility[b], cm (in.)	0.64 (0.25)	1.91 (0.75)
Compressive strength at 20% compression, MPa (psi)	0.096 (14)	0.096 (14)

[a]Produced by Zircar Products, Inc. [b]Can be wrapped around indicated diameter without damage.

Physical Properties of Alumina Papers (Zircar Products, Inc.)

Property	Type APA-1[a]	Type APA-2[a]	Type APA-3[a]
Color	White	White	White
Type of binder	Organic	None	Alumina
Thickness, in.:			
At 0 psi pressure	0.040	0.050	0.012
At 8 psi pressure	0.032	0.038	0.012
Weight, g/ft²	17.6	16.8	20.0
Density, lb/ft³	12	9	44
Breaking strength, lb/in. width	2.6	1.1	8.0
Organic content, %	5–6	0	0
Inorganic composition,[b] wt %:			
Al₂O₃	95	95	96
SiO₂	5	5	4

[a]Made from Saffil alumina fibers by Zircar Products, Inc. [b]Ashed basis.

Thermal Conductivity of Zirconia Felt (ZYF) in Various Atmospheres (Zircar Products, Inc.)

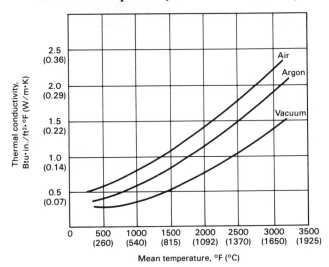

Typical Composition of a Precipitated Silica Filler (PPG Industries)

SiO₂, anhydrous basis	97.5%
SiO₂, as shipped	87.1%
Fe₂O₃	0.1%
Al₂O₃	0.5%
TiO₂	0.007%
CaO	0.3%
MgO	0.1%
NaCl or Na₂SO₄	1.5%

Properties of Commercial Ceramic Fiber Papers[a] (Standard Oil Engineered Materials)

Property	\u200bPaper grade\u200b 110	440	550	880	970	CG
Temperature limit						
°C	1260	740	1260	1427	1260	871
°F	2300	1300	2300	2600	2300	1600
Density						
kg/m³	320	208	208	192	160	640
lb/ft³	20	13	13	12	10	40
Organic binder, %	7–12	6–16	8–14	6–12	6–14	...
K-value						
At 205 °C (400 °F)	0.43	0.38	0.38	0.38	0.38	...
At 540 °C (1000 °F)	0.78	0.78	0.78	0.58	0.58	...
Tensile strength						
N/m²	1.2 × 10⁶	6.9 × 10⁵	6.0 × 10⁵	2.0 × 10⁵	6.0 × 10⁵	4.1 × 10⁶
lb/in.²	175	100	88	30	88	600
Burst strength						
N	40	17.8	13.4	8.9	8.9	44.4
lbf	9	4	3	2	2	10
Thickness						
mm	1.6, 3.2	1.6, 3.2	1.6, 3.2	3.2	0.80, 1.6, 3.2	0.40, 0.80, 1.6
in.	1/16, 1/8	1/16, 1/8	1/16, 1/8	1/8	1/32, 1/16, 1/8	1/64, 1/32, 1/16
Roll widths						
mm	1524[b]	610, 1220	610, 1220	305, 610, 1220	305, 610, 1220	914
in.	60[b]	24, 48	24, 48	12, 24, 48	12, 24, 48	36

[a]Fiberfrax fibers. [b]Sheets to any length.

5.2.2. Fillers: Extenders and Reinforcements

Typical Fillers for Plastics and Their Uses

Filler	Uses	Typical compatible resins
Alumina trihydrate	Fire-resistant filler	Polyesters
Carbon blacks	Electrical goods	Epoxies
Calcium carbonate		
Mineral	Tile, molded goods	Most resins
Precipitated	Pipe, putty	Most resins
Clays	Flooring tile	Most resins
Feldspar	Plastisols	Thermoplastics
Metals	Radiation shields, solders	Epoxies
Mica	Sheet molded goods	Most resins
Polymers		
Solid spheres	Molded goods	Most resins
Hollow spheres	Molded goods	Thermosets
Silica		
Diatomite	Films	Polyethylene
Novaculite	Electrical goods	Thermosets
Quartz flour	Molded goods	Thermosets
Tripolite	Molded goods	Thermosets
Wet process	Sheets, films	Thermoplastics
Vitreous	Electrical goods	Epoxies
Silicate glass		
Solid spheres	Molded goods	Most resins
Hollow spheres	Molded goods	Thermosets
Flakes	Electrical goods	Thermosets
Talc	Extruded and molded goods	PVC, polyalkenes
Wood and shell flour	Molded goods	Most resins

Classification of Fillers (Ref 5, p 513)

By particle morphology	By composition	
Crystalline	Inorganic	Organic
Fibers	Carbonates	Celluloses
Platelets	Fluorides	Fatty acids
Polyhedrons	Hydroxides	Lignins
Irregular masses	Metals	Polyalkenes
Amorphous	Oxides	Polyamides
Fibers	Silicates	Polyamines
Flakes	Sulfates	Polyaromatics
Solid spheres	Sulfides	Polyesters
Hollow spheres		Proteins
Irregular masses		

Types of Reinforcements or Fillers by Function (Ref 5, p 376)

Polymer matrix

Extenders—clay, sand, ground glass, wood flour
Rheology control—mica, asbestos, silica gel
Color—titanium dioxide, carbon, pigments
Flame/heat resistance—minerals
Heat distortion resistance—fibers, mica
Shrinkage resistance—particulate minerals, beads
Toughness—fibers, carbon black, dispersed rubber phase

Metal/inorganic matrix

Hardness, wear resistance—metal, interstitial powders
Creep resistance—oxide dispersions

Specific Gravity and Refractive Index for Some Inorganic Fillers

Fillers	Specific gravity	Refractive index
Carbonates		
Barium carbonate	4.43	1.6
Calcium carbonate		
Precipitated	2.65–2.68	1.63
Limestone	2.71	1.60
Whiting	2.71	1.60
Magnesium carbonate	2.20	1.5–1.71
Fluorides		
Calcium fluoride	3.18	1.43
Sodium aluminum fluoride	2.95	1.3
Hydroxides		
Aluminum hydroxide	2.4	1.58
Oxides		
Aluminum oxide	3.8	1.6
Magnesium oxide	3.5	
Silicon dioxide		
Colloidal sol	1.3	1.4
Diatomaceous	2.65	1.4
Novaculite	2.65	1.55
Pyrogenic	2.2	1.4
Quartz flour	2.65	1.55
Tripolite	2.65	1.55
Vitreous	2.18	1.4
Wet process	1.9–2.2	1.4
Titanium dioxide	3.9	2.55
Zinc oxide	5.6	2.0
Silicates		
Asbestos	2.56	1.57
Clay		
Kaolin	2.60	1.56
Calcined kaolin	2.63	1.62
Calcium silicate	2.33	1.4
Feldspar	2.6	1.53
Glass		
Ground glass	2.5	1.5
Flakes	2.5	1.5
Hollow spheres	0.2	1.5
Solid spheres	2.5	1.5
Mica		
Muscovite	2.75	1.59
Phlogopite	2.75	1.60
Vermiculite	2.25	1.60
Nepheline	2.6	1.6
Perlite	0.17	
Pyrophyllite	2.85	1.59
Talc	2.85	1.59
Wollastonite	2.9	1.59
Sulfates		
Barium sulfate		
Barytes	4.47	1.64
Blanc fixe	4.35	1.64
Calcium sulfate		
Gypsum	2.35	1.53
Anhydrite	2.95	1.59
Precipitated	2.95	1.59
Sulfides		
Lithopone	4.2	1.8
Zinc sulfide	4.0	2.37

Specific Gravities of Some Organic Fillers

Cellulose
  Cork 0.25
  Corn cob 1.2
  Shell flour..1.4
  Wood flour 0.65
Fatty acids and esters
  Sulfur-chlorinated vegetable oil 1.04
  Vulcanized vegetable oil.................................1.0
Polymers
  Polystyrene 1.05
  Phenol-formaldehyde......................................1.1
  Mineral rubber...1.0

Oil Absorption of Various Fillers

Filler	Oil absorption, wt %	Filler	Oil absorption, wt %
Carbonates		Calcium silicate	375
Barium carbonate	14	Feldspar	19
Calcium carbonate		Glass	
Precipitated	15–65	Ground glass	15
Limestone	1.5–15.0	Hollow spheres	35
Whiting	6.5–15.0	Solid spheres	30
Magnesium carbonate	80	Mica	
Oxides		Muscovite	47
Aluminum oxide	13	Phlogopite	24
Magnesium oxide	70	Nepheline	20
Silicon dioxide		Pyrophyllite	36
Diatomaceous	120	Talc	27
Novaculite	20	Wollastonite	26
Pyrogenic	150	Sulfates	
Quartz flour	32	Barium sulfate	
Tripolite	31	Barytes	6
Vitreous	15	Blanc fixe	14
Wet process	160	Calcium sulfate	
Titanium dioxide	24	Gypsum	21
Zinc oxide	13	Anhydrite,	25
Silicates		Precipitated	50
Asbestos	34	Sulfates	
Clay		Lithopone	14
Kaolin	36	Zinc sulfide	13
Calcined kaolin	25		

Properties of Some Filler Materials (Ref 72, p 6)

Filler	Nominal particle size, µm	Purity, wt %	Specific gravity	Melting or softening temperature, °F
Boron	44 (−325 M)	95 (as B)	2.34	3812
Boron carbide (B_4C)	44 (−325 M)	77.54 (as B)	2.52	4262
Aluminum	44 (−325 M)	99 (as Al)	2.70	1228
Solder glass (vitreous, contains lead oxide)	44 (−325 M)	...	5.38	824
Soda-lime glass	44 (−325 M)	...	2.5	1292

Conductive Additives for Thermoplastics (LNP Corp.)

There are numerous additives that can be compounded into a thermoplastic matrix to achieve conductivity. These can be divided into two groups — reinforcements and fillers:

Reinforcements	Fillers
PAN carbon fibers	Carbon powder
Pitch carbon fibers	Metal powders
Ni-plated carbon fibers	Aluminum flakes
Stainless steel fibers	Organic antistatic
Aluminum fibers	
Metallized glass fibers	

Reinforcing agents are fibrous materials that are either conductive themselves, or have been surface modified (generally metal plated) to achieve conductivity. Reinforcements increase the strength properties of the base resin, especially if they include sizing systems which chemically bond the fibers to the resin matrix. Fillers are particulate materials with aspect ratios (length/diameter) well below the critical level needed for reinforcement. Fillers generally reduce the strength properties of a base resin, although they do impart increased stiffness.

Hydrated Alumina (ATH) (ALCOA)

Hydrated alumina is used extensively in the production of aluminum chemicals, most commonly:

- Alum
- Zeolites
- Aluminum chlorhydrate
- Sodium aluminate
- Anhydrous aluminum fluoride
- Sodium aluminum sulfate
- Aluminum phosphate
- Alumina-based catalysts

All alumina trihydrate (ATH) grades are widely used as fire-retardant and smoke-suppressant fillers in polymer systems, including thermosets, thermoplastics, and elastomers. The particular grade required depends on the characteristics of the polymer material and the application.

ATH is also useful in the manufacture of crystal, glass, refractories, and in vitreous enamel and glazing mixtures.

Plastic systems using ATH as a filler are shown in the table below.

Thermoset	Thermoplastic	Elastomeric
Epoxy	Acrylic	Neoprene
Phenolic	PVA	SBR
Polyester	PVC	Natural rubber
Spray-up	Polybutylene	Butyl rubber
Hand lay-up	Polyethylene	Nitrile rubber
Foam	Polypropylene	EPDM
SMC		Silicone rubber
BMC		
Polyurethane		

Hollow Glass and Ceramic Microsphere Properties (Emerson and Cuming)

| | Standard industrial grade | Industrial grade | | High strength SMC/BMC | Electrical grade | | Hydrospace grade Buoyancy to 20 000-ft depths | Ceramic Low-cost filler |
		Higher strength	Lower density		Standard electrical grade	High temp, low loss		
Composition	←————— Sodium borosilicate glass —————→					Silica	Insoluble glass	Ceramic
Bulk density, lb/ft³	12.1	12.1	9.0	14.6	10.4	9.5	9.6	25.0
g/cm³	0.194	0.194	0.145	0.234	0.166	0.152	0.155	0.4
True particle density, lb/ft³	19.4	18.7	14.8	22.2	22.5	15.8	14.8	40.5
g/cm³	0.311	0.300	0.237	0.356	0.361	10.254	0.238	0.65
Particle size range, μm, % by weight >175	5	0	0	0	8	0	···	5
149–175	10	0	6	1	7	14	4	11
125–149	12	2	6	2	11	10	3	19
100–125	12	2	13	2	10	12	7	24
62–100	44	46	42	28	40	40	51	35
44–62	10	19	12	14	10	15	23	4
<44	7	31	21	53	14	9	12	2
Packing factor	0.624	0.647	0.614	0.657	0.46	0.559	0.651	0.620
Average wall thickness, μm	2	1.5	1.5	1.6	2	1.5	1.2	3.5
Softening temperature, °C	482	482	482	482	482	982	1093	1200
Strength under hydrostatic pressure, vol %, survivors at 110 kg/cm² (1500 psi)	47.0	76.6	44.0	96	···	55.0	79.4	83.0
Dielectric constant					1.3	1.2	···	···
Dissipation factor					0.002	0.0005	···	···

Applications for Hollow Microspheres (The P.Q. Corporation)

Polymer system	Products	Advantages
Polyester	Wood-simulated furniture Decorative pieces	Cost reduction Simulated density of wood Improved painting, nail and screw holding
Polyester	Autobody fillers	Cost reduction Easier filing and sanding Lower shipping costs
Polyester	Cultured marble	Weight reduction Resistance to thermal-shock cracking Easier installation
Polyester epoxy	Marine decks, hulls, and putties	Cost reduction Weight reduction, with improved stiffness Easier finishing Thermal insulation
Polyester	Bowling-ball cores	Density control
Polyester epoxy polyurethane	Plywood-patching compounds Pipe coverings Miscellaneous structural parts	Weight reduction Cost reduction Easier shaping and finishing Improved insulation
PVC plastisols	Automotive air filters Carpet backing, toys Coatings	Weight reduction No crease whitening Improved hand
Epoxy	Flotation devices	Buoyancy Cost reduction
Polyurethane	Foams and elastomers	Cost reduction Improved physical properties

Applications for Solid Microspheres (The P.Q. Corporation)

Polymer system	Products	Advantages
Nylon	Housewares Machine parts Bearings	Low heat distortion Improved stiffness and compressive strength Increased dimensional stability and reduced shrinkage
Polystyrene	Housewares Electronic cabinets Decorative frames	Low heat distortion Improved stiffness and compressive strength Shortened cycle time Improved paintability
ABS	Electronic cabinets Automotive trim Poultry-feeding cups	Reduced shrinkage Improved stiffness and compressive strength Increased abrasion resistance
SAN	Automotive glove compartment doors	Smooth surface Improved resistance to solvent crazing when painted Reduced stress from molding
Polyesters	Lantern handles Decorative frames Furniture parts Signs	Improved detail Increased rigidity and surface hardness Reduced warpage
Epoxies	Tooling molds Road markers	Improved detail Increased rigidity and surface hardness Reduced costs and warpage
Silicones	Molds	Cost reduction Improved detail
Urethane	Boats	Increased resistance to abrasion Reduced shrinkage and water absorption

Mohs' Hardness Index/Numbers for Alumina Trihydrates (ALCOA)

Hardness No.	Modified scale	Aluminas
1	Talc	
2	Gypsum	
		Hydrated
3	Calcite	
4	Fluorite, boehmite	Activated
5	Apatite	
6	Orthoclase	
7	Vitreous silica	
8	Quartz or stellite	
9	Topaz	
10	Garnet	
11	Fused zirconia	
12	Fused alumina	Calcined tabular
13	Silicon carbide	
14	Boron carbide	
15	Diamond	

Properties of a Precipitated Silica Filler (Hi-Sil T-690) (PPG Industries)

Median agglomerate size, μm 1.4

Average ultimate particle size, nm 21

Surface area, m^2/g 150

Oil adsorption (linseed oil), lb/100 lb 150

pH-5% in water at 25 °C 7

Ignition loss, %:
  At 1000 °C 10
  From 105 to 1000 °C 4
  At 105 °C 6

Equilibrium moisture content at room temperature
  and 50% relative humidity, % 7.2

Bulk density, tamped, lb/ft³ 3–4

Specific gravity 2.1

Refractive index 1.46

Wet sieve residue, 325 mesh, % 0.002

Hardness, Mohs' scale .... Not measurable — too soft

SECTION 6
Property Data: Polymer Matrix Composites

6.1. Data of General Interest

6.1.1. Materials Selection and Manufacturing Processes; Applications and Testing

Applications for Glass Fiber Reinforced Plastics*

Industry	Advantages of FRP	Applications
Automotive	High-volume production; fine finishes, reduced costs	Automobile body components; fender extenders; front ends; headlamp and taillamp housings; hoods; spoilers, instrument panels; shift consoles; under-the-hood components; truck hoods; fenders; cab and body components; insulated tanks; engine covers; housings; fender liners
Agricultural	Ruggedness; corrosion resistance	Farm tractor hoods, grilles, instrument housings, seating, fenders; garden tractor and lawn mower bodies and housings; fertilizer and pesticide tanks and sprayers; feed troughs
Appliances	Ability to produce complex molded parts without fasteners or welds	Room air conditioner cases, base pans, bulkheads; condenser and compressor fans; humidifier cases and blower wheels; dishwasher pump bodies; dryer ducts; home laundry tubs; water softener tanks, controls, piping; fan housings; gears; vacuum cleaner housings; iron handles; soap dispensers; microwave oven cook trays; television swivel stands; sump pump bases
Aviation/aerospace	High strength with light weight	Aircraft interior components for passengers and cargo; wing tips; antenna components; radomes; wing fuel tanks; ducting; rocket-motor cases; nozzles; nose cones; pressure vessels; instrument housings; launch tubes
Business machines	Excellent surface finish and dimensional stability at elevated temperatures; high strength	Machine covers and housings; access panels; keyboard caps; keys; printer heads; gears, cams, and levers; frames; mounting panels; printed circuit boards; fans and blowers
Chemical processing	Corrosion resistance	Chemical and fuel tanks, pipes, and ducting; storage tanks and hoppers; process pump and valve bodies, casings, impellers; pressure vessels; filters; fume-collection hoods and duct systems; scrubbing towers; electroplating racks and handling equipment; photographic processing equipment
Construction	Ruggedness; moderate cost; good appearance	Structural shapes; paneling; siding; skylighting; curtain wall components; glazing panels; patio covers; concrete pouring forms
Electrical/electronic	High dielectric strength with low moisture absorption	Electrical pole line hardware, crossarms, strain insulators, standoffs, brackets; shatterproof street lighting globes; switch control rods; hot sticks; electronic components; housings and backboards; utility line maintenance equipment

*From Owens-Corning Fiberglas.

(continued)

Applications for Glass Fiber Reinforced Plastics (continued)*

Industry	Advantages of FRP	Applications
House and home	Beauty; low maintenance; low cost	Architectural components; appliance and equipment components; furniture–chairs, tables, lawn furnishings; sinks; bathroom tub/shower units; skylights
Marine	Ease of repair; low maintenance; high performance	Pleasure, commercial, and military boat hulls and superstructures; barge covers; lighters; fuel tanks; water tanks; masts and spars; bulkheads; duct work; ventilation cowls; marker buoys; floating docks; outboard engine shrouds
Materials handling	High strength with light weight	Tote trays and bins; food-processing and delivery trays, boxes, bins; tanks and pipes; conveyor-system components; pallets and skids; cargo-handling equipment
Recreational	Low maintenance; good appearance	Motor homes; travel trailers; truck campers; camping trailers; pickup covers; water and snow skis; surfboards; golf clubs; hockey sticks; lacrosse sticks; archery bows; fishing rods; vaulting poles; recreational water craft, canoes; snowmobile and all-terrain vehicle bodies; golf carts; protective helmets; swimming pools; diving boards; playground equipment
Transportation	Toughness; lightness	Railway passenger and freight car components; transport seating; freight car roofs; hopper car covers; refrigerator car liners; air cargo "igloos"; motor truck and bus components; rapid transit car ends; third-rail covers; barges; truck trailer panels; refrigerated truck bodies

*From Owens-Corning Fiberglas.

Applications of Glass Fiber Composites by Manufacturing Method

Manufacturing method	Applications
Filament winding	Pressure bottles, silos, tubing, pipe, missile shells, nozzles
Press molding	Car bodies, housings, chemical equipment, electrical equipment, appliances, covers, lamp shades, sinks, tanks
Pultrusion	Rods, tubes, shapes, building components, electrical appliances, structural
Spray molding, contact lay-up	Boats, silos, car bodies, scooters, motorcycles, furniture, trucks, homes, chemical tanks, swimming pools, bathtubs, plumbing fixtures
Autoclave molding	Aircraft parts, propeller blades, high-strength parts, radomes

Selection of Bearing Materials for Various Conditions (Ref 29, p 415)

Operating requirement	Decreasing suitability[a]				
Low wear/long life	5	3	6	7	4
Low friction	11	9	5	3	4
High temperatures	10	9	11	7	8
Low temperatures	3	11	9	4	5
High loads	9	10	6	5	4
High speeds	11	9	8	5	7
High stiffness	11	9	6	4	5
Dimensional stability	10	11	9	7	8
Compatibility with fluid lubricants	7	10	4	8	2
Corrosive environments	10	7	3	4	2
Compatibility with abrasives	1	3	2	4	5
Tolerance to soft counterfaces	1	9	2	3	4
Compatibility with radiation	7	4	9	10	2
Space/vacuum	11	9	4	3	2
Minimum cost	1	2	3	4	5

[a]Key:
1 Unfilled thermoplastics
2 Filled/reinforced thermoplastics
3 Filled/reinforced ptfe
4 Filled/reinforced thermosetting resins
5 Ptfe impregnated porous metals
6 Woven ptfe/glass fibre
7 Carbons-graphites
8 Metal-graphite mixtures
9 Solid film lubricants
10 Ceramics, cermets, hard metals
11 Rolling bearings with self-lubricating cages

Main Plastics and Fillers of Interest for Bearings (Ref 29, p 390)

Polymers		Fillers and reinforcements		
Thermoplastics	Thermosetting resins	To improve mechanical properties	To reduce friction	To improve thermal properties
Polyethylene, high molecular weight	Phenolics	Asbestos	Graphite	Bronze
Acetal-homo-, and co-polymer	Polyesters	Glass	MoS_2	Silver
Polyamides (Nylon 6, 66 and 11)	Epoxies	Carbon	Ptfe	Carbon/graphite
Ptfe	Silicones	Textile fibres	(particles or fiber)	
Polyphenylene oxide	Polyimides	Mica		
Polycarbonate		Metals and oxides		

Manufacturing Processes for Polymer Matrix Composites (Ref 60, p 22–23)

Process	Tooling material	Tool life, number of parts Low	High	Relative dollars	Typical materials used	Advantages of process	Limitations of process	Applications
Structural foam (thermoplastic and RIM)	Steel	200 000	1 million	100	Polyurethanes, polycarbonates, polyphenylene oxide, phenylene ether copolymers, polybutylene terephthalate, ABS, polyethylene polystyrene, polypropylene	Large detailed parts, low production costs, rigidity, complex shapes, parts consolidation. High strength-to-weight ratio. Low density. Molded-in inserts. Low pressure molding due to low viscosity. Wide material selection. Minimizing or eliminating of sink marks. Low molded-in stresses. Improved chemical resistance.	Surfaces usually need secondary finishing and/or painting. Some sacrifice of physical properties relative to base resin. Longer cycle times than injection-molded parts.	Business machine housings, automotive fascias, medical and electronic cabinetry, furniture, materials-handling equipment.
	Machined aluminum	5 000	250 000	60 to 80				
	Cast aluminum; Kirksite	500	50 000	50 to 70				
	Cast aluminum-filled epoxy	10	500	20 to 30				
	Cast epoxy (100%)	2	25	5 to 20				
Injection molding	Steel	200 000	1 million	100	Broad range of thermoplastics and thermosets	High-volume production runs, close tolerances, molded-in color, low part cost, large material selection, parts consolidation.	High tooling investment. Long tooling lead times.	Wide range of applications in all industries.
	Machined aluminum	5 000	250 000	60 to 80				
Compression molding	Steel	200 000	1 million	100	Thermosets (polyesters, sheet molding compound, alkyds, ureas, phenolics, epoxies dialyl phthalates)	High-strength, heat-resistant parts. High modulus. Complex shapes. Parts consolidation. Excellent surface finish. Used for molding large parts, such as with polyester, as well as full range of sizes. Excellent fatigue resistance.	High tooling costs. Deflashing needed. Labor intensive. Parts generally need painting.	Automotive parts, electrical connectors, business-machine parts, power-tool housings.
Vacuum forming	Machined aluminum	5 000	250 000	60 to 80	ABS and ABS alloys, PVC, acrylic, polystyrene, polyethylene, polycarbonate, polypropylene	Excellent for complex contours, with minimum internal details. Large or small parts.	Often limited by large radii, shallow depths, large draft angles, loose tolerances. Exposed edges must be trimmed and buffed or milled.	Signage, business-machine housings, furniture, medical cabinetry, recreational products (boats, campers), transportation (interior and exterior parts), packaging (cups, plates). Recreational boating, materials handling, furniture, construction, transportation (truck hoods, bus seats).
	Cast aluminum; Kirksite	500	50 000	50 to 70				
	Cast aluminum-filled epoxy	10	500	20 to 30				
	Cast epoxy (100%)	2	25	5 to 20				
Hand layup sprayed glass fiber	Machined aluminum	5 000	250 000	60 to 80	Thermoset polyesters	Large parts, basically shells, can be molded with complex curves, excellent surface finish, and high rigidity.	Internal details (bosses, ribs) must be manually layed into inside wall and then overlayed with glass fiber. Labor intensive.	
	Cast aluminum; Kirksite	500	50 000	50 to 70				
	Wooden pattern	10	1 000	3 to 10				
Die casting	Steel	200 000	1 million	100	Zinc, magnesium, aluminum alloys	Complex shapes, intricate details, tight tolerances, excellent surface finishes direct from die. High-strength fatigue-resistant parts.	Usually requires costly secondary operations, such as deburring, deflashing, tapping, painting.	Small appliances, hardware, automotive parts, motors, power tools.

(continued)

Manufacturing Processes for Polymer Matrix Composites (Ref 60, p 22–23) (continued)

Process	Tooling material	Tool life, number of parts Low	Tool life, number of parts High	Relative dollars	Typical materials used	Advantages of process	Limitations of process	Applications
Matched metal die forming	Steel	200 000	1 million	100	Mild steel, aluminum	Effective for production of compound-surfaced parts which require high strength, rigidity, heat resistance.	High tooling costs due to need for progressive dies. Deburring required. Painting necessary.	Automotive body panels, major appliances, housings, large containers.
Sandcasting	Wooden pattern	10	1 000	3 to 10	Iron, bronze, brass, copper, aluminum	Complex shapes with low capital investment. Very rigid parts. High heat resistance and strength. Good surface finish.	Limited to bulky parts. Frequently require machining. Labor intensive. Secondary finishing operations can be extensive.	Wide range of industrial uses. Transportation components, materials-handling equipment, large generator parts.
Sheetmetal/brake formed	Fixtures				Steel, aluminum, brass	Excellent for punching and bending sheetmetal.	Compound curves, internal ribs should be avoided. Assembly intensive. Painting.	Electronic cabinetry, ducting, furniture, construction.

List of Composites Meeting Johnson Space Center Vacuum Stability Requirements for Polymeric Materials (Ref 4, p 738)

Material	Total weight loss, %	Volatile condensable materials, %
Thermosets		
Epoxy/glass (Hexcel F-161)[a]	0.30	0.02
Epoxy/glass (GE-101)	0.48	0.05
Epoxy/glass (Ferro 2209)[b]	0.53	0.00
Epoxy/glass (G-10)	0.10	0.01
Epoxy/glass (G-11)	0.61	0.03
Epoxy/glass (E-720)	0.54	0.04
Polyimide/glass (Hexcel F-174)	0.40	0.00
Graphite/epoxy (HY-E-1334)[c]	0.97	0.01
Epoxy/Kevlar 49 (F-164)[d]	0.00	0.00
Epoxy/graphite (Thornel-300/934)	0.62	0.00
Phenolic/glass	0.64	0.00
Epoxy/S-glass (Scotchply XP-251S)[e]	0.58	0.01
Polyimide/Kevlar (Skybond 703)	0.85	0.00
Silicone resin/silica fibers[f]	0.21	0.03
Thermoplastics		
Teflon FEP	0.06	0.06
Teflon TFE	0.10	0.03
Nylon 6/6-glass (70/30)	0.81	0.04
Acetal (Delrin)	0.48	0.07
KEL-F	0.03	0.01
Polycarbonate/glass laminated	0.10	0.01
Acrylic	0.57	0.01
Polypropylene/glass	0.13	0.04
Polyphenyle oxide	0.04	0.03
Polystyrene	0.26	0.01
Polysulfone	0.33	0.00
Polysulfone/glass (70/30)	0.24	0.01

[a]Preconditioning time, 2.75 h; preconditioning temperature, 163 °C (325 °F). [b]Preconditioning time–as received. [c]Preconditioning time, 1 h; preconditioning temperature, 177 °C (350 °F). [d]Preconditioning time, 3 h; preconditioning temperature, 177 °C (350 °F). [e]Preconditioning time, 0.5 h; preconditioning temperature, 140 °C (284 °F). [f]Preconditioning time, 16 h; preconditioning temperature, 204 °C (399 °F).

Factors Affecting the Conductivity of Thermoplastic Compounds*

Aspect ratio: A continuous pathway of particles is essential to electrical conductivity. The greater the aspect ratio, the less additive needed to make the resin conductive.

Loading level: The amount of additive used in a compound. The higher the loading, the greater the conductivity. The lowest loading of an additive required to initiate conductivity is called "critical concentration."

Resin matrix: Depending on the type of resin chosen, more or less additive may be required to meet conductivity requirements.

Processing: Improper processing can cause the fiber additives to break into shorter lengths and decrease their effectiveness. Conductivity is enhanced by proper tooling, gating, and processing.

Conductivity: The more conductive the additive is, the more conductive the finished product will be.

*From Wilson-Fiberfil International.

Process Comparisons for Glass Fiber Composites*

	Resin transfer molding	Injection molding	Pultrusion	RRIM	Hand lay-up spray-up	Filament winding	Compression: sheet molding compound	Compression: bulk molding compound	Preform molding
Factor limiting maximum size of part	Machine size	Machine size	Materials	Metering equipment	Mold size; part transport	Winding machine	Press rating and size	Press rating and size	Press rating and size
Maximum size to date, m² (ft²)	9.3 (100)	9.3 (100)	1.6 mm to 1.22 m × 2.44 m (1/16 to 48 × 96 in.)	4.6 (50)	279 (3 000)	93 (1 000)	4.6 (50)	4.6 (50)	18.6 (200)
Shape limitations	Moldable	Moldable	Round, rectangular	Moldable	None	Surface of revolution	Moldable	Moldable	Moldable
Usual useful production	Medium	High	Medium	Medium-high	Low-medium	Low-medium	High	High	High
Volume, No. parts/year	1000–20 000	50 000–1 000 000	3050 m (10 000 ft)	15 000–100 000	0–1 000	0–1 000	1 000–1 000 000	1 000–1 000 000	1 000–1 000 000
Production cycle time	10–20 min	15 s to 15 min	10 to 30 min	1–2 min	3 min to 24 h	5 min	1 1/2 to 5 min	1 1/2 to 5 min	1 1/2 to 5 min
Typical glass content, %	15–25	20–40	30–75	5–25	20–35	65–90	15–35	15–35	24–45
Strength orientation	Random	Random	Highly oriented	With flow	Random (usually)	Highly oriented	Random	Random	Random
Strength category	Low to medium	Low	High	Low	Medium	Very high	Low-medium	Low-medium	Medium-high
Wall thickness:									
Minimum, mm (in.)	0.76 (0.03)	0.76 (0.03)	1.6 (1/16)	2.03 (0.080)	0.76 (0.030)	0.25 (0.01)	0.76 (0.03)	1.5 (0.06)	0.76 (0.03)
Maximum, mm (in.)	25.4 (1.0)	12.7–25.4 (1/2–1)	12.7 (1/2)	12.7 (0.500)	38–up (1.5–up)	51 (2.0)	6.4 (0.25)	25.4 (1.0)	6.4 (0.25)
Tolerance, mm (in.)	...	...	+0.25 ± 25.4 (+0.01 ± 1.0)	±0.05 (±0.002)	+0.51 (+0.02)	+0.25 (+0.01)	+0.20 (+0.008)	+0.13 (+0.005)	+0.13 (+0.005)
Variations	Uniform	Uniform	Uniform	Uniform	As desired	As desired	Uniform desirable <3:1	As desired	Uniform desirable <2:1
Minimum draft:									
To depth 150-mm (6-in.)	1°	1°	0–2°	1°–3°	0–2°	3°	1°–3°	1°–3°	1°–3°
Over depth 150-mm (6-in.)	1°	1°+	0–2°	3°+	0–2°	3°+	3°+	3°+	3°+
Minimum inside radius, mm (in.)	1/2 part depth	1/2 part depth	1.5 (0.06)	1/2 part depth	6.4 (0.25)	3.18 (0.125)	1.5 (0.06)	1.5 (0.06)	3.18 (0.125)
Ribs	Yes	Yes	No	Yes	Yes	No	Yes	Yes	Not recommended
Bosses	Yes	Yes	No	Yes	Yes	No	Yes	Yes	Not recommended
Undercuts	Possible	Possible	No	Yes, with proper	Avoid	No	Avoid	No	No
Holes:									
Parallel	Yes	Yes	No	Yes	Yes	Yes	Yes	Yes	Not recommended
Perpendicular	Yes	Yes	No	Yes	Yes	—	Undesirable	Undesirable	Undesirable
Built-in cores	Yes	Yes	No	Yes	Possible	Possible	Possible	Possible	Possible
Metal inserts	Yes	Yes	No	Yes	Yes	Yes	Yes	Yes	Not recommended
Metal edge stiffeners	Yes	Yes	No	Yes	Yes	No	No	No	Yes
Surface finish:									
Number of finished surfaces	All	2	2	2	1	1	2	2	2
Quality of surface	Excellent	Excellent	Fair to good	Excellent	Excellent	Excellent	Very good	Excellent	Very good
Gel-coat surface	...	Yes	No	No	Yes	Yes	No	No	Yes
Surfacing mat.	Yes	No	No	No	Yes	Yes	No	No	Yes
Combination with thermoplastic line	Yes	Yes	No	Yes	Yes	Yes	Yes	Yes	No
Trim in mold	No	No	No	No	No	Yes	Yes	Yes	Yes
Molded-in labels	Yes	Yes	No	Yes	Yes	Yes	Difficult	Difficult	Difficult
Raised numbers	Yes	Yes	No	No	Yes	—	Yes	Yes	Yes
Translucency	Yes	Yes	Yes	No	Yes	Yes	No	No	Yes
Tool cost	High	High	Low	Low-medium	Low	Low	High	High	Medium
Capital equipment cost	High	High	Low	Low-medium	Low	Low	High	High	High

*Data from Owens-Corning Fiberglas.

Selection of Resins Used With Glass Fiber Reinforcements[a]

Selection of the resin system to be used is based on satisfying chemical, electrical, and thermal performance requirements of the finished product. There are two general classes of resins: *thermosets*, which become hard when heated; and *thermoplastics*, which are soft when heated and harden upon cooling.

Type of resin	Properties	Processes
Thermosets		
Polyesters[b]	Simplest, most versatile, most economical, and most widely used family of resins, having good electrical properties and good chemical resistance (especially to acids)	Compression molding; filament winding; hand lay-up; mat molding; pressure bag molding; continuous pultrusion; injection molding; spray-up; centrifugal casting; cold molding; comoform; encapsulation
Epoxies	Excellent mechanical properties, dimensional stability, and chemical resistance (especially to alkalis); low water absorption; self-extinguishing (when halogenated); low shrinkage; good abrasion resistance; very good adhesion properties	Compression molding; filament winding; hand lay-up; continuous pultrusion; encapsulation; centrifugal casting
Phenolics	Good acid resistance; good electrical properties (except arc resistance); high heat resistance	Compression molding; continuous laminating
Silicones	Highest heat resistance; low water absorption; excellent dielectric properties; high arc resistance	Compression molding; injection molding; encapsulation
Melamines	Good heat resistance; high impact strength	Compression molding
Diallyl phthalate	Good electrical insulation; low water absorption	Compression molding
Thermoplastics		
Polystyrene	Low cost; moderate heat distortion; good dimensional stability; good stiffness; good impact strength	Injection molding; continuous laminating
Nylon	High heat distortion; low water absorption; low elongation, good impact strength; good tensile and flexural strength	Injection molding; blow molding; rotational molding
Polycarbonate	Self-extinguishing; high dielectric strength; high mechanical properties	Injection molding
Styrene-acrylo-nitrile	Good solvent resistance; good long-term strength; good appearance	Injection molding
Acrylics	Good gloss, weather resistance, optical clarity, and color; excellent electrical properties	Injection molding; vacuum forming; compression molding; continuous laminating
Vinyls	Excellent weatherability; superior electrical properties; excellent moisture and chemical resistance; self-extinguishing	Injection molding; continuous laminating; rotational molding
Acetals	Very high tensile strength and stiffness; exceptional dimensional stability; high chemical and abrasion resistance; no known room-temperature solvent	Injection molding
Polyethylene	Good toughness; light weight; low cost; good flexibility; good chemical resistance; can be "welded"	Injection molding; rotational molding; blow molding
Fluorocarbons	Very high heat and chemical resistance; nonburning; lowest coefficient of friction; high dimensional stability	Injection molding; encapsulation; continuous pultrusion
Polyphenylene oxide modified	Very tough engineering plastic; superior dimensional stability; low moisture absorption; excellent chemical resistance	Injection molding
Polypropylene	Excellent resistance to stress or flex cracking; very light weight; hard, scratch-resistant surface; can be electroplated; good chemical and heat resistance; exceptional impact strength; good optical qualities	Injection molding; continuous laminating; rotational molding
Polysulfone	Good transparency; high mechanical properties, heat resistance, and electrical properties at high temperatures; can be electroplated	Injection molding

[a]From Owens-Corning Fiberglas.
[b]Properties shown also apply to some polyesters formulated for thermoplastic processing by injection molding.

Applications of Boron and SiC Filaments and Composites (Ref 4, p 193)

Product	Component and/or producer	Composite
Production		
F-14	Horizontal stabilizer	Boron/epoxy
F-15	Tail section, cabin floor, and stabilator	Boron/epoxy
Uttas helicopter	Structural beam reinforcement	Boron/epoxy
F-111	Wing pivot doubler	Boron/epoxy
Mirage 2000	Rudder	Boron/graphite/epoxy
Space shuttle	Fuselage	Boron/aluminum tubes
Fishing rods	Browning, Shakespeare, Rodon, and Phoenix	···
Tennis rackets	Head–Spalding, DuraFiber, Snauwaert & Depla, and Browning	···
Golf shafts	Aldilla	···
Research and Development		
F-14	Overwing and fairing	Boron + other filaments/fibers/epoxy
A-7	Outer wing	Boron/graphite/epoxy
C-130	Wing box	Boron/epoxy-reinforced aluminum
F-4	Rudder	Boron/epoxy
707	Foreflap	Boron/epoxy
F-100	Engine fanblades	Boron/aluminum
C-5A	Wing slat	Boron/epoxy
CH-54	Fuselage string and tail skid	Boron/epoxy
Avco B210	Bicycle tubes	Boron/aluminum
B-1	Horizontal and vertical stabilizers and wing slat	Boron/graphite/epoxy
F-111	Horizontal stabilizer	Boron/epoxy
CH-47	Rotor blade	Boron/epoxy
Compressor blades	···	SiC/titanium
Advanced structures	···	SiC/titanium

Alphabetical Index of Some ASTM Test Methods

Method	ASTM number[a]
Abrasion Resistance of Plastics	D1242-56 (1981), Vol 08.01
Arc Resistance of Solid Electrical Insulation	D495-73 (1979), Vol 08.01
Bearing Strength of Plastics	D953-80, Vol 08.01
Blocking of Plastic Film	D1893-67 (1978), Vol 08.02
Brittleness Temperature of Plastics and Elastomers by Impact	D746-79, Vol 08.01, 09.02
Coefficient of Linear Thermal Expansion	D696-79, Vol 08.01, 14.01
D-C Electrical Resistance or Conductivity	D257-78 (1983), Vol 08.01, 09.02, 10.02
Deflection Temperature Under Flexural Load	D648-82, Vol 08.01
Deformation of Plastics Under Load	D621-64 (1976), Vol 08.01
Dielectric Breakdown Voltage and Dielectric Strength at Commercial Power Frequencies	D149-81, Vol 08.01, 09.02, 10.02
Dielectric Constant and A-C Loss Characteristics	D150-81, Vol 08.01, 09.02, 10.02, 10.03
Diffuse Light Transmittance Factor of Reinforced Plastic Panels	D1494-60 (1980), Vol 08.01, 08.04
Flexural Fatigue of Plastics	D671-71 (1978), Vol 08.01
Flexural Properties of Plastics	D790-81, Vol 08.01
Flow Properties of Thermoplastic Molding Materials	D569-82, Vol 08.01
Impact Resistance of Plastics	D256-81, Vol 08.01, 09.02
Index of Refraction for Transparent Organic Plastics	D542-50 (1977), Vol 08.01
Light Transmission Factor of Reinforced Plastic Panels	D1494-60 (1980), Vol 08.01, 08.04
Luminous Transmittance and Haze of Transparent Plastics	D1003-61 (1977), Vol 08.01
Mar Resistance of Plastics	D673-70 (1982), Vol 08.01
Rate of Burning of Flexible Plastics in Vertical Position	D568-77, Vol 08.01
Rate of Burning of Self-Supporting Plastics in Horizontal Position	D635-81, Vol 08.01
Resistance of Plastics to Chemical Reagents	D543-67 (1978), Vol 08.01
Rockwell Hardness of Plastics	D785 65 (1981), Vol 08.01
Specific Gravity of Plastics by Displacement of Water	D792-66 (1979), Vol 08.01
Specular Gloss Test	D523-80, Vol 06.01, 08.01
Surface Abrasion of Transparent Plastics	D1044-82, Vol 08.01
Surface Irregularities of Transparent Plastic Sheets	D637-50 (1977), Vol 08.01
Tear Resistance of Film and Sheeting	D1004-66, Vol 08.01
Tensile Properties of Plastics	D638-82a, Vol 08.01
Tensile Properties of Plastics (Metric)	D638M-81, Vol 08.01
Tensile Properties of Thin Plastic Sheets	D882-83, Vol 08.01
Water Absorption of Plastics	D570-81, Vol 08.01
Water Vapor Permeability	E96-80, Vol 04.06, 08.03, 15.09

[a]Dates in parentheses are reissue dates.

Important Electrical Considerations for Engineering Thermoplastics

Property	Significance	Property	Significance
Dielectric constant	The ratio of the capacity of a condenser filled with the material to the capacity of an evacuated capacitor. It is a measure of the ability of the molecules to become polarized in an electric field. A low dielectric constant indicates low polarizability; thus the material can function as an insulator.	Volume resistivity	The electrical resistance of a unit cube calculated by multiplying the resistance in ohms between the faces of the cube by the area of the faces. The higher the volume resistivity, the better the material will function as an insulator.
Dissipation factor	A measure of the dielectric loss (energy dissipated) of alternating current to heat. A low dissipation factor indicates low dielectric loss, while a high dissipation factor indicates high loss of power to the material, which may become hot in use at high frequencies.	Surface resistivity	The resistance to electric current along the surface of a 1-cm^2 sample of material. Higher surface resistivity indicates better insulating properties.
		Dielectric strength	A measure of the voltage an insulating material can take before failure (dielectric breakdown). A high dielectric strength indicates that the material is a good insulator.

Typical Composite Efficiencies Attained in Reinforced Plastics (Ref 8, p 27)

Fiber configuration		Fiber length	Total fiber content (by volume), V_f	F_{long}[a], ksi (N/cm$^2 \times 10^3$) F_{theor}[b]	F_{test}[c]	Composite efficiency[d], %
Filament-wound (unidirectional)		Continuous	0.77	310 (214)	180 (124)	58.0
Cross-laminated fibers		Continuous	0.48	197 (136)	72.5 (50.0)	36.8
Cloth laminated fibers		Continuous	0.48	197 (136)	43.0 (29.6)	21.8
Mat laminated fibers		Continuous	0.48	197 (136)	57.2 (39.4)	29.0
Chopped fiber systems (random)		Noncontinuous	0.13	60.7 (41.8)	15.0 (10.3)	24.7
Glass flake composites		Noncontinuous	0.70	165.5 (114.1)	20.0 (13.8)	12.1

[a]F_{long} = Ultimate tensile strength in direction of greatest fiber content (longitudinal), if there is one.
[b]Theoretical strength based on "Rule of Mixtures":

$$F_{theor} = V_f S_f + (1 - V_f)S_m$$

where

S_f = 400 (275.8) ksi (N/cm$^2 \times 10^3$) − typical boron or carbon fiber strength

S_m = 10 (6.9) ksi (N/cm$^2 \times 10^3$) − typical resin strength

[c]F_{test} = typical experimental strength values.
[d]Composite efficiency = $(F_{test}/F_{theor}) \times 100$.

Physical Properties Relating to Reinforced Plastic Design Considerations

Mechanical Properties

Tensile properties
Compressive properties
Flexural properties
Shear properties
Impact strength
Properties at high rates of
  loading (dynamic properties)
Bearing strength
Surface hardness
Creep properties (creep-rupture
  and stress-relaxation)
Fatigue (cyclic properties)
Poisson's ratio
Notch sensitivity
Shatterproofness
Shockproofness
Tear resistance

Thermal Properties

Thermal conductivity
Thermal expansion
Specific heat
Flow temperature
Flammability (flame resistance)
Heat distortion temperature
  (deflection temperature under
  load)
Thermal shrinkage
Maximum safe operating
  temperature
Ignition properties
Brittleness temperature

Electrical Properties

Arc resistance
Electrical resistance (insulation
  resistance — volume and
  surface)
Dielectric strength and
  dielectric breakdown voltage
Dielectric constant and power
  factor

Optical Properties

Index of refraction (refractive
  index)
Light diffusion
Crazing resistance
Spectral transmission (haze)
Internal stress (transparent
  plastics)
Surface stability, optical
Optical uniformity and
  distortion

**Chemical and Permanence
Properties**

Water absorption
Water vapor permeability
  (diffusion) — gas
  transmission rate
Sunlight and weather exposure
  (aging)
Resistance to chemical reagents
Effects of radiation
Toxicity
Volatile loss (outgassing)
Stress-crazing
Impact sensitivity (LOX)
Accelerated service
  (temperature and humidity)

6.1.2. Fabrication and Molding Methods

Thermosetting and Thermoplastic Adhesives for Composite Bonding

Thermosetting Adhesives

Epoxy resins
  Epichlorohydrin-bisphenol A
  Cycloaliphatics
  Epoxy-novolacs
  Epoxy-nitriles
  Epoxy-phenolics
  Epoxy-polyamides
  Epoxy-polysulfides
Polyester resins
  Polyester-DAP
  Polyester-TAC
Phenolic resins
  Phenol-formaldehyde
  Vinyl-phenolics
  Nitrile-phenolics
  Polyamide-phenolics
  Resorcinol-formaldehyde
  Melamine-formaldehyde
Silicone resins
  Dimethyldichlorosilanes
  Phenyl-silicones
  Silicone-alkyds
  Silicone-epoxies
  Elastomerica silicones (RTV)

Polyimides
  Dianhydride-diamine (PI)
  Polybenzimidazole (PBI)
  Amide-imides
  Chain terminated imides

Thermoplastic Adhesives

Acrylic resins
  Methylmethacrylate
Cellulosics
  Cellulose acetate
  Acetate-butyrate
  Cellulose-nitrate
  Ethyl cellulose
Sulfones
  Polysulfone
  Polyethersulfone
  Polyarylsulfone
Vinyl resins
  Acetal
  Acetate
  Alcohol
  Chloride-acetate

Molding Processes Requiring a Release Agent

Compression
Transfer
Injection
Laminating

Reinforced plastic (FRP)
Polyester resin injection
Rotational

Characteristics of Some Epoxy Resin Curing Agents

Agent	Characteristics
Benzyl dimethyl amine (BDMA)	Used as accelerator; also controls B-staging
Borontrifluoro-mono-ethyl amine (BF₃MEA)	Used as accelerator; also controls B-staging (moisture sensitive)
Diamino, diphenyl sulfone (DDS)	Melts at 60 °C (140 °F); BF₃MEA used as accelerator
Dicyandiamide (DICY)	Used as accelerator; also controls B-staging
Diethylene triamine (DETA)	General-purpose agent for room-temperature curing; also high exotherm
Hexahydrophthalic anhydride (HHPA)	Melts at 35 °C (95 °F); soluble in liquid resin at room temperature
Metaphenylene diamine (MPDA)	Melts at 60 °C (140 °F); pot life is 4–6 h at room temperature
Methylene dianiline (MDA)	Melts at 85 °C (185 °F); pot life is 4–6 h at room temperature
Nadic methyl anhydride (NMA)	Liquid; long pot life; BDMA and others used as accelerators
Tridimethylaminomethyl phenol (DMP-30)	Used as accelerator
Triethylene triamine (TETA)	General-purpose agent for room-temperature curing; also high exotherm
2,6-diaminopyridene (DAP)	Cured properties similar to MDA, but slower reacting

Troubleshooting Guide for Gel Coats for Hand Lay-Up

Problem	Cause	Solution
Wrinkling of gel coat during lamination (alligatoring)	Uncured or thin gel coat. Gel coat swells and separates from mold surface in confined area because of insufficient cure and action of styrene in lay-up resin.	Check with wet film gauge for minimum 15-mil thickness: Apply gel coat evenly. Allow adequate gel coat cure time.
Waviness in gel coat	Too long a gel time in lay-up; too long a gel time for lay-up resin	Use more catalyst; adjust catalyst to weather conditions for 1-hour cure.
Streaks in gel coat (particularly pastel colors)	Draining of gel coat, causing color separation	Use heavier gel coat or lay molds flat.
Rough molded surface	Wax build-up	Wash off with styrene or buff with mold cleaner.
Glass pattern in mold	Soft mold gel coat	Use heat-resistant resin in future molds.
Star crazes in mold	Rough handling use of mallet in removing part from mold	1. Grind down to glass. 2. Apply mold gel coat. 3. Apply wax paper and tape. 4. Refinish.
Wrinkling of gel coat immediately after application	Trapped acetone; water in gel coat; insufficient catalyst in gel coat	Hold gun farther from mold; use higher atomization; use more catalyst; drain traps; check line; warm molds.
Dimples in gel coat (when using PVA film)	PVA-separating film not dried	Allow more drying time; clean line of moisture.
Cracking of gel coat	Too heavy a coat. Back-up layer not cured; shrinks later and cracks gel coat.	Use 25-mil maximum thickness; use fast cure on first layer.
Pits in gel coat	Foreign particles in film	Spray film in dust-free room.
Uneven color in gel	Air entrapment; poor hiding power; insufficient pigment	Use styrene for good flow; consult gel coat supplier; use 10% minimum pigment.
Dull surface	Rough mold	Refinish mold.
Difficulty in removing part from mold	Mold not broken in; rough mold; undercuts in mold; insufficient wax	Use PVA; repeat mold-prep process; fill undercuts; cover all areas.
Telegraphing of glass pattern in gel coat	Gel coat too thin; undercure	Use 15–20 mils; wait for full gel coat cure.
Patching does not match gel coat	Patch cured too fast	Use thinned gel coat; use low-catalyst concentrations; do not add filler.
Gel coat sticking to mold (brushed or sprayed)	Improper release agent or application	Apply release and let cure. If wax, allow to dry thoroughly and buff. If trouble persists, use PVA-sprayed film over wax.
Hazy or nonglossy surface	Entire part prematurely removed from mold; contamination of release prior to application of gel coat	Permit more complete cure of gel coat and lay-up.
Voids under gel coat	Small or large, flat blisters caused by separation of gel coat from lay-up. Gel coat should not cure tack-free in air but should remain sticky for better bond to lay-up.	Allow first lay-up application to cure prior to adding second and third (etc.). Inspect closely for blisters after lay-up. Cut out and putty mix: 1 part resin to 3 parts $CaCO_3$.
Open bubbles, blisters, and pinholes in gel coat surface	Trapped air, free solvent, dirt, or excessively high exotherm in gel coat or lay-up resin	Avoid mixing air into gel coat when introducing catalyst. Let stand for short period after mixing and before spraying. Keep containers and working area clean.
Soft areas	Uneven cure	More thorough mixing of catalyst into gel coat.
Cratering	Use of too high surface angle release, preventing gel coat from wetting in small spots, $1/16$–$1/4$ in., so that lay-up shows through gel coat	More careful selection and application of release agent.

Mold Release Methods for Hand Lay-Up Process

Release type	Characteristics
Fluorocarbons, silanes, and silicones—liquid or spray	More expensive; not very high gloss; low coefficient of friction
Internal releases—liquid (mixed into gel coat)	Eliminates need for mold waxing; provides paintable surface; high luster; good detail transfer
PVA—liquid (usually sprayed)	One-time use; excellent release; wash off part and mold; provides paintable surface; water-soluble
Release papers and release films—PVA film, coated paper, cellophane	One-offs; flat sheet molding
Wax—carnauba paste or liquid	Good detail transfer; multiple runs; high luster

Troubleshooting Guide for Hand Lay-Up

Problem	Cause	Solution
Cure in thickened rods or strings	Pregelation	Keep mixing containers clean and free of previously catalyzed gel coat. Use throw-away mixing containers.
Cracking and fissuring	Larger cracks caused by too thick areas of gel coat or excessive exotherm or thin point in lay-up; fissuring because of front or reverse impact blow	More uniform application of gel coat and better mixing with catalyst. Prevent accidental or injuring blows.
Fiber pattern: random fibers from mat or cross-hatch from woven roving weave	High exotherm; coarse weave material; too close to gel coat	Cure laminate in steps; use lower exotherm resin. Put more mat in front of woven roving. Best solution is application of an intermediate layer of more rigid resin-containing Vitro-Strand fibers.
Lay-up draining on vertical surfaces	Resin too low in viscosity; resin with insufficient thixotropic agent; mold or room too warm	Most probable correction is to increase thixotropic agent content of resin.
Bubbles	Air entrained in reinforcement after combination with resin	Add 0.2% green pigment to lay-up resin to see voids. Work lay-up more freely with brushes, squeegees, or serrated rollers. If possible, apply a liberal quantity of resin onto work before applying reinforcement, so that the resin forces air out from the bottom.
Bridging over small radius curves such as lap-strakes, etc.	Reinforcement too stiff; curves below design-allowables	Select more highly wettable or soluble mat or woven roving. Use loose-mixed putty to caulk small radii curvatures prior to lay-up. Redesign mold.
Thin areas	Gaps in lapping reinforcement caused by improper placement or short-cutting, etc.	Correct placement and cutting errors. Lay in patches to correct thin spots prior to removal from mold. Try pre-wetting of reinforcement by resin prior to placement in mold.
Fibers protruding from inner lay-up surface	Usually unavoidable if mat is sole reinforcement	For finish layer, apply woven fabric, woven roving, or veil mat on inside. After cure, sand and apply splatter paint.
Cracked or resin-rich areas usually at bottom or wellpoint	Drainage of resin in large lay-up to a low point and, because of high exotherm, results in cracking; possibly too high a resin-to-glass ratio	Introduce more thixotropic agent into resin. Continue to squeegee excess resin out of collection points until gelation occurs. Add additional reinforcement.
Warpage of part	Unbalanced laminate; flat surface	Use symmetric lay-up; design slight radius in surface.
Distortion of part	Undercured in mold	Allow full cure in mold.
Cracking next to stiffening members	Hard spot	Use fillet in corner where stiffener meets laminate.
Low impact strength	Insufficient glass; too much flexing	Use bag molding, more woven roving, roving; use stiffener or sandwich construction.
Slow curing laminating resin	Weather changes	Adjust catalyst to weather changes.
Roller picks up fibers when working on mat	Too close to gel time; styrene evaporation; rolling too fast	Adjust gel time, adjust fans; dip roller in styrene or fresh resin; more deliberate rolling.

Troubleshooting Guide for Injection Molding of Thermoplastic Polyester Resins*

Problem	Typical causes	Corrective actions
Cycle variations		
Erratic cycles	Mold held open for varying times	Use mold open timers to maintain constant cycles.
	Pressure variations	Insure sufficient pressure. Check system for leaks; nonreturn valve for seating.
	Cylinder temperature variations (cycling)	Check operation of temperature controls and use best controls available. Check consistency of line voltage. Check heater bands are working. Insure constant temperature of material in drums going into hopper. Check machine at equilibrium conditions. Check window or fan air flow.
	Mold temperature variations	Use mold temperature control. Adjust mold water lines. Check venting of mold. Check water hookup in mold.
	Feed rate variations	Check feed mechanism. Lower rear zone cylinder temperature.
Long cycles	Material temperature high	Reduce temperature. Reduce screw rpm or back pressure.
	Mold temperature excessive	Reduce temperature.
	Cycle time erratic	Maintain constant overall cycle.
	Heating capacity insufficient	Change mold to larger press.
	Delay in machine operation	Reduce machine dead time.
Dimensional variations	Molding conditions erratic	Adjust system for maximum uniformity. Maintain uniform gate-to-gate timing.
	Molding conditions different from previous runs	Check conditions for variations in material temperature, plastic pressure, mold temperature, cycle time, feed, line voltage, temperature regulators, pressure system, material temperature in hopper. Redry both virgin resin and regrind. Check condition and percent of regrind.
	Air temperature variation	Keep ambient air temperature constant during measurement. Close windows on P.M. shifts. Locate fan so air flows away from machine.
Discoloration		
Black or brown specks and streaks	Burned plastic flaking onto cylinder walls	Purge heating cylinder. Scour cylinder walls by purging with stiffer molding compound. Avoid long residence time at high temperature.
	Airborne dirt	Keep hopper covered. Keep virgin material covered.
	Cylinder overheating (general or local)	Reduce temperature of heater. Reduce rpm of screw. Reduce back pressure of screw.
	Burning due to cylinder or nozzle hang-up	Purge cylinder. Remove nozzle and clean.
Black spots	Burning due to entrapped air	Vent mold. Redesign part. Relocate gate. Reduce speed or injection pressure. Change cylinder and mold temperature to alter flow pattern of resin.
Flash	Material temperature excessive	Reduce temperature of material. Reduce temperature of mold. Reduce rpm of screw. Redry material properly.
	Pressure excessive	Reduce pressure. Reduce booster time.
	Feed excessive	Reduce feed.
	Feed erratic	Maintain constant cushion.
	Parting line or mating surfaces poor	Reface.
	Cycle time erratic	Maintain constant overall cycle.
	Clamp insufficient	Increase clamp pressure. Change to machine with greater clamping capacity.
Flow lines and folds	Mold cold	Raise temperature for glass-filled grades.
	Jetting due to small gates	Enlarge gates and reduce injection speed.
	Section thickness not uniform	Redesign part for greater uniformity. Eliminate heavy bosses and ribs.
Short shots	Pressure insufficient	Increase pressure.
	Injection speed slow	Increase injection speed.
	Entrapped air	Increase number and size of vents.
	Unbalanced plastic flow in multiple cavity molds	Correct balance.
	Injection forward time insufficient	Increase injection forward time.
	Gates small	Enlarge gates.
	Shot size too large for machine	Change to larger-capacity machine or reduce number of cavities.
	Feed insufficient	Increase feed and maintain a cushion.
	Mold cold	Raise mold temperature and zone control if required.
	Material cold	Increase temperature of material.

(continued)

Troubleshooting Guide for Injection Molding of Thermoplastic Polyester Resins (continued)*

Problem	Typical causes	Corrective actions
Sink marks (see also *Short shots*)	Thick sections, bosses, ribs, etc.	Rework mold and/or raise injection pressure.
Insufficient plastic in mold to allow for shrinkage	Feed insufficient	Increase speed.
	Injection pressure low	Raise injection pressure.
	Injection forward time short	Increase injection forward time.
	Gates unbalanced	Balance by restricting flow through gates nearest sprue.
	Injection speed slow	Increase speed. Increase gate size.
	Plastic too hot	Reduce cylinder temperature. Control mold temperature more strictly in those areas that heat up too much.
	Mold open time variations	Use timers.
	Piece ejected too hot	Cool mold or immerse piece in water. Lengthen cooling time.
Sticking		
Parts in mold	Injection pressure or cylinder temperature high	Lower pressure or temperature. Reduce screw rpm or back pressure.
	Feed excessive	Reduce feed.
	Injection forward time excessive	Reduce injection forward time.
	Unbalanced flow in multicavity mold	Use balanced runner system. Balance flow through gates.
	Mold undercuts	Remove, polish, and provide draft.
	Vacuum under deep draw part	Vent adequately.
	Mold open time variation	Maintain constant timing.
	Overpacking in gate area	Reduce injection time. Starve feed.
	Finish improper	Polish mold cavity.
	Core shifting	Align cores and provide taper locking. Balance flow into cavity with gating.
Sprue sticking	Pressure excessive	Reduce pressure and/or booster time.
	Material hot	Decrease material temperature.
	Sprue size excessive	Decrease sprue size by using extended nozzle and short sprue bushing.
	Draft insufficient	Increase draft angle.
	Improper fit between sprue bushing and nozzle	Make nozzle hole smaller than in sprue bushing.
	Undercuts or rough surface	Eliminate undercuts and polish surface.
	Injection dwell excessive	Reduce injection forward time.
	Feed excessive	Decrease feed.
Surface finish poor	Mold cold	Raise mold temperature.
	Injection pressure low	Raise injection pressure.
	Water condensed on mold face	Clean and repair water leaks or avoid condensation.
	Mold lubricant excessive	Clean mold and use lubricant sparingly.
	Plastic not contacting mold metal completely	Increase injection pressure. Increase feed. Increase mold temperature.
	Injection speed slow	Increase injection speed. Increase resin temperature. Use higher pressure; weigh feed or preplasticize feed without cushion of material ahead of injector. Increase back pressure of screw.
	Internal or external lubricants excessive	Check and obtain proper type.
	Mold surface poor	Polish mold.
	Jetting through gate	Relocate gate to impinge flow against pin or cavity wall. Use tab gate.
Warping	Ejected part too hot	Reduce plastic temperature. Reduce mold temperature. Lengthen mold closed time. Reduce screw rpm or back pressure.
	Plastic cold	Increase cylinder temperature. Raise mold temperature. Increase screw back pressure.
	Section thickness or contour of part varies	Operate mold halves at different temperatures. Make sections uniformly thick throughout part.
	Feed excessive	Reduce feed. Reduce injection pressure. Change stop position.
	Gates unbalanced on parts with multiple gates	Restrict flow of plastic through gates near sprue.
	Ejection system poorly designed or operated	Redesign.
	Mold temperature not uniform	Keep mold surface temperature uniform.
	Excessive material discharged from or packed into area around gate	Regulate injection forward time. Change gate size.

(continued)

Troubleshooting Guide for Injection Molding of Thermoplastic Polyester Resins (continued)*

Problem	Typical causes	Corrective actions
Warping (continued)	Length and/or direction of flow improper	Use multiple gates to shorten length of flow and nullify effect of differential shrinkage.
	Differential shrinkage in glass-filled grades	Use low-warp grades.
Weld lines poor or obvious	Air slow in escaping from mold	Provide adequate vents. Provide overflow well.
	Gas trap	Vent down a knockout pin.
	Lubricant on mold excessive	Remove lubricant and use sparingly. Use mold lubricant only when absolutely required.
	Weld line too far from gate	Relocate gate or use balanced multiple gates.
	Section thickness varies within part	Redesign part. Relocate gates.
	Mold cold	Raise mold temperature.
	Pressure insufficient	Increase injection pressure.
	Injection speed slow	Increase injection speed.

*Data from GAF.

Matched Mold Methods for Fiberglass-Reinforced Plastics*

COMPRESSION MOLDING

A high-volume, high-pressure method suitable for molding complex, high-strength fiberglass-reinforced plastic parts. Fairly large parts can be molded with excellent surface finish. Thermosetting resins are normally used.

Process Description

Matched molds are mounted in a hydraulic or modified mechanical molding press. A weighed charge of sheet or bulk molding compound, or a "preform" or fiberglass mat with resin added at the press, is placed in the open mold. (In the case of preform or mat molding, the resin may be added either before or after the reinforcement is positioned in the mold, depending on part configuration.) The two halves of the mold are closed, and heat (105 to 160 °C; 225 to 320 °F) and pres-

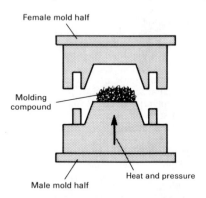

Compression molding

sure (1.0 to 13.8 MPa; 150 to 2000 psi) are applied. Depending on thickness, size, and shape of the part, curing cycles range from less than a minute to about five minutes. The mold is opened and the finished part is removed. Typical parts: automobile front ends, appliance housings and structural components, furniture, electrical components, business machine housings and parts.

Resin Systems

Polyesters (combined with fiberglass reinforcement as bulk or sheet molding compound, preform, or mat), general-pur-

*From Owens-Corning Fiberglas.

pose, flexible or semirigid, chemical resistant, flame retardant, high heat distortion; also phenolics, melamines, silicones, diallyl phthalate, some epoxies.

Molds

Single- or multiple-cavity hardened and chrome plated molds, usually cored for steam or hot oil heating; sometimes electric heat is used. Side cores, provisions for inserts, and other refinements are often employed. Mold materials include cast or forged steel, cast iron, and cast aluminum.

Major Advantages

Highest volume and highest part uniformity of any thermoset molding method. The process can be automated. Great part design flexibility, good mechanical and chemical properties obtainable. Inserts and attachments can be molded in. Superior color and finish are obtainable, contributing to lower part finishing cost. Subsequent trimming and machining operations are minimized.

TRANSFER MOLDING

Similar to compression molding, except that the charge of sheet or bulk molding compound is heated in a separate chamber, then transferred under heat and pressure into a closed mold where the shape of the part is determined and the final cross-linking reaction takes place. The process is used for small intricate parts that would be difficult to mold using the conventional compression molding process.

INJECTION MOLDING

Highest-volume method of any of the fiberglass-reinforced plastic processes, using single- or multiple-cavity molds to produce very large volumes of complex parts at very high production rates, with a wide variety of mechanical, chemical, electrical, and thermal properties provided by the wide range of thermoplastic resins available.

Process Description

The molding compound — in pellet form, compound concentrate, or dry blend — is heated in the injection chamber of the molding machine. The material is then injected, either by plunger or reciprocating screw under high pressure and in hot, fluid form, into a relatively cold closed mold. After a short cooling cycle, usually 15 to 90 s, the compound is solidified to a degree sufficient to enable the part to be removed from the mold without distortion. A wide variety of small to medium size parts are routinely injection molded; household appliance parts, gears, pump housings, valve bod-

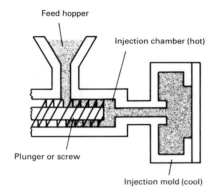

Injection molding of thermoplastics

ies, and large parts such as home laundry tubs, automobile instrument panels, fender liners, and other components requiring high dimensional stability and complexity of design.

Resin Systems

All thermoplastics — notably nylon, acetal, vinyl, polycarbonate, polyethylene, polystyrene, polypropylene, polysulfone, modified polyphenylene oxide, fluorocarbons, ABS (acrylonitrile-butadiene-styrene), and SAN (styrene-acrylonitrile). (Some thermosets can also be injection molded.)

Molds

Single- or multiple-cavity, hardened tool steel molds.

Major Advantages

Highest volume and highest uniformity of part of any molding method. The process is usually highly automated. Great part precision and design flexibility are obtainable. Parts range from very small to quite large, with complex detail, and other design features not possible to mold in any other process. There is versatility with the many resin systems and compounds available, as well as pigmentation and additives for color and many special chemical or thermal properties.

PRESSURE BAG MOLDING

Suited primarily to the production of seamless FRP tanks for water softening, water handling and storage applications, filtration tanks, and fire extinguisher tanks.

Process Description

A fiberglass preform is made by spray-up, mat fabrication, or a combination of methods. An inflatable elastic pressure bag is positioned within the preform and the assembly placed into a cold closed mold. Resin is injected into the mold and the pressure bag is inflated to about 275 kPa (40 psi). Heat is applied and the part is cured within the mold. When cure

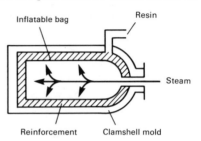

Pressure bag molding

is complete, the bag is deflated and pulled through an opening at one end of the molded part. The part is then removed from the mold.

Molds

Steel "clamshell" molds, hollow, with provisions for injection of resin. Comparatively inexpensive for metal closed molds.

Resin Systems

Polyesters, general-purpose and semirigid, chemical-resistant, low-moisture-absorption types.

Major Advantages

Specialized process yields pressure vessels of one-piece construction at moderately high production rates. Large vessels are possible with the process. Mold cost is low to moderate. Process can use low-cost continuous strand roving, chopped in preform processing. Color can be molded in. Translucent containers which permit visual inspection of their contents can be produced by this process.

Cold Molding Methods for Fiberglass-Reinforced Plastics*

Several fiberglass-reinforced plastic production methods do not employ external heat to effect part cure, the input material having already been heated, or requiring no heat for cure. Two of these methods combine the best properties of fiberglass reinforcements having continuous or long strands with the fine finish, color, and chemical and abrasion resistance of thermoplastic sheets.

COLD PRESS MOLDING

This process is an economical press molding method for manufacturing intermediate volumes of products (200 to 8000), using low-pressure, room-temperature cure and inexpensive molds.

*From Owens-Corning Fiberglas.

Process Description

Fiberglass reinforcement (preform or mat) is applied between matching molds, along with thermosetting resin. Woven roving may be added for strength. Molds may be gel coated. The molds are closed under moderate pressure (138 to 345 kPa; 20 to 50 psi) and the part is cured without heat other than the exothermic heat generated. Elevated molding pressures are not necessary.

Molds

These may be inexpensive metal, plaster, or fiberglass-reinforced plastic molds. Such molds will not have shear edges, and therefore parts must be trimmed after molding. Ribs, bosses, and other shape complexities cannot be produced easily using this process.

Resin Systems

Polyesters, fast curing at room temperature, low-viscosity types.

Major Advantages

Without the necessity for high molding temperatures and pressures, and with relatively inexpensive tooling, cold press molding offers a simple, low-cost way to produce parts hav-

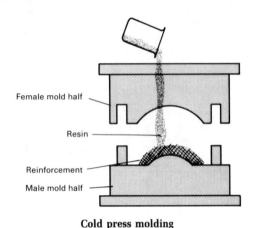

Cold press molding

ing good surface appearance, molded-in color, and dimensionally accurate surfaces, with good mechanical properties. The method could be an excellent way to produce small runs of parts which might later be produced by a higher-volume method, such as compression molding with mat or preform, with no design change and little or no alteration of the material formulation.

COLD STAMPING

A process employing reinforced thermoplastic sheets and typical sheet metal press and die equipment, offering good production rates and parts with good surface finishes.

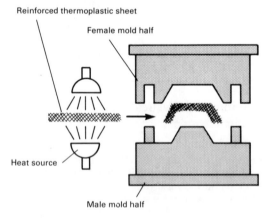

Cold stamping

Process Description

A reinforced-thermoplastic sheet or blank is preheated and placed into a metal mold mounted in a sheet metal stamping press. The molds are closed rapidly to form the part. Molds are held at, or slightly above, room temperature. A cooling cycle of 10 to 12 s requires the press to "dwell" at the bottom of its stroke.

Molds

Matched metal. Heaters are not required, although cooling is required for optimum cycle times.

Resin Systems

Thermoplastic sheet, acrylic or other suitable type.

Major Advantages

The process closely approaches that of standard sheet metal stamping as widely used in the automotive and appliance industries in terms of production rates and low labor costs. Also, pigmentation can be molded in, finishing costs can be reduced, and molded-in bosses and ribs are possible.

COMOFORMING

A special process which combines vacuum-formed thermoplastic shapes with cold molded fiberglass-reinforced plastic laminates to produce parts with excellent surface appearance, weatherability, and strength.

Process Description

A standard thermoformer is used to form a thin thermoplastic skin, usually of acrylic. This skin is placed in a cold mold and a fiberglass-reinforced laminate applied by preform, mat, or spray-up within the thermoformed sheet. The mold is closed, and the part is cured at room temperature by chemical means and is removed when the mold is opened. An alternative method allows the thermoforming and cold molding to be done in the same molds and press, producing the same types of laminate but with less equipment.

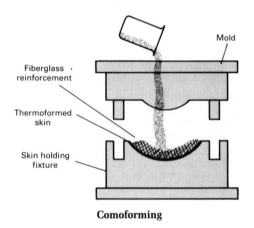

Comoforming

Molds

Matched metal, plaster, or FRP, as used in cold press molding.

Resin Systems

Fast-cure polyesters, plus acrylic thermoformed sheet.

Major Advantages

The method produces laminates having excellent mechanical properties as normally found in FRP moldings with long fibers. They also have the superior surface finish, appearance, and weatherability obtainable with thermoplastic sheet (especially acrylic). The thermoplastic skin considerably enhances the impact strength of the composite. The process offers controlled dimensions, low-cost equipment and molds, and short lead time. Thermoforming and laminating can be done on the same press affording shorter cycle times than with standard cold molding.

Contact Molding Methods for Fiberglass-Reinforced Plastics*

HAND LAY-UP

A low-to-medium-volume contact mold method suitable for making boats, tanks, housings and building panels for prototypes and for other large parts requiring high strength.

Process Description

A pigmented gel coat is first applied to the mold by spray gun for a high-quality surface. When the gel coat has become tacky, fiberglass reinforcement (usually mat or cloth) is manually placed on the mold. The base resin is applied by pouring, brushing, or spraying. Squeegees or rollers are used to densify the laminate, thoroughly wetting the reinforcement with the resin, and removing entrapped air. Layers of fiberglass mat or woven roving and resin are added for thickness.

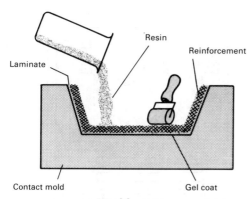

Hand lay-up

Catalysts and accelerators are added to the resin system to allow the resin to cure without external heat. The amounts of catalyst and accelerator are dictated by the working time necessary and overall thickness of the finished part.

The laminate may be cored or stiffened with honeycomb, balsa, foam plastic, or other materials to provide weight reduction or flotation.

Resin Systems

General-purpose, room-temperature curing polyesters which will not drain or sag on vertical surfaces. Also, certain epoxies.

Molds

Simple, single-cavity, one-piece, either male or female, of any size. (Vacuum bag, pressure bag, or autoclave methods may be used to speed cure, increase glass content, and improve off-mold surface finish.)

Major Advantages

Simplest method offering low-cost tooling, simple processing, and wide range of part size. Design changes are readily made. There is a minimum investment in equipment. With good operator skill, good production rates and consistent quality are obtainable.

SPRAY-UP

A low-to-medium-volume, open mold method similar to hand lay-up in its suitability for making boats, tanks, tub/shower units, and other simple medium-to-large-size shapes such as truck roofs, vent hoods, and commercial refrigeration

*From Owens-Corning Fiberglas.

display cases. Greater shape complexity is possible with spray-up than with hand lay-up.

Process Description

Fiberglass continuous strand roving is fed through a combination chopper and spray gun. This device simultaneously deposits chopped roving and catalyzed resin onto the mold. The laminate thus deposited is densified with rollers or squeegees to remove air and thoroughly work the resin into

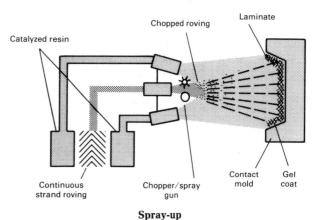

Spray-up

the reinforcing strands. Additional layers of chopped roving and resin may be added as required for thickness. Cure is usually at room temperature or may be accelerated by moderate application of heat.

As with hand lay-up, a superior surface finish may be achieved by first spraying gel coat onto the mold prior to spray-up of the substrate. Woven roving is occasionally added to the laminate for specific strength orientation. Also, core materials are easily incorporated.

Resin Systems

General-purpose, room-temperature curing polyesters, low-heat-curing polyesters.

Molds

Simple, single-cavity, usually one-piece, either male or female, as with hand lay-up molds. Occasionally molds may be assembled in several pieces, then disassembled when removing the part. This technique is useful when part complexity is great.

Major Advantages

Simple, low-cost tooling, simple processing; portable equipment permits on-site fabrication; virtually no part size limitations. The process may be automated.

FOR HIGHER MECHANICAL PROPERTIES FROM CONTACT MOLDS

The following techniques are available to improve off-mold surface and to speed the cure of parts produced by contact molding methods.

These contact molding processes are characterized by simple tooling of easily worked material such as plaster, wood, sheet metal, or fiberglass-reinforced plastic. Production parts have one highly finished side (i.e., that against the mold). There can be a very wide range of part sizes. Extremely high physical properties are attainable, since very high glass-to-resin ratios can be achieved using woven reinforcements.

Vacuum Bag Molding

A flexible film (PVA or cellophane) is placed over the completed lay-up or spray-up, its joints sealed, and a vacuum drawn. Atmospheric pressure eliminates voids in the laminate, and forces excess resin and air from the mold. The addition of pressure further results in higher glass concentration, and provides better adhesion between layers of sandwich construction.

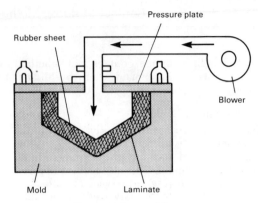

Pressure bag molding

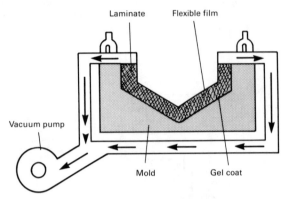

Vacuum bag molding

Pressure Bag Molding

A tailored rubber sheet is placed against the finished lay-up or spray-up, and air pressure is applied between the rubber sheet and a pressure plate. (Steam may be applied to heat the resin to accelerate cure.)

Pressure eliminates voids and drives excess resin and air out of the laminate, densifying it and improving the off-mold surface finish. This process is not compatible with male molds.

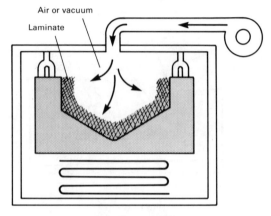

Autoclave molding

Autoclave Molding

Either vacuum bag or pressure bag process can be further modified by the use of an autoclave, which provides additional heat and pressure, producing greater laminate densification and faster cure. This process is usually employed in the production of high-performance laminates using epoxy-resin systems in aircraft and aerospace applications.

Other Molding Methods for Fiberglass-Reinforced Plastics*

The following fiberglass-reinforced plastic production methods offer specialized approaches to the manufacture of parts involving unusual properties such as very large size, extremely high strength, highly directional fiber orientation, unusual shape, or constant cross section. These production methods will commonly be the only ones suitable for most applications for which they were designed.

FILAMENT WINDING

A process resulting in a high degree of fiber orientation and high glass loading to provide extremely high tensile strengths in the manufacture of hollow, generally cylindrical products such as chemical and fuel storage tanks and pipe, pressure vessels, and rocket motor cases.

Process Description

Continuous strand reinforcement is utilized to achieve maximum laminate strength. Reinforcement is fed through a resin bath and wound onto a suitable mandrel (pre-impregnated roving may also be used). Special winding machines lay down continuous strands in a predetermined pattern to provide maximum strength in the directions required. When

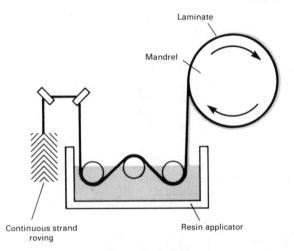

Filament winding

sufficient layers have been applied, the wound mandrel is cured at room temperature or in an oven. The molding is then stripped from the mandrel. Equipment is available to perform filament winding on a continuous basis.

*From Owens-Corning Fiberglas.

Molds

Mandrels of suitable size and shape, made of steel or aluminum, which form the inner surface of the hollow part. Some mandrels are designed to be collapsible, to facilitate part removal.

Resin Systems

Polyesters and epoxies.

Major Advantages

The process affords highest strength-to-weight ratio of any fiberglass-reinforced plastic manufacturing process, and provides highest degree of control over uniformity and fiber orientation. Filament-wound structures can be accurately machined. The process may be automated when high volume makes this economically feasible. The reinforcement used is low in cost. Integral vessel closures and fittings may be wound into the laminate.

CONTINUOUS PULTRUSION

A continuous process for the manufacture of products having a constant cross section, such as rod stock, structural shapes, beams, channels, pipe, tubing, and fishing rods.

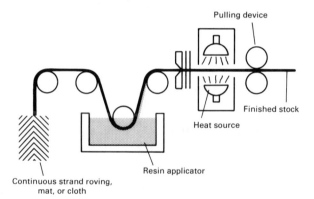

Continuous pultrusion

Process Description

Continuous strand fiberglass roving, mat or cloth is impregnated in a resin bath, then drawn through a steel die which sets the shape of the stock and controls the resin/glass ratio. A portion of the die is heated to initiate the cure. With rod stock, cure is effected in an oven. A pulling device establishes production speed.

Molds

Hardened steel dies.

Resin Systems

General-purpose polyesters and epoxies.

Major Advantages

The process is a continuous operation and can be readily automated. It is adaptable to shapes with small cross-sectional areas, and uses low-cost reinforcement. Very high strengths are possible due to high fiber concentration and orientation parallel to the length of the stock being drawn. There is no practical limit to the length of stock produced by continuous pultrusion.

CONTINUOUS LAMINATING

A continuous process for the production of sheet form output such as glazing panels, corrugated or flat construction panels, and electrical insulation material.

Process Description

Fiberglass chopped rovings and reinforcing mat or fabric are combined with resin and sandwiched between two cellophane skins or other suitable carrier sheets. The material is then fed through rollers to eliminate entrapped air and determine finished laminate thickness. By being passed

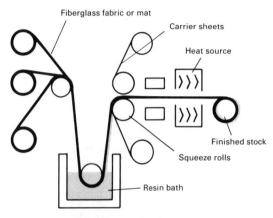

Continuous laminating

through a heating zone (88 to 120 °C; 190 to 250 °F), the material is cured. This process is adaptable to various part thicknesses, and yields good production volume.

Resin Systems

Thermosets, usually weather- and flame-retardant polyesters, or phenolics and melamines for electrical applications. Also, some thermoplastics, primarily acrylics.

Molds

Steel rolls which set the thickness and shape of the stock.

Major Advantages

Unlimited panel lengths are attainable. The process can be automated. Very low tooling cost. A wide variety of surface finishes and textures can be obtained. Wall thicknesses can be closely controlled. Corrugations may be produced by molds or special rollers just prior to the curing stage.

CENTRIFUGAL CASTING

A process for the production of cylindrical hollow shapes: water tanks, pipe, tubing, and containers.

Process Description

Chopped strand mat is positioned inside a hollow, cylindrical mold; or continuous strand roving is chopped and directed onto the inside walls of the mold, which is heated (82 to 93 °C; 180 to 200 °F) and is rotating. Resin is applied to the inside of the rotating mandrel and centrifugal force distributes the resin thoroughly, impregnating the reinforcement. To accelerate cure, hot air is blown through the rotating mold. The mold is stopped and the finished part is removed.

Resin Systems

General-purpose and chemical-resistant polyesters.

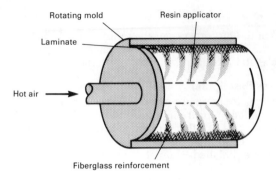

Centrifugal casting

Molds

Cylindrical metal tubing.

Major Advantages

The method can be automated to produce high volumes of pipe or tanks at low tooling cost. Good surfaces are provided inside and out. Low material waste and uniform, void-free laminates are achieved. External threads may be molded in.

ENCAPSULATION

A simple method for the embedment of electronic or electrical parts (such as coils, transformers, connectors) in a protective FRP capsule.

Process Description

Milled fibers or short chopped fiberglass strands are combined with catalyzed resin, and the mixture is poured into small open molds to cast terminal blocks, electrical casings,

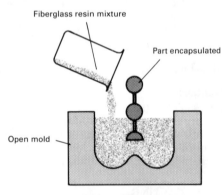

Encapsulation

and the like. The fibers decrease shrinkage and crazing, and increase the useful temperature range of the resin system. Cure may be at room temperature.

Resin Systems

Low exotherm polyesters, epoxies, and silicones.

Molds

Simple open molds of rubber, plaster, wood, or FRP.

Major Advantages

An extremely simple process which lends itself to high volume and automation at very low cost. Inserts of any size, shape, and number can be encapsulated. There is little or no material waste.

ROTATIONAL MOLDING

A versatile method for the production of large and irregularly shaped hollow parts, such as tanks, containers, furniture, and toys.

Process Description

A thermoplastic powder and chopped fiberglass strands or milled fibers are introduced into a hollow mold which is heated while being rotated in two planes. Rotation tumbles the molding powder and fiberglass reinforcement throughout the inner surface of the mold. After the material is fully dispersed and melted, the mold is chilled and the finished part removed from inside the opened mold.

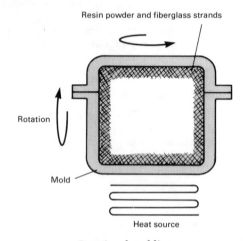

Rotational molding

Resin Systems

Most thermoplastics, in powder form, especially polyethylene, polypropylene, and vinyls.

Molds

Relatively inexpensive, cast aluminum or sheet metal hollow molds.

Major Advantages

The process can be automated to produce medium-to-large-size parts in high volumes. There is no material waste.

FIBERGLASS BACKED VACUUM FORMING

The process consists of vacuum forming a thermoplastic sheet into the desired shape, and then applying a fiberglass-

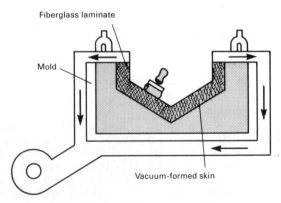

Fiberglass backed vacuum forming

reinforced thermosetting resin to the vacuum-formed part. The vacuum-formed part acts as the mold for the reinforcement and becomes the appearance surface of the finished part.

Process Description

The process is comparable to hand lay-up and spray-up with a gel coat, except that the vacuum-formed thermoplastic unit takes the place of the gel coat. The thermosetting spray-up or lay-up takes place in a simple holding fixture.

Resin Systems

General-purpose polyesters, formulated for good adhesion to thermoplastic sheet and for fast room-temperature cure.

Molds

None, since the vacuum-formed part acts as the mold. Only a simple holding fixture is required, to help the skin hold its shape as the spray-up or lay-up is being accomplished. Conventional vacuum forming molds are used to produce the thermoplastic skin.

Major Advantages

The process eliminates the gel coat operation in spray-up, and affords an opportunity to inspect the outer part surface before the application of the fiberglass laminate takes place — an opportunity obviously impossible with a gel coat operation. Production rates are slightly higher than with gel coating, filling the production gap between spray-up and closed-mold methods. The system is ideal for producing bathroom components, all-terrain vehicle bodies, small boats, snowmobile shrouds, camper trailer bodies, concrete pouring forms, and many other products.

Sheet Molding Compound Systems*

PROCESS DESCRIPTION

Sheet molding compounds for structural molding composites are made by depositing a chemically thickenable, low-viscosity polyester resin uniformly on a continuous polyethylene carrier film. Continuous fiberglass rovings are then chopped to a predetermined length and randomly deposited on the resin. Others may be continuously deposited in the machine direction without chopping.

Simultaneously, a second polyethylene carrier film is covered with a given amount of thickenable catalyzed polyester resin, and brought into contact with the first carrier film. The resultant resin-glass-resin sandwich is passed under kneading and/or compaction rolls to assist in wetting of the reinforcements to form a uniform sheet. The sheet is then wound onto a take-up roll. To ensure that the loss of styrene monomer is held at the lowest possible level, the rolls are wrapped and sealed with barrier film or foil.

Reinforcement content is controlled by the speed of the carrier film passing under the deposition area, the quantities of rovings being continuously deposited (if any) and chopped, and by the chopper speed.

*From PPG Industries.

Rolls of compound are aged for a predetermined period of time to allow uniform viscosity rise before being prepared for molding. Since the compounds are nearly tack-free as they thicken, they can be readily cut into various shapes for mold charging and stripped of the polyethylene film by the press operator.

Molds

Following placement in the matched metal die and mold closing, isotropic sheet compounds flow uniformly throughout the integral contours of the mold to form the final part. Those containing continuous roving are limited to cross-directional flow.

Major Advantages

Sheet molding compounds offer the molder many advantages. They can either be made by the molder himself or be purchased from a compounder, depending on volume. They are stable for 3 to 9 months if properly wrapped and stored at 16 °C (60 °F). Reuse of most edge and charge trim from the compound minimizes waste without adverse effect on the molded product. Parts have not only good physical properties, weatherability, and corrosion resistance with improved surfaces, but also a quite uniform glass/resin ratio throughout.

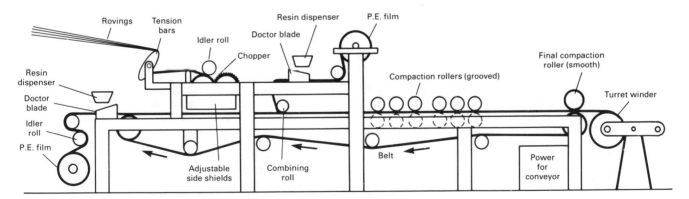

Sheet molding compound (SMC) machine

Thermoset Molding Systems: Premix, BMC, and Injection*

PREMIX COMPOUNDS

In premix molding compounds, the fiberglass chopped strand along with resin, pigment, filler and catalyst are thoroughly mixed for use in bulk form. The premixed material is then formed, in some cases by extrusion, into accurately weighed premold charges, placed into the cavity of the mold, and molded under heat and pressure. Pressure used may be from 690 kPa to 10.3 MPa (100 to 1500 psi). Curing temperature may vary from 105 to 150 °C (225 to 300 °F). Curing time depends on the compound and the configuration of the molded part. Under most circumstances it lasts from 30 s to 5 min.

BMC COMPOUNDS

BMC or bulk molding compounds are formulated with chopped strand fiberglass, resin, thickener, pigment, filler and a catalyst as ingredients. Low-profile agents may also be used when surface finish is critical. The compound is taken in measured quantities by weight and placed into a heated cavity of a mold, where pressure and additional heat are applied. Curing time, temperature, and pressure are dependent on the type of mix and the shape of the product being molded.

*From PPG Industries.

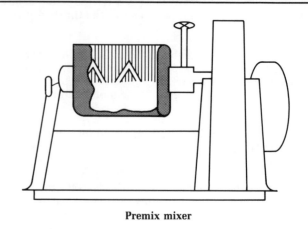

Premix mixer

INJECTION MOLDING THERMOSET COMPOUNDS

Injection molding thermoset compounds are prepared much in the same way as premix or bulk molding compounds. The thermosetting material is placed in the injection molding unit, where accurate amounts are injected into the heated mold. Cycle is controlled by temperature and pressure applied, and may vary from 45 s to 2 min, depending on the product being molded.

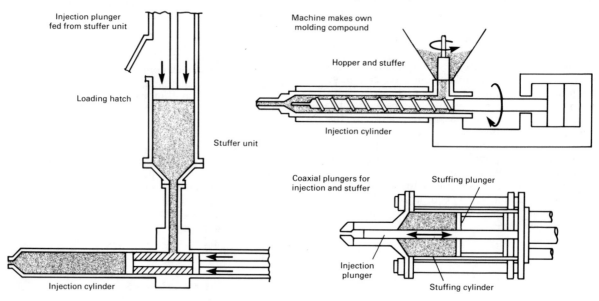

Thermoset injection molding machines

Advantages of Prepregs*

"Prepregs" is the industry term for high-quality reinforcing fabrics (usually fiberglass) which are pre-impregnated with a resin system and partially cured ("B" stage). This pre-impregnated material is delivered to the manufacturer, who molds his products directly from the fabric without the necessity of adding resin. (Prior to the introduction of prepregs, resin was applied manually, by wet lay-up, during lamination.) The material can be produced either dry or tacky to fit

*From Hexcel-Trevarno.

specific applications. Advantages of prepregs are: (a) elimination of formulation problems and mess of "wet lay-up"; (b) consistent quality and resin content; (c) adaptability to assembly line technique for increased production at lower unit costs; (d) fewer rejects; (e) less variance in mechanical properties; (f) finest quality materials; (g) design freedom through prepreg's "workability" in producing irregular shapes, walls of varying thickness, undercuts, flanges, protrusions, etc.; (h) lower inventory levels since no resins or catalysts need be stocked.

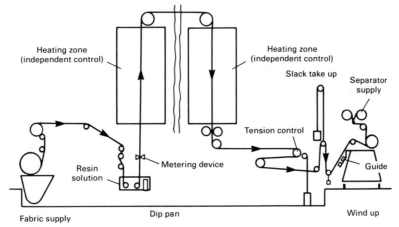

Schematic diagram of impregnating process

ADDITIONAL ILLUSTRATIONS OF PREPREG MOLDING METHODS

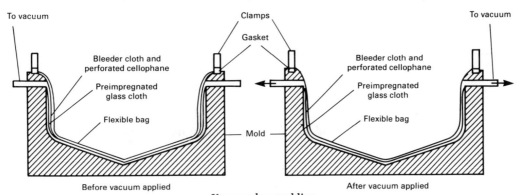

Vacuum bag molding

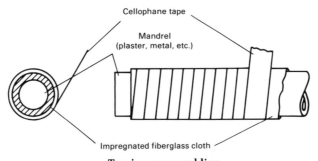

Tension wrap molding

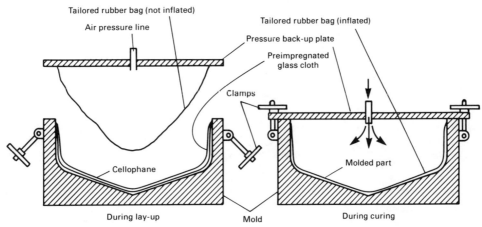

Pressure bag molding

(continued)

ADDITIONAL ILLUSTRATIONS OF PREPREG MOLDING METHODS (continued)

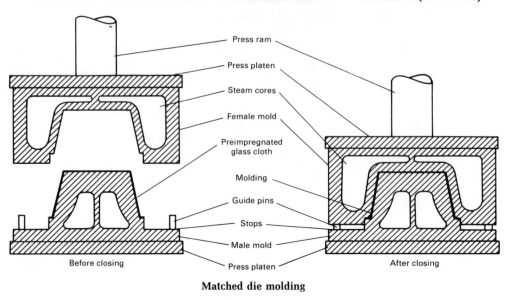

Press ram
Press platen
Steam cores
Female mold
Preimpregnated glass cloth
Molding
Guide pins
Stops
Male mold
Press platen

Before closing

After closing

Matched die molding

Fabrication of Tubular Products From Preimpregnated Fiberglass Fabrics (Hexcel-Trevarno)

In the basic fabrication method, the fabric is cut to a pattern and rolled onto a mold or mandrel (much as one would roll a sheet of paper around a pencil). After curing under heat and pressure, the mandrel is removed, leaving the hollow-core part.

Typical methods of rolling prepreg onto a mandrel

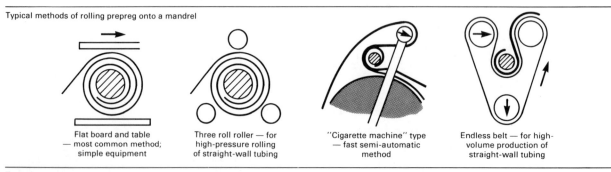

| Flat board and table — most common method; simple equipment | Three roll roller — for high-pressure rolling of straight-wall tubing | "Cigarette machine" type — fast semi-automatic method | Endless belt — for high-volume production of straight-wall tubing |

Typical methods of applying pressure

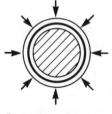   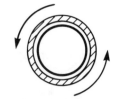

Flexible film — shrinkage of cellophane, Mylar Tedlar film applies pressure

Matched metal mold — high pressure; even sizing of part

Pressure bag — high-presure molding where consistency of outside diameter is not important

Centrifugal casting — low-pressure molding where control of outside diameter is important

Internal bag — high-pressure molding where consistency of inside diameter is not important

6.1.3. Machining of Composites

Design Guides for Fiberglass Composites[d]

| Characteristic | | Compression molding | | | Injection molding (thermoplastics) | Cold press molding | Spray-up and hand lay-up |
		Sheet molding compound	Bulk molding compound	Pre-form molding			
Minimum inside radius, mm (in.)		1.59 ($^1/_{16}$)	1.59 ($^1/_{16}$)	3.18 ($^1/_8$)	1.59 ($^1/_{16}$)	6.35 ($^1/_4$)	6.35 ($^1/_4$)
Molded-in holes		Yes[a]	Yes[a]	Yes[a]	Yes[a]	No	Large
Trimmed in mold		Yes	Yes	Yes	No	Yes	No
Core pull and slides		Yes	Yes	No	Yes	No	No
Undercuts		Yes	Yes[b]	No	Yes	No	Yes[b]
Minimum recommended draft, °/in.		Depths of 6.35–152 mm ($^1/_4$–6 in.), 1–3°; depths of 152 mm (6 in.) and over, 3° or as required				2° 3°	0°
Minimum practical thickness, mm (in.)		1.27 (0.50)	1.52 (0.060)	0.76 (0.030)	0.89 (0.035)	2.03 (0.080)	1.52 (0.060)
Maximum practical thickness, mm (in.)		25.4 (1)	25.4 (1)	6.35 (0.250)	12.7 (0.500)	12.7 (0.500)	No limit
Normal thickness variation, mm (in.)		±0.13 (±0.005)	±0.13 (±0.005)	±0.20 (±0.008)	±0.13 (±0.005)	±0.25 (±0.010)	±0.50 (±0.020)
Maximum thickness build-up, heavy build-up and increased cycle		As required	As required	2-to-1 maximum	As required	2-to-1 maximum	As required
Corrugated sections		Yes	Yes	Yes	Yes	Yes	Yes
Metal inserts		Yes	Yes	NR[c]	Yes	No	Yes
Bosses		Yes	Yes	Yes	Yes	NR[c]	Yes
Ribs		As required	Yes	NR[c]	Yes	NR[c]	Yes
Molded-in labels		Yes	Yes	Yes	No	Yes	Yes
Raised numbers		Yes	Yes	Yes	Yes	Yes	Yes
Finished surfaces (reproduces mold surface)		Two	Two	Two	Two	Two	One

[a]Parallel or perpendicular to ram action only. [b]With slides in tooling, or split mold. [c]Not recommended. [d]Owens-Corning Fiberglas Corp.

Evaluation of Conventional Tools for Machining Aramid Fiber Composites (Ref 100)

Tool description	Size	Speed	Cutting rate	Evaluation	
				Quality	Comments
Circular saw, carbide tipped ....7^1/$_2$ in., 40 tooth		3500 sfpm	4 mat'l 5 fpm	Good	Light fuzz downside only
Remington grit-edge circular ....10-in. diam		4500 sfpm	6 fpm	Good	Not recommended due to grit loading
Cutoff blade-composition ....10-in. diam		4500 sfpm	...	Poor	Generates heat and swells edges
Band saw, wavy set ....^{1}/$_4$ in., 32 tooth		650 sfpm	4 mat'l 2 fpm	Good	...
Band saw, knife edge..................^{1}/$_2$ in.		1000 sfpm	10 fpm	Good	32 plies 281 style dry woven Kevlar
Band saw, scalloped knife edge ....^{1}/$_2$ in.		1000 sfpm	10 fpm	Poor	32 plies 281 style dry woven fabric
Saber saw–Bosch T118A ....1^5/$_8$ in., 20 TPI		191 sfpm	2 fpm	Good	...
Saber saw–Bosch T101P.............	...	191 sfpm	4 fpm	Good	...
Saber saw–Airtech AR3925	...	191 sfpm	2 fpm	Good	Special blade high cost
Saber saw–Bosch T130 Course	...	191 sfpm	1 fpm	Good	Grit coated blade for special applications
Saber saw, knife edge–Bosch 118A......2^1/$_4$ in.		191 sfpm	10 fpm	Good	2 plies 281 style dry woven fabric
Scroll saw	...	...	...	...	Not available
Drill, HSS, standard twist ....^{1}/$_8$ in., 135°		600 rpm	...	Poor	Rejected
Drill, solid, carbide twist ....^{3}/$_8$ in., 135°		600 rpm	...	Poor	Rejected
Carbro single-flute drill................^{3}/$_8$ in.		600 rpm	...	Poor	Rejected
Turboflute drill ....^{3}/$_8$ in.		600 rpm	...	Poor	Rejected
Countersink 2-flute standard	...	600 rpm	...	Poor	Rejected
Weldon countersink....................	...	600 rpm	...	Good	Overheats fast
Router, opposed helical ....^{1}/$_4$ in.		24,000 rpm	6 in./min.	Good	Carbide bit broke-diameter too small
Router, diamond cut carbide...........^{1}/$_4$ in.		24,000 rpm	...	Poor	Rejected
Hog mill, Omcut......................1 in.		4,000 rpm	6 in./min	Good	Suitable for heavy sections
Bosch Nibbler Model 7501.............14 GA		1460 strokes per min.	2 fpm	Excellent	Clean and economical
Circular saw, carbide teeth (modified) Mod #1 ....10 in., 60 tooth		4500 sfpm	6 fpm	Good	No advantage over standard blade
Countersink serrated	...	600 rpm	N/A	Poor	Unsatisfactory finish
HSS twist drill modified ....^{1}/$_4$, ^{5}/$_{16}$, ^{3}/$_8$, ^{1}/$_2$ in.		600 rpm	24 s/in.	Good	Excellent with backup
Solid carbide twist drill (modified)......^{3}/$_8$ in.		600 rpm	20 s/in.	Good	Excellent for graphite composite
Turboflute or Paraflute drill (modified) ....^{3}/$_8$ in.		750 rpm	24 s/in.	Good	Excellent with backup
Circular saw, carbide teeth (modified) Mod #2 ....10 in., 60 tooth		4500 sfpm	6 fpm	Good	More effective than standard

Suggested Operating Conditions for Tools Used in Machining Aramid Fiber Composites[a] (Ref 101)

Type of tool	Suggested speed[g]	Feed rate	Remarks
Single-flute drill ....60–182 smpm (200–600 sfpm)		0.5–0.13 mm/rev (0.002–0.005 in./rev)	Drill locater bushing required. Controlled feed rate recommended. Backup support required.
Brad-point drill ....1800–7600 smpm (6000–25 000 sfpm)		1.5–0.15 mm/rev (0.002–0.006 in./rev)	Controlled feed rate required. Cannot be used to drill metals.
Spade drill ....60–135 smpm (250–450 sfpm)		0.13–0.25 mm/rev (0.005–0.010 in./rev)	Primarily used on thin materials, 0.51 mm (0.020 in.) thick. Backup may be required. Drill locater bushing required.
Self-centering drill ....7–60 smpm (25–200 sfpm)		0.13–0.25 mm/rev (0.005–0.010 in./rev)	Special operational techniques required.
Core drill ....10–20 smpm (30–60 sfpm)		0.15–0.25 mm/rev (0.006–0.010 in./rev)	Backup required. Used on low-resin-content laminates only.
Hole saw ....50–100 smpm (150–300 sfpm)		0.13–0.25 mm/rev (0.005–0.010 in./rev)	Backup required.
Countersinks ....50–250 rpm		...	Microstop tooling required.
Band saws: Conventional ....900–1800 smpm (3000–6000 sfpm)		Up to 1 m/min[c] (up to 36 in./min)	Raker or straight set. 5–9 teeth/cm (14–22 teeth/in.). Hone.
Carbide grit ....900–1800 smpm (3000–6000 sfpm)		Up to 1 m/min[c] (up to 36 in./min)	...

(continued)

Suggested Operating Conditions for Tools Used in Machining Aramid Fiber Composites[a] (Ref 101)
(continued)

Type of tool	Suggested speed[g]	Feed rate	Remarks
Circular saw	900–1800 smpm (3000–6000 sfpm)	Up to 1 m/min[c] (up to 36 in./min)	Coolants[d] may be required, particularly for heavy cuts.
Saber saw	2500 strokes/min	Up to 1 m/min[c] (up to 36 in./min)	Specialty designs available.
Router bits	20 000–27 000 rpm	Up to 1.5 m/min[c] (up to 60 in./min)	Some secondary edge finishing may be required, depending on end-use application.
Edge sander (power)	1200–1500 smpm (4000–5000 sfpm)	...	80–180 grit aluminum oxide or silicon carbide. Belt sanding preferred over disk sanding.
Lathe tools	75–150 smpm (250–500 sfpm)	0.04–0.06 mm/rev (0.0015–0.0024 in./rev)	Depth of cut, 0.25–0.50 mm (0.01–0.02 in.).
Grinding machine	1350–2000 smpm (4500–6500 sfpm)	23–53 smpm (75–175 sfpm)	Use aluminum oxide or silicon carbide wheel. Depth of cut: rough, 0.025–0.075 mm (0.001–0.003 in.); finish, 0.013–0.026 mm (0.0005–0.001 in.). Traverse, $^1/_4$–$^1/_2$ wheel width/pass.
Milling machine	Up to 400 rpm	Up to 0.3 m/min (up to 12 in./min)	Secondary edge finishing required.
Knibbler	2500 strokes/min	Up to 1.5 m/min[c] (up to 60 in./min)	Maximum material thickness, 2.54 mm (0.1 in.).
Water jet	———	For specific recommendations, contact manufacturer. ———	
Laser	———	For specific recommendations, contact manufacturer. ———	
Power shears	...	760–1015 mm/min[c] (30–40 in./min)	Maximum material thickness, 2.54 mm (0.1 in.).

[a]The above machining information should be considered to be only a general guide. Variations within and/or beyond listed suggestions may be required to optimize tool life and/or cut-edge quality for each different composite.
[b]smpm = surface meters per minute; sfpm = surface feet per minute. [c]Depending on material thickness. [d]Suggest water-soluble coolant – 150–200 parts water, 1 part coolant concentrate.

Evaluation of Water-Jet Router for Machining of Aramid Fiber Composites (Ref 100)

Thickness		Jewel hole diameter		Pressure		Cutting time, s	Area of cut		Cutting time for 645 cm² (100 in.²), min	Quality
mm	in.	mm	in.	MPa	ksi		cm²	in.²		
3.18	0.125	0.30	0.012	310	45	41	38.7	6.0	11.4	Fair
3.18	0.125	0.30	0.012	310	45	28	38.7	6.0	7.77	Fair
3.18	0.125	0.30	0.012	310	45	57	38.7	6.0	15.83	Fair
4.57	0.180	0.30	0.012	310	45	100	55.7	8.64	19.29	Poor
4.57	0.180	0.30	0.012	310	45	90	55.7	8.64	17.36	Poor
4.57	0.180	0.30	0.012	310	45	105	55.7	8.64	20.25	Poor
1.52	0.060	0.30	0.012	310	45	14	18.6	2.88	8.1	Good
1.52	0.060	0.30	0.012	310	45	11	18.6	2.88	6.3	Good
1.52	0.060	0.30	0.012	310	45	8.5	18.6	2.88	4.9	Good
1.52	0.060	0.41	0.016	310	45	14	18.6	2.88	8.1	Good
1.52	0.060	0.41	0.016	310	45	9	18.6	2.88	5.2	Good
1.52	0.060	0.41	0.016	310	45	12	18.6	2.88	6.9	Good
4.57	0.180	0.41	0.016	310	45	134	55.7	8.64	25.84	Fair to good
4.57	0.180	0.41	0.016	310	45	131	55.7	8.64	25.27	Good
4.57	0.180	0.41	0.016	310	45	66	55.7	8.64	12.73	Good
3.18	0.125	0.41	0.016	310	45	45	38.7	6.0	12.49	Good
3.18	0.125	0.41	0.016	310	45	47	38.7	6.0	13.05	Good
3.18	0.125	0.41	0.016	310	45	44	38.7	6.0	12.22	Good
7.87	0.310	0.30	0.012	310	45	140	96.0	14.88	15.68	Poor
7.87	0.310	0.41	0.016	310	45	186	96.0	14.88	21.04	Poor

Factors for Drilling of Boron/Epoxy Composites (Ref 3, p 389)

Composite	Speed[a]		Feed (equivalent)		Tool recommendation	Drilling time, min	Tool life, holes
	smpm	sfpm	mm/rev	in./rev			
Boron-epoxy, 2.0 mm (0.08 in.)	91–183	300–600	0.013	0.0005	Core with end-set diamonds	0.1	300–400
Boron-epoxy, 25.4 mm (1.00 in.)	91–183	300–600	0.013	0.0005	Core with end-set diamonds	1.0	75–100
Boron-epoxy/titanium multilayer, 12.7 mm (0.50 in.) total	61–152	200–500	0.0013	0.00005	Core	4.0	30–50

[a]smpm = surface meters per minute; sfpm = surface feet per minute.

Evaluation of Fixed-Head Water-Jet for Machining of Aramid Fiber Composites (Ref 100)

Thickness		Working pressure		Orifice size		Cutting speed		
mm	in.	MPa	ksi	mm	in.	mpm	fpm	Results
1.5	0.06	345	50	0.18	0.007	3.0	10	Good
9.7	0.38	345	50	0.30	0.012	0.3	1	Good

6.1.4. Sandwich Construction

Characteristics and Geometry of Honeycombs*

The basic geometry of honeycomb provides six primary characteristics:

1. Highest strength-to-weight ratio as a sandwich core
2. Highest stiffness-to-weight ratio as a sandwich core
3. High ratio of exposed surface area to total volume
4. Exposure of surface area in parallel cells
5. Variable ratio of honeycomb material area to volume
6. Uniform crushing strength under compression.

Each characteristic can be combined with the qualities inherent in the material selected. Because honeycomb can be produced from almost any material available in continuous web or roll form, the extent to which honeycomb's characteristics can be used to advantage is unlimited.

*Data from Hexcel-Trevarno.

Use of Honeycomb's Properties*

STRUCTURAL APPLICATIONS

Thermal

Honeycomb is available for a wide range of temperature exposures. Aluminum honeycomb, made from alloys such as 5052, 5056, and 2024, has been used for service temperatures up to 215 °C (420 °F). Stainless steel is useful in even higher-temperature environments. Fiberglass reinforced phenolic honeycomb has excellent thermal stability and has been used from −253 to +205 °C (−423 to +400 °F).

Polyimide resins extend the upper limit to about 315 °C (600 °F), with short-time exposures up to 370 °C (700 °F). Good mechanical properties result in extensive use of nonmetallic honeycomb in aircraft and missile structures operating within such temperature ranges. New products under development, such as ceramics, can extend this upper range.

Electrical

Electrical and mechanical properties of glass-fiber reinforced plastic honeycomb have led to standard use of the material in both airborne and ground radomes. These structures must withstand high loads which require considerable core strength and they must serve as efficient radar windows without attenuating or distorting transmissions. The dielectric constants of these materials are quite low — 1.04 to 1.28 — with corresponding loss tangents of 0.001 to 0.004.

Fatigue Resistance

Bonded honeycomb sandwich has replaced riveted structures which were subject to high-energy, sonic vibration in jet aircraft. Stress concentration is minimized when loads are distributed evenly in this type of bonded structure, and, as a result, operating life is increased by several orders of magnitude. While fatigue resistance is relative to operating con-

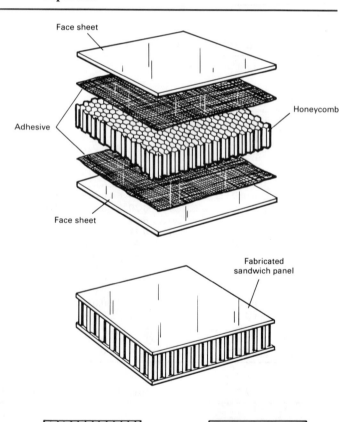

*Data from Hexcel-Trevarno.

Honeycomb Sandwich Construction

ditions and design, all studies show that honeycomb structures resist fatigue loading to extents which are far superior to alternative design methods.

Rigidity

As with fatigue resistance, rigidity is a design-sensitive characteristic. However, the nature of a honeycomb sandwich allows design of low-deflection structures at minimum weight. Where smooth reflective surfaces of extremely low deflection and high accuracy are required, honeycomb sandwich is a widely used design approach. Applications range from mammoth ground support radar equipment through solar energy concentrators to small, specialized reflectors for spacecraft.

Environmental Stability

Because honeycomb core exhibits all of the physical properties of the material from which it is made, the choice of material dictates its performance in structural or nonstructural exposure.

A protective coating is applied to aluminum honeycomb to give excellent corrosion resistance. The effectiveness of the coating has been documented by extensive testing under conditions of high humidity and salt spray. Several types of paper, films, and fibrous sheet materials are available in honeycomb form which are resistant to a wide range of chemicals, including mild acid and caustic solutions.

NONSTRUCTURAL APPLICATIONS

Energy Absorption

An extremely effective mechanical-energy absorber, honeycomb may be used to control forces exerted on decelerating objects. Materials such as sponge, solid rubber, foams, cork, and paper wadding generally exhibit spring characteristics with the attendant rebound problem. Aluminum, ar-

amid, and stainless steel honeycomb, however, have the unique property of failing at a constant load while completely dissipating energy otherwise released in rebound.

The threshold at which compressive failure begins can be eliminated by prestressing honeycomb core to produce slight initial compressive failure. Exposed to further loading, the prestressed core carries the crushing load at a near-linear rate. Such control cushions the impact of air-dropped supplies, provides earthquake-damage restraints for above-ground pipelines, protects instrumental missile assemblies, safeguards human occupants of rapid-transit vehicles, and acts as a safety system in fuel-cask transfer pools in nuclear power plants.

Directionalizing

The parallel cell orientation of expanded honeycomb allows it to directionalize air and fluid flow. Honeycomb's slight cell edge exposure minimizes pressure loss when liquids or gases pass through it. Typical directionalizing applications include grilles, registers, and wind-tunnel straightening vanes where honeycomb, properly placed, encourages higher velocities while reducing turbulence.

RF Shielding

Metal honeycomb has been used extensively in grilles and registers of shielded radio equipment. The electrical continuity of the foil ribbons permits use of honeycomb as an effective RF shield at all frequencies below the cut-off frequency of the cell size and thickness used.

Heat Exchange

The extremely high ratio of surface area to unit volume provided by honeycomb offers many possibilities in heat-exchange applications ranging from air conditioners to large industrial cooling towers. Materials available for heat exchangers include aluminum and stainless steel alloys.

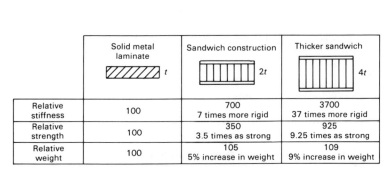

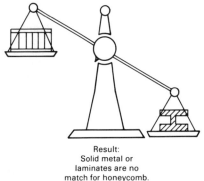

	Solid metal laminate	Sandwich construction	Thicker sandwich
	t	$2t$	$4t$
Relative stiffness	100	700 7 times more rigid	3700 37 times more rigid
Relative strength	100	350 3.5 times as strong	925 9.25 times as strong
Relative weight	100	105 5% increase in weight	109 9% increase in weight

Result: Solid metal or laminates are no match for honeycomb.

Honeycomb Configurations (Hexcel-Trevarno)

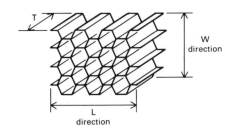

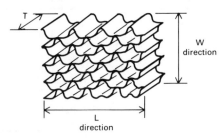

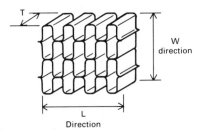

Honeycomb Property Data[a]

Honeycomb type[b]	Nominal density, lb/ft³	Compressive Bare Strength, psi Typ	Min	Stabilized Strength, psi Typ	Min	Modulus (typ), ksi	Crush strength (typ), psi	Plate shear "L" direction Strength, psi Typ	Min	Modulus (typ), ksi	"W" direction Strength, psi Typ	Min	Modulus (typ), ksi
Aluminum Alloy Honeycombs													
5052-1/8-0.0007	3.1	270	200	290	215	75	130	210	155	45.0	130	90	22.0
5052-3/16-0.0015	4.4	500	360	525	385	145	250	330	280	68.0	215	160	30.0
5052-1/4-0.001	2.3	165	120	175	130	45	75	140	100	32.0	85	57	16.2
5052-3/8-0.0007	1.0	30	20	45	20	10	25	45	32	12.0	30	20	7.0
5056-1/8-0.0007	3.1	340	250	360	260	97	170	250	200	45.0	155	110	20.0
5056-3/16-0.0015	4.4	600	460	650	490	180	310	410	340	68.0	245	198	27.5
5056-1/4-0.001	2.3	205	145	210	155	58	100	170	130	32.0	105	62	15.0
5056-3/8-0.0007	1.0	35	25	50	35	15	35	60	45	15.0	35	25	9.0
2024-1/8-0.0015	5.0	700	525	780	620	200	425	500	400	82.0	315	250	33.0
2024-3/16-0.0015	3.5	330	250	370	290	86	200	290	230	55.0	180	143	23.0
Glass Fabric Reinforced Plastic Resin Honeycombs													
Phenolic-3/16	4.0	500	350	600	480	57	···	260	210	11.5	140	110	5.0
Phenolic-1/4	3.5	350	260	500	400	46	···	230	170	9.0	120	110	3.5
Phenolic-3/8	3.2	320	245	440	350	38	···	200	160	8.0	105	85	3.0
NP[c]/polyester-3/16	4.5	520	365	670	470	80	···	280	195	13.5	130	90	5.2
NP[c]/polyester-1/4	4.0	420	295	560	390	68	···	260	180	13.0	120	85	5.0
Aramid Fiber Paper[d] Honeycombs													
1/8 (1.5 gauge)	1.8	110	70	130	85	···	···	90	65	3.7	50	36	2.0
1/8 (2 gauge)	3.0	300	180	330	270	20	···	180	162	7.0	95	85	3.5
3/16 (2 gauge)	2.0	150	90	170	105	11	···	110	72	4.2	55	40	2.2
3/16 (5 gauge)	6.0	650	580	700	650	···	···	390	330	14.5	185	150	6.0
1/4 (2 gauge)	1.5	90	45	95	55	6	···	75	45	3.0	35	23	1.5
1/4 (5 gauge)	3.1	275	180	285	240	···	···	170	135	7.0	85	60	3.0
3/8 (2 gauge)	1.5	90	45	95	55	6	···	75	45	3.0	35	23	1.5
3/8 (5 gauge)[e]	3.0	285	···	300	···	17	···	170	···	5.6	95	···	3.0

[a]Data from Hexcel-Trevarno. [b]Materials designated as follows: Material/alloy–cell size–foil thickness for aluminum alloy honeycombs. All cell configurations are hexagonal. [c]NP = initial impregnation of fabric is nylon-modified phenolic resin; final dip coats are polyester. [d]Du Pont's Nomex, an aramid-fiber paper treated with phenolic resin. [e]Preliminary data.

Common Honeycomb Adhesives[a]

Nitrile phenolic films
Modified epoxy liquid and pastes
Modified epoxy film, 120 °C (250 °F) cure and below
Modified epoxy film, 175 °C (350 °F) cure
Epoxy/polyimide film
Polyimide films
Modified urethane pastes and liquids
Core splicing adhesives, pastes, and tapes

[a]From Hexcel-Trevarno.

Comparison of Design Considerations for Various Honeycombs[a]

Consideration	5052 and 5056 aluminum alloys	2024 aluminum alloy	Commercial grade aluminum	Phenolic resin impregnated glass fabric	Phenolic resin impregnated bias weave glass fabric	Nylon-modified phenolic and polyester impregnated glass fabric	Aramid fiber phenolic treated paper
Maximum service temperature	175 °C (350 °F)	215 °C (420 °F)	175 °C (350 °F)	175 °C (350 °F)	175 °C (350 °F)	82 °C (180 °F)	260 °C (500 °F)
Flammability	E	E	E	E	E	E	E
Impact resistance	G	G	G	F	G	F	F
Moisture resistance	G[b]	F[b]	G[b]	E	E	E	E
Fatigue strength	G	G	G	G	G	G	G
Heat transfer	H	H	H	L	L	L	L

[a]Data from Hexcel-Trevarno. Key: E = excellent; G = good; F = fair; H = high; L = low. [b]Moisture and corrosion resistance are excellent for corrosion-resistant grades.

Honeycomb Core Products*

Description	Advantages	Applications	Approvals/specifications
Nomex cells with phenolic resin matrix	High strength-to-weight ratio, corrosion and chemical resistant, low radar image, good electrical properties	Aircraft, missile, and space-vehicle parts	BMS 8-124, DMS-1974, LCM-28-1041, MIL-C-81986
Same as above but with industrial grade Nomex; not for use in aircraft	As above	Marine panels, racing cars, mass transportation, industrial applications	
Fiberglass cells with phenolic resin matrix	Higher heat, moisture, and corrosion resistance, high stiffness-to-weight ratio, radar transparency	Aircraft, missile and space-vehicle parts	BMS 8-124, MIL-C-8073D, FMS 1022
Expansion is in the transverse direction, yielding rectangular cells	Flexible, formable core in ribbon direction, useful where a single curvature is needed	Aircraft, missile, and space-vehicle parts where moldability is needed	BMS 8-124, DMS-1974, LCM-28-1041, MIL-C-81986
Matrix resin partially cured, allowing a later, final cure	Adds flexibility to core in all directions, i.e., formability to compound curvatures	Same as above	MIL-C-8073A, BMS 8-124, DMS-1974, LCM-28-1041
Partially cured, over-expanded rectangular cells	Maximum flexibility, formability, and moldability	Same as above	MIL-C-8073A, BMS 8-124, DMS-1974, LCM-28-1041

Structural and Nonstructural Panels

Description	Advantages	Applications	Approvals/specifications
Nomex/phenolic core, with two plies of woven fiberglass/phenolic on each side	High stiffness, low flammability and smoke, low creep; low radar image, noncorroding	Marine bulkheads, joiner panels, doors, false decks, sanitary partitions; firewalls and floors for railroad cars	U.S. Navy, U.S. Coast Guard — approved for use on Coast Guard Fleet; not for use in aircraft
Nomex/phenolic core, with three plies on one side, two plies on the other side	Extra stiffness, stronger surface for fasteners	Same as above	Same as above
Same as above, with phenolic foam in the honeycomb cells	Improved fire resistance in some tests; sound absorption	Same as above	Same as above
Fiberglass/epoxy skins with unidirectional Nomex/phenolic core	Lightweight, stiff and strong	Aircraft flooring, interiors, luggage racks, galleys, racing car shells, boat hulls	Boeing
Carbon fiber/epoxy skins with Nomex/phenolic core	Ultralight, stiff and strong	Aircraft flooring, interiors	Boeing, Canadair
Fiberglass/phenolic skins with Nomex/phenolic core	Low FST, lightweight, high impact resistant	Aircraft cargoliners and floors, galleys	
Carbon fiber/phenolic skins with Nomex/phenolic core	Ultralight, stiff, strong, low FST	Aircraft flooring, decking, containers, furniture, boats, trailers, mobile homes	
Unidirectional fiberglass/resin skins with Nomex/phenolic core	Low FST, lightweight, high impact resistance	Aircraft galleys and cargo floors	Airbus
Aluminum core, with woven fiberglass/epoxy skins (various aluminum grades available)	High dimensional stability (similar to steel). Can be worked with standard industrial tools	Helicopter platforms, computer housings, exterior and interior panels for buildings, transportation; aircraft cargoliners and floors	Boeing, McDonnell Douglas
High temperature resistant Aerolam F	Withstands autoclave conditions	Tooling boards	
Can be tailored to meet specific needs. Core can be fiberglass, foam, Nomex, metal honeycomb, etc., with skins of Kevlar, veneers, vinyls, fiberglass, carbon, etc.			

Fabricated Products

Description	Advantages	Applications	Approvals/specifications
Filament-wound tubular structures of carbon, glass or Kevlar fibers and fiber hybrids of these fibers in an epoxy matrix	High strength and stiffness for torque translation, lightweight; high critical speed, longer shaft capability, parts reduction. Proprietary end-fitting	Automotive, machine and aircraft drive shafts, push/pull rods, struts, ship stanchions, towers, windmills	F.A.A.
Armor programs for various military agencies involving advanced composite technology	Weight savings coupled with improved fire protection	Various military programs, vehicles, shelters	

(continued)

Honeycomb Core Products* (continued)

Description	Advantages	Applications	Approvals/specifications
Both woven and unidirectional laminates from fiberglass and/or aramid fibers Leafsprings, aircraft seats and trays, custom-made panels, military and medical devices, etc.	High impact strength coupled with burn-through protection Lightweight, corrosion resistant, provides noncatastrophic failure; high stiffness, reduced parts count, more easily assembled	Cargoliners	

*Data from Ciba-Geigy.

Properties of Typical Honeycomb Sandwich Facing Materials[a]

Facing material	Yield strength[b] MPa	Yield strength[b] ksi	Modulus of elasticity GPa	Modulus of elasticity 10^6 psi	λ_f,[h] $1 - \mu^2$	Weight per mil thickness kg/m²	Weight per mil thickness lb/ft²	Comments
Aluminum:								
1100-H14	89.6C	13C	68.9	10.0	0.89	0.068	0.014	Moderate cost, workable, excellent chemical resistance, scars easily
3003-H16	124C	18C	68.9	10.0	0.89	0.068	0.014	Fair strength, good weather resistance, moderate cost
5052-H34	165C	24C	69.6	10.1	0.89	0.068	0.014	Better strength, good weather resistance, moderate cost
6061-T6	241C, T	35C, T	68.9	10.0	0.89	0.068	0.014	Good strength, workable, only heat treatable alloy easily welded
2024-T3	290T	42T	72.4	10.5	0.89	0.068	0.014	Excellent strength, heat treatable, soft stage for working, fair corrosion resistance
7075-T6	455T	66T	71.0	10.3	0.89	0.068	0.014	High strength, fair corrosion resistance
Mild carbon steel	345	50	200	29	0.91	0.20	0.040	Low cost, high weight, good availability
Stainless steel:								
316	414	60	193	28	0.94	0.20	0.040	High cost, corrosion resistant
17-7 PH 1380	1380	200	200	29	0.94	0.20	0.040	High strength, heat treatable
Graphite, woven	586	85	55.8	8.1	0.99	0.039[c]	0.008[c]	High cost, strength, and modulus
Graphite, unidirectional 1290	1290	187	126	18.3	0.99	d	d	High cost, strength, and modulus
Fiberglass prepreg:								
Epoxy F155[e]	427	62	22.8	3.3	0.98	0.046	0.0095	Excellent strength, low-temperature cure
Epoxy F161[e]	399	57.9	25.5	3.7	0.98	0.044	0.0090	Heat resistant, good strength
Phenolic F120[e]	331	48	24.1	3.5	0.98	0.043	0.0088	High temperature, good strength
Polyester F141[e]	331	48	24.1	3.5	0.98	0.044	0.0090	Good strength, low cost
Polyimide F174[e]	414	60	24.1	3.5	0.98	0.042	0.0085	High temperature resistant
Epoxy unidirectional F155	690	100	44.8	6.5	0.98	d	d	Highest strength
Kevlar F155[f]	221C	32C	28.3C	4.1C	1.00	0.034	0.007	High cost, high tensile
Kevlar F155[f]	483T	70T	28.3T	4.1T	...	...	...	Low weight, tough, low compression
Fiberglass mat, polyester resin	96.5	14	6.3	0.92	0.98	0.034	0.007	Very low cost
Woven roving, polyester resin	262	38	12.8	1.85	0.98	0.034	0.007	Very low cost
Ext fir plywood[g]	18.3	2.65	12.4	1.8	0.99	0.015	0.003	...
Pine plywood[g]	20.7	3.0	12.4	1.8	0.99	0.015	0.003	...
Luan plywood[g]	15.5	2.25	12.4	1.8	0.99	0.010	0.002	...
Tempered hardwood, 1.12 Mg/m³ (70 lb/ft³)	24.8	3.6	4.5	0.65	0.99	0.024	0.005	Good hardwood, low cost, smooth surface
Gypsum board	0.83	0.12	2.1	0.30	0.98	0.020	0.004	Fire resistant, low cost

[a]Data from Hexcel-Trevarno. [b]Yield strength is the lower of tensile (T) or compressive (C). [c]3K-70-PU ($12^1/_2 \times 12^1/_2$ Thornel 300/3000 Tow 7 mil/ply). [d]Dependent on fiber areal weight of 145 g/m² (4.28 oz/yd²). [e]All values are for 1581 fabric and autoclave cure. [f]Both compressive (C) and tensile (T) shown for 285 styles. [g]For calculations involving plywood, "effective thickness" should be used except for locating centroid: actual thicknesses of 6.35, 9.52, and 12.7 mm ($^1/_4$ or 0.250, $^3/_8$ or 0.375, and $^1/_2$ or 0.500 in.) correspond to effective thicknesses of 3.56, 3.56, and 6.60 mm (0.14, 0.14, and 0.26 in.), respectively. [h]μ = Poisson's ratio of facing; $\lambda = 1 - \mu^2$.

Honeycomb Manufacturing Process*

Two processes are used: expansion and corrugation. Aluminum foils are cleaned and treated with a clear protective coating that interacts with the aluminum surface to provide superior protection for honeycomb exposed to corrosive environments.

Expansion. Material in web or roll form is fed into high-precision machines where continuous ribbons of structural

*Data from Hexcel-Trevarno.

adhesives are applied. Sheets are cut and stacked layer upon layer. Horizontal slices are sawed to required thickness. Slices are then expanded into honeycomb panels. The expansion method is most generally used in the manufacture of aramid, aluminum, and fiberglass-reinforced cores.

Corrugation. Corrugation is used primarily for metallic core with higher densities than can be obtained using the expansion method. Densities of 190 to 880 kg/m³ (12 to 55 lb/ft³) are available from this adhesive bonded corrugated process.

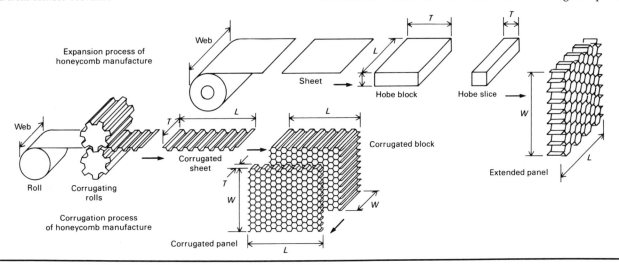

Honeycomb Used in Helicopter Blade To Stabilize Thin Skin

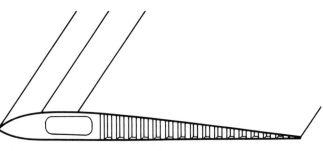

"Green Cure" or Partially Cured Honeycomb Is Flexible Enough To Form Parts of Complex Shape

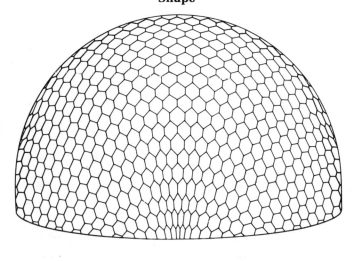

Overexpanded Honeycomb Used for Simple Curvature

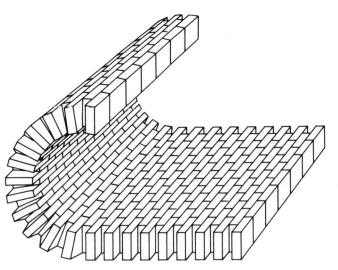

6.2. Property Data for Polymer Matrix Composites

6.2.1. Comparative Properties of Fiber Reinforced and Filled Resin Matrix Composites

Comparative Properties of Neat-Resin and Carbon- and Glass-Reinforced Engineering Thermoplastics
(Ref 33, p 38–41)

Resin type	Tensile strength MPa	ksi	Flexural modulus GPa	10^6 psi	Impact strength,[a] notched/unnotched J/cm	ft·lb/in.	Heat-deflection temperature °C	°F
Amorphous Resins								
Acrylonitrile-styrene-butadiene (ABS):								
Base resin	55	8	2.8	0.4	1.9/>21	3.5/>40	95	205
30% glass fiber	100	14.5	7.6	1.1	0.75/3.5	1.4/6.5	105	220
30% carbon fiber	130	18.8	12.4	1.8	0.59/2.4	1.1/4.5	105	220
Nylon:								
Base resin	66	9.5	2.6	0.38	>21/NB	>40/NB	125	255
30% glass fiber	148	21.5	7.9	1.15	0.64/3.7	1.2/7.0	140	285
30% carbon fiber	207	30	15.2	2.2	0.64/4.3	1.2/8.0	145	290
Polycarbonate:								
Base resin	62	9	2.3	0.33	1.4/>21	2.7/>40	130	265
30% glass fiber	128	18.5	8.3	1.2	2.0/9.34	3.7/17.5	150	300
30% carbon fiber	165	24	13.1	1.9	0.96/5.34	1.8/10.0	150	300
Polyetherimide:								
Base resin	105	15.2	2.8	0.4	0.53/13	1.0/25	200	392
30% glass fiber	197	28.5	8.6	1.25	0.75/5.60	1.4/10.5	215	420
30% carbon fiber	234	34	17.2	2.5	0.75/6.67	1.4/12.5	215	420
Polyphenylene oxide (PPO):								
Base resin	66	9.5	2.5	0.36	1.4/>21	2.7/>40	85–150	185–300
30% glass fiber	145	21	9.0	1.3	1.2/5.1	2.3/9.5	155	310
30% carbon fiber	159	23	11.7	1.7	0.53/3.0	1.0/5.6	155	310
Polysulfone:								
Base resin	69	10	2.8	0.4	0.64/>21	1.2/>40	170	340
30% glass fiber	124	18	8.3	1.2	0.96/7.5	1.8/14	185	365
30% carbon fiber	159	23	14.5	2.1	0.64/3.5	1.2/6.5	185	365
Styrene-maleic-anhydride (SMA):								
Base resin	54	7.8	3.2	0.47	0.2/NB	0.4/NB	115	235
30% glass fiber	103	15	9.0	1.3	0.59/2.4	1.1/4.5	120	250
Thermoplastic polyurethane:								
Base resin	14	2	0.1	0.015	>21/NB	>40/NB	32	90
30% glass fiber	57	8.2	1.3	0.19	5.1/15	9.5/28	170	340
Crystalline Resins								
Acetal:								
Base resin	61	8.8	2.8	0.4	0.69/>21	1.3/>40	110	230
30% glass fiber	134	19.5	9.7	1.4	0.96/4.8	1.8/9.0	165	325
20% carbon fiber	81	11.8	9.3	1.35	0.53/1.6	1.0/3.0	160	320
Nylon 66:								
Base resin	80	11.6	2.8	0.41	0.48/>21	0.9/>40	77	170
30% glass fiber	179	26	9.0	1.3	1.5/11	2.9/20	255	490
30% carbon fiber	241	35	20.0	2.9	0.80/6.4	1.5/12	257	495
Polybutylene terephthalate (PBT):								
Base resin	59	8.5	2.3	0.34	0.48/>21	0.9/>40	85	185
30% glass fiber	134	19.5	9.7	1.4	1.4/9.1	2.6/17	210	410
30% carbon fiber	152	22	15.9	2.3	0.64/3.5	1.2/6.5	210	410
Polyethylene terephthalate (PET):								
30% glass fiber	159	23	9.0	1.3	1.0/···	1.9/···	225	435
Polyphenylene sulfide (PPS):								
Base resin	74	10.8	4.1	0.6	0.16/1.9	0.3/3.5	140	280
30% glass fiber	138	20	11.0	1.6	0.75/4.5	1.4/8.5	260	500
30% carbon fiber	186	27	16.9	2.45	0.59/2.9	1.1/5.5	265	505

[a]NB = no break.

Effects of Various Reinforcements and Fillers/Additives on Properties of Engineering Thermoplastic Resins (Ref 33, p 34–35)

Modifying agent	Typical loading, %	Tensile strength	Flexural modulus	Impact strength	Heat-deflection temperature:	Flammability resistance	Electrical conductivity	Wear resistance	Chemical resistance	Dimensional stability	Molding precision	Creep resistance	Cost effectiveness
Reinforcements													
Glass fiber	10–50	↑↑↑	↑↑↑	↓	↑↑↑↑	↑↑	—	↑↑	↑↑	↑↑↑	↑↑	↑↑↑	↑↑
Carbon fiber (chopped strand)	10–40	↑↑↑↑	↑↑↑↑	↓	↑↑↑	↑↑	↑↑↑↑	↑↑	↑↑	↑↑↑↑	↑↑	↑↑↑↑	↑↑↑↑
Aramid fiber (chopped strand)	5–20	↑↑	↑↑	↓	↑↑↑	↑	—	↑↑	↑↑	↑↑	↑↑	↑↑	↑↑↑
Mineral fiber (wollastonite, PMI, calcium, sulfate)	10–40	↑↑	↑↑	↓↓	↑↑	↑↑	—	↑	↑	↑↑	↑↑	↑↑	↑
Fillers/Additives													
Minerals (talc, clay, mica, calcium carbonate, silica)	40	↑	↑↑	↓↓	↑↑	↑	—	↑	↑↑[a]	↑↑	↑↑↑↑	↑↑	→
Metals (flake/fibers)	10–40	↓	↑↑	↓↓	↑↑	↑	↑↑↑	↑	↑↑[a]	↑	↑↑	↑↑	↑↑↑
Carbon powder	10–20	↓↓	↑	↓↓	↑	↓	↑↑↑↑	→	↑	↑	↑	↑	→
Flame retardants: Organic	5–20	↓	→	↓	→	↑↑↑	—	→	↑↑[a]	→	—	→	↑↑
Inorganic	5–40	↓	↑↑	↓↓	↑	↑↑↑	—	→	↑	↑	↑↑	↑	↑
Internal lubricants (polytetrafluoroethylene, molybdenum disulfide, silicone)	5–15	↓	↑	↓↓	↑	↑	—	↑↑↑↑	↑	↑	↑↑	↑↑	↑↑
Glass beads	10–40	↓↓	↑↑	↓↓	↑	↑	—	↑	↑↑	↑↑	↑↑↑↑	↑↑	→
Impact modifiers	5–15	↓↓	→	↑↑↑↑	↓↓	→	—	→	↑↑[a]	↓↓	→	↓↓	↓↓[a]
Antistatic agents	1–5	↓	→	→	→	—	↑↑	—	—	—	—	—	↑↑↑
UV stabilizers	Up to 1	↓	→	→	→	—	—	—	—	—	—	—	↑↑↑

Legend: Effects shown by arrow designations are ranked relatively for the particular property, rather than by absolute values. For arrows followed by an "a," the effect depends on the specific additive/filler.

↑ → ↓ } Negligible effect
↑↑ →→ ↓↓ } Some effect
↑↑↑ →→→ ↓↓↓ } Significant effect
↑↑↑↑ →→→→ ↓↓↓↓ } Very substantial effect
—— No effect

Comparative Properties of Selected Commercial Reinforced and Nonreinforced Engineering Plastics[a]

Property	Material: Polyphenylene sulfide Trade name(s): Ryton (R-4) Filler content: Glass	Polyphenylene sulfide Ryton (R-10) Glass and mineral	Polyamide-imide Torlon Unfilled	Polysulfone Udel Unfilled	Polysulfone Thermocomp Sulfil 30% glass	Polyetherimide Ultem Unfilled	Polyetherimide Ultem 30% glass	Polyethersulfone Victrex 20% glass	Polyamide Thermocomp Zytel Nylon 66 33% glass
Mechanical[b]									
Tensile strength, ksi	17.5	10.0–13.5	27.0	10.2	14.5	15.0	24.5	18.0	28.0 (dry)
Elongation, %	0.9	0.7	12–18	50–100	1.5	60	2.7	3	3 (dry)
Flexural strength, ksi	26.0	16.0–20.0	30.0	15.4	20.0	21.0	33.0	25.0	41.0 (dry)
Flexural modulus, 10^6 psi	1.7	2.0	0.664	0.39	1.05	0.48	1.2	0.856	1.3 (dry)
Compressive strength, ksi	26.0	18.5	...	40.0	...	20.3	...	...	29.4 (dry)
Rockwell hardness	R123	R120	E78	M69	M90–100	M109	...	M98	M100 (dry)
Izod impact at 75 °F, ft·lbf/in:									
Notched	1.3	0.7–1.0	2.5	1.2	1.1	1.0	2.0	1.5	2.2 (dry)
Unnotched	4.5	1.7–2.8	...	...	...	25	8.0	...	...
Thermal									
Heat-deflection temperature[c] at 264 psi, °F	500	500	525	345	350	392	410	420	485
UL temperature index, °C	200/220	220/240	...	140	150	170	170	180	140
Coefficient of linear thermal expansion, in./in.·°F $\times 10^{-5}$	1.2	1.2	1.9	3.1	1.4	3.5	1.1	1.5	0.9–1.1
Thermal conductivity, Btu·in./h·ft²·°F	2.0	3.9	1.7	1.8	1.5	1.5	...	2.3	2.7
Oxygen index, %	46.5	53	42	38	32	47	...	40	24
UL94 flammability rating	94 V-0/94-5V	94 V-0/94-5V	94 V-0	94 V-0	94 V-9	94 V-0	94 V-0	94 V-0	94 HB to 94 V-0[d]
Electrical									
Dielectric strength, V/mil	450	320–400	600	425	475	620	...	375	300–525
Dielectric constant at 1 MHz	3.8	4.8–6.1	3.8–4.0	3.0	3.5	3.0	...	3.5	3.7
Dissipation factor at 1 MHz	0.0014	0.01–0.02	0.005	0.003	0.005	0.001	...	0.006	0.02
Physical									
Density,[e] g/cm³	1.6	1.9	1.4	1.25	1.46	1.27	1.51	1.5	1.38
Water absorption, %	0.05	0.07	0.28	0.3	0.3	0.25	0.18	0.35	1.0
Chemical									
Chemical resistance	Affected by some oxidizing acids, some amines, and halogens; resistant to all solvents below 400 °F	Affected by some oxidizing acids, some amines, and halogens; resistant to all solvents below 400 °F	Affected by most alkalis and organic solvents	Affected by some aromatic and chlorinated hydrocarbons	Affected by some aromatic and chlorinated hydrocarbons	Affected by phenol, methylene chloride, and 1,1,2-trichloroethane	Affected by phenol, methylene chloride, and 1,1,2-trichloroethane	Affected by ketones and some chlorinated hydrocarbons	Affected by strong alkalis, phenols, formic acid, and hydrofluoric acid

[a]Data from Phillips Chemical Company. [b]All PPS specimens prepared using 275 °F mold temperature. Mechanical performance can be improved by using cold mold. [c]PPS specimens annealed 2 h at 500 °F. [d]Rating determined by amount of flame-retardant additive which can affect mechanical and electrical properties. [e]For cost comparison, calculate cost per cubic inch as follows: ¢/in.³ = resin cost (¢/lb) × density (g/cm³) × 0.0361.

Comparative Properties of Selected Commercial Reinforced and Nonreinforced Engineering Plastics[a] (continued)

Property	Thermoplastic polyester PBT — Valox Celanex Tenite — 30% glass	Thermoplastic polyester PET — Rynite — 30% glass	Phenolic — Durez Genal Plenco — Mineral	Phenolic — Durez Genal Plenco — 30% glass	Epoxy — Polyset Plaskon — 60% glass	Polyimide — Kinel Vespel Skybound — 50% glass	Diallyl Phthalate — Durez Plaskon Diall — 30% glass	Unsaturated polyester — Reichold Glastic — Glass and mineral
Mechanical								
Tensile strength, ksi	17.0–19.0	23.0	6.0–9.7	7.0–18.0	5.0–20.0	6.4	6.0–11.0	5.0–6.0
Elongation, %	2–4	2–3	0.1–0.5	0.2	4	...	3–5	...
Flexural strength, ksi	26.0–29.0	33.0	11.0–14.0	15.0–60.0	18.0–40.0	21.3	9.0–20.0	10.0–12.0
Flexural modulus, 10^6 psi	1.2	1.4	1.2	2.0–3.3	2.0–4.5	2.0	1.5	1.3
Compressive strength, ksi	18.0–23.5	25.0	22.5–35.0	26.0–70.0	18.0–40.0	37.0	25.0–35.0	24.0
Rockwell hardness	M90	...	E88	E54–101	M100–112	M118	E80–87	...
Izod impact at 75 °F, ft·lbf/in.								
Notched	1.3–1.6	1.6–1.9	0.26–0.36	0.5–18.0	0.3–10	5.6	0.4–15.0	0.5
Unnotched	...	...	...	...	...	...	...	...
Thermal								
Heat-deflection temperature at 264 psi, °F	428	435	360–475	350–600	225–500	660	330–550	330–500
UL temperature index, °C	140	...	170	170	130	180 (30% glass)	130	...
Coefficient of linear thermal expansion, in./in.·°F × 10^{-5}	1.4	1.6	1.1–1.5	0.5–1.2	0.6–2.8	0.7	0.6–2.0	...
Thermal conductivity, Btu·in./h·ft²·°F	1.3	2.0	...	5.1–9.0	6.0	3.0	2.1	...
Oxygen index, %	20	20	50–55	50–55	34	40	35	...
UL94 flammability rating	94HB to 94 V-0[d]	94HB	94 V-1 to V-0	94 V-0	94HB to 94 V-0[d]	94 V-0	94HB to 94 V-0[d]	94HB to 94 V-0
Electrical								
Dielectric strength, V/mil	460–550	550	200–350	140–400	250–400	450	400–450	390
Dielectric constant at 1 MHz	3.6	3.5	9.0–15.0	4.0–7.0	4.5	4.7	4.0	5.0
Dissipation factor at 1 MHz	0.02	0.014	0.7–0.2	0.1–0.2	0.01	0.007	0.01	0.01
Physical								
Density,[e] g/cm³	1.52	1.6	1.42–1.84	1.7–2.0	1.6–2.0	1.6–1.7	1.7–2.0	1.8–2.0
Water absorption, %	0.06–0.08	0.05	0.1–0.3	0.03–1.2	0.04–0.20	0.7	0.12–0.35	0.1–0.3
Chemical								
Chemical resistance	Affected by alkalis and some polar solvents	Affected by alkalis and some polar solvents	Affected by some strong acids and most strong alkalis	Affected by some strong acids and most strong alkalis	Affected by some strong acids	Affected by strong alkalis	Affected slightly by strong acids and alkalis	Affected by some strong acids and most strong alkalis

[a] Data from Phillips Chemical Company. [b] All PPS specimens prepared using 275 °F mold temperature. Mechanical performance can be improved by using cold mold. [c] PPS specimens annealed 2 h at 500 °F. [d] Rating determined by amount of flame-retardant additive which can affect mechanical and electrical properties. [e] For cost comparison, calculate cost per cubic inch as follows: ¢/in.³ = resin cost (¢/lb) × density (g/cm³) × 0.0361.

Applications and Properties of Filled Resins[a]

Material formulation[b]	Resin type[b]	Application	Melt flow rate,[c] g/10 min	Density,[d] g/cm³	Mold shrinkage,[e] in./in.	Tensile strength,[f] ksi	Elongation break[g] (2 in. per min), %	Water absorption[h] (24 h at 73 °F), %	Flexural strength,[i] ksi	Flexural modulus,[i] 10⁶ psi	Izod impact strength[k] at 73 °F, ft·lb/in. Notched 1/8 in.	Notched 1/4 in.	Unnotched 1/8 in.	Unnotched 1/4 in.	Gardner impact (2 lb) at 73 °F, in.·lb	Heat deflection[m] (1/8 in.), °F 66 psi	264 psi	Shore hardness[n] A	D
Calcium Carbonate Filled																			
10% HP	PP	Sheeting	0.5–2.5	0.975	0.016–0.025	4.8	100	0.02	6.6	0.250	1.10	0.90	12.80	10.15	30	190	140	95	74
20% HP	PP	Lawn and patio equipment	3.0–7.0	1.04	0.010–0.017	4.2	50	0.02	6.0	0.270	0.78	0.65	13.30	8.20	8	195	145	95	72
40% HP	PP	Food containers	18.0–25.0	1.28	0.006–0.012	3.8	40	0.02	6.5	0.370	0.70	0.60	10.20	5.05	35	220	150	94	74
40% HP	PP	Housewares	25.0–35.0	1.25	0.006–0.012	3.5	40	0.02	6.2	0.350	0.60	0.50	17.60	8.00	40	220	160	96	75
20% CP	PP	Outdoor equipment	12.0–18.0	1.06	0.010–0.016	3.3	>100	0.02	5.0	0.200	1.30	1.10	22.10	20.00	>130	175	125	96	63
40% CP	PP	Toys, housewares	6.0–10.0	1.23	0.006–0.012	3.5	25	0.02	6.0	0.350	1.20	0.55	9.65	6.50	80	210	145	93	70
Hi 40% CP	PP	Housewares	18.0–25.0	1.23	0.008–0.015	2.4	80	0.02	4.0	0.200	1.50	0.90	14.10	9.15	80	190	135	97	67
Talc Filled																			
20% HP	PP	Medical instruments	3.0–7.0	1.04	0.010–0.015	5.3	18	0.01	7.0	0.400	0.70	0.60	8.05	7.65	10	250	150	87	72
30% HP	PP	Medical instruments	8.0–12.0	1.13	0.005–0.012	4.9	12	0.02	7.5	0.475	0.65	0.55	6.75	5.80	9	255	158	90	74
40% HP	PP	AC housings, fan shrouds	3.0–7.0	1.24	0.004–0.010	5.0	7.5	0.02	8.5	0.625	0.60	0.50	5.33	4.35	8	268	180	94	75
20% CP	PP	Cabinets	6.0–10.0	1.03	0.010–0.018	4.1	40	0.02	5.8	0.280	1.30	1.10	21.00	19.50	60	220	130	85	70
40% CP	PP	Downspouts	3.0–7.0	1.24	0.008–0.014	4.5	20	0.02	7.0	0.510	0.75	0.60	5.85	4.15	10	245	150	93	78
Mica Filled																			
30% Unc	PP	Pool equipment	3.0–7.0	1.13	0.005–0.008	4.7	5	0.04	7.5	0.590	0.60	0.53	3.05	2.55	6	263	195	92	75
40% Unc	PP	Instrument panels	3.0–7.0	1.23	0.003–0.008	5.5	4	0.05	8.5	0.900	0.55	0.50	2.20	1.70	4	278	210	97	78
12% Cou	PP	Automotive components	6.0–10.0	0.990	0.008–0.015	5.3	10	0.03	7.0	0.420	0.80	0.60	10.10	7.70	12	250	170	86	66
20% Cou	PP	Automotive components	4.0–10.0	1.03	0.005–0.008	5.8	5	0.01	10.0	0.520	0.85	0.55	3.50	3.90	24	275	203	90	77
30% Cou	PP	Pool equipment	3.0–7.0	1.13	0.005–0.008	6.3	5	0.04	10.9	0.730	0.60	0.55	3.25	2.95	7	275	195	93	76
40% Cou	PP	Automotive components	3.0–7.0	1.23	0.003–0.008	6.5	3	0.05	11.5	1.000	0.55	0.53	3.10	2.15	6	285	220	96	76
50% Cou	PP	Automotive components	6.0–10.0	1.35	0.002–0.006	5.5	3	0.06	11.0	1.300	0.52	0.45	2.25	1.75	4	310	260	94	76
Glass Filled																			
10% Unc	PP	Automotive components	…	0.970	0.005–0.008	6.0	4	0.05	8.5	0.350	0.80	0.60	7.00	6.00	6	275	250	98	74
20% Unc	PP	Appliances	…	1.04	0.003–0.006	6.5	3	0.05	9.0	0.525	1.1	1.0	7.10	7.35	6	290	260	89	74
30% Unc	PP	Appliances	…	1.13	0.002–0.006	8.0	2	0.05	10.0	0.650	1.1	1.0	10.80	7.75	8	295	270	97	78
10% Cou	PP	Automotive components	…	0.970	0.005–0.008	7.2	5	0.05	10.5	0.470	1.2	0.60	7.45	6.70	7	285	265	96	73
20% Cou	PP	Automotive components	…	1.04	0.003–0.006	10.0	3	0.05	11.0	0.600	1.3	1.0	7.80	7.45	8	310	275	92	75
30% Cou	PP	Automotive components	…	1.13	0.002–0.006	10.5	2	0.05	13.0	0.750	1.6	1.4	11.50	7.90	10	320	295	95	77

(continued)

Applications and Properties of Filled Resins[a] (continued)

Material formulation[b]	Resin type[b]	Application	Melt flow rate,[c] g/10 min	Density,[d] g/cm³	Mold shrinkage,[e] in./in.	Tensile strength,[f] ksi	Elongation break[g] (2 in. per min), %	Water absorption[h] (24 h at 73 °F), %	Flexural strength,[i] ksi	Flexural modulus,[j] 10⁶ psi	Izod impact strength[k] at 73 °F, ft·lb/in. Notched 1/8 in.	Notched 1/4 in.	Unnotched 1/8 in.	Unnotched 1/4 in.	Gardner impact (2 lb) at 73 °F, in.·lb	Heat deflection[m] °F, 1/8 in. 66 psi	264 psi	Shore hardness[n] A	D
Combination/Special																			
FR	PP	Electrical appliances	1.0–1.5	0.950	0.015–0.020	4.4	25	0.03	6.6	0.190	0.90	0.70	...	...	8	190	130	...	...
TF/FR	PP	Industrial appliances	7.0–11.0	1.18	0.007–0.012	5.0	10	0.02	9.0	0.500	0.80	0.60	6.35	6.75	5	265	175	92	77
CC/GF	PP	Closures	6.0–10.0	1.24	0.008–0.013	4.0	6	0.03	7.3	0.400	0.85	0.50	7.35	5.30	8	270	185	93	75
MF/TF	PP	Trays	0.5–2.5	1.17	0.004–0.010	4.4	8	0.08	8.4	0.600	0.90	0.80	4.70	3.55	8	257	194	94	77
CP/Blk	PP	Containers, automotive	8.0–12.0	0.910	0.010–0.020	4.0	100	0.02	5.0	0.450	1.5	1.3	20.90	17.20	120	190	150	94	66
Other Materials																			
MDPE	PE	Wires, cables, conduits	1.0	0.938	0.020–0.033	2.2	500	0.02	...	0.050	NB[p]	14.5	NB[p]	NB[p]	>160	...	...	89	52
FR/Gray 30% CC	PE	Pipe conduits	1.1	1.00	0.015–0.026	2.1	>450	0.02	...	0.060	NB[p]	2.15	NB[p]	NB[p]	145	...	...	90	53
10% MF	PE	Housewares	1.0–2.0	1.20	0.012–0.025	1.3	>50	0.02	1.1	0.025	>7.9	>7.0	NB[p]	NB[p]	100	...	...	90	51
HDPE	PE	Stereo equipment	3.0–5.0	1.02	0.015–0.025	3.4	40	0.02	4.5	0.200	3.25	1.60	15.35	14.00	75	178	124	88	60
20% GF	ABS	Automotive components	1.5–2.5	1.17	0.000–0.002	9.0	2.5	0.02	13.0	0.650	1.25	1.0	4.40	5.30	10	207	194	98	81

[a]Data from Washington Penn Plastic Co. [b]Key: PP–polypropylene; PE–polyethylene; CC–calcium carbonate filled; TF–talc filled; MF–mica filled; GF–glass filled; HP–homopolymers; CP–copolymers; Hi–high impact; HDPE–high-density polyethylene; MDPE–medium-density polyethylene; FR–flame retardant; Cou–coupled; Unc–uncoupled; Blk–black. [c]ASTM D-1238. [d]ASTM D-729. [e]ASTM D-953. [f]ASTM D-638. [g]ASTM D-570. [h]ASTM D-350. [i]ASTM D-790. [j]ASTM D-256. [k]ASTM D-648. [m]ASTM D-648. [n]ASTM D-2240. [p]NB = no break.

Properties of Filled and Unfilled Ptfe Composites (Ref 29, p 393)

Property	Filler 15 wt % glass	12 1/2 wt % glass, 12 1/2 wt % MoS₂	15 wt % graphite	20 wt % carbon, 5 wt % graphite	55 wt % bronze, 5 wt % MoS₂	None[a]
Specific gravity	2.19	2.3	2.12	2.1	3.9	2.2
Tensile strength, MPa	17.5	13	9.5	11.6	13.0	9
Elongation, %	300	230	130	70	90	400
Flexural modulus, GPa	1.1	1.1	1.4	1.2	1.5	0.6
Deformation under load at 25 °C and 14 MPa, %	11	4	8.1	2.9	4.6	CF
Expansion coefficient from 25 to 100 °C, 10^{-5} /°C	12.1	11	12.5	8.4	10.1	17
Thermal conductivity, W/m·°C	0.43	0.51	0.45	0.44	0.72	0.25
Specific wear rate, 10^{-7} mm³/N·m	1.4	1.2	6.8	1.2	1.0	4000
Friction coefficient on steel at 0.01 m/s	0.09	0.09	0.12	0.12	0.13	0.1
Limiting PV, MPa × m/s, at:						
0.05 m/s	0.33	0.5	0.35	0.53	0.44	0.04
0.5 m/s	0.39	0.62	0.6	0.53	0.44	0.06
5 m/s	0.5	0.62	0.95	0.42	0.44	0.09

[a]CF = continuous flow.

Mechanical Properties of a Bulk Molding Compound Reinforced With Glass Fibers and With Glass and Ceramic Fibers[a]

Property	Glass fiber[b]	Ceramic fiber[c,d]	Coated ceramic fiber[c,d]
Tensile strength:			
MPa	23.6	27.4	30.8
ksi	3.42	3.98	4.47
Flexural strength:			
MPa	64.4	51.2	62.1
ksi	9.34	7.43	9.00
Flexural modulus:			
GPa	17.2	15.7	15.7
10⁶ psi	2.50	2.28	2.28
Izod impact strength, notched:			
J/m	224	264	262
ft·lb/in.	4.20	4.94	4.91
Izod impact strength, unnotched:			
J/m	257	227	271
ft·lb/in.	4.82	4.25	5.07

[a]Data from Standard Oil Engineered Materials. [b]6.35-mm ($^1/_4$-in.) chopped fiberglass, 15.6 wt %. [c]Fiberfrax. [d]6.35-mm ($^1/_4$-in.) chopped fiberglass, 11.7 wt %, and chopped Fiberfrax, 3.9 wt %.

Typical Properties of Sheet Molding Compounds[a] (Ref 7, p 2.48)

Fiber reinforcement	Specific gravity	Elongation, %	Tensile strength MPa	ksi	Tensile modulus GPa	10⁶ psi
33% glass	1.67	1.4	103	15	11.7	1.7
35% carbon	1.66	0.6	103	15	26.9	3.9
24% glass, 26% carbon	1.65	0.5	117	17	24	3.5

[a]Data generated from flat plates from an SMC formulation using an isophthalic polyester matrix resin. Molding conditions: 7 MPa (1 ksi) at 149 °C (300 °F) for 3 min. Reinforcing fibers ranged in length from 12.7 to 50.8 mm (0.5 to 2 in.).

Properties of Poly(amide-imide) Engineering Resins With Various Reinforcements and Additives[a]

Nominal composition	Properties/characteristics	Applications
High-Strength Composites		
3% TiO₂ ¹/₂% fluorocarbon	Best impact resistance, most elongation, and good mold-release and electrical properties	Connectors, switches, relays, thrust washers, spline liners, valve seats, poppets, mechanical linkages, bushings, wear rings, insulators, cams, picker fingers, ball bearings, rollers, thermal insulators
30% glass fiber 1% fluorocarbon	High stiffness, good retention of stiffness at elevated temperatures, very low creep, and high strength	Burn-in sockets, gears, valve plates, fairings, tube clamps, impellers, rotors, housings, back-up rings, terminal strips, insulators, brackets
30% glass fiber 1% fluorocarbon	Can be molded to greater thickness with some sacrifice in mechanical properties	Same as above, but for parts requiring thicker cross sections
30% glass fiber 4% TiO₂ 1% fluorocarbon	High stiffness, good retention of stiffness at elevated temperatures, very low creep, and high strength	Structural, electrical, valve plates, metal replacement
30% graphite fiber 1% fluorocarbon	Best retention of stiffness at high temperatures, best fatigue resistance; electrically conductive	Metal replacement, housings, mechanical linkages, gears, fasteners, spline liners, cargo rollers, brackets, valves, labyrinth seals, fairings, tube clamps, standoffs, impellers, shrouds, potential use for EMI shielding
33% carbon fiber 1% fluorocarbon	Can be molded to greater thickness with some sacrifice in mechanical properties; electrically conductive	Injection molding of normally troublesome thick cross sections
Proprietary blend of carbon fibers and fluorocarbons	High stiffness and lubricity	Service requiring high stiffness and some lubricity, especially sliding vanes; potential use for EMI shielding
Wear-Resistant Composites		
12% graphite powder 8% fluorocarbon	Good for reciprocating motion or bearings subject to high loads at low speed; best wear resistance	Bearings, thrust washers, wear pads, strips, piston rings, seals
12% graphite powder 8% fluorocarbon	Designed for bearing use; good wear resistance, low coefficient of friction, and high compressive strength	Bearings, thrust washers, wear pads, strips, piston rings, seals, vanes, valve seats
20% graphite powder 3% fluorocarbon	Better wear resistance at high speeds	Bearings, thrust washers, wear pads, strips, piston rings, seals, vanes, valve seats
High-Performance Composite		
40% glass fiber 1% fluorocarbon	Best cost-to-performance ratio	Switches, relays, terminal strips, wear bands, back-up rings, housings, impellers, brackets, thermal insulators

[a]From Amoco Chemicals Corp.

Properties of Thermoplastics With Lubricating Fillers (Ref 29, p 396)

Property	Polycarbonate 22% ptfe	Polycarbonate 15% ptfe, 30% glass	Acetal 22% ptfe	Acetal 15% ptfe, 30% glass	Nylon 66 44% ptfe	Nylon 66 15% ptfe, 30% glass	Polyimide 15% graphite	Polyimide 15% MoS$_2$
Specific gravity	1.33	1.55	1.5	1.75	1.43	1.49	1.51	1.59
Tensile strength, MPa	45	120	40	107	38	163	45	41
Flexural modulus, GPa	1.3	8.3	2.1	9.7	2.1	9.3	3.8	3.5
Heat-distortion temperature at 1.8 MPa, °C	130	145	100	160	82	250	>260	>260
Izod notched impact strength, J/cm	1.1	1.1	0.27	0.38	0.27	0.98	···	···
Moisture absorption in 24 h, %	0.14	0.06	0.25	0.2	0.55	0.5	0.32	0.32
Specific wear rate, 10^{-7} mm^3/N·m	···	5.8	3.2	38	2.3	3.1	5	50
Friction coefficient on steel	0.15	0.2	0.15	0.28	0.18	0.26	0.1–0.3	0.1–0.3
Limiting PV, MPa × m/s, at:								
0.05 m/s	···	0.97	>1.4	0.44	>1.4	0.61	6	5
0.5 m/s	0.06	1.05	0.4	0.42	0.95	0.7	···	···
5 m/s	···	0.46	0.17	0.28	0.28	0.46	···	···

Properties of Electrically Conductive Metal-Flake/Fiber Composites[a]

ASTM test	Property	Aluminum flakes SMA 40 wt % flakes	Aluminum flakes ABS 40 wt % flakes	Nylon 5 wt % fibers	Stainless steel fibers Polycarbonate 5 wt % fibers	Stainless steel fibers Polycarbonate 5 wt % fibers	Stainless steel fibers Polycarbonate 10 wt % fibers	Stainless steel fibers PES 10 wt % fibers	Stainless steel fibers ABS 7 wt % fibers	Stainless steel fibers PPS[b] 5 wt % fibers	Stainless steel fibers Mod PPO 3 wt % fibers
Mechanical Properties											
D638	Tensile strength:										
	MPa	44.8	22.7	58.6	93.7	68.0	75.8	72.5	41.3	73.0	55.1
	ksi	6.5	3.3	8.5	13.6	10.0	11.0	10.5	6.0	10.6	8.0
D638	Elongation, %	1.8	1.5	2.0	3.0	5.0	4.0	1.7	3.8	1.3	5.0
D638	Tensile modulus:										
	GPa	···	···	···	4.8	3.1	3.4	5.1	···	···	2.4
	10^5 psi	···	···	···	7.0	4.5	5.0	7.3	···	···	3.5
D790	Flexural strength:										
	MPa	72.3	42.7	117	13.8	110	117	130	68.9	114	82.7
	ksi	10.5	6.2	17.0	20.0	16.0	17.0	18.9	10.0	16.5	12.0
D790	Flexural modulus:										
	GPa	6.9	4.1	4.4	5.2	3.1	3.4	4.1	3.0	10.0	2.5
	10^5 psi	10.0	6.0	6.4	7.5	4.5	5.0	6.0	4.3	14.5	3.6
D256	Izod impact strength, notched[c]:										
	J/m	64.0	74.8	42.7	80.1	69.4	74.8	64.0	53.4	42.7	80.1
	ft·lb/in.	1.2	1.4	0.8	1.5	1.3	1.4	1.2	1.0	0.8	1.5
Physical Properties											
D792	Specific gravity	1.42	1.54	1.22	1.28	1.27	1.33	1.55	1.12	1.50	1.10
D570	Water absorption in 24 h, %	···	···	···	0.12	0.12	0.12	0.4	0.4	0.03	0.08
···	Linear mold shrinkage in 3 mm ($^1/_8$ in.), %	···	···	···	0.15	0.4	0.3	0.5	0.4	0.3	0.5
Thermal Properties											
D648	Deflection temperature under load[d]:										
	°C	130	88	232	143	141	141	215	93	252	124
	°F	266	190	450	290	285	285	420	200	486	255
D648	Deflection temperature under load[e]:										
	°C	···	···	···	149	146	146	221	···	···	···
	°F	···	···	···	300	295	295	430	···	···	···
···	Flammability at minimum thickness (UL 94)	···	V-0	HB	V-1	V-0	HB	V-0	···	V-0	HB
Electrical Properties											
D257	Volume resistivity, Ω·cm	10^0	10^0	10^1	10^1	10^1	10^0	10^0	10^1	10^2	10^3
D257	Surface resistivity, Ω/sq.	···	···	10^2	10^2	···	···	···	10^6	10^3	···
···	Shielding effectiveness,[f] dB	50	50	40	40	40	50	50	···	···	···

[a]Data from Wilson-Fiberfil International. All tests conducted at 23 °C (73 °F). [b]Also contains 5 wt % PAN carbon fiber. [c]Izod impact test bars 6.35 × 12.7 mm ($^1/_4$ × $^1/_2$ in.). [d]At 1.82 MPa (264 psi). [e]At 0.45 MPa (66 psi). [f]Attenuation at 1000 MHz for 3-mm ($^1/_8$-in.) section.

Shielding Effectiveness Comparison of Various Fiber Composites (Wilson-Fiberfil International)

Depending on the conductivity of the specific composite, shielding attenuation equivalent to that of nickel paint can be achieved.

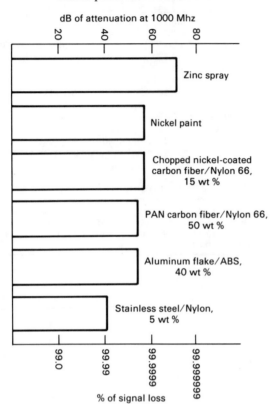

Comparison of Injection-Moldable Polycarbonates for EMI Shielding[a]

Property	Un-filled	Propri-etary system[b]	25% carbon fiber	25% metal-lized glass	30% Al flake
Tensile strength:					
MPa	62	55	138	82.7	41
ksi	9.0	8.0	20.0	12.0	6.0
Elongation, %	6–8	7	1.1	2.0	2.9
Izod impact:					
J/cm	1.3	0.9	1.1	0.9	1.1
ft·lb/in.	2.4	1.7	2.0	1.6	2.0
Flexural strength:					
MPa	89.6	93.1	193	145	75.8
ksi	13.0	13.5	28.0	21.0	11.0
Flexural modulus:					
GPa	2.3	2.3	13.8	6.8	4.4
10^6 psi	0.33	0.33	2.00	0.98	0.64
DTUL[c]:					
°C	132	142	146	143	141
°F	270	288	295	290	285
Specific gravity	1.20	1.27	1.31	1.40	1.44
Attenuation at 1.0 GHz, dB	0	40	30	20	30

[a]Data from Wilson-Fiberfil International. [b]Conductive additives include stainless steel fibers. [c]At 1.82 MPa (264 psi).

Comparison of Properties of Metal- and Resin-Matrix Composites (Ref 21, p 68)

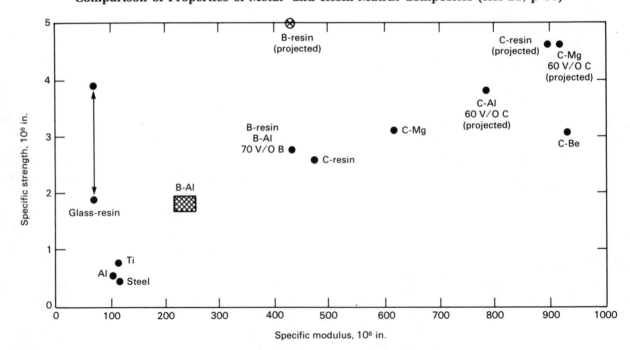

Effects of Fiber Type on Properties of SMC's (Ref 51, p 46–48)

(a) Effect of fiber type on density of unfilled SMC's. (b) Effect of fiber type on Young's modulus of filled SMC's. (c) Effect of fiber type on Poisson's ratio of filled SMC's. (d) Effect of fiber type on density of filled SMC's.

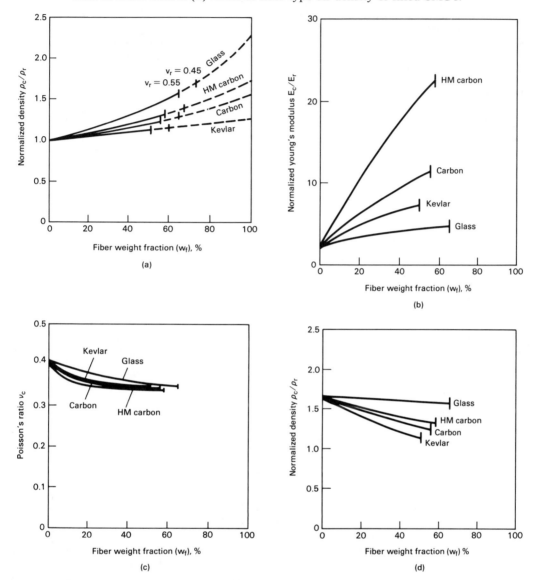

6.2.2. Comparison of Various Epoxy Matrix Composites

Room-Temperature Tensile and Specific Strengths of Various Fiber-Reinforced Epoxies (Ref 11, p 70)

Type	Fiber[a] Density, lb/in.^{3}	Volume %	Composite Density[b] (D), lb/in.^{3}	Strength (S), ksi	Modulus (E), 10^6 psi	S/D 10^3 in.	E/D 10^6 in.	Source
S-glass	0.090	60	0.072	290	7.4	4000	100	Owens-Corning Fiberglas Corp.
B/W	0.095	61.8	0.075	133.3	20.1	1780	268	Air Force Materials Laboratory
Be	0.067	73.3	0.061	150	3.3	2500	54	Naval Ordnance Laboratory
C	0.072	NR[c]	NR[c]	90	40	1500	660	Rolls-Royce Ltd.
C	0.072	40	0.055	105	22	1900	400	Royal Aircraft Est.
C	0.054	64	0.051	62	4.8	1210	94	Union Carbide Corp.
Al$_2$O$_3$	0.143	44	0.063	72	24	810	380	GE-SSL
Al$_2$O$_3$	0.143	14.2	0.059	113	6.0	1920	102	GE-SSL

[a]Unidirectional orientation. [b]Density of resin $\approx$ 0.046 lb/in.^{3}. [c]NR = not reported.

Properties of Conventional Structural Materials and Bidirectional (Cross-Ply) Fiber/Epoxy Composites
(Ref 27, p 2.39)

Material	Fiber volume fraction (V_f), %	Tensile modulus (E), GN/m²	Tensile strength (σ_u), GN/m²	Density (ρ), g/cm³	Specific modulus (E/ρ)	Specific strength (σ_u/ρ)
Mild steel	...	210	0.45–0.83	7.8	26.9	0.058–0.106
Aluminum:						
2024-T4	...	73	0.41	2.7	27.0	0.152
6061-T6	...	69	0.26	2.7	25.5	0.096
E-glass/epoxy	57	21.5	0.57	1.97	10.9	0.26
Kevlar-49/epoxy	60	40	0.65	1.40	29.0	0.46
Carbon fiber/epoxy	58	83	0.38	1.54	53.5	0.24
Boron/epoxy	60	106	0.38	2.00	53.0	0.19

Impact Characteristics of Epoxy Composites
(3M)

Comparison of Dielectric Properties of Quartz/Epoxy and Fiberglass/Epoxy[a]

Material[b]	Property[c] Dielectric constant	Loss tangent
Astroquartz	3.52	0.017
"E" fiberglass	4.38	0.020

[a]Data from J. P. Stevens. [b]Resin content, 35%. [c]Measured at 9.375 GHz and room temperature.

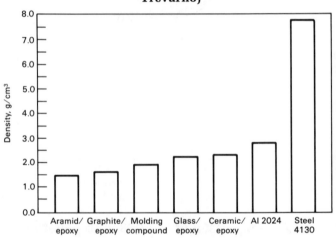

Comparison of Typical Densities of Epoxy Laminates and Other Materials (Hexcel-Trevarno)

Comparison of Average Properties of Epoxy Sheet Molding Compound and Competitive Materials[a]

Material	Flexural strength MPa	ksi	Tensile strength MPa	ksi	Izod impact strength J/cm	ft·lb/in.	Specific gravity
Epoxy SMC[b]	400	58	228	33	19	35	1.85
Glass mat polyester	200	29	172	25	6	11	1.63
ABS	76	11	55	8	2	4	1.07
Polycarbonate	90	13	69	10	9	17	1.2
Structural steel	...	...	414–1380	60–200	...	...	7.8
Aluminum	...	...	138–414	20–60	...	...	2.7
Beryllium copper	...	...	758	110	...	...	8.94
Magnesium	...	...	138–345	20–50	...	...	1.8
Zinc	...	...	138–172	20–25	...	...	7.14

[a]Data from Quantum Composites, Inc. [b]Lytex brand epoxy sheet molding compound, approximately 60% glass fiber content.

Physical Properties of Epoxy Preimpregnated Unidirectional Tapes[a]

Type	Ultimate tensile strength MPa	ksi	Tensile modulus GPa	10⁶ psi	Poisson's ratio (V₁₂)	Ultimate compressive strength MPa	ksi	Compressive modulus GPa	10⁶ psi	Transverse shear strength MPa	ksi	Shear modulus MPa	ksi	Interlaminar shear MPa	ksi	Specific gravity, g/cm³	Cured ply range μm	mils
Standard graphite/epoxy	1515	220	131	19.0	0.3	1310	190	131	19.0	66	9.5	4.1	0.6	110	16.0	1.58	51–254	2–10
1.5%-strain graphite/epoxy	1895	275	134	19.4	0.3	1585	230	131	19.0	66	9.5	4.1	0.6	110	16.0	1.60	102–203	4–8
1.8%-strain graphite/epoxy	2585	375	138	20.0	0.3	1585	230	134	19.5	69	10.0	4.1	0.6	110	16.0	1.61	102–203	4–8
Intermediate-modulus graphite/epoxy	2760	400	165	24.0	0.3	1380	200	145	21.0	66	9.5	4.1	0.6	110	16.0	1.60	102–203	4–8
High-modulus graphite/epoxy	780	113	239	34.7	0.3	345	50.0	228	33.0	34	5.0	4.8	0.7	35	5.1	1.80	64–254	2.5–10
Ultrahigh-modulus graphite/epoxy	760	110	314	45.6	0.3	338	49.0	316	45.9	37	5.4	4.8	0.7	66	9.5	1.83	64–254	2.5–10
Pitch-100/epoxy	1035	150	421	61.0	0.3	255	37.0	310	45.0	34	5.0	4.8	0.7	31	4.5	1.45	64–127	2.5–5
Kevlar 49/epoxy	1365	198	46	6.7	0.3	207	30.0	41	6.0	59	8.5	2.1	0.3	52	7.5	1.90	127–254	5–10
E-glass/epoxy	1035	150	41	6.0	0.3	827	120	41	6.0	...	...	...	...	76	11	1.90	102–305	4–12
S-2 glass/epoxy	1690	245	52	7.6	0.3	827	120	60	8.7	...	...	...	...	76	11	2.02	102–305	4–12
Nicalon SiC/epoxy	1380	200	110	16.0	0.3	1655	240	...	...	...	...	...	...	97	14.0	1.90	76–305	3–12

[a]Data from Fiberite.

Comparison of Fatigue Strengths of Graphite/Epoxy, Steel, Fiberglass/Epoxy, and Aluminum (Hercules)

Note greater strength of graphite/epoxy after 10⁷ cycles.

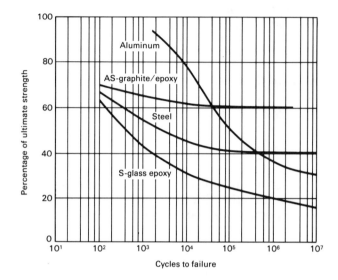

Prepreg Characteristics, Cure Conditions, and Mechanical Properties of SiC/Epoxy Laminates[a]

Prepreg Characteristics

Resin content . . . . 26 wt %
Volatile content 1.0%
Tack (at 23 °C; 73 °F) . . . . Good
Width tolerance . . . . ±0.51 mm (±0.02 in.)
Filament count (nominal) 5.51/mm (140/in.)
Storage temperature . . . . −18 °C (0 °F)

Typical Cure Conditions

Temperature . . . . 177 °C (350 °F)
Pressure . . . . 345–585 kPa (50–85 psi)
Time . . . . 90 min[b]

Mechanical Properties

Tensile strength:
    At room temperature . . . . 160 kg/mm² (229 ksi)
    At 127 °C (260 °F) . . . . 133 kg/mm² (190 ksi)
Tensile modulus:
    At room temperature . . . . 23 Mg/mm² (33 × 10⁶ psi)
    At 127 °C (260 °F) . . . . 23 Mg/mm² (33 × 10⁶ psi)
Compressive strength:
    At room temperature . . . . 228 kg/mm² (326 ksi)
    At 127 °C (260 °F) . . . . 162 kg/mm² (232 ksi)
Horizontal shear strength:
    At room temperature . . . . 10.5 kg/mm² (15.0 ksi)
    At 127 °C (260 °F) . . . . 6.3 kg/mm² (9.0 ksi)
Flexural strength:
    At room temperature . . . . 220 kg/mm² (314 ksi)
    At 127 °C (260 °F) . . . . 221 kg/mm² (316 ksi)
Flexural modulus:
    At room temperature . . . . 22.4 Mg/mm² (32.0 × 10⁶ psi)
    At 127 °C (260 °F) . . . . 21 Mg/mm² (30.0 × 10⁶ psi)
Density . . . . 2.325 Mg/m³ (0.084 lb/in.³)

[a]Data from Avco. Laminates were fabricated and tested by the University of Dayton under AFML Contract (F33615-78-C-5172). [b]Postcure, 4 h at 190 °C (375 °F).

Typical Properties of Boron/Epoxy Laminates[a] (Ref 8, p 52)

Temperature °C	°F	0° Ultimate tensile strength MPa	ksi	0° Tensile modulus GPa	10⁶ psi	0° Failure strain, %	90° Ultimate tensile strength MPa	ksi	0° Ultimate compressive strength MPa	ksi
Unidirectional Construction (0°)										
−55	−67	1240−1380	180−200	205−220	30−32	0.59−0.64	69−90	10−13	3105−3310	450−480
24	75	1275−1450	185−210	205−215	30−31	0.62−0.69	55−83	8−12	2585−3070	375−445
125	260	1275−1345	185−195	200−205	29−30	0.63−0.66	41−55	6−8	1860−2070	270−300
190	375	1105−1170	160−170	195−200	28−29	0.56−0.61	21−28	3−4	585−1000	85−145
Cross-Plied Construction (0°/90°)										
−55	−67	690−760	100−110	130	18.9	0.54−0.56	690−760	100−110	...	...
24	75	690−760	100−110	125	18.0	0.57−0.61	690−760	100−110	...	...
125	260	655−690	95−100	120	17.5	0.58	655−690	95−100	...	...
190	375	580−655	84−95	105−115	15.5−16.5	0.56−0.58	580−655	84−95	...	...

[a]Nominal fiber content, 55 vol %.

Tension-Tension Fatigue Behavior of Unidirectional E-Glass and Kevlar Epoxy Composites and Aluminum (Du Pont)

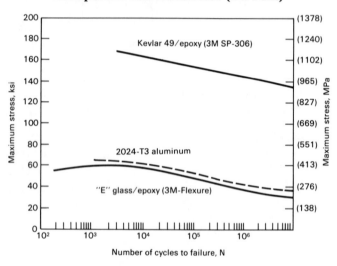

Comparison of Kevlar-Reinforced Vs. Glass-Reinforced Epoxy Molding Compound

Property	Kevlar 49	Glass[a]
Fiber denier	380	...
Fiber content, vol %	50	47
Fiber length:		
mm	12.7	12.7
in.	0.5	0.5
Specific gravity of CFMC	1.32	1.86
Tensile strength:		
MPa	197	99.3
ksi	28.5	14.4
Tensile modulus:		
GPa	20.4	15.4
10⁶ psi	2.96	2.23
Flexural strength:		
MPa	245	201
ksi	35.5	29.1
Flexural modulus:		
GPa	18.4	16.1
10⁶ psi	2.67	2.33
Izod impact strength:		
J/cm	9.29	10.1
ft·lb/in.	17.4	19

[a]Glass-reinforced epoxy compound is U.S. polymeric EM-7302.

Shear Stress-Strain Curve for Boron/Epoxy (Ref 27, p 8.28)

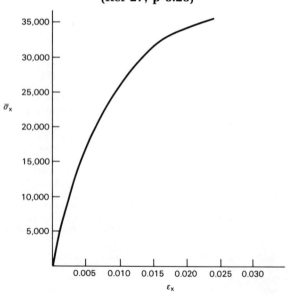

Stress-Strain Curve for ±45° Boron/Epoxy Laminate (Ref 27, p 8.26)

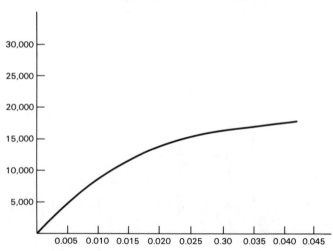

Variation in Coefficient of Thermal Expansion of Epoxy Resin With Addition of Various Filler Materials (Ref 103, p 16)

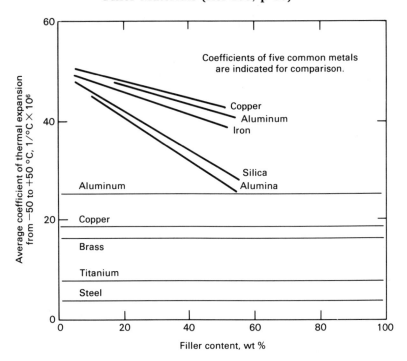

Constant-Amplitude-Fatigue Design Data for Boron/Epoxy Laminates at Room Temperature and 177 °C (350 °F) (Ref 4, p 192)

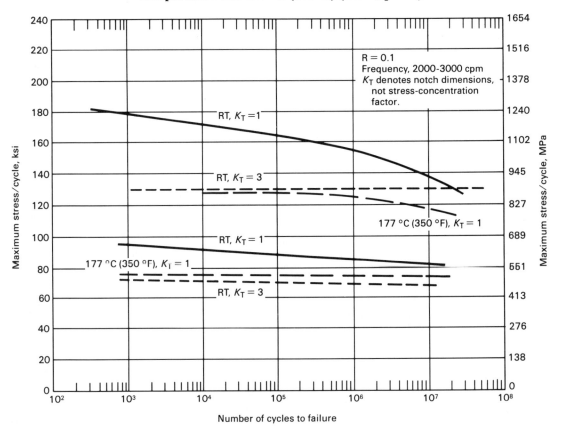

Comparison of Whisker-Reinforced and Fiber-Reinforced Epoxy (Ref 75, p 18)

Property	Epoxy plus 35% Si_3N_4 whiskers	Epoxy plus 14% Al_2O_3 whiskers	Epoxy plus 70% S-glass fibers	Epoxy plus 14% S-glass fibers
Density, g/cm^3	1.90	1.64	2.11	1.38
Elastic modulus:				
GPa	103	41	61	...
10^6 psi	15	6	8.9	...
Strength:				
MPa	276	779	2070	517
ksi	40	113	300	75
Specific elastic modulus	7.9×10^6	3.65×10^6	4.2×10^6	...
Specific strength	22.2×10^3	69×10^3	140×10^3	54.5×10^3

6.2.3. Carbon/Graphite Fiber Reinforced Resin Matrix Composites

Design Properties of Carbon

Base resin	PAN carbon fiber content, %	Specific gravity[b]	Physical — Water absorption,[c] %	Mold shrinkage,[d] in./in.	Tensile strength,[e] ksi	Flexural strength,[f] ksi	Mechanical — Flexural modulus,[f] 10^6 psi	Shear strength,[g] ksi	Izod impact strength,[h] ft · Notched	Unno
Nylon 66	10	1.18	0.80	0.0040	20.0	30.0	1.00	...	1.0	
	20	1.23	0.60	0.0025	28.0	42.0	2.40	12.0	1.1	
	30[t]	1.28	0.50	0.0020	35.0	51.0	2.90	13.0	1.5	1
	40	1.34	0.40	0.0020	40.0	60.0	3.40	14.0	1.6	1
	30[q]	1.38	0.48	0.0025	30.5	43.0	2.30	12.5	1.4	1
	30[r]	1.36	0.45	0.0025	27.0	36.0	1.75	11.5	1.4	1
	30[u]	1.27	0.80	0.0020	32.0	46.5	2.50	12.5	2.1	1
Nylon 6	30	1.28	0.80	0.0020	32.0	46.0	2.40	12.5	1.8	1
Nylon 6/12	30	1.22	0.15	0.0020	29.0	42.0	2.30	12.0	1.8	1
Nylon 6/10	30	1.23	0.12	0.0020	28.0	40.5	2.20	12.0	1.8	1
Super tough nylon	30	1.22	0.35	0.0025	24.0	34.0	2.00	11.0	3.0	1
Amorphous nylon	30	1.27	0.12	0.0015	30.0	47.5	2.20	12.5	1.2	
Polycarbonate	30	1.33	0.08	0.0015	24.0	36.0	1.90	10.0	1.8	1
Polysulfone	30	1.37	0.15	0.0015	23.0	32.0	2.05	9.5	1.2	
Polyethersulfone	30	1.48	0.30	0.0015	26.0	37.5	2.05	10.5	1.0	
	30[q]	1.57	0.20	0.0020	23.5	33.5	1.90	9.3	1.0	
PEEK	15[q]	1.46	0.10	0.0020	19.5	24.0	1.50	7.8	0.9	
	30	1.39	0.10	0.0010	31.0	36.0	2.20	12.4	1.2	1
Acetal	20	1.46	0.50	0.0050	11.8	13.7	1.35	8.2	1.0	
Polyester (PBT)	30	1.41	0.04	0.0020	22.0	29.0	2.30	8.0	1.2	
Polyphenylene sulfide	30	1.45	0.04	0.0010	27.0	34.0	2.45	9.5	1.1	
	30[q]	1.57	0.03	0.0015	25.5	28.0	2.40	8.0	0.8	
ETFE[s]	20	1.72	0.02	0.0025	12.0	16.0	1.20	6.5	5.5	
PVDF	15	1.77	0.02	0.0060	13.5	18.0	1.15	7.5	1.0	

[a]Data from LNP Corp. This information is based on experience and is intended for use as a guide only. [b]ASTM D792. [c]ASTM D570. [d]ASTM D955. [e]ASTM D638. [f]ASTM D790. [g]ASTM D732. [h]ASTM D256. [j]ASTM D648. [k]ASTM C177. [m]ASTM D696. [n]ASTM D257. [p]F = fair; G = good; E = excellent. [q]Plus PTFE lubricant. [r]Plus 15% PTFE/silicone lubricant. [s]Based on Du Pont Tefzel fluoropolymer. [t]Grade RC 1006. [u]Grade RC 1006 HI.

Effect of Resin Matrix on Conductivity

Because nylon is crystalline, it becomes surface conductive at lower loadings than does amorphous polycarbonate with addition of pitch carbon fiber.
(Wilson-Fiberfil International)

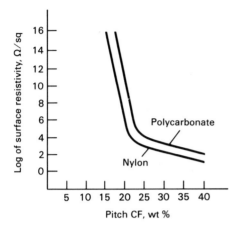

Effect of Loading Level on Conductivity

Because PAN carbon fiber has a higher aspect ratio than that of pitch, less is needed to achieve conductivity. Stainless steel fibers provide conductivity at lowest loadings.
(Wilson-Fiberfil International)

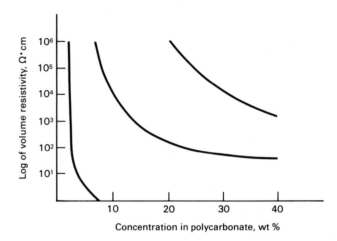

...orced Thermoplastic Composites[a]

Thermal												
Thermal conductivity,[k] Btu·in./ ft²·h·°F	Coefficient of linear thermal expansion,[m] 10^{-5} in./ in.·°F	Flamma-bility (UL94)	Coefficient of friction		Wear factor, 10^{-10} in.³·min/ ft·lb·h	Limiting PV			Electrical: Surface resistivity,[n] Ω/sq	Chemical[p]		
			Static	Dynamic		10 fpm	100 fpm	1000 fpm		Acids	Bases	Solvents
4.0	2.0	HB	0.17	0.21	60	10 000	16 000	5 000	10^7	F	G	E
5.5	1.4	HB	0.16	0.20	40	19 000	25 000	7 000	1300	F	G	E
7.0	1.1	HB	0.16	0.20	20	21 000	27 000	8 000	150	F	G	E
8.5	0.8	HB	0.13	0.18	14	22 000	27 500	8 500	75	F	G	E
7.0	1.1	HB	0.11	0.15	10	29 000	42 000	19 000	200	F	G	E
7.0	1.1	HB	0.10	0.11	6	29 000	43 000	20 000	1000	F	G	E
6.8	1.1	HB	0.18	0.22	22	20 000	26 500	7 000	150	F	G	E
7.0	1.0	HB	0.18	0.21	30	18 000	22 000	7 500	150	F	G	E
6.5	0.9	HB	0.19	0.23	25	18 000	20 000	17 000	250	F	G	E
6.5	0.9	HB	0.20	0.25	25	18 000	21 000	7 500	250	F	G	E
5.8	1.2	HB	0.20	0.25	25	20 000	26 000	7 500	1100	F	G	E
7.0	1.1	HB	0.19	0.24	90	10 000	11 000	6 000	150	F	F	F
4.9	0.9	Vl	0.18	0.17	85	8 000	8 500	5 500	3500	F	F	P
5.5	0.6	VO	0.17	0.14	75	8 500	8 500	6 000	200	G	E	P
6.0	0.8	VO	0.17	0.15	80	10 000	10 000	7 000	100	G	E	P
5.7	0.8	VO	0.13	0.17	40	35 000	33 000	16 000	100	G	E	P
4.5	1.8	VO	0.18	0.20	60	42 000	40 000	22 000	5000	E	E	G
7.1	0.7	VO	0.19	0.13	60	...	...	...	120	E	E	G
4.6	2.2	HB	0.11	0.14	40	13 000	20 000	15 000	2000	F	G	E
4.6	0.5	HB	0.12	0.15	24	18 000	22 000	10 000	500	F-G	P-F	E
5.2	0.6	VO	0.23	0.20	160	12 000	20 000	10 000	250	E	E	E
5.4	0.8	VO	0.13	0.15	75	27 000	35 000	30 000	150	E	E	E
6.0	1.0	VO	0.16	0.18	28	...	...	...	1200	E	E	E
2.2	2.5	VO	0.25	0.25	14	15 000	11 000	<5 000	700	E	E	E

Properties of Nickel-Coated Carbon Fiber Reinforced Composites[a]

ASTM test	Property	Nylon 66 15 wt % fiber	Nylon 66 40 wt % fiber	PPS 15 wt % fiber	PPS 40 wt % fiber	Polycarbonate 10 wt % fiber	Polycarbonate 15 wt % fiber	Polycarbonate 20 wt % fiber
Mechanical Properties								
D638	Tensile strength:							
	MPa	96.5	138	75.8	138	72.4	82.7	96.5
	ksi	14.0	20.0	11.0	20.0	10.5	12.0	14.0
D638	Elongation, %	1.6	2.5	1.6	2.5	4.0	3.0	3.0
D638	Tensile modulus:							
	GPa	7.6	...	6.9	9.6	...	...	...
	10^5 psi	11.0	...	10.0	14.0	...	...	...
D790	Flexural strength:							
	MPa	145	186	131	207	103	110	124
	ksi	21.0	27.0	19.0	30.0	15.0	16.0	18.0
D790	Flexural modulus:							
	GPa	6.9	13.8	9.6	17.2	4.4	6.7	8.3
	10^5 psi	10.0	20.0	14.0	25.0	6.4	9.7	12.0
D256	Izod impact strength, notched:[b]							
	J/m	37.4	53.4	32.0	37.4	...	...	...
	ft·lb/in.	0.7	1.0	0.6	0.7	...	...	...
D256	Izod impact strength, unnotched:[b]							
	J/m	294	...	134	...	...	...	...
	ft·lb/in.	5.5	...	2.5	...	...	...	...
Physical Properties								
D792	Specific gravity	1.20	1.46	1.45	1.68	...	...	...
D570	Water absorption in 24 h, %	1.0	0.8	0.04	0.03	...	...	...
...	Linear mold shrinkage in 3 mm ($^1/_8$ in.), %	0.5	0.1	0.1	0.05	...	...	...
Thermal Properties								
D648	Deflection temperature under load:[c]							
	°C	238	243	243	253	138	143	149
	°F	460	470	470	488	280	290	300
...	Flammability at minimum thickness (UL 94)	HB	HB	V-O	V-O	V-O	V-O	V-O
Electrical Properties								
D257	Volume resistivity, Ω·cm	10^0	10^{-2}	10^2	10^0	...	...	...
D257	Surface resistivity, Ω/sq	10^1	10^0	10^3	10^1	...	...	...
...	Shielding effectiveness,[d] dB	55	>60	20	40	...	...	...

[a]Data from Wilson-Fiberfil International and American Cyanamid. All tests conducted at 23 °C (73 °F). [b]Izod impact test bars 6.35 × 12.7 mm ($^1/_4$ × $^1/_2$ in.). [c]At 1.82 MPa (264 psi). [d]Attenuation at 1000 MHz for 3-mm ($^1/_8$-in.) section.

EMI Shielding Data for Nickel-Coated Carbon Fiber in Polyester Bulk Molding Compounds[a]

Sample	Fiber content, wt %	Shielding effectiveness, dB, at frequency of: 0.5 MHz	1.5 MHz	5 MHz	15 MHz	50 MHz	250 MHz	500 MHz	960 MHz
A	3.0	41	41	41	41	41	41	41	42
B	4.0	49	49	49	49	49	51	55	56
C	5.5	54	54	54	55	55	59	62	62
D	7.0	56	57	57	57	57	61	61	61

[a]Data from American Cyanamid.

Impact Strength of Unidirectional Composites With 60 Vol % AS4-12 Carbon Fibers, Compared With Impact Strength of the Pure Resins (Ref 94, p 388)

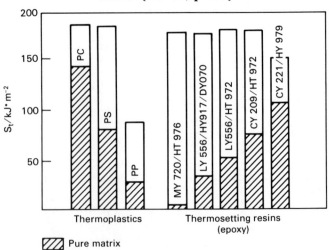

Properties of Pitch Carbon Fiber Reinforced Composites[a]

ASTM test	Property	Nylon 20 wt % chopped fiber	Nylon 30 wt % chopped fiber	Nylon 40 wt % chopped fiber	Nylon 30 wt % chopped fiber[f]	Poly-carbonate 25 wt % chopped fiber	Poly-propylene 40 wt % chopped fiber
Mechanical Properties							
D638	Tensile strength:						
	MPa	89.6	107	121	138	72.3	31.7
	ksi	13.0	15.5	17.5	20.0	10.5	4.60
D638	Elongation, %	2.5	2.0	1.5	1.9	2.0	1.1
D638	Tensile modulus:						
	GPa	10.3	13.8	17.9	17.2	11.4	...
	10^5 psi	15.0	20.0	26.0	25.0	16.5	...
D790	Flexural strength:						
	MPa	152	179	193	227	68.9	23.4
	ksi	22.0	26.0	28.0	33.0	10.0	3.4
D790	Flexural modulus:						
	GPa	6.9	10.3	13.8	13.1	...	7.6
	10^5 psi	10.0	15.0	20.0	19.0	...	11.0
D256	Izod impact strength, notched:[b]						
	J/m	32.0	37.4	42.7	64.1	58.7	37.4
	ft·lb/in.	0.6	0.7	0.8	1.2	1.1	0.7
D256	Izod impact strength, unnotched:[b]						
	J/m	...	694	...	...	320	214
	ft·lb/in.	...	13.0	...	...	6.0	4.0
D695	Compressive strength:						
	MPa	138	138	158	172	...	...
	ksi	20.0	20.0	23.0	25.0	...	...
D732	Shear strength:						
	MPa	65.5	65.5	68.9	68.9	...	...
	ksi	9.5	9.5	10.0	10.0	...	...
D785	Rockwell hardness	E44	E53	E54	E55	...	...
D621	Deformation under load,[c] %	0.54	0.38	0.2	...	...	...
Physical Properties							
D792	Specific gravity	1.24	1.30	1.36	1.40	1.35	1.15
D570	Water absorption in 24 h, %	0.75	0.6	0.5	0.5	...	0.03
...	Linear mold shrinkage in 3 mm (1/8 in.), %	0.4	0.3	0.2	0.2	0.1	0.1
Thermal Properties							
D696	Coefficient of linear thermal expansion:						
	10^{-5} m/m·°C	2.3	1.6	0.9	1.3	...	...
	10^{-5} in./in.·°F	1.3	0.9	0.5	0.7	...	...
D648	Deflection temperature under load:[d]						
	°C	232	241	246	243	143	127
	°F	450	465	475	470	290	260
D648	Deflection temperature under load:[e]						
	°C	254	254	259	254	...	...
	°F	490	490	498	490	...	...
...	Flammability at minimum thickness (UL 94)	HB	HB	HB	HB	V-O	HB
Electrical Properties							
D257	Volume resistivity, Ω·cm	10^6	10^4	10^3	10^4	10^5	10^2
D257	Surface resistivity, Ω/sq	10^6	10^4	10^3	10^4	10^5	10^2

[a]Data from Wilson-Fiberfil International. All tests conducted at 23 °C (73 °F) unless otherwise noted. [b]Izod impact test bars 6.35 × 12.7 mm (1/4 × 1/2 in.). [c]Deformation at 27.6 MPa and 50 °C (4 ksi and 122 °F). [d]At 1.82 MPa (264 psi). [e]At 0.45 MPa (66 psi). [f]Contains 15 wt % glass fiber.

Room-Temperature Mechanical Properties of As-Fabricated Unfilled and Filled Graphite/PMR-15 Laminates

Property	Graphite/PMR-15, unfilled	Graphite/PMR-15, filled with:			
		Boron	B_4C	Lime glass	Aluminum
Interlaminar shear strength:					
MPa	101.8	82.7	90.3	51.0	95.1
ksi	14.765	12.00	13.09	7.40	13.80
Flexural strength:					
MPa	1840	928	1080	1541	1729
ksi	266.8	134.6	156.6	223.5	250.8
Flexural modulus:					
GPa	120	117	112	115	108
10^6 psi	17.4	17.0	16.3	16.7	15.7

A potential problem in the use of graphite fiber reinforced resin matrix composites is the dispersal of graphite fiber during accidental fires. Airborne electrically conductive fibers originating from burning composites could enter and cause shorting in electrical equipment located in surrounding areas. A variety of matrix fibers have been tested for their ability to prevent loss of fiber from graphite fiber/PMR polyimide (see table above) and graphite fiber/epoxy composites (see table below) in a fire. The fillers tested included powders of boron, boron carbide (B_4C), lime glass, and aluminum. Of these fillers, boron was the most effective and prevented any loss of graphite fiber during burning. Mechanical properties of composites containing boron filler were measured and compared with those of composites containing no filler (Ref 72, p 6–7).

Room-Temperature Properties of Unfilled and Boron-Filled Graphite/Epoxy Laminates

Property	Graphite/epoxy, unfilled			Graphite/epoxy, boron-filled		
	As-fabricated	1032 h at 205 °C (400 °F)	1000 h at 60 °C (140 °F); 95% RN	As-fabricated	1032 h at 205 °C (400 °F)	1000 h at 60 °C (140 °F); 95% RN
Interlaminar shear strength:						
MPa	85.48	84.61	81.78	80.65	70.50	78.99
ksi	12.399	12.272	11.861	11.698	10.225	11.457
Flexural strength:[a]						
MPa	1530.5	1544.1	1488.8	1618.6	1610.3	1596.9
ksi	221.978	223.957	215.932	234.760	233.548	231.611
Weight change,[b] %	...	−0.90	+1.50	...	−1.00	+2.00

[a]Average of three determinations. Normalized to 60 vol % fiber. [b]Average of four determinations.

Properties of Electrically Conductive Carbon Black Composites (18–20 Wt %)[a]

ASTM test	Property	HDPE	EVA	Nylon	Polypropylene Grade 1	Grade 2
Mechanical Properties						
D638	Tensile strength:					
	MPA	22.0	13.1	58.6	28.9	22.0
	ksi	3.2	1.9	8.5	4.2	3.2
D638	Elongation, %	10.0	50	2.0	7.0	10.0
D638	Tensile modulus:					
	GPa	...	0.34	...	2.8	...
	10^5 psi	...	0.5	...	4.0	...
D790	Flexural strength:					
	MPa	20.7	11.0	103	44.1	24.1
	ksi	3.0	1.6	15.0	6.4	3.5
D790	Flexural modulus:					
	GPa	0.82	0.27	3.4	2.3	1.0
	10^5 psi	1.2	0.4	5.0	3.3	1.5
D256	Izod impact strength, notched:[b]					
	J/m	80.1	NB[c]	26.7	26.7	160
	ft·lb/in.	1.5	NB[c]	0.5	0.5	3.0
D785	Rockwell hardness	...	R32	...	...	...

(continued)

Properties of Electrically Conductive Carbon Black Composites (18–20 Wt %)[a] (continued)

ASTM test	Property	HDPE	EVA	Nylon	Polypropylene Grade 1	Polypropylene Grade 2
Physical Properties						
D792	Specific gravity	1.03	1.02	1.22	1.08	1.02
D570	Water absorption in 24 h, %	0.02	0.05	0.8	0.01	0.02
···	Linear mold shrinkage in 3 mm (1/8 in.), %	1.5	1.4	1.8	1.5	1.2
Thermal Properties						
D648	Deflection temperature under load:[d]					
	°C..........................	49	63	93	69	49
	°F..........................	120	145	200	156	120
D648	Deflection temperature under load:[e]					
	°C..........................	89	···	···	···	···
	°F..........................	192	···	···	···	···
···	Flammability at minimum thickness (UL 94)..............	···	···	HB	HB	HB
Electrical Properties						
D257	Volume resistivity, $\Omega \cdot$cm	10^3	10^2	10^3	10^4	10^3
D257	Surface resistivity, Ω/sq..........................	10^4	10^3	10^4	10^5	10^4

[a]Data from Wilson-Fiberfil International. All tests conducted at 23 °C (73 °F). [b]Izod impact test bars 6.35 × 12.7 mm (1/4 × 1/2 in.). [c]NB = nonbreaking. [d]At 1.82 MPa (264 psi). [e]At 0.45 MPa (66 psi).

Variations Among Reported Unidirectional Properties for a Widely Used Graphite/Epoxy System
(Ref 27, p 8.73)

Property	Source[a] 1	2[b]	3[b]	4	5	Maximum Δ
Elastic Constants						
Longitudinal tensile modulus:						
GPa	143	125	145	142	128	
10^6 psi..........................	20.8	18.1	21	20.6	18.5	16%
Longitudinal compressive modulus:						
GPa	128	100	145	137	128	
10^6 psi..........................	18.6	14.5	21	19.8	18.5	45%
Transverse tensile modulus:						
GPa	13.1	12.4	11.7	9.0	11.0	
10^6 psi..........................	1.9	.8	1.7	1.3	1.6	46%
Shear modulus:						
GPa	5.9	···	4.5	5.5	4.5	
10^6 psi..........................	0.85	···	0.65	0.8	0.65	31%
Poisson's ratio	0.30	···	···	0.32	0.25	28%
Strength Properties						
Longitudinal tensile strength:						
MPa..........................	1890	1310	1240	1130	1165	
ksi	274	190	180	164	169	67%
Longitudinal compressive strength:						
MPa..........................	1930	869	1240	869	1115	
ksi	280	126	180	126	162	122%
Transverse tensile strength:						
MPa..........................	65.5	35.9	55.2	37.2	41.4	
ksi	9.5	5.2	8	5.4	6.0	83%
Transverse compressive strength:						
MPa..........................	269	···	207	145	172	
ksi	39	···	30	21	25	86%
In-plane shear strength:						
MPa..........................	119	···	82.7	51	···	
ksi	17.3	···	12	7.4	···	106%
Interlaminar shear strength:						
MPa..........................	···	93.1	90	···	49	
ksi	···	13.5	13	···	7.1	90%

[a]From major airframe company reports. [b]From divisions of same company.

Physical Properties of Graphite Fabric Prepreg[a]

Type	Ultimate tensile strength MPa	ksi	Tensile modulus GPa	10⁶ psi	Poisson's ratio (V₁₂)	Ultimate compressive strength MPa	ksi	Compressive modulus GPa	10⁶ psi	Transverse shear strength MPa	ksi	Shear modulus GPa	10⁶ psi	Inter-laminar shear MPa	ksi	Specific gravity, g/cm³	Cured ply range μm	mils
Standard bidirectional graphite/epoxy	586	85	68.9	10.0	0.09	565	82	62.1	9.0	93.1	13.5	4.8	0.7	67.6	9.8	1.59	127–381	5–15
1.5%-strain bidirectional graphite/epoxy	690	100	68.9	10.0	0.09	586	85	62.1	9.0	93.1	13.5	4.8	0.7	66.2	9.6	1.60	178–381	7–15
High-modulus bidirectional graphite/epoxy	345	50	117	17.0	0.09	152	22	110	16.0	34.5	5.0	4.8	0.7	31.0	4.5	1.80	203–381	8–15
Ultrahigh-modulus bidirectional graphite/epoxy	345	50	152	22.0	0.09	152	22	152	22.0	34.5	5.0	4.8	0.7	34.5	5.0	1.80	152–330	6–13
Standard woven unidirectional graphite/epoxy	1310	190	129	18.7	0.25	1105	160	124	18.0	62.1	9.0	4.1	0.6	82.7	12.0	1.60	178–254	7–10
Standard bidirectional hybrid graphite/S-2 glass	483	70	51.7	7.5	0.09	565	82	48.3	7.0	34.5	5.0	4.1	0.6	65.5	9.5	1.80	254–381	10–15
Standard bidirectional hybrid graphite/Kevlar 49	448	65	51.0	7.4	0.09	276	40	41.4	6.0	...	...	...	...	34.5	5.0	1.55	254–381	10–15

[a]Data from Fiberite Corp.

Typical Tensile Properties of Carbon-Fiber Composites at Room Temperature

Fiber description	Matrix	Laminate orientation, degrees	Ultimate tensile strength MPa	ksi	Tensile modulus GPa	10⁶ psi
HT	Epoxy	0	1295	188	138	20
		90	16.5	2.4	7.6	1.1
		0/±45	793	115	90	13.1
HM	Epoxy	0	717	104	165	24
		90	46.9	6.8	7.6	1.1
		±45	145	21	17.2	2.5
		0/90	365	53	100	14.5
A	Epoxy	0	1160	168	136	19.7
GY-70	Epoxy	0	621	90	290	42
		90	21	3	5.9	0.86
Fortafil 5Y	Polyimide	0	627	91	163[a]	23.6[a]
		90	20.0	2.9	...	...
HMG-50	Polyimide	0	481	69.7	178	25.8
		90	17.9	2.6	...	...

[a]In flexure.

Selected Physical and Mechanical Properties of High-Strength Graphite/Epoxy Composites With Various Fiber Orientations (Ref 2, p 214)

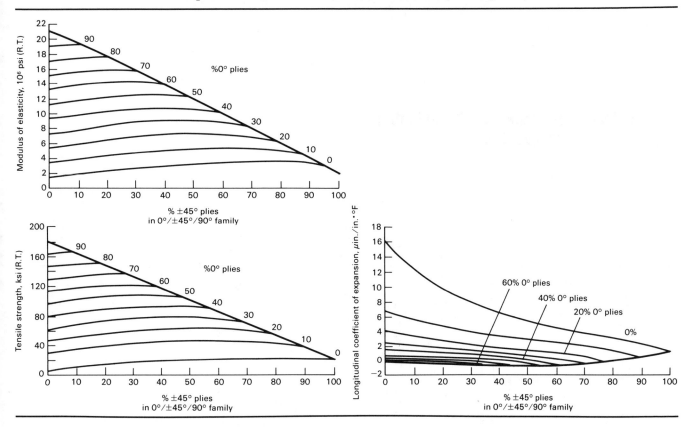

Typical Mechanical Properties of Graphite/Epoxy Composites With Various Fiber Orientations (Ref 21, p 47)

Fiber type	Ultimate tensile strength		Tensile modulus (E)		Ultimate compressive strength		Compressive modulus (E)	
	MPa	ksi	GPa	10⁶ psi	MPa	ksi	GPa	10⁶ psi
Chopped fiber:								
Molding compound	352	51	108	15.7	469	68	...	...
Unidirectional (nonwoven):								
High-strength fiber (0°)	1625	236	138	20.0	993	144	113	16.4
High-strength fiber (0°/±45°)	496	72	57	8.3	503	73	50	7.3
Medium-strength fiber (0°)	1455	211	147	21.3	1405	204	131	19.0
Medium-strength fiber (0°/±45°)	503	73	65	9.4	503	73	64	9.3
High-modulus fiber (0°)	1240	180	215	31.2	758	110	177	25.6
Fabric (woven):								
Medium-strength fiber........................	510	74	70	10.2	510	74	63	9.2

Resin-Dependent Properties of Graphite Composites

Resin type	Continuous use temperature		Maximum use temperature		Interlaminar shear strength		Comments
	°C	°F	°C	°F	MPa	ksi	
Thermosets							
Epoxy, 120 °C (250 °F) cure	70	160	105	225	55–103	8–15	690 kPa (100 psi) mold pressure
Epoxy, 175 °C (350 °F) cure	120	250	150	300	103	15	690 kPa (100 psi) mold pressure
Polyimide, 315 °C (600 °F) cure	290	550	370	700	103	15	690 kPa (100 psi) mold pressure with postcure
Phenolic.............................	260	500	315	600	...	...	...
Thermoplastics							
Polysulfone	150	300	175	350	97	14	Moisture resistant
Polyphenylsulfone	180	360	205	400	97	14	Moisture resistant

6.2.4. Glass Fiber Reinforced Resin Matrix Composites

Comparison of Engineering Properties of Fiberg

Material	Glass fiber content, wt %	Flexural strength, ksi	Flexural modulus, 10^5 psi	Tensile strength at yield, ksi	Tensile modulus, 10^5 psi	Compressive strength, ksi	Ultimate tensile elongation, %	Izod impact strength, ft·lb/in. of notch	Thermal conductivity, Btu·in./ft²·h·°F (K value)	Bt
Glass Fiber Reinforced Thermosets										
Sheet molding compound (SMC)	15–30	18–30	14–20	8–20	16–25	15–30	0.3–1.5	8–22	1.3–1.7	0.
Bulk molding compound (BMC)	15–35	10–20	14–20	4–10	16–25	20–30	0.3–.5	2–10	1.3–1.7	0.
Preform/mat (compression molded)	25–50	10–40	13–18	25–30	9–20	15–30	1–2	10–20	1.3–1.8	0.
Cold press molding—polyester	20–30	22–37	13–19	12–20	...	...	1–2	9–12	1.3–1.8	0.
Spray-up—polyester	30–50	16–28	10–12	9–18	8–18	15–25	1.0–1.2	4–12	1.2–1.6	·0.
Filament wound—epoxy	30–80	100–270	50–70	80–250	40–90	45–70	1.6–2.8	40–60	1.92–2.28	0.
Rod stock—polyester	40–80	100–180	40–60	60–180	40–60	30–70	1.6–2.5	45–60	1.92–2.28	0.
Molding compound—phenolic	5–25	18–24	30	7–17	26–29	14–35	0.25–0.6	1–6	1.1–2.0	0.
Glass Fiber Reinforced Thermoplastics										
Acetal	20–40	15–28	8–13	9–18	8–15	11–17	2	0.8–2.8	...	
Nylon	6–60	7–50	2–26	13–33	2–20	13–24	2–10	0.8–4.5	...	0.
Polycarbonate	20–40	17–30	7.5–15	12–25	7.5–17	14–24	2	1.5–3.5	...	
Polyethylene	10–40	7–12	2.1–6	6.5–11	4–9	4–8	1.5–3.5	1.2–4.0	...	
Polypropylene	20–40	7–11	3.5–8.2	5.5–10.5	4.5–9	6–8	1–3	1–4	...	
Polystyrene	20–35	10–17	8–12	10–15	8.4–12.1	13.5–19	1.0–1.4	0.4–4.5	...	0.
Polysulfone	20–40	21–27	8–15	13–20	15	21–26	2–3	1.3–2.5	...	
ABS (acrylonitrile butadiene styrene)	20–40	23–26	9.2–15	11–16	6–10	12–22	3–3.4	1–2.4	...	
PVC (polyvinyl chloride)	15–35	20–25	9–16	14–18	10–18	13.4–16.8	2–4	0.8–1.6	...	
Polyphenylene oxide (modified)	20–40	17–31	8–15	15–22	9.5–15	18–20	1.7–5	1.6–2.2	...	
SAN (styrene acrylonitrile)	20–40	15–21	8.0–18	13–18	9–18.5	12–23	1.1–1.6	0.4–2.4	...	
Thermoplastic polyester	20–35	19–29	8.7–15	14–19	13–15.5	16–18	1–5	1.0–2.7	1.3	
Unreinforced Thermoplastics										
Acetal	N.Ap.	13–14	4	8–10	4–5	5	25–60	1.2–2.3	...	0.
Nylon	N.Ap.	5–18	2–4	9	2–5	7–10	29	1–4	...	0.
Polycarbonate	N.Ap.	13	3	9–11	3.5	12	100–130	16	...	0.
Polyethylene (high density)	N.Ap.	...	0.7–2.6	4	0.6–1.5	2.7–3.6	30–900	0.6–20.0	...	
Polypropylene	N.Ap.	5–8	1.2–2.7	3–5	1.2	3.7–8	200–700	0.5–20.0	...	
Polystyrene (high impact)	N.Ap.	3–10	1–5	3–5	2–4	4–9	15–30	0.7–3.6	...	
Polysulfone	N.Ap.	1.5	4	10	3.6	14	50–100	1.3	...	
ABS (high heat)	N.Ap.	9	3–4	6.6	2.8–4.1	6.8–12.5	10–20	2.5	...	
PVC	N.Ap.	13–16	4	6–7	4	...	...	2–20	...	
Polyphenylene oxide (modified)	N.Ap.	15	4	10	3.7	15	50–100	1.5–1.9	...	
SAN	N.Ap.	9.7–17.5	5	9–11	5	14–17	2.5–3.7	0.4	...	
Metals										
Gray cast iron	N.Ap.	10	N.Av.	15–30	120	25	1	4–4.4	288–408	0.
Low-carbon steel (cold rolled)	N.Ap.	28	300	29–33	300	28	38–39	N.Ap.	260–460	0.
Stainless steel	N.Ap.	30–35	280	30–35	280	30	50–60	8.5–11.0	96–185	0.
Aluminum, wrought	N.Ap.	20	100	6–27	100	N.Av.	30–40	N.Ap.	810–1620	0.
Aluminum, die cast	N.Ap.	8–26	100	8–26	100	9	6–8	N.Ap.	610–1100	0.
Magnesium, die cast	N.Ap.	14	65	8–30	65	10–14	4–6	3	288–960	0.2
Zinc, die cast	N.Ap.	N.Av.	N.Av.	10–25	N.Av.	N.Av.	10	4.3	764–792	0.
Brass, plain yellow wrought	N.Ap.	14	150	14	150	N.Av.	60–65	N.Ap.	804	

[a]Data from Owens-Corning Fiberglas. N.Av. = not available; N.Ap. = not applicable. [b]Classification shown is highest obtainable rating. Less-critical applications may permit use of materials with low sifications. [c]Key: E = excellent (outstanding); G = good (acceptable); F = fair (test before using); P = poor (not recommended). [d]Attacked by oxidizing acids. [e]Disintegrates in sulfuric acid. [f]So aromatic and chlorinated hydrocarbons. [g]Soluble in ketones and esters, and in aromatic and chlorinated hydrocarbons. [h]Below 176 °F (80 °C). [j]Softens in some aromatic and chlorinated aliphatics. P to alcohol. [k]Dissolves or swells in some aromatic and chlorinated aliphatics. Resistant to alcohol.

...forced Plastics and Competitive Materials[a]

Rockwell hardness	Dielectric strength V/mil	Specific gravity	Density, lb/in.^{3}	Heat distortion at 264 psi, °F	Continuous heat resistance, °F	Coefficient of thermal expansion, 10^{-6} in./in.·°F	Chemical resistance[c] Weak acids	Strong acids	Weak alkalis	Strong alkalis	Organic solvents
H50–H112	300–450	1.7–2.1	0.061–0.075	400–500	300–400	8–12	G to E	F	F	P	G to E
H80–H112	300–450	1.8–2.1	0.065–0.075	400–500	300–400	8–12	G to E	F	F	P	G to E
H40–H105	300–600	1.5–1.7	0.054–0.061	350–400	150–400	10–18	G to E	F	F	P	G to E
H40–H105	300–600	1.5–1.7	0.054–0.061	350–400	150–400	10–18	G to E	F	F	P	G to E
H40–H105	200–400	1.4–1.6	0.050–0.058	350–400	150–350	12–20	G to E	F	F	P	G to E
M98–120	300–400	1.7–2.2	0.061–0.079	350–400	500	2–6	E	F	E	G	E
H80–112	200–400	1.6–2.0	0.058–0.072	325–375	150–500	3–8	G to E	F	F	F	G to E
M90–99	150–370	1.7–1.9	0.061–0.069	400–500	325–350	4.5–9	F	P	F	P	F
M78–M94	500–600	1.55–1.69	···	315–335	185–220	19–35	F	P	F	P	E
···	400–500	1.47–1.7	0.049	300–500	300–400	11–21	G	P	E	F	G
M75–M100	450	1.24–1.52	···	285–300	275	12–18	E	G^d	G	F	P^f
···	450–500	1.16–1.28	···	150–200	280–300	17–27	E	G^d	E	E	G^h
R95–R115	500–600	1.04–1.22	···	230–300	300–320	16–24	E	G^d	E	E	G^h
M70–M95	350–425	1.20–1.29	0.045–0.048	200–220	180–200	17–22	E	G^d	G	G	P^f
M85–M92	···	1.38–1.55	···	333–350	···	12–17	E	E	E	E	G
M75–M102	···	1.23–1.38	···	215–240	200–230	16–20	E	G^d	E	E	P^g
M80–M88	500–550	1.45–1.62	···	155–165	···	12	E	G	E	E	P^g
M95	···	1.20–1.38	···	220–315	240–265	10–20	E	E	E	E	G^j
M77–M103	···	1.22–1.40	···	210–230	200–220	16–21	G	G^e	G	G	P^g
R118–M70	560–750	1.45–1.61	···	380–470	275–375	24–33	F	P	P	P	E
M78–M94	465–500	1.42	0.052	230–255	185–220	45	F	P	F	P	E
R108–118	300–470	1.12–1.14	0.039–0.041	120–150	250–300	55–63	G	P	E	F	G
M70	400–425	1.20	0.043	265–290	275	39	E	G^d	G	F	P^f
···	450–500	0.95	···	100–130	180–230	6	E	G^d	E	E	P^f
R50–R110	500–600	0.9	···	125–140	190–240	38	E	G^d	E	E	G^h
M12–M45	300–600	1.05	0.039	175–205	150–180	22–56	E	G^d	G	G	P^f
M69–R12.0	425	1.24	···	345	300–345	31	E	E	E	E	G
R113	350–500	1.05	···	215–245	190–230	41–52	G	G^e	G	G	P^g
D80	···	1.4	···	155–165	···	···	E	E	E	E	P^g
M75	···	1.06	···	375	···	30	E	E	E	E	G^k
M80	400–500	1.08	···	190–220	140–205	36	G	G^e	G	G	P^g
B93	C	7.19	0.26	N.Ap.	N.Av.	6	Rusted by water, oxygen, and salt solutions; poor acid resistance; good alkaline resistance.
B72	C	7.8	0.28	N.Ap.	N.Av.	6–8	
B90	C	7.92	0.29	N.Ap.	N.Av.	9–10	Rusted by water, oxygen, and salt solutions; poor acid resistance; good alkaline resistance.
B1–B5	C	2.6–2.8	0.10	N.Ap.	N.Av.	12–13	Poor acid resistance (especially hydrochloride and sulfuric); poor chloride solution resistance; good alkaline and organic resistance.
E59	C	2.57–2.96	0.09	N.Ap.	N.Av.	12–13	
E50–E59	C	1.81	0.07	N.Ap.	N.Av.	14–16	Poor acid resistance (especially hydrochloric and sulfuric); poor chloride salt resistance (must be chemically treated for appearance when exposed to weather).
B44	C	6.6	0.24	N.Ap.	N.Av.	15–16	Poor acid resistance (especially hydrochloric and sulfuric); poor chloride salt resistance (must be chemically treated for appearance when exposed to weather).
F58–F64	C	8.5	0.31	N.Ap.	N.Av.	11–12	Poor acid resistance (except hydrofluoric); good alkaline resistance, corrodes in presence of salt, salt spray, or industrial atmospheres.
							Poor resistance to strong acids and bases; poor resistance to atmosphere.
							Good resistance to atmosphere; poor resistance to soft water and high-salinity water.

Properties of Fiberglass-Reinforced Nylon 6[a]

ASTM method	Property	Unrein-forced	10% glass	20% glass	30% glass	40% glass	50% glass	60% glass
General								
D792	Specific gravity	1.13–1.15	1.21	1.28	1.37	1.46	1.57	1.70
...	Specific volume:							
	cm³/kg	...	827	784	730	686	636	589
	in.³/lb	...	22.9	21.7	20.2	19.0	17.6	16.3
D570	Water absorption in 24 h, %	1.8	1.4	1.3	1.1	0.9	0.8	0.7
D570	Equil. cont. immersion, %	9.5	8.8	8.0	6.5	4.6	4.3	4.0
D955	Mold shrinkage, mm/mm (in./in.):							
	3.2-mm (⅛-in.) avg section	0.013	0.0060	0.0045	0.0035	0.0030	0.0025	0.0020
	6.4-mm (¼-in.) avg section	0.016	0.0100	0.0065	0.0045	0.0040	0.0035	0.0030
Mechanical								
D638	Tensile strength:							
	MPa	81.4	93.1	128	159	179	214	228
	ksi	11.8	13.5	18.5	23.0	26.0	31.0	33.0
D638	Tensile elongation, %	80	3–4	3–4	3–4	2–3	2–3	1–2
D790	Flexural strength:							
	MPa	103	117	197	234	248	310	338
	ksi	15.0	17.0	28.5	34.0	36.0	45.0	49.0
D790	Flexural modulus:							
	GPa	2.76	4.48	5.86	8.27	10.3	13.8	19.3
	10⁶ psi	0.40	0.65	0.85	1.20	1.50	2.00	2.80
D695	Compressive strength:							
	MPa	60.7	110	152	159	159	165	172
	ksi	8.8	16.0	22.0	23.0	23.0	24.0	25.0
D732	Shear strength:							
	MPa	66.2	66.8	68.9	82.7	86.2	89.6	93.1
	ksi	9.6	9.7	10.0	12.0	12.5	13.0	13.5
D256	Izod impact strength, notched:[b]							
	J/m	53.4	53.4	74.7	123	160	160	160
	ft·lb/in.	1.0	1.0	1.4	2.3	3.0	3.0	3.0
D256	Izod impact strength, unnotched:[b]							
	J/m	...	267–320	587–641	1070	1070	1070	1070
	ft·lb/in.	...	5–6	11–12	20	20	20	20
D785	Rockwell hardness	R119	R118	R119	R121	R121	R121	R121
			M86	M88	M92	M92	M98	M101
Thermal								
D621	Deformation under load,[c] %	1.6[d]	1.4	1.0	0.9	0.4	0.3	0.2
D648	Deflection temperature at 0.45 MPa (66 psi):							
	°C	175	210	216	218	218	218	218
	°F	347	410	420	425	425	425	425
D648	Deflection temperature at 1.82 MPa (264 psi):							
	°C	75	191	210	216	216	216	216
	°F	167	375	410	420	420	420	420
Cenco	Thermal conductivity:							
	W/m·K	0.19	0.36	0.39	0.48	0.49	0.50	0.52
	Btu·in./ft²·h·°F	1.3	2.5	2.7	3.3	3.4	3.5	3.6
D696	Coefficient of linear thermal expansion:							
	10⁻⁵ m/m·K	8.3	4.5	4.0	3.1	2.2	1.6	1.4
	10⁻⁵ in./in.·°F	4.6	2.5	2.2	1.7	1.2	0.9	0.8

[a]Data from LNP Corp. Test specimens dry, as molded. [b]Izod impact test bars 6.35 × 12.7 mm (¼ × ½ in.). [c]In 24 h at 28 MPa and 50 °C (4 ksi and 122 °F). [d]At 14 MPa (2 ksi).

Properties of Fiberglass-Reinforced Nylon 66[a]

ASTM method	Property	Unrein-forced	10% glass	20% glass	30% glass	40% glass	50% glass	60% glass
General								
D792	Specific gravity	1.13–1.15	1.21	1.28	1.37	1.46	1.57	1.70
...	Specific volume:							
	cm³/kg	...	827	784	730	686	636	589
	in.³/lb	...	22.9	21.7	20.2	19.0	17.6	16.3
D570	Water absorption in 24 h, %	1.5	1.1	0.9	0.9	0.6	0.5	0.4
D570	Equil. cont. immersion, %	8.0	7.8	5.6	3.8	3.0	2.6	2.3
D955	Mold shrinkage, mm/mm (in./in.):							
	3.2-mm (1/8-in.) avg section	0.015	0.0065	0.0050	0.0040	0.0040	0.0030	0.0020
	6.4-mm (1/4-in.) avg section	0.018	0.0150	0.0060	0.0055	0.0050	0.0040	0.0030
Mechanical								
D638	Tensile strength:							
	MPa	81.4	96.5	131	186	214	221	228
	ksi	11.8	14.0	19.0	27.0	31.0	32.0	33.0
D638	Tensile elongation, %	60	3–4	3–4	3–4	2–3	2–3	1–2
D790	Flexural strength:							
	MPa	103	138	200	262	290	321	345
	ksi	15.0	20.0	29.0	38.0	42.0	46.5	50.0
D790	Flexural modulus:							
	GPa	2.83	4.48	5.86	8.96	11.0	15.2	19.3
	10⁶ psi	0.41	0.65	0.85	1.30	1.60	2.20	2.80
D695	Compressive strength:							
	MPa	33.8 (1%)	124	159	165	172	186	207
	ksi	4.9 (1%)	18.0	23.0	24.0	25.0	27.0	30.0
D732	Shear strength:							
	MPa	66.2	68.9	73.1	86.2	87.6	91.7	95.1
	ksi	9.6	10.0	10.6	12.5	12.7	13.3	13.8
D256	Izod impact strength, notched:[b]							
	J/m	48.0	42.7	64.1	107	139	139	139
	ft·lb/in.	0.9	0.8	1.2	2.0	2.6	2.6	2.6
D256	Izod impact strength, unnotched:[b]							
	J/m	...	267–320	427–480	907	1015	1070	1070
	ft·lb/in.	...	5–6	8–9	17	19	20	20
D785	Rockwell hardness	R118	R121 M92	R121 M93	R121 M96	R121 M96	R121 M100	R121 M104
Thermal								
D621	Deformation under load,[c] %	1.4[d]	0.9	0.7	0.6	0.4	0.3	0.2
D648	Deflection temperature at 0.45 MPa (66 psi):							
	°C	182	260	260	260	260	260	260
	°F	360	500	500	500	500	500	500
D648	Deflection temperature at 1.82 MPa (264 psi):							
	°C	66	252	252	254	260	260	260
	°F	150	485	485	490	500	500	500
Cenco	Thermal conductivity:							
	W/m·K	0.25	0.39	0.42	0.49	0.52	0.55	0.58
	Btu·in./ft²·h·°F	1.7	2.7	2.9	3.4	3.6	3.8	4.0
D696	Coefficient of linear thermal expansion:							
	10⁻⁵ m/m·K	8.1	4.9	4.1	3.2	2.5	1.8	1.6
	10⁻⁵ in./in.·°F	4.5	2.7	2.3	1.8	1.4	1.0	0.9

[a]Data from LNP Corp. Test specimens dry, as molded. [b]Izod impact test bars 6.35 × 12.7 mm (1/4 × 1/2 in.). [c]In 24 h at 28 MPa and 50 °C (4 ksi and 122 °F).
[d]At 14 MPa (2 ksi).

Properties of Fiberglass-Reinforced Nylon 6/10[a]

ASTM method	Property	Unrein-forced	10% glass	20% glass	30% glass	40% glass	50% glass	60% glass
General								
D792	Specific gravity	1.07–1.09	1.15	1.22	1.30	1.41	1.50	1.65
...	Specific volume:							
	cm³/kg	...	871	820	770	715	658	607
	in.³/lb	...	24.1	22.7	21.3	19.8	18.2	16.8
D570	Water absorption in 24 h, %	0.40	0.24	0.22	0.20	0.18	0.16	0.14
D570	Equil. cont. immersion, %	3.50	3.20	2.00	1.85	1.80	1.40	0.85
D955	Mold shrinkage, mm/mm (in./in.):							
	3.2-mm (1/8-in.) avg section	0.013	0.0045	0.0040	0.0035	0.0030	0.0025	0.0020
	6.4-mm (1/4-in.) avg section	0.016	0.0060	0.0050	0.0045	0.0040	0.0035	0.0030
Mechanical								
D638	Tensile strength:							
	MPa	58.6	82.7	124	145	179	200	214
	ksi	8.5	12.0	18.0	21.0	26.0	29.0	31.0
D638	Tensile elongation, %	85	3–4	3–4	3–4	2–3	2–3	1–2
D790	Flexural strength:							
	MPa	82.7	117	179	221	241	296	317
	ksi	12.0	17.0	26.0	32.0	35.0	43.0	46.0
D790	Flexural modulus:							
	GPa	1.93	5.17	6.20	7.58	8.96	13.1	15.9
	10⁶ psi	0.28	0.75	0.90	1.10	1.30	1.90	2.30
D695	Compressive strength:							
	MPa	20.7 (1%)	96.5	117	138	159	172	179
	ksi	3.0 (1%)	14.0	17.0	20.0	23.0	25.0	26.0
D732	Shear strength:							
	MPa	57.9	59.3	64.8	75.8	82.7	86.2	89.6
	ksi	8.4	8.6	9.4	11.0	12.0	12.5	13.0
D256	Izod impact strength, notched:[b]							
	J/m	32.0	48.0	58.7	128	171	171	171
	ft · lb/in.	0.6	0.9	1.1	2.4	3.2	3.2	3.2
D256	Izod impact strength, unnotched:[b]							
	J/m	...	320–374	641–694	1070	1070	1070	1070
	ft · lb/in.	...	6–7	12–13	20.0	20.0	20.0	20.0
D785	Rockwell hardness	R111	R117 M87	R119 M89	R120 M93	R121 M93	R122 M99	R123 M102
Thermal								
D621	Deformation under load,[c] %	1.6[d]	2.0	1.4	0.9	0.5	0.3	0.2
D648	Deflection temperature at 0.45 MPa (66 psi):							
	°C	149	210	216	221	224	224	224
	°F	300	410	420	430	435	435	435
D648	Deflection temperature at 1.82 MPa (264 psi):							
	°C	57	204	210	216	216	218	218
	°F	135	400	410	420	420	425	425
Cenco	Thermal conductivity:							
	W/m · K	0.22	0.29	0.43	0.50	0.53	0.58	0.61
	Btu · in./ft² · h · °F	1.5	2.0	3.0	3.5	3.7	4.0	4.2
D696	Coefficient of linear thermal expansion:							
	10⁻⁵ m/m · K	9.0	4.5	4.0	2.7	2.2	1.6	1.4
	10⁻⁵ in./in. · °F	5.0	2.5	2.2	1.5	1.2	0.9	0.8

[a]Data from LNP Corp. Test specimens dry, as molded. [b]Izod impact test bars 6.35 × 12.7 mm (1/4 × 1/2 in.). [c]In 24 h at 28 MPa and 50 °C (4 ksi and 122 °F).
[d]At 14 MPa (2 ksi).

Properties of Fiberglass-Reinforced Nylon 6/12[a]

ASTM method	Property	Unrein-forced	10% glass	20% glass	30% glass	40% glass	50% glass	60% glass
General								
D792	Specific gravity	1.06–1.08	1.14	1.21	1.30	1.40	1.49	1.64
...	Specific volume:							
	cm³/kg	...	878	827	777	719	661	611
	in.³/lb	...	24.3	22.9	21.5	19.9	18.3	16.9
D570	Water absorption in 24 h, %	0.40	0.24	0.22	0.20	0.18	0.14	0.13
D570	Equil. cont. immersion, %	3.00	3.20	2.00	1.85	1.80	1.40	0.85
D955	Mold shrinkage, mm/mm (in./in.):							
	3.2-mm (1/8-in.) avg section	0.012	0.0045	0.0030	0.0035	0.0030	0.0025	0.0020
	6.4-mm (1/4-in.) avg section	0.016	0.0060	0.0040	0.0045	0.0040	0.0035	0.0030
Mechanical								
D638	Tensile strength:							
	MPa	60.7	82.7	124	152	179	200	214
	ksi	8.8	12.0	18.0	22.0	26.0	29.0	31.0
D638	Tensile elongation, %	150	6–7	5–6	4–5	3–4	2–3	1–2
D790	Flexural strength:							
	MPa	82.7	124	193	221	269	296	317
	ksi	12.0	18.0	28.0	32.0	39.0	43.0	46.0
D790	Flexural modulus:							
	GPa	2.00	5.17	6.21	7.58	8.96	13.1	15.9
	10⁶ psi........................	0.29	0.75	0.90	1.10	1.30	1.90	2.30
D695	Compressive strength:							
	MPa	16.5 (1%)	96.5	131	152	159	172	179
	ksi	2.4 (1%)	14.0	19.0	22.0	23.0	25.0	26.0
D732	Shear strength:							
	MPa	59.3	59.3	64.8	75.8	82.7	86.2	89.6
	ksi	8.6	8.6	9.4	11.0	12.0	12.5	13.0
D256	Izod impact strength, notched:[b]							
	J/m	53.4	48.0	58.7	128	171	171	171
	ft·lb/in.	1.0	0.9	1.1	2.4	3.2	3.2	3.2
D256	Izod impact strength, unnotched:[b]							
	J/m	...	320–374	641–694	1070	1070	1070	1070
	ft·lb/in.	...	6–7	12–13	20.0	20.0	20.0	20.0
D785	Rockwell hardness	R114	R117 M87	R119 M89	R120 M93	R121 M93	R122 M99	R123 M102
Thermal								
D621	Deformation under load,[c] %	1.6[d]	2.0	1.4	0.9	0.5	0.3	0.2
D648	Deflection temperature at 0.45 MPa (66 psi):							
	°C	149	210	216	221	224	224	224
	°F	300	410	420	430	435	435	435
D648	Deflection temperature at 1.82 MPa (264 psi):							
	°C	57	204	210	213	216	218	218
	°F	135	400	410	415	420	425	425
Cenco	Thermal conductivity:							
	W/m·K	0.22	0.29	0.43	0.50	0.53	0.58	0.61
	Btu·in./ft²·h·°F	1.5	2.0	3.0	3.5	3.7	4.0	4.2
D696	Coefficient of linear thermal expansion:							
	10⁻⁵ m/m·K	9.0	4.5	4.0	2.7	2.2	1.6	1.4
	10⁻⁵ in./in.·°F	5.0	2.5	2.2	1.5	1.2	0.9	0.8

[a]Data from LNP Corp. Test specimens dry, as molded. [b]Izod impact test bars 6.35 × 12.7 mm (1/4 × 1/2 in.). [c]In 24 h at 28 MPa and 50 °C (4 ksi and 122 °F). [d]At 14 MPa (2 ksi).

Comparative Fatigue Strength of Nonwoven Unidirectional Glass Fiber Reinforced Plastic Laminates

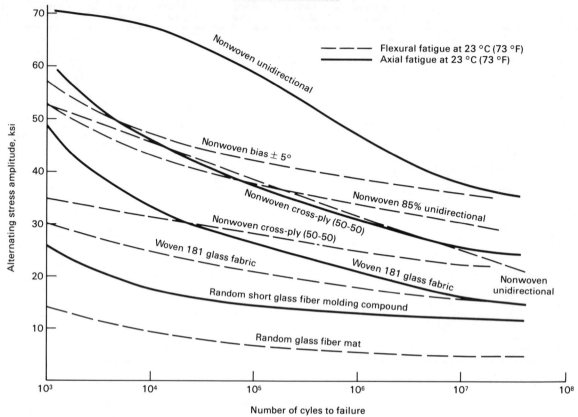

Properties of Fiberglass-Reinforced Nylon 11 and Nylon 12[a]

ASTM method	Property	Nylon 11 Unrein-forced	Nylon 11 30% glass	Nylon 12 Unrein-forced	Nylon 12 30% glass	Nylon 12 50% glass
General						
D792	Specific gravity	1.05	1.24	1.02	1.24	1.45
...	Specific volume:					
	cm³/kg	...	806	...	806	690
	in.³/lb	...	22.3	...	22.3	19.1
D570	Water absorption in 24 h, %	0.40	0.20	0.25	0.07	0.06
D955	Mold shrinkage, mm/mm (in./in.), 3.2-mm (1/8-in.) avg section	0.012	0.003	0.075	0.004	0.0030
Mechanical						
D638	Tensile strength:					
	MPa	53.8	96.5	51.0	119	152
	ksi	7.8	14.0	7.4	17.2	22.0
D638	Tensile elongation, %	300	3–4	8–9	4–6	4–6
D790	Flexural strength:					
	MPa	53.8	138	48.3	162	221
	ksi	7.8	20.0	7.0	23.5	32.0
D790	Flexural modulus:					
	GPa	0.98	6.03	1.21	6.89	8.96
	10⁶ psi	0.142	0.875	0.175	1.00	1.30
D256	Izod impact strength, notched:[b]					
	J/m	96.1	117	107	176	224
	ft·lb/in.	1.8	2.2	2.0	3.3	4.2
D256	Izod impact strength, unnotched:[b]					
	J/m	...	374–427	...	747–854	961–1070
	ft·lb/in.	...	7–8	...	14–16	18–20
D648	Deflection temperature at 1.82 MPa (264 psi):					
	°C	54	168	52	171	177
	°F	130	335	125	340	350

[a]Data from LNP Corp. Test specimens moisture conditioned at 50% R.H. [b]Izod impact test bars 6.35 × 12.7 mm (1/4 × 1/2 in.).

Electrical Properties of Fiberglass-Reinforced Nylon Resins[a]

ASTM method	Property	Nylon 6 30% glass	Nylon 6 40% glass	Nylon 66 30% glass	Nylon 66 40% glass	Nylon 6/10 30% glass	Nylon 6/10 40% glass
D149	Dielectric strength, V/mil	450	420	440	400	440	420
D257	Volume resistivity, $\Omega \cdot$ cm	2.8×10^{14}	7×10^{13}	5×10^{14}	1.8×10^{14}	7.6×10^{15}	1.5×10^{15}
D257	Surface resistivity, Ω	9×10^{13}	2×10^{13}	2.9×10^{14}	7.8×10^{13}	1.5×10^{15}	6.8×10^{14}
D495	Arc resistance, s	135	135	130	135	125	130
D150	Dielectric constant:						
	100 Hz	4.2	4.4	4.2	4.4	4.2	4.3
	10^3 Hz	3.9	4.0	3.9	4.4	3.8	3.9
	10^6 Hz	3.6	3.9	3.5	4.0	3.5	3.9
D150	Dissipation factor:						
	100 Hz	0.009	0.010	0.009	0.009	0.013	0.015
	10^3 Hz	0.014	0.016	0.012	0.012	0.016	0.018
	10^6 Hz	0.018	0.021	0.018	0.019	0.015	0.016

[a]Data from LNP Corp. Test specimens dry, as molded.

Physical Property Comparison of Fiberglass-Reinforced Nylons and Die-Casting Alloys[a]

ASTM method	Property	Nylon 6, 40% glass	Nylon 66, 60% glass	Nylon 6/10, 40% glass	Nylon 66, 40% glass	Zinc, ASTM XXIII	Aluminum, ASTM S9	Magnesium, ASTM AZ91B
Mechanical								
D638	Tensile strength:							
	MPa	179	228	179	214	283	117–207	228
	ksi	26.0	33.0	26.0	31.0	41.0	17.0–30.0	33.0
D790	Flexural strength:							
	MPa	248	345	241	290	···	121	···
	ksi	36.0	50.0	35.0	42.0	···	17.5	···
D790	Flexural modulus:							
	GPa	10.3	19.3	9.0	11.0	41.4	68.9	46.9
	10^6 psi	1.5	2.8	1.3	1.6	6.0	10.0	6.8
···	Charpy impact strength, unnotched:[b]							
	J/m	···	···	···	···	2300	32–187	107
	ft·lb/in.	···	···	···	···	43	0.6–3.5	2.0
D256	Izod impact strength, notched:[b]							
	J/m	160	139	171	139	347	···	···
	ft·lb/in.	3.0	2.6	3.2	2.6	6.5	···	···
D256	Izod impact strength, unnotched:[b]							
	J/m	1070	1070	1070	1015	···	···	···
	ft·lb/in.	20	20	20	19	···	···	···
D696	Coefficient of thermal expansion:							
	10^{-5} m/m·K	2.2	1.6	2.2	2.5	2.7	2.2	2.5
	10^{-5} in./in.·°F	1.2	0.9	1.2	1.4	1.5	1.2	1.4
D785	Rockwell hardness	M92	M104	M93	M96	M103	···	···
General								
D792	Specific gravity	1.46	1.70	1.41	1.46	6.61	2.89	1.81
···	Density:							
	kg/m³	1520	1680	1440	1520	6600	2880	1810
	lb/ft³	95	105	90	95	412	180	113
···	Strength-to-weight ratio	0.87	1.00	0.92	1.00	0.32	0.31	0.95

[a]Data from LNP Corp. [b]Test bars, 6.35 mm ($^1/_4$ in.).

Properties of High-Strength Molding Compound[a]
(Ref 53, p 6)

Tensile strength, MPa (ksi):
  At 22 °C (72 °F) 268 (38.9)
  At 120 °C (250 °F) 193 (28.7)
  At 150 °C (300 °F) 159 (23.4)
  At 175 °C (350 °F) 121 (17.5)
Flexural strength, MPa (ksi):
  At 22 °C (72 °F) 396 (57.4)
  At 120 °C (250 °F) 248 (36.2)
  At 150 °C (300 °F) 131 (19.0)
  At 175 °C (350 °F) 48.3 (7.0)
Flexural modulus, GPa (10^6 psi):
  At 120 °C (250 °F) 11.9 (1.73)
  At 150 °C (300 °F) 10.8 (1.56)
  At 175 °C (350 °F) 2.96 (0.43)
Elongation, % 1.45
Strain energy density, in.·lb/in.^{3} 224.2
Tensile modulus, GPa (10^6 psi) 16.4 (2.38)
Izod impact (notched), J (ft·lb) 24.4 (18)
Izod impact (unnotched), J (ft·lb) 47.5 (35)
Glass-transition temperature, °C (°F) 192 (378)
Specific gravity 1.9
Linear shrinkage, mils/in. .... −1.42
Water absorption in 24 h at 23 °C (73 °F) 0.04

[a]Source: Premix. New high-strength molding compounds feature excellent mechanical properties. They are usually formulated with tough resin systems reinforced by continuous-strand glass fibers or glass cloth, or a hybrid that includes aramid fibers, and are generally compression molded. Values above apply to Premix Premi-Glas 1200E HSMC, an epoxy resin reinforced with 65% glass.

Comparison of Dielectric Properties of Quartz/ Polyimide and Fiberglass/Polyimide[a]

Material[b]	Property[c]	
	Dielectric constant	Loss tangent
Astroquartz:		
At room temperature	3.30	0.0054
At 315 °C (600 °F)	3.25	0.0030
"E" fiberglass:		
At room temperature	3.87	0.009
At 315 °C (600 °F)	4.20	0.0085

[a]Data from J. P. Stevens. [b]Resin content, 35%. [c]Measured at 9.375 GHz.

Typical Properties of Glass and Quartz Composites (Epoxy Resin)

Property	E-glass	S-glass	Quartz
Density, g/cm^3	0.065	0.060	0.065
Tensile strength, unidirectional:			
MPa	1105	1380	···
ksi......................	160	200	···
Tensile strength, fabric:			
MPa	414	586	483
ksi	60	85	70
Tensile modulus, unidirectional:			
GPa	41.4	55.2	···
10^6 psi	6.0	8.0	···
Tensile modulus, fabric:			
GPa	23.4	27.6	20.7
10^6 psi	3.4	4.0	3.0
Compressive strength, fabric:			
MPa	345	414	276
ksi	50	60	40
Dielectric constant	4.3	3.8	3.2
Loss factor	0.19–0.24	0.25	0.01
Coefficient of thermal expansion:			
10^6 cm/cm·°C	8.6	6.3	4.0
10^6 in./in.·°F	4.8	3.5	2.2

Tensile Strengths of Filled and Unfilled Heat-Resistant Resins after Thermal Aging at 260 °C (500 °F)[a]
(Ref 124)

Base resin	Fiberglass content, wt %	Tensile strength, ksi (MPa), after aging for:						
		0 h	100 h	250 h	500 h	750 h	1000 h	1500 h
ETFE	20	.... 11.3 (77.9)	11.5 (79.3)	10.0 (69.0)	7.0 (48.3)	5.0 (34.5)	3.8 (26.2)	2.3 (15.9)
FEP	20	.... 5.0 (34.5)	5.1 (35.2)	4.8 (33.1)	4.7 (32.4)	4.7 (32.4)	4.6 (31.7)	4.5 (31.0)
Polyphenylene sulfide	40	.... 23.2 (160)	16.4 (113)	16.0 (110)	15.5 (107)	15.0 (103)	14.5 (100)	13.8 (95.2)
Polyethersulfone	40	.... 22.7 (157)	15.6 (108)	14.8 (102)	14.3 (98.6)	13.7 (94.5)	12.2 (84.1)	10.5 (72.4)
Polyimide	30	.... 13.0 (89.6)	15.0 (103)	14.3 (98.6)	13.4 (92.4)	12.8 (88.3)	12.0 (82.7)	11.2 (77.2)
Polyamide-imide	0	.... 27.4 (189)	27.2 (188)	26.6 (183)	26.0 (179)	24.5 (169)	23.5 (162)	22.0 (152)
Polyarylsulfone	0	.... 13.1 (90.3)	11.5 (79.3)	10.5 (72.4)	10.0 (69.0)	9.5 (65.5)	8.4 (57.9)	7.6 (52.4)
Poly-p-oxybenzoate	0	.... 23.0 (159)	18.0 (124)	16.3 (112)	16.0 (110)	15.5 (107)	15.1 (104)	13.0 (89.6)
Nylon 66	50	.... 31.0 (214)	17.6 (121)	10.3 (11.0)	9.4 (64.8)	···	···	···
Polyester	40	.... 22.1 (152)	Melted	···	···	···	···	···
Polysulfone	40	.... 20.3 (140)	Melted	···	···	···	···	···

[a]Tested at 23 °C (73 °F).

Tensile Strengths of Filled and Unfilled Heat-Resistant Resins at Elevated Temperatures (Ref 124)

Base resin	Fiberglass content, wt %	23 °C (73 °F)	93 °C (200 °F)	149 °C (300 °F)	177 °C (350 °F)	204 °C (400 °F)	232 °C (450 °F)
ETFE	20	11.3 (77.9)	6.85 (47.2)	4.0 (27.6)	2.0 (13.8)	(a)	(a)
FEP	20	5.0 (34.5)	4.2 (29.0)	2.3 (15.9)	1.2 (8.3)	(a)	(a)
Polyphenylene sulfide	40	23.2 (160)	11.2 (77.2)	8.1 (55.8)	4.8 (33.1)	1.1 (7.6)	(a)
Polyethersulfone	40	22.7 (157)	19.4 (134)	13.1 (90.3)	4.9 (33.8)	3.1 (21.4)	(a)
Nylon 66	50	31.2 (215)	16.0 (110)	12.4 (85.5)	7.3 (50.3)	2.2 (15.2)	(a)
Polyester	40	19.4 (134)	7.4 (51.0)	4.1 (28.3)	0.6 (4.1)	(a)	(a)
Polysulfone	40	17.3 (119)	14.9 (103)	2.3 (15.9)	1.1 (7.6)	(a)	(a)
Polyimide	30	13.0 (89.6)	6.2 (42.7)	4.8 (33.1)	3.1 (21.4)	2.3 (15.9)	1.8 (12.4)
Polyamide-imide	0	27.4 (189)[b]	19.9 (137)	16.2 (112)	11.4 (78.6)	8.2 (56.5)	6.9 (47.6)
Polyarylsulfone	0	11.1 (76.5)	10.4 (71.7)	8.7 (60.0)	7.4 (51.0)	5.7 (39.3)	3.2 (22.1)
Poly-p-oxybenzoate	0	13.9 (95.8)	11.2 (77.2)	9.3 (64.1)	7.8 (53.8)	6.4 (44.1)	3.9 (26.9)

Tensile strength, ksi (MPa), at above temperatures.

[a]Could not sustain load. [b]Annealed prior to test.

Properties of Fiberglass-Reinforced Polyester Resins

ASTM test method	Property	Woven cloth	Chopped roving	Sheet molding compound
Mechanical Properties				
D792	Specific gravity	1.5–2.1	1.35–2.30	1.65–2.60
D790	Flexural yield strength:			
	MPa	276–552	68.9–276	68.9–248
	ksi	40–80	10–40	10–36
D790	Flexural modulus of elasticity:			
	GPa	···	6.9–20.7	6.9–15.2
	10^5 psi	···	10–30	10–22
D695	Compressive strength:			
	MPa	172–345	103–207	103–207
	ksi	25–50	15–30	15–30
D638	Tensile strength:			
	MPa	207–345	103–207	55.2–138
	ksi	30–50	15–30	8–20
D638	Elongation, %	0.5–2.0	0.5–5.0	···
D638	Tensile modulus of elasticity:			
	GPa	10.3–31.0	5.5–13.8	···
	10^5 psi	15.0–45.0	8.0–20.0	···
D256	Izod impact strength, notched:[a]			
	J/m	267–1600	107–1070	374–1175
	ft·lb/in.	5.0–30.0	2.0–20.0	7.0–22.0
···	Barcol hardness	60–80	50–80	50–70
Electrical Properties				
D257	Volume resistivity,[b] $\Omega\cdot$cm	10^{14}	10^{14}	10^{14}–10^{15}
D149	Dielectric strength,[c] V/mil:			
	Short-time test	350–500	350–500	380–450
	Step-by-step test	300–400	300–450	350–400
D150	Dielectric constant:			
	At 60 Hz	4.1–5.5	3.8–6.0	4.4–6.3
	At 10^3 Hz	4.2–6.0	4.0–6.0	4.4–6.1
	At 10^6 Hz	4.0–5.5	3.5–5.5	4.2–5.8
D150	Dissipation (power) factor:			
	At 60 Hz	0.01–0.04	0.01–0.04	0.007–0.021
	At 10^3 Hz	0.01–0.06	0.01–0.05	0.007–0.015
	At 10^6 Hz	0.01–0.03	0.01–0.03	0.016–0.024
D495	Arc resistance, s	60–120	120–180	120–200
Resistance Characteristics				
···	Heat resistance (continuous):			
	°C	149–177	149–177	149–204
	°F	300–350	300–350	300–400
D570	Water absorption[d] in 24 h, %	0.05–0.50	0.1–1.0	0.10–0.15
···	Effect of sunlight	Slight	Slight	Slight

[a]Test bar 13 × 13 mm ($^1/_2$ × $^1/_2$ in.). [b]At 50% R.H. and 23 °C (73 °F). [c]Material thickness, 3.2 mm ($^1/_8$ in.). [d]Material thickness, 3.1 mm ($^1/_8$ in.).

Properties of Glass Fiber Vs. Microsphere Reinforced Polypropylene Compounds (Ref 97, p 24)

Filler/reinforcement[a]	Warpage		Flexural modulus		Flexural strength		Notched Izod impact strength at 23 °C (73 °F)		Shrinkage, mm/mm	Heat-deflection temperature		Tensile yield strength		Elongation at break, %
	mm	in.	GPa	ksi	MPa	ksi	J/m	ft·lb/in.	(in./in.)	°C	°F	MPa	ksi	
Control	0.0	0.00	1.75	254	56.5	8.2	42.7	0.8	0.018	122	252	37.2	5.40	425
20% GF	4.0	0.16	3.07	445	70.3	10.2	74.7	1.4	0.002	160	320	44.5	6.45	10
30% GF	3.2	0.13	4.39	636	77.2	11.2	80.1	1.5	0.001	160	320	46.5	6.75	10
40% GF	2.2	0.09	6.00	870	80.7	11.7	74.7	1.4	0.001	161	322	42.1	6.11	10
30% GF; 10% Z-600	0.0	0.00	4.69	680	66.9	9.7	58.7	1.1	0.001	157	315	35.4	5.13	10
25% GF; 15% Z-600	0.0	0.00	3.64	528	60.0	8.7	53.4	1.0	0.001	159	318	33.1	4.80	10
20% Z-600	0.0	0.00	2.09	303	51.0	7.4	37.4	0.7	0.017	123	253	31.0	4.50	269
30% Z-600	0.0	0.00	2.12	308	48.3	7.0	42.7	0.8	0.016	126	259	26.7	3.87	225
40% Z-600	0.0	0.00	2.52	366	49.0	7.1	42.7	0.8	0.014	129	264	23.8	3.45	114

[a]GF = glass fiber; Z-600 = Zeeospheres Type 600 (alumina/alumina ceramic).

6.2.5. Hybrid Composites

Effect of Various Ratios of Carbon to Glass Fibers on Typical Properties of Hybrid Polyester Resin Matrix Composites (Ref 99, p 184)

Property	Carbon to glass fiber ratio[a]			
	All glass	1 carbon, 3 glass	1 carbon, 1 glass	3 carbon, 1 glass
Tensile strength:				
MPa	607	641	676	793
ksi	88	93	98	115
Tensile modulus:				
GPa	36.5	62.1	85.5	112
10^6 psi	5.3	9.0	12.4	16.3
Flexural strength:				
MPa	931	1070	1205	1275
ksi	135	155	175	185
Flexural modulus:				
GPa	35.2	62.7	78.0	110
10^6 psi	5.1	9.1	11.3	16.0
Interlaminar shear strength:				
MPa	65.5	74.5	76	83
ksi	9.5	10.8	11	12
Density, g/cm^3	1.95	1.88	1.82	1.69

[a]Fiber volume, 65%.

Properties of Various Hybrid Composites[a] (Ref 98, p 37)

Material	Configuration[b]	Tensile failure stress		Compressive failure stress		Modulus of elasticity							
						Tension				Compression			
						Longitudinal		Transverse		Longitudinal		Transverse	
		MPa	ksi	MPa	ksi	10^4 MPa	10^6 psi	10^4 MPa	10^6 psi	10^4 MPa	10^6 psi	10^4 MPa	10^6 psi
S-glass/graphite[c]	0°$_4$/±45°$_2$	975	141.5	665	96.5	4.75	6.9	2.1	3.1	3.9	5.7	2.2	3.3
Graphite[c]/boron	0°$_4$/±45°$_2$	1085	157.5	680	98.7	14.7	21.4	3.0	4.4	11.7	17.0	1.8	2.6
Boron/graphite[c]/graphite[c]	0°$_3$/±45°/90°	856	124.2	655	95.0	15.1	22.0	5.7	8.4	2.4	3.5	1.5	2.2
S-glass/boron	0°$_5$/±45°	1665	241.5	517	75.1	4.96	7.2	2.96	4.3	4.9	7.1	2.3	3.4
Aramid[d]/graphite[c]/aramid	0°$_2$/±45°$_2$/90°	496	72.0	176	25.5	4.8	7.0	2.7	4.0	3.9	5.7	2.0	3.0
Graphite[c]/graphite[e]	0°$_4$/±45°$_3$	634	92.0	545	89.1	7.4	10.8	2.48	3.6	7.1	10.4	2.1	3.1
Graphite[f]/boron	0°$_5$/±45°	800	116.0	625	90.7	7.4	10.8	2.48	3.6	8.8	12.8	1.7	2.5
S-glass/graphite[e]	0°$_4$/±45°	751	109.0	399	57.9	1.93	2.8	0.75	1.1	3.6	5.3	1.9	2.7

[a]Data from Boeing Aerospace Corp. and Material Sciences Corp. [b]All configurations are symmetrical about the last ply in the sequence; subscripts indicate the number of plies. [c]Union Carbide Thornel 300. [d]Du Pont Kevlar 49. [e]Hercules HMS, high modulus. [f]Hercules HTS, intermediate tensile strength and modulus.

Properties of Carbon/Glass/Polyester Hybrid Composites[a] (Ref 98, p 37)

Property	0% carbon, 100% glass[b]	25% carbon, 75% glass	50% carbon, 50% glass	75% carbon, 25% glass
Tensile strength:				
MPa	604	641	689	807
ksi	87.7	93.0	100	117.0
Tensile modulus:				
10^4 MPa	4.00	6.39	8.96	12.3
10^6 psi	5.81	9.27	13.0	17.9
Flexural strength:				
MPa	945	1062	1220	1262
ksi	137	154	177	183
Flexural modulus:				
10^4 MPa	3.54	6.34	7.8	11.2
10^6 psi	5.14	9.2	11.4	16.3
Interlaminar shear strength:				
MPa	65	74	76	83
ksi	9.5	10.8	11.0	12.0
Density:				
kg/m^3	1932	1873	1814	1688
lb/in.^{3}	0.069	0.067	0.065	0.060

[a]Data from Celanese Corp., PPG Industries. All fiber content by volume. Matrix resin 48% thermoset polyester plus 52% continuous, unidirectionally oriented fiber by volume equivalent to 30% resin and 70% glass by weight. Properties apply to longitudinal fiber direction. [b]Less than optimum data most recently reported.

Typical Properties of Nylon 66[a] Glass Fiber and Carbon Fiber Hybrid (Ref 99, p 184)

Property	Reinforcement, wt %		
	20% carbon fiber, 25% glass fiber[b]	25% carbon fiber, 20% glass fiber[b]	30% carbon fiber, 15% glass fiber[b]
Specific gravity	...	1.42	1.41
Ultimate elongation, %	3.9	3.8	3.5
Ultimate tensile strength:			
MPa	125	114	101
ksi	18.1	16.5	14.7
Tensile modulus:			
GPa	...	11.8	...
10^6 psi	...	1.71	...
Flexural strength:			
MPa	236	234	214
ksi	34.3	34.0	31.0
Flexural modulus:			
GPa	11.4	13.2	13.0
10^6 psi	1.65	1.92	1.89
Compressive strength:			
MPa	...	117	103
ksi	...	17.0	15.0
Izod impact strength, notched:			
J/m	85.4	80.1	64.1
ft·lb/in.	1.6	1.5	1.2
Heat-deflection temperature:[c]			
°C	...	250	250
°F	...	480	580
Mold shrinkage, mm/mm (in./in.)	...	0.003	0.003
Volume resistivity, Ω·cm	10^5	10^3	10^3

[a]Injection molding compound. [b]6.35-mm (0.25-in.) chopped glass fibers. [c]At 1.82 MPa (264 psi).

Density and Strength of Hybrid/Epoxy Composites

Construction	Density		Tensile strength		Compressive strength	
	Mg/m^3	lb/in.^{3}	MPa	ksi	MPa	ksi
T-300/boron, 0/±45°, 75%/25%	1.66	0.06	1086	157.5	681	98.7
Boron/AS, 0/±45°, 52%/48%	1.77	0.064	856	124.2	655	95.0
S-glass/boron, 0/±45°, 86%/14%	1.85	0.067	1665	241.5	518	75.1
HTS/boron, 0/±45°	1.61	0.058	800	116.0	625	90.7

Typical Properties of Hybrid Sheet Molding Compounds (Ref 99, p 182)

Compound	Glass fiber content, wt %	Carbon fiber content, wt %	Specific gravity, g/cm³	Tensile strength MPa	Tensile strength ksi	Tensile modulus GPa	Tensile modulus 10⁶ psi	Flexural strength MPa	Flexural strength ksi	Flexural modulus GPa	Flexural modulus 10⁶ psi
XMC[a]	70	...	1.9	483	70	...	...	931	135	34.5	5
	26	35	1.83	689	100	...	...	1140	165	93.1	13.5
HMC[b]	60–65	...	1.88	241	35	11.7–14.5	1.7–2.1	379	55	...	...
	28	30	1.60	NA[c]	NA[c]	24.1	3.5	NA[c]	NA[c]	...	...
SMC	30	...	1.70	83–103	12–15	...	...	165–207	24–30	10.3–13.1	1.5–1.9
	...	35	1.66	103	15	26.9	3.9	NA[c]	NA[c]	...	...

[a]XMC (trademark of PPG Industries, Inc.) = directionally reinforced molding compound. [b]HMC (trademark of PPG Industries, Inc.) = high-strength molding compound. [c]NA = not available.

Strength and Cost Factor of Hybrid/Epoxy Composites

Construction	Tensile strength MPa	Tensile strength ksi	Compressive strength MPa	Compressive strength ksi	Cost factor ¢/cm³	Cost factor $/in.³
S-glass/T-300, 0/±45°, 64%/36%	976	141.5	665	96.5	11.3	1.86
S-glass/boron, 0/±45°, 86%/14%	1665	241.5	518	75.1	15.7	2.58
S-glass/HMS, 0/±45°, 52%/48%	752	109.0	399	57.9	17.3	2.84

Ceramic and Kevlar Fiber Hybrids[a]

Nextel 312	Kevlar	HTS graphite	Test	Ultimate stress MPa	Ultimate stress ksi	Ultimate strain	Modulus GPa	Modulus 10⁶ psi
100	0	0	Tension	855	124	0.01428	59.3	8.6
			Tension	827	120	0.01411	58.6	8.5
			Shear	73.1	10.6	...	3.45	0.50
			Shear	73.1	10.6	...	3.59	0.52
			Compression	931	135	0.0134	69.6	10.1
			Compression	786	114	...	...	...
25	75	0	Tension	951	138	0.0150	60.7	8.8
			Tension	958	139	0.0155	59.3	8.6
			Compression	483	70	0.0154	62.7	9.1
50	50	0	Tension	848	123	0.0136	61.4	8.9
			Tension	841	122	0.0134	61.4	8.9
			Compression	703	102	0.0148	68.3	9.9
75	25	0	Compression	745[b]	108[b]	0.0113	71.7	10.4
25	0	75	Tension	1062	154	...	110	16.0
			Tension	1027	149	...	109	15.8
			Compression	1062[b]	154[b]	0.0100	105	15.3
50	0	50	Tension	1062	154	0.0115	91.0	13.2
			Tension	...	...	...	86.2	12.5
			Compression	1138[b]	165[b]	0.0117	104	15.1
			Compression	1076[b]	156[b]	...	...	...
75	0	25	Compression	752[b]	109[b]	0.0094	79.3	11.5
			Compression	683[b]	99[b]	...	...	...

[a]Data from 3M. Hybrid fiber composites were fabricated by drum-winding fibers onto epoxy resin film. In this manner, two types of fibers were interspersed within the lamina. Three hybrid systems were examined: 25% Nextel 312/75% Kevlar, 50% Nextel 312/50% Kevlar, and 75% Nextel 312/25% Kevlar. [b]Bond failure between composite and sandwich beam core.

SECTION 7
Property Data: Metal Matrix Composites

7.1. Data of General Interest; Fabrication Methods; Applications

Schematic Representation of Three Primary Fabrication Methods: (a) Diffusion Bonding Processes; (b) Continuous Casting; and (c) Plasma Spraying (Ref 8, p 148)

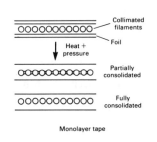
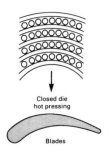
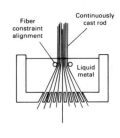
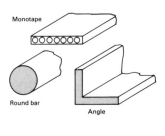
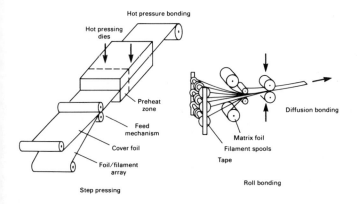
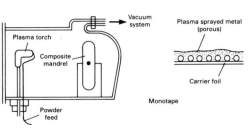

(a)

(b)

(c)

Factors Affecting the Realizable Strength of Fiber-Reinforced Metals (Ref 74, p 186, 187)

Factor	Ideal case	Factor	Ideal case
Fiber Parameters–Strength		Fiber-Fiber ....Fibers must be separated by matrix; otherwise, contact points serve as stress concentrators.	
Initial strength ....Should be at a maximum.		**Matrix Parameters**	
Scatter in strength ....Variation in strength between individual fibers should be minimized.		Mechanical ....Rate of work hardening should be optimized, properties relative to fibers — i.e., E_m E_f.	
Strength retension ....Should be maximized during handling, fabrication, etc.		Chemical ....Should wet fibers, but not react to weaken fibers, particularly at high temperatures; oxidation and corrosion resistance are important.	
Fracture statistics ....All fibers should be loaded evenly. Also depends on points discussed above for initial strength and scatter in strength.		Thermal ....Coefficient of expansion should be similar to that of fiber, recrystallization, temperature.	
Fiber Parameters–Shape and Size		**Fabrication Parameters**	
Taper ....Should be minimized.		Fiber production ....Factors include cost to obtain optimum values cited under Fiber Parameters.	
Diameter ....Should be minimized. Fiber strength usually increases with decreasing diameter. Statistically, more fibers can be packed in per unit volume for small diameters.		Fiber handling ....Technique depends on length and length-to-diameter ratio, and on properties desired in composite (see Fiber Parameters–Relation to Matrix); surface treatment (metallizing).	
Length ....Should be optimized in terms of length greater than critical length. Note: strength also increases with decreasing length.		Incorporation into matrix .....Many techniques — powder metallurgy, infiltration, impact forming, etc. Fibers must be wetted and bonded to matrix. Must form composite of theoretical density.	
Variation in fiber diameter ....Should be as uniform as possible.			
Cross section ....Not all fibers are circular, hence cross-sectional shape will affect maximum packing density and stress distribution.		Composite shaping ....Extrusion, forging, pressing — depends on fiber brittleness (and size).	
Fiber Parameters–Relation to Matrix		Joining ....Chemical, pressure, electron welding, and diffusion bonding techniques are more effective than mechanical joining.	
Orientation ....Fibers should be parallel to tensile axis. However, properties can be tailor-made by varying orientation.			
Distribution ....Fibers should be uniformly distributed in the matrix.			
Overlap ....Short fibers must have a minimum overlap distance; otherwise, crack in matrix will propagate through regions between fiber ends.			

Fabrication Methods and Applications for Composites of the Metal Fiber/Metal Matrix System (Ref 43, p 85, 87)

Fiber	Matrix	Fabrication method	Field of application	Source/author
Al_5Co_2	Al-1.5Cu alloy	Directional solidification	High-tensile-strength electrical conductors	United Technologies
Al_4Ce	Al-Ce alloy (8 to 12% Ce)	Directional solidification	Nuclear industry	Pechiney
Stainless steel	Al	Powder metallurgy; lamination	Aircraft industry	Tumanov *et al*; Pechiney
Be ribbons	Al; Ti	Al- or Ti-clad Be rods are inserted in drilled Al or Ti preforms; mechanical deformation of preform	Shafts for high-speed rotating engines — e.g., helicopter rotor shafts, aircraft control rods	R. Schmidt
Ta	Mg	Infiltration technique	Aircraft industry	I. Ahmad *et al*
Mo	Ti or Ti alloy	Powder metallurgy; fiber alignment by extrusion, rolling, etc.	Supersonic aircraft; rocket propulsion	A. Schwope *et al*
Be or Ti-clad Be	Ti alloy; pure Ti	Explosive welding of interposed layers of Ti and Be fibers	Aircraft construction	R. Trabocco

(continued)

Fabrication Methods and Applications for Composites of the Metal Fiber/Metal Matrix System (Ref 43, p 85, 87) (continued)

Fiber	Matrix	Fabrication method	Field of application	Source/author
Be	Ti; Ti-6Al-4V; Ti-6Al-6V-2Sn; Ti-5Al-2.5Sn	Hot extrusion of mixed precursor of Ti and Be; the latter can be present as fiber preforms.	Aircraft industry	Brush Wellmann, Inc.
Ni$_3$Al	Ni−2 to 10% Al	Powder metallurgy; the fiberlike phase is formed *in situ* during mechanical working.	Oxidation resistant; high strength at high and low temperatures	Cabot Corp.
Ni$_3$Ta	Ni-Ta alloy	Unidirectional solidification	⋯	Euratom
Ni-Cr-Al-Y	Ni alloy	Powder metallurgy	Sealing elements in turbines, compressors	Brunswick Corp.
W-1ThO$_2$	Superalloy	Investment casting	Jet engines	U.S. Army
W	W-Ni-Fe alloy	Liquid phase sintering; the W fibers are recrystallized to avoid dissolution.	⋯	E. Zukas; NASA
Stainless steel	Ni alloys	Electroforming	Rocket engines	NASA
Mo, Ti, Nb	Ni superalloys	Powder metallurgy	⋯	U.S. Army Material Development
W	Cu	Melt impregnation	Electrical machinery	Hitachi
W filament coated with successive layers of B and Ti or Al$_2$O$_3$	Fe or Fe alloys	Not defined	⋯	U.S. Composites Corp.
Ti or Mo band coated with SiC	Ti, Mg	Various processes	⋯	Armines
Nb filaments	Ni, Cu, Ag	Nb filaments imbedded in a Cu, Ni, or Ag matrix are passed through a molten bath of Sn, until formation of Nb$_3$Sn.	Superconductors	Imperial Metal Industries

Fabrication Methods and Applications for Composites of the Ceramic Oxide or Glass Fiber/Metal Matrix System (Ref 43, p 92)

Fiber	Matrix	Fabrication method	Field of application	Source/author
Al$_2$O$_3$; SiC; Al-oxy-nitride	Al-Cu alloy	Mixing of minute filaments in molten matrix; after solidification, the filaments penetrate through grain-boundary regions.	⋯	R. Rosenberg
Al$_2$O$_3$	Al-Li alloy	Infiltration with a molten 1−8% Li alloy; reaction occurs between Al$_2$O$_3$ fiber and the Li.	⋯	Du Pont de Nemours
Al$_2$O$_3$-SiO; Y-Al$_2$O$_3$	Al; Al-Zn alloy	Melt impregnation; powder metallurgy; etc.	Aeronautics	Sumitomo
Ni-coated glass-ceramic fibers	Al	Powder metallurgy	Sliding parts	Honda Motor Co.
Al$_2$O$_3$ (continuous and polycrystalline)	Mg or Mg alloy	Melt impregnation of aligned fibers; preferred composites contain at least 50% fibers.	Turbine blades, springs, shafts, etc.	Du Pont de Nemours
C, B, glass, ceramic, metal	Mg alloy	Powder metallurgy; the composite contains two different Mg alloy phases.	Multiple applications	Dannohl
Glass	Light metals	Melt impregnation	Constructional parts	F. A. W. Schade
Metal-coated glass	Fe, Be, Ti, Al, Sn	Chopped metal-coated glass fibers are mixed with glass filaments and metal powder; hot pressing	Dimensionally stable machine parts — e.g., friction elements	Owens-Corning
Glass	Pb	Melt impregnation	Battery plates; bearing materials; acoustic insulation	D. M. Goddard *et al*

Fabrication Methods for Various Fiber/Metal Matrix Systems (Ref 43, p 82, 83)

Fiber	Matrix	Fabrication method	Fiber	Matrix	Fabrication method
Be	Ag, Ag-Al-Ge	Powder metallurgy	Al₂O₃	Ag	Liquid infiltration
	Al	Extrusion		Al	Liquid infiltration
Be, Ni-B	Be, Ni-B	Melting, directional solidification		2-S and Al alloys	Casting
				Ni and Ni alloys	Liquid infiltration, powder metallurgy, electroforming
Cr, Al₃Ni	Al-Cu, Cu-Cr	Melting			
Cu₆Sn₅	Cu-Sn	Directional solidification		Fe, Ti, Nichrome	Powder metallurgy, melting
NiAl₃	Al-Ni	Extrusion, melting			
Mo	Ti-6Al-4V	Cold pressing		Refractory metals	Slurry
	Ti	Sintering, extrusion	Boron	Al, Cu, Ni	Casting, powder metallurgy
Steel	Al	Hot pressing, diffusion bonding			
				Al alloys	Liquid infiltration
	Al-10%Si	Casting		Ni	Electroforming
	Ag	Vacuum infiltration		Al	Powder metallurgy plus laminated sheets
	Mg	Casting			
Steel, W	Al alloy	Hot rolling		Ni, Fe, Ti, Cu	Powder metallurgy, diffusion bonding
Steel, Mo, W, and Ge	Ag, Cu, Al	Slip casting, vacuum infiltration, extrusion, swaging			
				Ti	Diffusion bonding
W	Cu	Liquid infiltration	E glass, SiO₂-MgO-Al₂O₃	Cu, brass, steel	Liquid infiltration
	Cu, Al	Liquid infiltration			
	Cu alloys, L605, Ni-Fe	Vacuum infiltration	Fiberfrax	Ni-Sn	Vacuum infiltration
			Glass + UO₂	Al alloys	Rolling
	Ni	Pneumatic impaction	Graphite	Al-4%Ni-Cr	Powder metallurgy
	U	Vacuum infiltration	NbC	Nb	Directional solidification
	Ni, Ti, Al, Cu	Molecular forming	SiC	Ni-Cr	Directional solidification
	W	Vapor deposition	Si₃N₄	Ag, Ag + 1% Si	Hot pressing
W, Mo	Ni, Co, L605	Powder metallurgy	SiO₂	Al	Hot pressing
	Nichrome steel	Forging, rolling	ZrO₂, Y₂O₃, HfO₂, ThO₂, HfB₂, etc.		
Al₂O₃	Ag	Liquid infiltration		W	Extrusion
	Al	Liquid infiltration			

Methods Used To Fabricate Fiber-Reinforced Metals (Ref 11, p 80)

Matrix state	Method used to combine matrix with fibers	Method used to form and shape composite	Composite system: matrix (fiber)
Molten	Infiltration into bundles of parallel fibers	No additional working	Cu (W); Cu-alloy (W); Al (glass); Ag, Al, Ni (Al₂O₃)
	Infiltration into fiber mats	Rolling and machining	Ag (steel, Mo)
	Fiber solidified directly from melt	Some specimens rolled	Al (NiAl); Ta (Ta₂C); Nb (Nb₂C)
Powder	Mix	Hot press	Al (steel)
	Blend, cold press, sinter	Hot extrusion, anneal	Ti, Ti-alloy (Mo)
	Mix, press	Hot extrusion, swage	Al (W)
	Mix, hot press, cold press, sinter	Hot roll, swage, or forge	Ni/Cr, Co-alloy, Co, steel (W)
	Mix, melt matrix, blend	Extrude, sinter, cold roll	Ni/Cr, Fe (Al₂O₃)
	Blend, dry	Extrude, sinter, roll	Ag, Ni (Si₃N₄)
Molecular	Electroplating	No additional working	Ni (Al₂O₃, W)
	Electroplating	No additional working	Ni (Al₂O₃)
	Vapor plating, electrocodeposition, hot press	Hot roll	Ni (Al₂O₃)
Sheet or foil	Alternating layers of sheet and fibers	Diffusion bond	Al (steel, Be); Al (B, Be)
	Alternating layers of sheet and fibers	Diffusion bond	Ti, Ti-alloy (B); Ti (Be)
	Alternating layers of sheet and wire mesh	Hot roll	Al (stainless steel)
Combination	Draw fibers through molten matrix	Hot press coated fibers	Al (SiO₂)
	Electroplate fibers	Encapsulate in can, draw, sinter	Ag (steel)

Fabrication Methods and Applications for Composites of the Carbon, Boron, Carbide Fiber/Metal Matrix System (Ref 43, p 89, 91)

Fiber	Matrix	Fabrication method	Field of application	Source/author
Borsic	Al	Powder metallurgy; the composite article is clad with a sheet of Ti by diffusion bonding.	Turbine blades	United Aircraft; Pratt & Whitney
C coated with Ag-Al alloy	Al-4.5M-0.6Mn-1.5Mg	Powder metallurgy	High-temperature-resistant components; arms, aircraft	Union Carbide
Continuous B or Borsic plus β SiC discontinuous fibers	Al	Liquid phase hot pressing	⋯	R. Hermann *et al*; F. Swindels
C (graphite, amorphous C)	Ni/Co aluminide	Coating C fibers with Ni or Co; mixing with Ni-Co-Al powder; hot pressing	⋯	United Technologies
C coated with boride of Ti, Zr, Hf	Al or Al alloys; Mg; Pb; Sn; Cu; Zn	Melt impregnation	⋯	Fiber Materials, Inc.
C pretreated with molten NaK alloy	Al	Melt impregnation	⋯	A. P. Levitt; Anvar
C	Al alloy containing carbide-forming metal — e.g., Ti, Zr, etc.	Melt impregnation	⋯	Hitachi
SiC with W core	Al-Cu alloy	Coating the filaments with Cu; passing the Cu-coated filaments through an Al melt	⋯	Thomson CSF
C	Mg or Mg alloy	Hot pressing of alternating layers of matrix metal and fibers; small amounts of Ti, Cr, Ni, Zr, Hf, or Si are added to promote wetting.	Aircraft industry	A. P. Levitt
C	Mg or Mg alloy	Melt impregnation; the molten Mg matrix contains small amounts of Mg nitride to enhance wetting of the fibers.	Turbine fan blades; pressure vessels; armor plates	I. Kalnin
C coated with Ti	Mg	Melt infiltration; liquid phase hot pressing	⋯	A. P. Levitt
SiC	Be or Be alloys with Ca, W, Mo, Fe, Co, Ni, Cr, Si, Cu, Mg, Zr	Vacuum impregnation with molten Be or plasma spraying of fibers with Be and consolidation by metallurgical process	Aerospace and nuclear industry	Research Institute for Iron & Steel of the Tohoku University
B plus stainless steel; Borsic plus Mo fibers	Al; Ti	Impregnation; spraying; etc.; combination of high-strength ductile and brittle fibers	Aerospace industry	Inst. Fiziki Fverdogo Tela Akademii Nauk SSSR
SiC	Ti or Ti-3Al-2.5V alloy	Hot pressing of interposed layers of fibers and matrix sheets; SiC fibers are previously coated with Zr diffusion barrier layer.	Compressor blades; airfoil surfaces	General Motors
Carbides of Nb, Ta, W	Ni-Co and Fe-Cr alloys	Unidirectional solidification	Aircraft industry	Onera; General Electric
SiC containing 0.01-20% free carbon	Cr-base alloys	Powder metallurgy; the free carbon reacts with the chromium to form carbides, thus improving bonding ability.	High-strength, heat-resistant material — e.g., turbine vanes and blades; rocket nozzles	The Research Institute for Special Inorganic Materials, Japan

(continued)

Fabrication Methods and Applications for Composites of the Carbon, Boron, Carbide Fiber/Metal Matrix System (continued) (Ref 43, p 89, 91)

Fiber	Matrix	Fabrication method	Field of application	Source/author
SiC containing 0.01-30% free carbon	Co or Co-base alloys	Powder metallurgy or melt impregnation; carbide formation between the fibers and the Co matrix	High-strength, heat-resistant material — e.g., turbine vanes and blades; rocket nozzles	The Research Institute for Special Inorganic Materials, Japan
SiC containing 0.01-20% free carbon	Mo-base alloys	Powder metallurgy	High-strength, heat-resistant material — e.g., turbine vanes and blades; rocket nozzles	The Research Institute for Special Inorganic Materials, Japan
C coated with carbides	Ni or Ni alloys	Melt impregnation	Aeronautic industry	Union Carbide
B	Cu-Ti-Sn alloy	Liquid phase sintering	Cutting tools	General Dynamics
C	Bronze	Various processes	Bearing materials	UK Atomic Energy Authority
C	Cu alloy	Powder metallurgy; fibers are mixed with a slurry of Cu powder and 2% of a carbide-forming metal powder (Ti, Cr, . . .)	High-strength, electrically conductive materials	Hitachi
C coated with Ti boride	Al, Cu, Sn, Pb, Ag, Zn, Mg	Matrix contains alloying elements of Ti and B to prevent deterioration of the TiB coating of the fibers.	Aeronautic industry	Aerospace Corp.
C coated with Ni	Metals with melting points lower than that of Ni	Melt impregnation	...	Brown-Bover; NASA
C coated with SiO_2 + SiC	Al, Mg, Ti, Ni	Melt impregnation; powder metallurgy	...	Fiber Materials
Monocarbides of Ta, Ti, W	Al, Al-Si alloy, Ag or Ag alloys, Cu or Cu alloys	Melt impregnation	Abrasion-resistant materials	Union Carbide
SiC	Si	Melt impregnation	...	General Electric; S. Yajima
β SiC	Ag or Ag alloys	...	Electric conductors, contacts, etc.	Research Institute for Iron, Steel and Other Metals of the Tohoku University
C	Si	Powder metallurgy	Abrasive materials	General Electric
SiC whiskers	Ag	Hot pressing (in zero-gravity conditions)	...	S. Takahashi
C coated with TiB	Mg, Pb, Sn, Cu, Al, Zn	Melt impregnation	...	Fiber Materials

Classification of Most Composite System Interfaces (Ref 110, p 4)

Class I[a]	Class II[b]	Class III[c]
Copper/tungsten	Copper (chromium)/ tungsten	Copper (titanium)/ tungsten
Copper/alumina		
Silver/alumina	Eutectics	Aluminum/carbon
Aluminum/BN-coated boron	Niobium/tungsten	(<700 °C)
	Nickel/carbon	Titanium/alumina
Magnesium/boron[d]	Nickel/tungsten[e]	Titanium/boron
Aluminum/boron[d]		Titanium/silicon carbide
Aluminum/stainless steel[d]		Aluminum/silica
Aluminum/silicon carbide[d]		

[a]Filament and matrix do not react and are mutually insoluble. [b]Filament and matrix do not react, but exhibit some solubility. [c]Filament and matrix react to form surface coating. [d]Pseudo–Class I system. [e]Becomes reactive at lower temperatures with formation of Ni_4W.

Reactivity of Boron With Various Metals (Ref 121)

Metal	Definite reaction			Little or no reaction		
	Temperature		Time[a],	Temperature		Time[a],
	°C	°F	h	°C	°F	h
Fe	700	1292	50	600	1112	(b)
	900	1652	1	...	...	...
Co	700	1292	50	600	1112	(b)
	900	1652	1	...	...	...
Al	700[c]	1292[c]	0.2	600	1112	1
Mg	...	...	...	600	1112	100
	...	...	...	700[c]	1292[c]	0.2
Be	...	...	...	1000	1832	24
Ti	...	...	...	600	1112	100
Cr	900	1652	1	...	...	...
Ag	...	...	...	900	1652	2
Ni	600	1112	100	...	...	...
	700	1292	100	...	...	...
	900	1652	1	...	...	...

[a]Time required for consolidation of metal powders and fibers by hot pressing. [b]Reaction time so short that the reaction is finished as soon as fibers have been hot pressed. [c]Molten.

Reactivity of Metals With SiC Fibers

Metal	Melting temperature, °C	Reactivity[a] in 1 h in H_2 gas at temperature, °C, of:									
		450	530	620	650	750	850	950	1000	1100	1200
Al	660	○	○	○							
Ag	961				○						
Cu	1083					○	○	○			
Ni	1453								○	x	x
Co	1495								○	△	x
Fe	1537								○	○	○
Ti	1668								△	△	△
Cr	1875								○	○	△
Mo	2610								○	○	○

[a]Key: ○ = no reaction ; △ = slight reaction; x = significant reaction.

Some Fiber/Metal Combinations (Ref 112)

Matrix	C	B	Al_2O_3	SiC	Steel	Be	W	Ni_3Nb
Ti		+	+	+		+		
Al	+	+	+	+	+	+		
Ni	+		+	+			+	+

Factors Influencing the Choice of Constituents and Fabrication Process for Whisker Composites (Ref 77, p 301)

Selection type	Factors affecting decision	
	Constituent factors	Composite factors
Matrix	Density Melting point Ductility (toughness) Shear strength Corrosion resistance	Design requirements Application Environment Constituent functions
Whiskers	Strength Modulus Density Melting point Size and aspect ratio Availability (and cost)	Chemical and mechanical compatibility of whisker coating/ matrix Economics
Whisker coating	Wetting Bonding Stability	Performance
Whisker configuration (orientation)		Volume fraction Spacing Overlap Alignment direction(s)
Fabrication process	Damage to whiskers Ease of process Automation	Methods which optimize constituent properties with composite performance

Actual and Potential Composite Fabrication Methods (Ref 18, p 274)

General method	First step	Consolidation
Powder metallurgy ...	Pack fibers in matrix, or slip cast matrix about fibers, or use fugitive binder to hold fibers together	Sinter, hot press, or hot isostatic press
Foil metallurgy	Use fugitive binder or "glue" to bond fibers to foil, or press foils (flat or grooved) about fibers, or flash bond or continuous bond (roll) foils about fibers	Diffusion bond in hot press, or hot isostatic press braze
Casting	Cast entire part, or cast matrix about fibers (individual units or continuous tapes)	None when casting entire part; for continuous tapes, hot press or hot isostatic press
Electrodeposition	Electroplate or form shapes or tapes; electrophoresis	Hot press or hot isostatic press
Vapor deposition	CVD; ion plate; vaporion beam	Hot press or hot isostatic press
Metal spraying	Plasma spray; molten metal spray	Hot press or hot isostatic press

Methods Used for Fabrication of Metallic-Matrix Whisker Composites (Ref 77, p 328, 329)

Method used to combine constituents	Matrix	Whisker[a] Composition	Whisker[a] Alignment	Coating	Consolidation process[b]	Forming or shaping process	Remarks
Deposition (molecular)							
Chemical vapor deposition	Ni or NiCr	Al_2O_3, SiC	R	C	H.P. or L.P.H.P.	Hot roll	SiC whiskers reacted chemically with matrix
Electrocodeposition	Ni	Al_2O_3	R	C	H.P.	Hot roll	Problem of voids and low whisker content
	Ni	Al_2O_3, SiC	R	C, U	...	Hot roll	Problem of voids and low whisker content
Electroplate	Ni	Al_2O_3	A	C, U	As deposited or hot forged	Hot roll	Align whiskers via flow in electroplating bath
	Ni	Al_2O_3, SiC	A	C	H.P.	Hot roll	Problem of consolidation and whisker breakage
Electroform	Ni	Al_2O_3	A	C	Cold and hot rolling, H.P.	Hot roll	SiC whiskers reacted with matrix, also whisker breakage
			R		...	Hot roll	Excellent properties, limited to very small specimens
Liquid (Matrix)							
Alloy (eutectic)	Al, Nb	$NiAl_3$, Nb_2C	A	U	Unidirectional solidification	Extrude, roll	Whisker content fixed by alloy composition
	Resin	Al_2O_3	A	C	Polymerize matrix	Extrude, roll	Achieved high-strength composites
	Ag	Al_2O_3	A	C	Solidification of matrix	Extrude, roll	Achieved high-strength composites
	Cu, Ni-alloys	Al_2O_3	A	C	Solidification of matrix	Extrude, roll	Achieved varying degrees of success; coating stability a problem
Infiltration	Al	B_4C	A	C	Solidification of matrix	Extrude, roll	Achieved varying degrees of success; coating stability a problem
	Al-alloys	Al_2O_3	A	C	Solidification of matrix	Extrude, roll	Achieved varying degrees of success; coating stability a problem
Melting of powdered matrix	Fe, NiCr	Al_2O_3, SiC	R	C, U	L.P.H.P.	Extrude, roll	Problems of interfacial reactions and whisker segregation
Casting of melt and whiskers	Co-alloys	Al_2O_3, SiC	R	C, U	Solidification of melt	Extrude, roll	Whiskers concentrated at grain boundaries, low whisker content
	Ag, Fe, Ni	Si_3N_4	A	U	Dry, burn off organic carrier, H.P.	Extrude, roll	Excellent whisker alignment and packing
Spin, extrude, draw slurry of whisker, matrix powder, and carrier solution	Al	SiC	A	U	Dry, burn off organic carrier, H.P.	Extrude, roll	Excellent whisker alignment and packing
	Al-alloy	SiC	A	U	Dry, burn off organic carrier, L.P.H.P.	Extrude, roll	Excellent whisker alignment, fabrication of complex shapes
	Cu, Al, Mg	SiC	A	U	Dry, burn off, sinter, H.P.	Extrude, roll	Excellent whisker alignment
	Al-alloy	SiC	A	U	Dry, burn off, H.P.	Extrude, roll	Excellent whisker alignment, problem of carbon burn-off
Filter slurry or settle out whiskers and matrix	Ag-alloy	Si_3N_4	R	U	L.P.H.P.	Extrude, roll	Problem of matrix porosity and wetting
	Cu, Al-alloy	Al_2O_3, SiC	R	U	H.P.	Hot roll	Little whisker alignment, much whisker breakage
	NiCr, Al-alloy	SiC	A	C, U	L.P.H.P.	Clad and roll	Ni-coated whiskers were aligned magnetically during settling in slurry
	NiCr, Al-alloy	Al_2O_3, SiC	A, R	C, U	L.P.H.P.	Clad and roll	Ni-coated whiskers were aligned magnetically during settling in slurry
Impregnation into whisker strands and tapes	Mg-alloy	Al_2O_3, SiC	A, R	C, U	L.P.H.P.	Hot extrusion	Successfully rolled composites with aligned whiskers
	Resin	SiC	A	U	H.P. and squeeze out excess resin	Hot extrusion	Achieved excellent alignment, little whisker breakage
	Resin	Al_2O_3	A	U	H.P. and squeeze out excess resin	Hot extrusion	Achieved excellent alignment, little whisker breakage
	Resin	Si_3N_4	A	U	H.P. and squeeze out excess resin	Hot extrusion	Achieved excellent alignment, little whisker breakage
Solid State							
Powder matrix and whiskers	Ni	Al_2O_3	R	C	H.P.	Hot extrusion	Many whiskers broken
	Al	Si_3N_4	A	U	Extrusion	Hot extrusion	Extended long rods, but many whiskers broken
	Ni, Ti	Al_2O_3, SiC	R	U	High-energy-rate forming	Hot extrusion	Chemical reactions minimized, but whiskers still broken
Deposit whiskers on Al-foil	Al	SiC	A	U	Diffusion bonding	Hot extrusion	Achieved excellent alignment of whiskers

R = random whisker alignment; A = whiskers unidirectionally aligned; C = coated whiskers; U = uncoated whiskers. [b]H.P. = hot pressing; L.P.H.P. = liquid phase hot pressing.

7.2. Comparative Properties of Various Fiber Reinforced Metal Matrix Composites

Typical Mechanical Properties of Some Metal-Matrix Composites (Ref 5, p 621)

Fiber	Matrix	Reinforcement, vol %	Density, g/cm³[a]	Longitudinal tensile strength, MPa[b]	Longitudinal modulus, GPa[b]	Transverse tensile strength, MPa[b]	Transverse modulus, GPa[b]
G T50	201 Al	30	2.380	620	170	50	30
G T50	201 Al	49		1120	160	...	...
G GY70	201 Al	34	2.380	660	210	30	30
G GY70	201 Al	30	2.436	550	160	70	40
G HM pitch	6061 Al	41	2.436	620	320	...	...
G HM pitch	AZ31 Mg	38	1.827	510	300	...	...
B on W, 142-μm fiber	6061 Al	50	2.491	1380	230	140	160
Borsic	Ti	45	3.681	1270	220	460	190
G T75	Pb	41	7.474	720	200	...	...
G T75	Cu	39	6.090	290	240	...	...
FP	201 Al	50	3.598	1170	210	(140)	140
SiC	6061 Al	50	2.934	1480	230	(140)	140
SiC	Ti	35	3.931	1210	260	520	210
SiC whisker	Al	20	2.796	340	100	340	100
B₄C on B	Ti	38	3.737	1480	230	>340	>140
G T75	Mg	42	1.799	450	190	...	...
G HM	Pb	35	7.750	500	120	...	...
G T75	Al-7% Zn	38	2.408	870	190	...	...
G T75	Zinc	35	5.287	770	120	...	...
G T50	Nickel	50	5.295	790	240	...	...
G T75	Nickel	50	5.342	828	310	30	40
G (81.3 μm)	2024 Al	50	2.436	760	140	...	...
G (142 μm)	2024 Al	60	2.436	1100	180	...	...
Superhybrid	Grafitic	60	2.048	860	120	220	60
Superhybrid	S-glass	60	2.159	740	80	190	30
Superhybrid	Kevlar	60	1.799	700	80	190	10

[a]To convert g/cm³ to lb/in.³, divide by 27.68. [b]To convert MPa to psi, multiply by 145; to convert GPa to psi, multiply by 145 000.

Strength Properties of Several Types of Fiber-Reinforced Metals (Ref 74, p 209)

Filament	Matrix	Fabrication technique[a]	Test temperature, °C	V_f	L/d_f	S_c MPa	S_c ksi	$\bar{S}_f$ MPa	$\bar{S}_f$ ksi	$\bar{S}_f^*$ MPa	$\bar{S}_f^*$ ksi	β
Metallic Filaments												
Stainless steel	Al	HPF	RT	0.11	250	179	26.9	1515	220	1405	204.0	0.93
W	Cu	VI	RT	0.63	∞	772	112.0	1240	180	1215	176.0	0.98
W	Cu	VI	250	0.69	∞	855	124.0	1240	180	1225	178.0	0.99
W	Cu	VI	250	0.56	40	558	81.0	1240	180	972	141.0	0.78
W	Cu	VI	250	0.51	20	490	71.0	1240	180	938	136.0	0.76
W	Cu	VI	250	0.57	10	386	56.0	1240	180	655	95.0	0.53
W	Cu	VI	RT	0.768	∞	1755	254.9	2255	327	2250	326.0	~1.00
W	Cu	VI	RT	0.357	75	827	120.0	2255	327	2220	322.0	0.99
Oxide Filaments												
Al-coated E-glass	Al	HP	RT	0.50	∞	310	45.0	896	130	593	86.0	0.66
Al-coated E-glass	Al	HP	480	0.50	∞	178	25.85	731	106	342	49.6	0.47
Al-coated E-glass	Al	HP	540	0.50	∞	135	19.63	710	103	263	38.1	0.37
Al-coated E-glass	Al	VI	RT	0.50	∞	121	17.5	896	130	230	33.4	0.26
Al-coated E-glass	Al	VI	480	0.50	∞	96	13.89	731	106	189	27.48	0.26
Al-coated SiO₂	Al	HP	RT	0.48	∞	871	126.3	3035	440	1725	250.0	0.57
Al-coated SiO₂	Al	HP	100	0.48	∞	958	139.0	3035	440	1935	281.0	0.65
Al-coated SiO₂	Al	HP	500	0.48	∞	288	41.8	1795	260	593	86.0	0.33
Pb-coated E-glass	Pb	VI	RT	0.08	∞	103	15.0	1515	220	1055	153.0	0.70

[a]HPF = hot pressed filaments in powdered matrix; VI = vacuum infiltration; HP = hot pressed metal-coated filaments. [b]S_c = composite tensile strength; $\bar{S}_f$ = average strength of fibers tested individually; $\bar{S}_f^*$ = average strength of fibers in the matrix during fracture; β = effective fiber strength factor = $\bar{S}_f^*/\bar{S}_f$.

Potential Specific Strength and Specific Modulus of Conventional Alloys and Advanced Fiber-Reinforced Composites at Room Temperature (Ref 11, p 7)

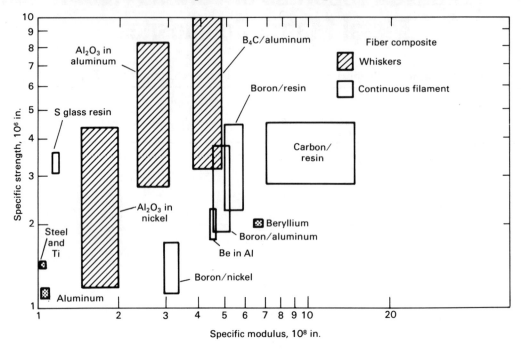

Room-Temperature Tensile and Specific Strengths of Various Fiber-Reinforced Composites[a] (Ref 10, 11)

Composition Matrix	Fiber	Density, lb/in.³ Matrix	Density, lb/in.³ Fiber	Fiber content,[b] vol %	Composite Tensile strength,[c] ksi	Composite Strength/ density, 10³ in.	Source
Al	B/W	0.097	0.095	50	110	1150	United Aircraft Research Laboratory
	B	···	···	10	43	453	GE-AETD
	Steel	0.097	0.282	25	173	1210	Harvey Aluminum Co.
	Be	0.097	0.067	40	80	830	North American
	SiO₂	0.097	0.079	48	118	1340	Rolls-Royce Ltd.
	Al₂O₃	0.097	0.143	35	161	1425	GE-SSL
	B₄C	0.097	0.091	10	29	302	GE-SSL
	CuAl₂	0.097	0.157	50	39	307	United Aircraft Research Laboratory
	Al₃Ni	0.097	0.143	10	48	470	United Aircraft Research Laboratory
Nb	Nb₂C	0.299	0.286	31	172	570	United Aircraft Research Laboratory
Ni	B/W	0.320	0.095	75	384	1470	General Technology, Inc.
	W	0.320	0.695	9.4	61.4	173	Battelle N.W.
	W	···	···	40	161	344	NGTE (England)
	Al₂O₃	0.320	0.143	19	171	600	GE-SSL
	C	0.320	0.054	56	80	467	Union Carbide Corp.
Al-10.2Si	Al₂O₃	0.096	0.143	15[d]	40.7	395	Melpar, Inc.
Ni-20Cr	Al₂O₃	0.308	0.143	9	255	870	Horizons, Inc.
NiCr	W	···	···	22	73	185	Clevite Corp.
Fe	Al₂O₃	0.284	0.143	36	237	1017	Horizons, Inc.
Ta	Ta₂C	0.598	0.544	29	155	267	United Aircraft Research Laboratory, twice solidified
	Ta₂C	0.598	0.544	29	118	203	United Aircraft Research Laboratory, cold swaged, 67% reduction
Ag	Si₃N₄	0.378	0.115	15[d]	40	119	Explosive Research Development Est. (England)
	Al₂O₃	0.378	0.143	24	232	720	GE-SSL
	Steel	···	···	44	65	191	MIT
Ti	Mo[e]	···	···	20	96	457	Clevite Corp.
Cu	W	···	···	77	255	420	NASA-Lewis
Co	W	···	···	30	107	244	Clevite Corp.
	Mo	···	···	17	52	156	Clevite Corp.
316 stainless steel	W	···	···	18	58.6	175	Clevite Corp.

[a]Early and developmental metal-matrix composite work. [b]Unidirectional fiber orientation except where otherwise noted. [c]Highest reported values. [d]Random fiber orientation. [e]Short, discontinuous fibers and whiskers.

Reinforcement Efficiencies of Continuous Aligned Fiber-Reinforced Metals[a] (Ref 13, p 199)

Property	Matrix: Fiber: Boron	Aluminum Carbon[b]	Silica	Nickel alloy Carbon	Tungsten	Titanium Boron
Longitudinal strength	0.70[c]	1.05	0.27	0.63	0.90	0.78
Longitudinal modulus	0.97[c]	0.91	1.00	0.80	...	0.94
Transverse strength	1.00	0.25	1.00	...	...	0.60

[a]Values given are fractions of the rule-of-mixtures values for longitudinal strength and modulus, and fractions of the matrix strength in the transverse direction. [b]Matrix was 0.88Al-0.12Si. [c]Boron was coated with SiC (Borsic).

Specific Moduli of Advanced Composite Materials Compared With Titanium

Material[a]	Longitudinal modulus (E_{11}), in. $\times 10^7$ (m $\times 10^7$)	Transverse modulus (E_{22}), in. $\times 10^7$ (m $\times 10^7$)	Shear modulus (G_{12}), in. $\times 10^7$ (m $\times 10^7$)
Borsic/aluminum	35 (0.89)	20 (0.51)	10 (0.25)
Boron/epoxy	43 (1.09)	4 (0.10)	1.4 (0.04)
Carbon/epoxy	42 (1.07)	1.9 (0.05)	1.3 (0.03)
Titanium	10 (0.25)	10 (0.25)	4 (0.10)

[a]All composites, 50 vol % fiber.

Stress–Plastic Strain Curves for Pure Silver and Ag-30 Wt % Composites With Various Types of Fibers (Ref 74, p 123)

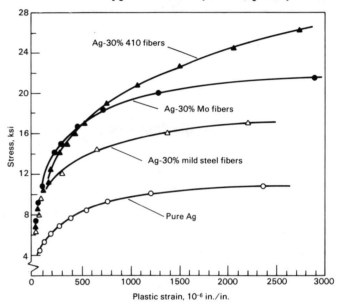

Tensile Strength of *In Situ* Extruded ZrO_2 Ceramic-Fiber-Reinforced Niobium Matrix Composites as a Function of Temperature (Ref 88, p 522)

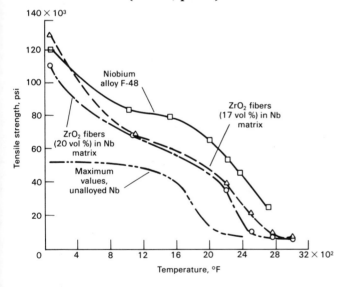

Comparison of Densities and Maximum Use Temperatures of Composites (Ref 6, 8, 13)

Composite system (fiber/matrix)	Density, g/cm³	Maximum use temperature, °C
Graphite/epoxy	1.6	200
Graphite/polyimide	1.6	300
Graphite/glass	2.0	500–650
Graphite/aluminum	2.3	325
Boron/aluminum	2.7	325
Silicon carbide/glass	2.8	>650
FP alumina/glass	3.1	>650
Boron/titanium	3.7	650
Beryllium/titanium	...	315
Silicon carbide/titanium	...	650

Comparative Properties of Various Graphite Fiber Reinforced Metals (Ref 14)

Composite	Fiber content, vol %	Strength, ksi	Modulus of elasticity, 10⁶ psi	Density, lb/in.³	Strength/density, 10⁶ in.	Modulus/density, 10⁶ in.
Graphite[a]/lead	41	104	29.0	0.270	0.385	107.0
Graphite[b]/lead	35	72	17.4	0.280	0.260	62.3
Graphite[a]/zinc	35	110.9	16.9	0.191	0.580	88.5
Graphite[a]/magnesium	42	65	26.6	0.064	1.016	393.7

[a]Thornel 75 fiber. [b]Courtaulds HM fiber.

Coefficients of Thermal Expansion of Matrices and Reinforcing Fibers (Ref 5, 14)

Matrix	Expansion, $10^{-6}/°C$	Filament	Expansion, $10^{-6}/°C$
Aluminum	23.9	Boron	6.3
Titanium	8.4	Borsic	6.3
Iron	11.7	SiC	4.0
Nickel	13.3	Alumina	8.3
Silicon	2.5	Carbon[a]	~0
Copper	17	SiO_2	~0
Silver	19	Tungsten	4.5
Magnesium	26	Al_2O_3	8
		Beryllium	12

[a]Parallel to basal plane.

Stress–Plastic Strain Curves for Pure Silver and Composites of Silver–Mild Steel Fiber of Various Fiber Concentrations (By Weight) (Ref 74, p 122)

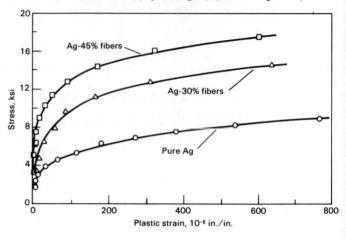

Tensile Properties of FP Alumina/Magnesium (Al_2O_3/Mg) Composites With 70 Vol % FP (Ref 7, p 2.81)

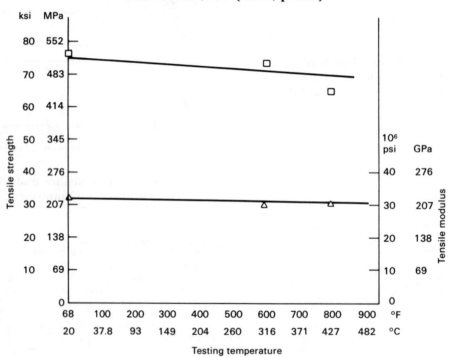

7.3. Fiber Reinforced Aluminum Matrix Composites

7.3.1. Boron Fiber Reinforced Aluminum

Mechanical Properties of Boron/Aluminum Composites[a]

Matrix	Fiber orienta- tion		Ultimate tensile strength		Elastic modulus	
			MPa	ksi	GPa	10^6 psi
Al-6061	0°	....	1515	220	207	30
	90°	....	138	20	138	20
Al-2024	0°	....	1550	225	207	30
	90°	....	214	31	145	21

[a]These samples contain 48% Avco 5.6-mil (142-μm) boron. The longitudinal tensile specimens are 6 in. (152 mm) by 5/16 in. (7.9 mm) by 6 ply, and the transverse tensile bars are 6 in. (152 mm) by 1/2 in. (12.7 mm) by 6 ply.

Properties of B$_4$C-Coated Cast Boron/ Aluminum (A-357) Composites[a] (Avco Specialty Materials Div.)

Ultimate tensile strength		Average tensile strength		Elastic modulus	
MPa	ksi	MPa	ksi	GPa	10^6 psi
1345−1620	195−235	1480	215	214	31

[a]Preliminary data. Process, vacuum infiltration casting. Temperature, 630–660 °C (1165–1220 °F). B$_4$C coating on filaments permits immersion in molten Al alloys at 675 °C (1250 °F) for time in excess of 30 min without degradation of tensile strength.

Impact Energy of Full-Size Notched Charpy Specimens Containing 55% Boron Fibers (Ref 17, p 552)

Fiber diameter, μm	Matrix	Total impact energy, J
200	Aluminum 1100........................	95
145	Aluminum 1100........................	55
145	Al alloy 6061	35
Unreinforced	Ti-6Al-4V.............................	25
Unreinforced	Al alloy 6061-T6	15

Properties of Hot-Molded B$_4$C-Coated Boron/ Aluminum (6061) Composites[a] (Avco Specialty Materials Div.)

Orientation		Ultimate tensile strength		Elastic modulus	
		MPa	ksi	GPa	10^6 psi
0°	....	1450	210	241	35
90°	....	152	22	131	19

[a]Preliminary data. Fiber volume, 49%. Process, vacuum bag or argon. Time, 20 min. Pressure, 1.4–2.8 MPa (200–400 psi). Temperature, 610 °C (1130 °F).

Flow Chart of a Typical Fabrication Scheme for a Boron/Aluminum Composite

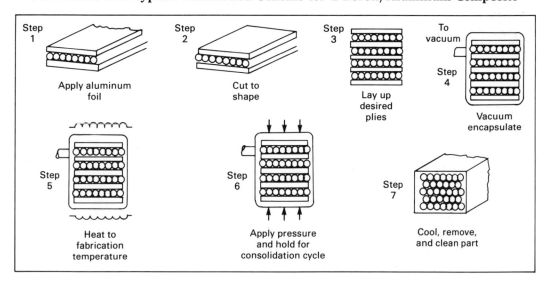

Step 1 — Apply aluminum foil

Step 2 — Cut to shape

Step 3 — Lay up desired plies

Step 4 — To vacuum / Vacuum encapsulate

Step 5 — Heat to fabrication temperature

Step 6 — Apply pressure and hold for consolidation cycle

Step 7 — Cool, remove, and clean part

Stress-Strain Behavior of 5.6-Mil Boron/ Al 6061 Composites (Ref 19)

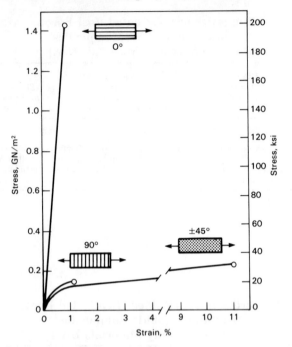

Failure Strain of T6-Treated Vs T6M-Treated 0–90° Cross-Ply B/Al 6061 Composites (Ref 119)

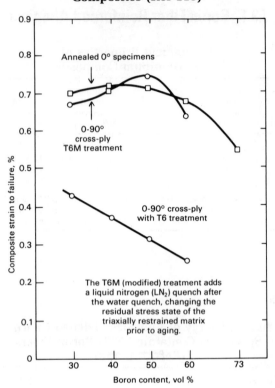

Effect of Transverse Rolling on Tensile Strength of B/Al Composites (Ref 119, p 37–42)

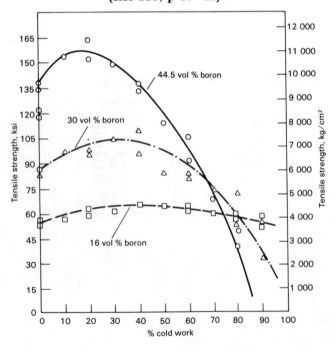

Typical Mechanical Properties of 50 Vol % Unidirectional Reinforced B/Al 6061 Composites (Ref 33, p 7-2)

Tensile strength:[a]
Longitudinal................................. 1490 MPa (216 ksi)
Transverse.................................138 MPa (20 ksi)
Tensile modulus:[a]
Longitudinal............................. 214 GPa (31×10^6 psi)
Transverse............................. 138 GPa (20×10^6 psi)
Poisson's ratio:
Longitudinal...0.23
Transverse...0.13
Compressive strength:[a]
Longitudinal............................. 1725 MPa (250 ksi)
Transverse.................................207 MPa (30 ksi)
Compressive modulus:[a]
Longitudinal............................. 221 GPa (32×10^6 psi)
Transverse............................. 138 GPa (20×10^6 psi)
Longitudinal shear strength[a] ....159 MPa (23 ksi)
Longitudinal shear modulus[a]................. 41 GPa (6×10^6 psi)
Longitudinal bearing strength[a] ....827 MPa (120 ksi)
Unnotched fatigue strength:[a]
Longitudinal.............................1035 MPa (150 ksi)[b]
Transverse............................. 41 MPa (6 ksi)[b]
Creep[c]................................. At 1105 MPa (160 ksi),
total strain averages 0.06% in 100 h.

[a]At 24 °C (75 °F). [b]At runout (10^7 cycles). [c]At 370 °C (700 °F).

Comparison of S-N Curves for Boron/Aluminum Composites (Ref 9, p 114)

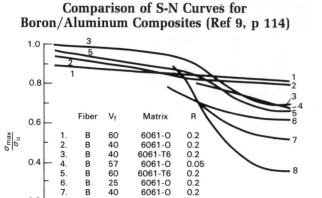

	Fiber	V_f	Matrix	R
1.	B	60	6061-O	0.2
2.	B	40	6061-O	0.2
3.	B	40	6061-T6	0.2
4.	B	57	6061-O	0.05
5.	B	60	6061-T6	0.2
6.	B	25	6061-O	0.2
7.	B	40	6061-O	0.2
8.	BSiC	30	Ti-6Al-4V	0.4

Comparison of Fatigue Behavior of Boron/ Aluminum Composite and Boron Fiber Bundle (Same Cross-Sectional Area) (Ref 9, p 110)

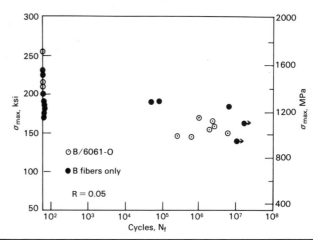

⊙ B/6061-O

● B fibers only

R = 0.05

7.3.2. Borsic Fiber Reinforced Aluminum

Notched Charpy Impact Energy of Borsic/Al and B/Al Composites[a] (Ref 16, p 449)

Specification number	Matrix[b]	Fiber	Fabrication conditions Temperature, °C	Pressure, MPa	Time, min	Orientation	Impact Energy[c] J	ft·lb
1	6061-F	Borsic	565	145.0	60	LT	6.7	4.9
2	6061-F	Borsic	565	145.0	60	TT	2.0	1.5
3	6061-F	Borsic	565	145.0	60	TL	1.3	1.0
4	6061-F	Borsic	490	72.5	30	LT	7.8	5.8
5	6061-F	Borsic	450	72.5	30	LT	9.4	6.9
6	6061-F	Boron	450	145.0	30	LT	17.7	13.1
7	6061-F	Boron	480	145.0	30	LT	13.2	9.7
8	6061-F	Boron	450	72.5	30	LT	18.5	13.6
9	1100	Borsic	450	72.5	30	LT	18.4	13.6
10	1100	Boron	450	72.5	30	LT	26.0	19.2
11	1100	Boron	450	72.5	30	LT	22.8	16.8
12	1100	Boron	450	72.5	30	LT	28.4	20.9
13	1100	Boron	450	72.5	30	LT	30.0	22.1
14	1100	Boron	450	72.5	30	LT	>30.0	>22.1
15	1100	Boron	450	145.0	30	LT	26.1	19.2
16	1100	Boron	450	145.0	30	LT	21.5	15.9
17	1100	Boron	450	145.0	30	LT	28.2	20.8
18	2024-F	Boron	450	72.5	30	LT	8.1	6.0
19	2024-F	Boron	450	72.5	30	LT	15.4	11.4
20	5052/56	Boron	450	145.0	30	LT	8.0	5.9
21	6061/1100-F	Boron	450	145.0	30	LT	26.6	19.6
22	6061/1100-F	Boron	450	145.0	30	LT	25.6	18.9
23	6061/1100-T6	Boron	450	145.0	30	LT	22.3	16.4

[a]All composites contain 50–60 vol % fiber. [b]F and T6 indicate the as-fabricated and heat treated conditions, respectively. [c]At 23 °C (73 °F).

Variation in Impact Energy per Unit Area With the Parameter $V_f d_f (\sigma_{uf})^2/24\tau_{my}$ for Borsic/Aluminum and Boron/Aluminum Composites of the LT Type

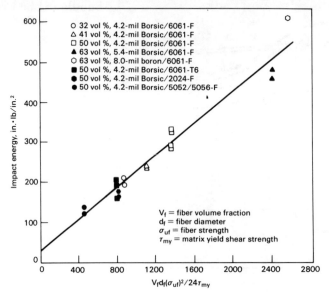

○ 32 vol %, 4.2-mil Borsic/6061-F
△ 41 vol %, 4.2-mil Borsic/6061-F
□ 50 vol %, 4.2-mil Borsic/6061-F
▲ 63 vol %, 5.4-mil Borsic/6061-F
◎ 63 vol %, 8.0-mil boron/6061-F
■ 50 vol %, 4.2-mil Borsic/6061-T6
● 50 vol %, 4.2-mil Borsic/2024-F
● 50 vol %, 4.2-mil Borsic/5052/5056-F

V_f = fiber volume fraction
d_f = fiber diameter
σ_{uf} = fiber strength
τ_{my} = matrix yield shear strength

Thermal Expansion Behavior of Borsic/Al 2024 (a and b) and Borsic/Al 1100 (c and d) Composites

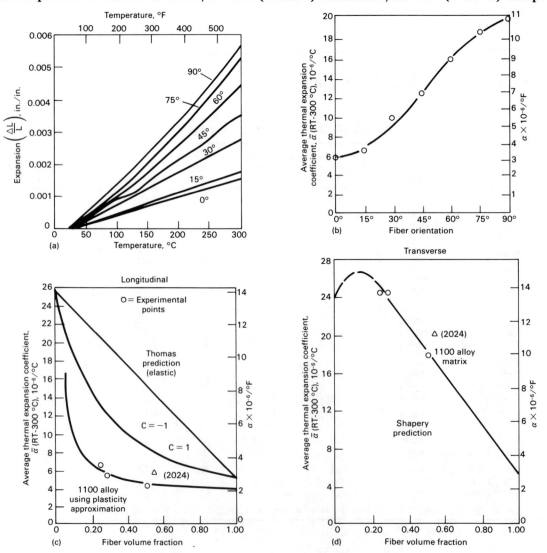

Axial Tensile Stress–Strain Curves for 64 Vol % Borsic Fiber Reinforced Al 2024 in As-Fabricated (Equivalent to Annealed) and T6 (Solution Treated and Quenched) Conditions (Ref 18)

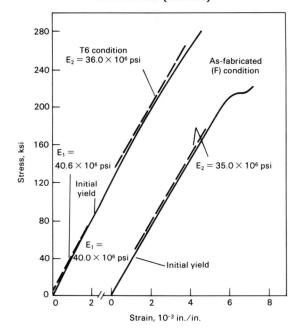

Composite Tensile Strength as a Function of Orientation for Borsic Fiber Reinforced Al 6061 in the T6 Condition (150-μm or 6-Mil Fibers) (Ref 18)

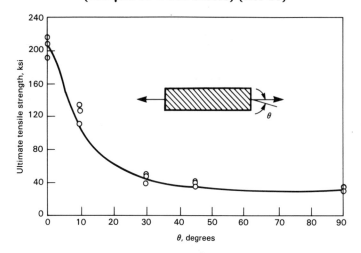

Tensile Stress–Strain Curve for ±45° Borsic-Reinforced Al 6061 Tested at Room Temperature (Ref 16, p 443)

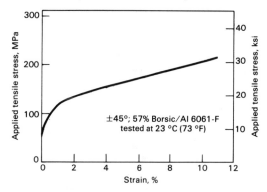

Composite Transverse Tensile Strength as a Function of Matrix Strength for 60 Vol % Borsic Fiber Reinforced Al 6061 (Ref 18)

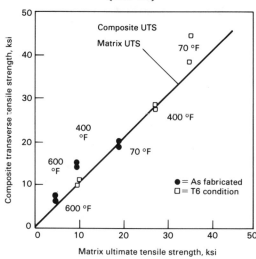

Composite Axial and Transverse Elastic Moduli for Borsic Fiber Reinforced Al 6061 (100-μm or 4-Mil Fibers) (Ref 18)

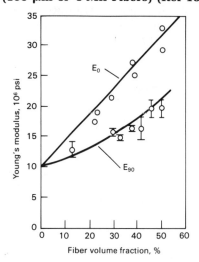

7.3.3. Alumina and FP Alumina Fiber Reinforced Aluminum

Hardness at 25 °C of Al-9Si-3Cu Alloy Reinforced With Al₂O₃ Fibers[a] (Ref 87)

Volume fraction (V_f)	Vickers Hardness Number, HV 10
0	131
0.12	179
0.18	190
0.24	212

[a]"Saffil" fiber, RF grade.

Coefficients of Thermal Expansion for Al₂O₃ Fiber/Al Alloy Composites (Imperial Chemical Industries, PLC)

Volume fraction (V_f)	Coefficient of thermal expansion[a] (α), 10^{-5} °C^{-1}	
	In-plane	Normal
0	2.03	2.03
0.12	1.66	1.76
0.18	1.54	1.66
0.24	1.55	1.57

[a]Coefficients (at 20–200 °C) of composites containing "Saffil" fiber RF grade in Al-9Si-3Cu alloy measured parallel and normal to planes of fiber orientation.

Axial and Transverse Tensile Strengths of FP Alumina Fiber Reinforced Aluminum (Ref 18)

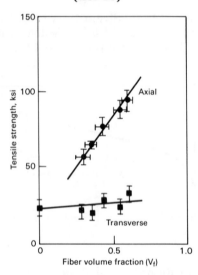

Modulus of Al₂O₃-Reinforced Aluminum Alloy (Imperial Chemical Industries, PLC)

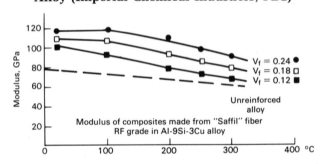

Tensile Strength of Al₂O₃-Reinforced Aluminum Alloy (Imperial Chemical Industries, PLC)

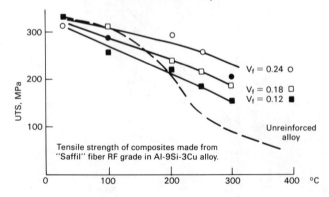

Specific Strength of FP/Al Composites Vs Other Fiber/Al Materials (DWA Composite Specialties, Inc.)

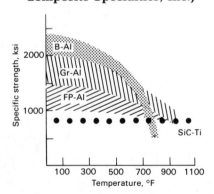

7.3.4. Stainless Steel Reinforced Aluminum

Tensile Strength as a Function of Fiber Volume for Stainless Steel/Aluminum Composites (Ref 120)

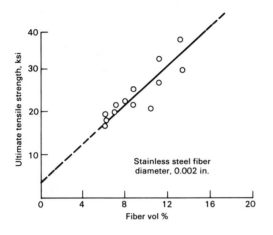

Effect of Aspect Ratio on Strength/Composition Relationship for Stainless Steel/Aluminum Composites (Ref 120)

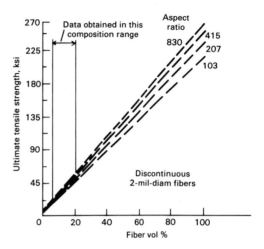

Comparison of 500 °F Stress-Rupture Properties of Stainless Steel/Aluminum and Boron/Aluminum Composites and Al Alloy Matrix (Ref 89, p 78)

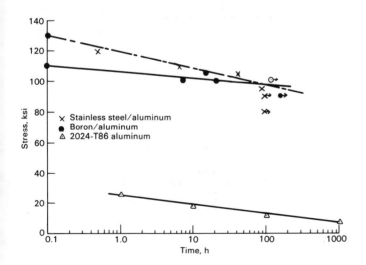

Semiplastic Strain Range for Silica/Aluminum and Stainless Steel/Aluminum Composites (Ref 89, p 95)

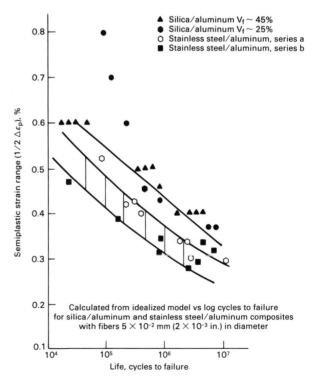

Effect of Temperature on Elastic Modulus of Stainless Steel/Aluminum Composites
(Ref 89, p 70)

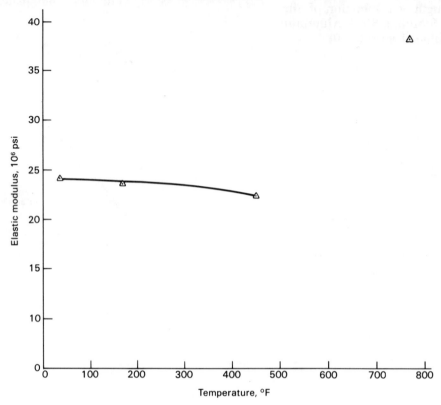

7.3.5. Other Aluminum Matrix Composite Data

Tensile Strengths and Moduli for Two Graphite/Al Composites (Ref 2, p 395)

Composite	Fiber loading, vol %	Tensile strength		Tensile modulus		Wire diameter	
		MPa	ksi	GPa	10⁶ psi	mm	in.
VS0054/201 Al	48–52	1035–1070	150–155	345	50	0.64[a]	0.025[a]
GY70SE/201 Al	37–38	793–827	115–120	207	30	0.71[b]	0.028[b]

[a]Two strand. [b]Eight strand.

Property Data for a Commercial (DWG-A1) Graphite/Aluminum Composite
(DWA Composite Specialties, Inc.)

Fiber[a]	Young's modulus		Modulus transverse		Tensile strength			
					Longitudinal		Transverse	
	GPa	10⁶ psi	GPa	10⁶ psi	MPa	ksi	MPa	ksi
P55	207–221	30–32	28–41	4–6	517–621	75–90	28–48	4–7
P75	276–296	40–43	28–41	4–6	621–724	90–105	28–48	4–7
P100	379–414	55–60	28–41	4–6	552–834	80–121	28–48[b]	4–7[b]

[a]Union Carbide Thornel fibers. [b]Estimated.

Mechanical Properties of Aramid/Aluminum/Laminate (Arall 1) Vs Other Materials (Ref 32, p 36)

Property		Alcoa/3M Arall 1,[a] 0.052 in. thick	7075-T6,[b] 0.06 in. thick	2024-T3,[b] 0.06 in. thick	Graphite/epoxy composites[c] (60% fiber content) Unidirectional (100% in 0°)		Typical structure (42% in 0°, 50% in 45° and 8% in 90°)
Average Mechanical Properties							
Tensile ultimate strength, ksi	L[d]	111	83	86	0°	180	95
	LT	60	83	67	90°	8	40
Tensile yield strength, ksi	L	86	74	52	0°	NA	NA
	LT	51	72	46	90°	NA	NA
Compressive yield strength, ksi	L	54	73	43	0°	180	95
	LT	57	76	49	90°	30	42
Elongation, %	L	1.0	11	18	0°	NA	NA
	LT	9.0	11	18	90°	NA	NA
Elastic tensile modulus, 10^6 psi	L	9.9	10.4	10.5	0°	NA	NA
	LT	7.6	10.4	10.5	90°	NA	NA
Ultimate strain, μin./in.	L	21 000			0°	8 700	4 750
	LT	100 000			90°	4 750	4 750
Density, lb/in.^{3}		0.085	0.101	0.101		0.056	0.056
Blanking shear, ksi		38	50	42		···	···
Bearing ultimate (e/D = 2.0), ksi	L	93	166	137	0°	···	···
	LT	88	168	140	90°	···	···
Bearing yield (e/D = 2.0), ksi	L	92	126	96	0°	NA	NA
	LT	87	126	96	90°	NA	NA
Specific Mechanical Properties							
TUS/density, $\times 10^3$ in.	L	1 306	822	673	0°	3 214	1 696
	LT	706	822	663	90°	143	714
TYS/density, $\times 10^3$ in.	L	1 011	733	515	0°	NA	NA
	LT	600	713	455	90°	NA	NA
CYS/density, $\times 10^3$ in.	L	635	723	426	0°	NA	NA
	LT	671	752	485	90°	NA	NA
E/density, $\times 10^6$ in.	L	116	103	104	0°	375	196
	LT	89	103	104	90°	30	89

[a]The Arall 1 properties are from limited tests of laboratory-produced material. [b]Taken from Aluminum Standards and Data, 1982, and from internal Alcoa test results. [c]Taken from "DOD/NASA Advanced Composite Design Guide," Vol 1-A, "High Strength Graphite/Epoxy, F180" (the generic description for Hercules AS/3501-6 and Celanese T300/5208). [d]The L direction for Arall 1 is defined to be parallel to the fiber direction.

Impact Energy Dissipated by Full-Size Notched Impact Specimens as a Function of Aluminum Composite Material Properties (Ref 18)

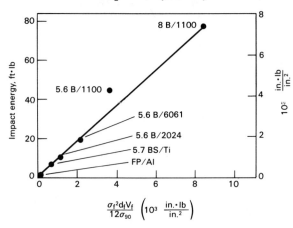

Creep Properties of Al/SiO₂ Composites and Other Materials at 400 °C (Ref 89, p 73)

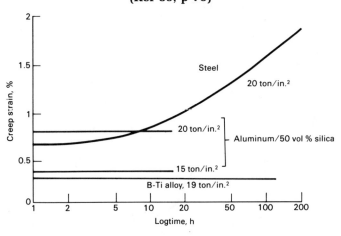

Composite Density as a Function of Fiber Reinforcement Level for Various Aluminum Matrix Composite Systems (Ref 119)

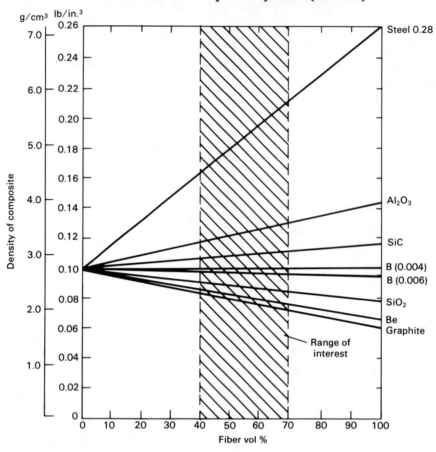

Strength-to-Density Ratio as a Function of Temperature for Specimens of Plasma-Sprayed Boron Filament Reinforced Aluminum and Other Alloys (Ref 11, p 92)

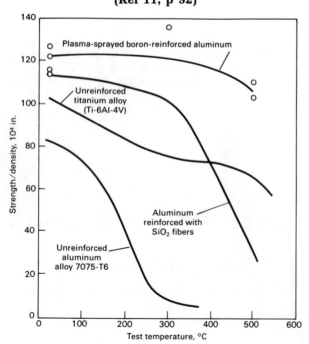

Tensile Strength Vs Vol % Loading for SiC/Al 2024 and B/Al Composites (Ref 89, p 47)

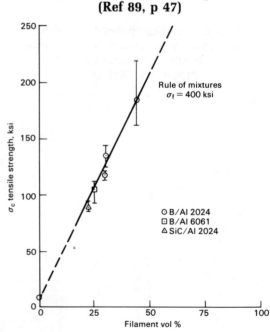

Effect of Aluminum Alloy Heat Treatment on Off-Axis Tensile Strength of Be-Reinforced Aluminum Composites (Ref 89, p 56)

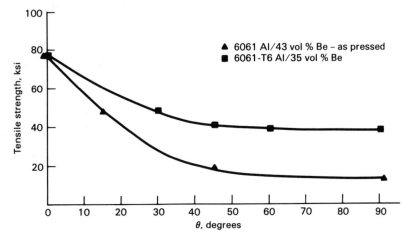

Young's Modulus of Be Wire Reinforced Aluminum Composites (Ref 89, p 42)

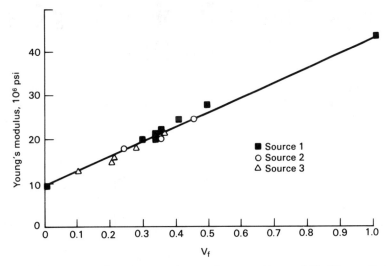

Data sources: 1. Herman, M., "Beryllium Wire–Metal Matrix Composites Program," Final Report, Contract N00019-67-C-0532, Allison Report EDR 5950, July 31, 1968.
2. Hoffmanner, A. L., "Study of Methods to Produce Composite Beryllium Blades," Final Report, Contract NOw-65-0281-f, August 12, 1966.
3. Toy, A., Atteridge, G. D., and Sinizer, D. I., "Development and Evaluation of the Diffusion Bonding Process," AFML-TR-66-350, November, 1966.

Full-Size Charpy Specimen Test Data for Some Aluminum Composites (Ref 18)

| Material | Energy dissipated | | | |
| | Notched | | Unnotched | |
	J	ft · lb	J	ft · lb
6061-T6 Al	13.6	10	>203	>150
Ti-6Al-4V	17.6	13	>353	>260
30 vol % SiC$_p$/6061 Al, annealed	2.7	2	24.4	18
50 vol % B/6061 Al (0°), annealed	27.1	20	54.2	40
60 vol % FP/Al (0°)	0.41	0.3	...	...

Modulus of Elasticity Vs Wire Orientation for Be-Reinforced Al* (Ref 89, p 60)

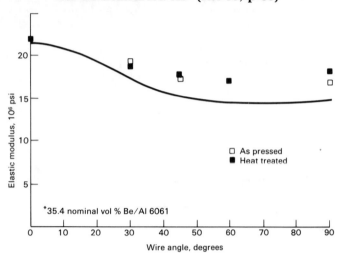

Off-Axis Tensile Strength of Al/50 Vol % B and Al/34 Vol % Be Orthogonal Cross-Ply Composites (Ref 89, p 57)

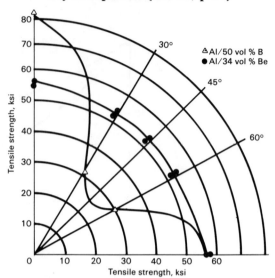

7.4. Fiber Reinforced Titanium Alloy Composite Property Data

Room-Temperature Notched Izod Impact Strength of SiC/A-40 Ti Composites Containing Various Amounts of Fiber (Ref 118)

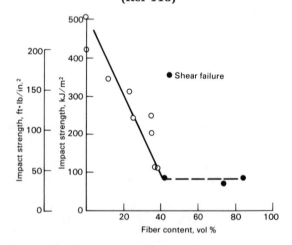

Axial Elastic Modulus of SiC/Ti-6Al-4V Composite as a Function of Fiber Content (Ref 9, p 279)

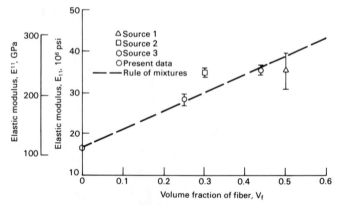

Data sources: 1. Brentnall, W. D., and Toth, I. J., "High Temperature Titanium Composites," Technical Report AFML-TR-73-223, Air Force Materials Laboratory, Dayton, OH, 1973.

2. Tsareff, T. C., Sippel, G. R., and Herman, M., "Metal-Matrix Composites," DMIC Memo 243, Defense Metals Information Center, Battelle Memorial Institute, May, 1969.

3. Mayakuth, D. J., Monroe, R. E., Favor, R. J., and Moon, D., "Titanium Base Alloys/6Al-4V, DMIC Processes and Properties Handbook," Battelle Memorial Institute, Feb, 1971.

Tensile Strength and Elastic Modulus of B₄C-Coated Boron/Titanium Composites[a] (Avco Specialty Materials Div.)

Orientation	Ultimate tensile strength		Elastic modulus	
	MPa	ksi	GPa	10^6 psi
0°	1340–1485	194.4–215.4	219–225	31.7–32.7
90°	476–488	69.1–70.8	...	...

[a]Preliminary data. Fiber volume, 36%. Process: hot pressed for 30 min at 41–55 MPa (6–8 ksi) and 925 °C (1700 °F).

Room- and Elevated-Temperature Fatigue Curves for SiC/Ti-6Al-4V System at a Stress Ratio of R = 0.0 (Ref 9, p 283)

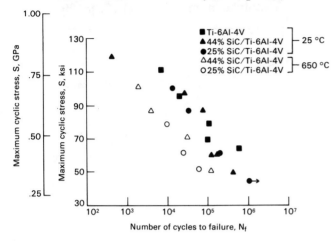

Room-Temperature Fatigue Data From Previous Figure for Ti-6Al-4V and for SiC/Ti-6Al-4V Composites Plotted as Maximum Cylic Matrix Strain Vs Cycles to Failure (Ref 9, p 284)

The matrix strains in the composite are calculated using the composite moduli determined in this study.

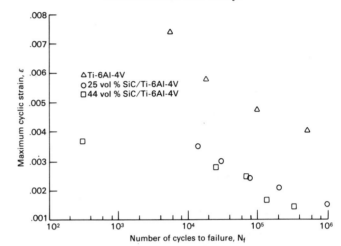

Axial and Transverse Tensile Strengths of SiC/Ti-6Al-4V Composite as a Function of Fiber Content (Ref 9, p 280)

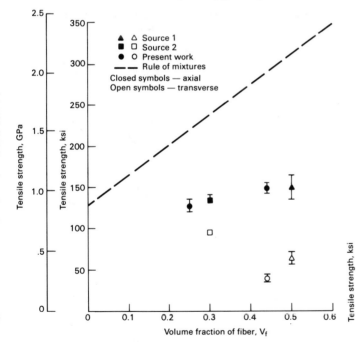

Data sources: 1. Brentnall, W. D., and Toth, I. J., "High Temperature Titanium Composites," Technical Report AFML-TR-73-223, Air Force Materials Laboratory, Dayton, OH, 1973.
2. Tsareff, T. C., Sippel, G. R., and Herman, M., "Metal-Matrix Composites," DMIC Memo 243, Defense Metals Information Center, Battelle Memorial Institute, May, 1969.

Room- and Elevated-Temperature Tensile Strengths of 36 Vol % SiC/A-70 Ti Composites, Hot Pressed A-70 Foil, and A-75 and Ti 6-4 Bar Stock (Ref 118)

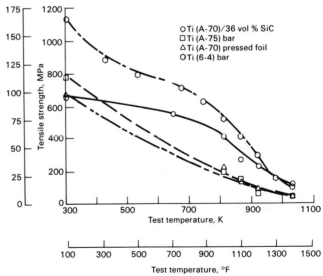

Elevated-Temperature Properties of Borsic/Ti-6Al-4V Composite Vs Monolithic Materials (Ref 8, p 250)

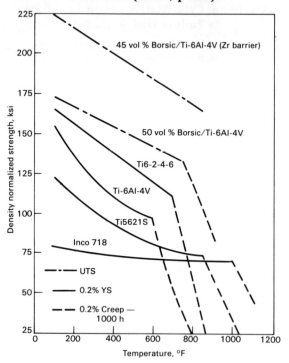

Tensile-Strength-to-Density Ratio Vs Temperature for Unreinforced Titanium Alloy and Alloy Reinforced With 20 Vol % Molybdenum Fiber (Ref 22, p 129)

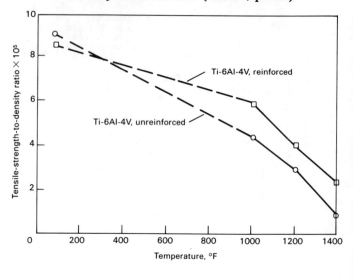

Tensile Strength Vs Temperature for Unreinforced Titanium Alloy and Alloy Reinforced With 20 Vol % Molybdenum Fiber (Ref 22, p 128)

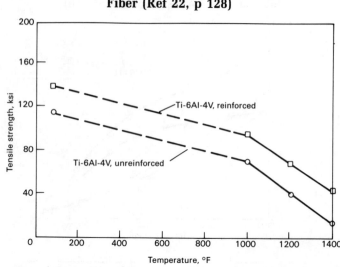

Modulus Values for Reinforced Titanium Composites (Ref 89, p 44)

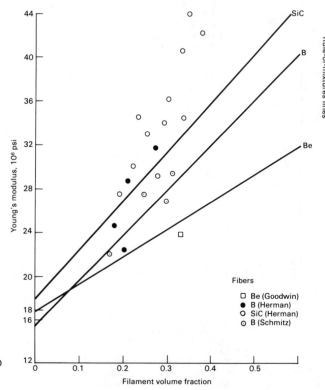

7.5. Tungsten Fiber Reinforced Copper Composites

Room-Temperature Tensile Properties of Tungsten Fiber/Copper Alloy Composites (Ref 10, p 151)

Binder material	Maximum solubility of alloying element in tungsten	Alloying element content wt %	at. %	Specimen	Fiber content, vol %	Tensile strength MPa	ksi	Reduction in area, %	Type of fracture
Pure copper	Insoluble in tungsten	0	0	···	65	1556	225.7	···	Ductile
				···	70.2	1641	238.0	···	Ductile
				···	75.4	1722	249.8	···	Ductile
Copper-nickel	0.3	5	5.4	1	79	1700	246.6	34	Ductile
				2	78.4	1724	250.0	37	Ductile
				3	76	1509	218.9	32	Ductile
		10	10.9	4	74.1	908	131.7	Nil	Brittle
				5	75.5	750	108.8	Nil	Brittle
				6	79.5	354	51.3	Nil	Brittle
Copper-cobalt	0.3	1	1.1	7	77.3	1513	219.4	···	Semiductile
		5	5.4	8	76	1470	213.2	1.5	Semiductile
				9	74.8	1581	229.3	2.3	Ductile
				10	74.7	1015	147.2	···	Brittle
				11	74.9	1187	172.1	···	Brittle
Copper-aluminum	2.6	5	11.3	12	63.4	682	98.9	Nil	Brittle
				13	72.4	1060	153.8	Nil	Semiductile
				14	76.1	1065	154.5	Nil	Semiductile
		10	20.8	15	76.7	955	138.5	···	Brittle
Copper-titanium	8	10	12.8	16	78.2	1542	223.7	···	Semiductile
				17	71.7	1518	220.1	10	Semiductile
		25	30.7	18	76.3	1287	186.7	···	Brittle
Copper-zirconium	3	10	7.2	19	72.8	1489	216.0	Nil	Brittle
				20	78.5	1760	255.3	Nil	Ductile
				21	75.6	1564	226.9	Nil	Semiductile
				22	64.7	1190	172.6	Nil	Brittle
				23	64.3	1349	195.7	Nil	Semiductile
		33	25.5	24	75.9	736	106.7	Nil	Brittle
Copper-chromium	Complete solid solubility (miscibility gap)	1	1.2	25	78.7	1541	223.5	7.4	Semiductile
				26	77.5	1572	228.0	25.8	Ductile
				27	77.2	1558	225.9	7.5	Semiductile
		2	2.4	28	76.4	1666	241.7	16.4	Ductile
Copper-niobium	Complete solid solubility	1	0.6	29	75.4	1635	237.1	20.6	Ductile
				30	75.1	1538	223.1	24.7	Ductile

Modulus of Elasticity of Tungsten-Fiber-Reinforced Copper Composites (Ref 49, p 25)

Specimen number	Fiber content, vol %	Modulus of elasticity GPa	10⁶ psi	Specimen number	Fiber content, vol %	Modulus of elasticity GPa	10⁶ psi
Continuous Reinforcement				E-10	66.6	305	44.3
				E-11	67.3	314	45.6
E-01	18.7	182	26.4	E-12	75.5	333	48.3
E-02	24.3	184	26.7	E-13	Copper	122	17.7
E-03	26.7	195	28.3	E-14	Tungsten	405	58.8
E-04	29.0	207	30.0	**Discontinuous Reinforcement**			
E-05	32.3	205	29.8				
E-06	35.9	208	30.2	ED-01	23.0	180	26.1
E-07	55.0	287	41.6	ED-02	25.0	199	28.9
E-08	57.6	288	41.7	ED-03	25.7	194	28.2
E-09	64.9	306	44.4	ED-04	26.9	192	27.9

Comparison of 100-Hour Rupture Strengths of Copper and Copper Composites (Ref 48, p 54)

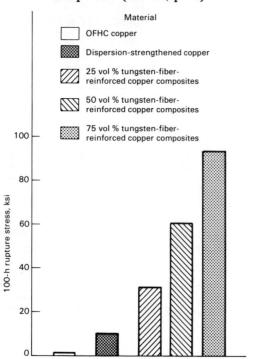

Comparison of 100-Hour Rupture Stress (a) and 100-Hour Rupture Stress/Density (b) of Tungsten-Fiber-Reinforced Copper Composites and Superalloys at 1500 °F (Ref 48, p 55)

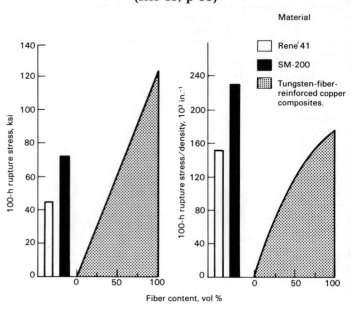

Stress for Given Creep Rate as a Function of Fiber Content for Tungsten-Fiber-Reinforced Composites at (a) 1200 °F and (b) 1500 °F (Ref 48, p 45)

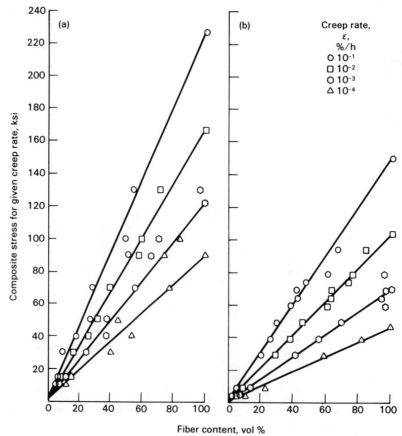

7.6. Fiber Reinforced Nickel, Nickel-Base, Cobalt, Iron-Base, and Superalloy Composites Property Data

Rupture Strengths of Tungsten and Tungsten Alloy Wire Reinforced Nickel-, Cobalt-, and Iron-Base Superalloy Composites (Ref 92)

Matrix alloy[a]	Wire	Wire diameter		Wire content, vol %	Density		100-h rupture strength[b]		Stress-density for 100-h rupture	
		mm	in.		g/cm³	lb/in.³	MPa	ksi	m	in.
ZhS6	VRN (W)	0.3–0.5	0.012–0.020	40	12.5	0.45	138	20	1125	44 300
EPD-16	···	···	···	···	8.3	0.3	51	7.4	635	25 000
	Tungsten	0.25	0.010	40	12.7	0.46	131	19	1040	41 000
Nimocast 713C	···	···	···	···	8.0	0.29	48	7	613	24 000
	Tungsten	1.27	0.050	20	10.3	0.37	93	13.5	927	36 500
MARM322E	···	···	···	···	···	···	48	7	···	···
	W-2ThO₂	0.08	0.003	40	···	···	207	30	···	···
Ni-Cr-W-Al-Ti	···	···	···	···	9.15	0.33	23	3.3	254	10 000
	218CS (W)	0.38	0.015	40	13.3	0.48	138	20	1058	41 700
	W-2ThO₂	0.38	0.015	40	13.0	0.47	193	28	1513	59 600
	W-Hf-C	0.38	0.015	40	13.3	0.48	324	47	2491	98 000
Fe-Cr-Al-Y[c]	W-1ThO₂	0.38	0.015	56	12.5	0.45	242[d]	35[d]	1957	77 000
	W-Hf-C	0.38	0.015	35	11.3	0.41	242	35	2147	84 500

[a]Nominal compositions, wt %: ZhS6, Ni-12.5Cr-4.8Mo-7W-2.5Ti-5Al; EPD-16, Ni-6Al-6Cr-2Mo-11W-1.5Nb; Nimocast 713C, Ni-12.5Cr-6Al-1Ti-4Mo-2Nb-2.5Fe; MARM322E, Co-21.5Cr-25W-10Ni-0.8Ti-3.5Ta; Ni-Cr-W-Al-Ti, Ni-15Cr-25W-2Al-2Ti; Fe-Cr-Al-Y, Fe-24Cr-5Al-1Y. [b]At 1100 °C (2010 °F). [c]Because of its extreme oxidation resistance, Fe-Cr-Al-Y is referred to here as a superalloy. [d]831-h rupture strength.

Comparison of 100-Hour Rupture Strength at 1093 °C (2000 °F) for Some Tungsten-Wire Reinforced Composites Vs Superalloys (Ref 92)

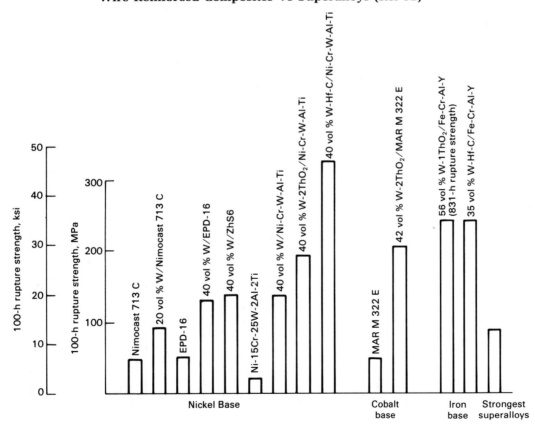

Comparison of Ratio of 100-Hour Rupture Strength to Density for Tungsten-Wire-Reinforced Nickel- and Iron-Base Superalloy Composites at 1093 °C (2000 °F) (Ref 92)

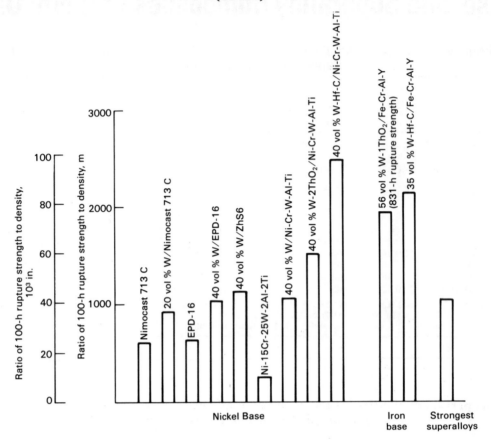

Miniature Izod Impact Strengths of Superalloys and Tungsten-Wire-Reinforced Superalloy and Nichrome Composites (Ref 92)

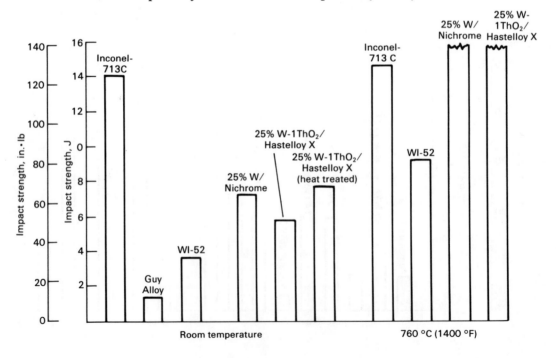

Thermal Cycling Data for Tungsten/Superalloy Composites (Ref 92)

Composite material	Heat source	Cycle	No. of cycles	Remarks
40 vol % W/Nimocast 258	Fluidized bed	RT–1100 °C (2010 °F)	400	No apparent damage to interface
13 vol % W/Nimocast 713C	Fluidized bed	20–600 °C (70–1110 °F)	200	No cracks
		550–1050 °C (1020–1920 °F)	2–12	Cracks at interface
		20–1050 °C (70–1920 °F)	2–25	Cracks at interface
W/EI435 (14, 24, and 35 vol %)	Electric resistance furnace	RT–1100 °C (2010 °F) 2.5 min to temperature Water quench	100	Number of cycles for fiber/ matrix debonding: 14 vol %, 90 to 100; 24 vol %, 60 to 70; 35 vol %, 35 to 50
W/EI435 (15 and 32 vol %)	Self resistance	30 s to heat and cool 480–700 °C (900–1290 °F) 500–800 °C (930–1470 °F) 530–900 °C (980–1650 °F) 570–1000 °C (1050–1830 °F) 600–1100 °C (1110–2010 °F)	1000	All 15 vol % specimens warped and decreased in length; cracks at interface. 32 vol % specimens did not deform externally, but matrix cracks between fibers were observed.
W/Ni-W-Cr-Al-Ti (35 and 50 vol %)	Self resistance	1 min to heat, 4 min to cool	100	35 vol %, warpage and shrinkage; 50 vol %, no damage
W/Ni-Cr-Al-Y (35 and 50 vol %)	Self resistance	RT–1093 °C (2000 °F)	100	35 vol %, warpage; 50 vol %, no damage
W/2IDA (35 and 50 vol %)	Self resistance	RT–1093 °C (2000 °F)	100	35 vol %, warpage and shrinkage; 50 vol %, no damage
50 vol % W/Ni-Cr-Al-Y	Self resistance	427–1093 °C (800–2000 °F)	1000	Internal microcracking
30 vol % W-1ThO$_2$/Fe-Cr-Al-Y	Passage of electric current	1 min to heat, 4 min to cool RT–1204 °C (2200 °F)	1000	No damage. Surface roughening but no cracking.

Resistance of Fiber-Free Hastelloy X and Hastelloy X Composites to Single and Repeated Tension (Ref 115)

Temperature °C	°F	Type of wire	Wire content, vol %	Tensile (σ_c) kN/cm^2	ksi	Fatigue (σ_{cf}) (1 × 10^6) kN/cm^2	ksi	σ_{cf}/σ_c
RT	RT	None	0	79	114	41	60	0.53
		TZM molybdenum	36	86	125	60	87	0.70
		NF tungsten	37	79	115	48	70	0.61
816	1500	None	0	36	52	17	24	0.46
		NF tungsten	23	41	60	37	54	0.90
899	1650	None	0	30	43	12	18	0.41
		NF tungsten	30	47	68	38	55	0.81
982	1800	None	0	19	27	4	6	0.25
		NF tungsten	30	36	52	24	35	0.67

Effect of Thermal Cycling on Residual Room-Temperature Tensile Properties of W-1ThO$_2$/Fe-Cr-Al-Y (Ref 117, p 132)

Exposure	Observations	Ultimate tensile strength MPa	ksi	Modulus GPa	10^6 psi	Strain to failure μin./in.
As-fabricated	...	655	95.0	179	26.0	3400
As-fabricated	...	581	84.2	201	29.2	3700
100 cycles, 29–1095 °C (85–2000 °F)	No visual change	563	81.7	259	37.6	2400
100 cycles, 29–1095 °C (85–2000 °F)	No visual change	618	89.6	219	31.7	4300
1000 cycles, 29–1095 °C (85–2000 °F)	Surface roughening	590	85.5	258	37.4	3200
1000 cycles, 29–1095 °C (2000 °F)	Surface roughening	557	80.8	177	25.6	3300
100 cycles, 29–1205 °C (85–2200 °F)	No visual change	624	90.5	250	36.3	3100
100 cycles, 29–1205 °C (85–2200 °F)	No visual change	587	85.2	228	33.1	3600
1000 cycles, 29–1205 °C (85–2200 °F)	Surface roughening	503	72.9	158	22.9	3200
1000 cycles, 29–1205 °C (85–2200 °F)	Surface roughening	487	70.6	170	24.6	3000

Elevated-Temperature Tensile Properties of W/Fe-Cr-Al-Y (Ref 117, p 129)

Test temperature °C	°F	Specimen	Filament orientation	Ultimate tensile strength MPa	ksi	Total elongation, %
648	1200	P34-15-2	±15°	776	112.6	25.2
		P36-15-2	±15°	717	104.0	12.8
		P34-45-2	±45°	564	81.8	35.5
		P36-45-2	±45°	539	78.2	24.6
		P34-90-2	90°	189	27.4	3.5
		P36-90-2	90°	180	26.2	3.2
760	1400	P34-15-1	±15°	571	82.8	12.0
		P36-15-1	±15°	534	77.4	14.2
		P34-45-1	±45°	185	26.9	23+[a]
		P36-45-1	±45°	163	23.6	24.6
		P36-90-1	90°	111	16.1	6.5
648	1200	P21-1-1[b]	0°	737	106.9	3.0
		P21-1-2[c]	0°	768	111.4	2.9

[a]Test terminated before failure. [b]Creep specimen after 1077-h creep test at 1037 °C (1900 °F). [c]Creep specimen after 990-h creep test at 1093 °C (2000 °F).

Comparison of Typical Creep Behavior at 1100 °C (2010 °F) of Nimocast 713C With and Without Tungsten Reinforcement (Ref 114)

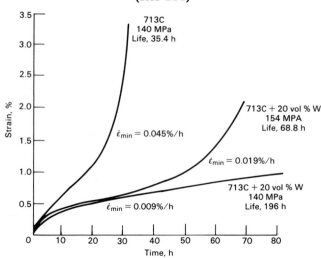

Ratio of Fatigue Strength (σ_F) to Ultimate Tensile Strength (UTS) for W/Ni Continuously Reinforced Composites (Ref 116)

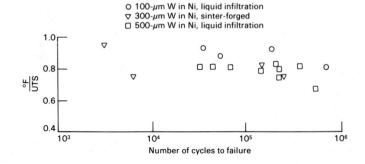

Notched Impact Strength Vs Temperature for W/Fe-Cr-Al-Y Composites (Ref 117, p 131)

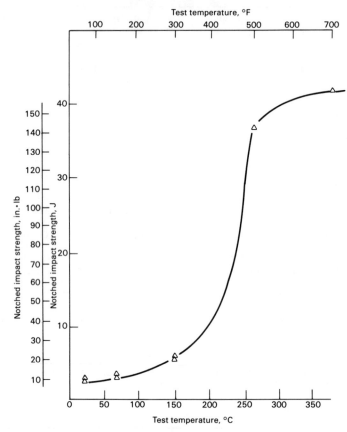

Comparative, Density-Normalized Larson-Miller Stress-Rupture Curves (Longitudinal Properties) (Ref 117, p 131)

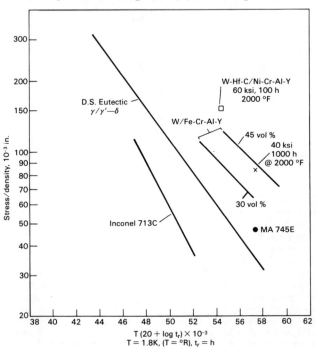

Izod Impact Strengths of Unnotched Tungsten Fiber/Nickel-Base Superalloy Composites Compared With Minimum Impact Criterion for Turbine Blades and Vanes (Ref 8, p 348)

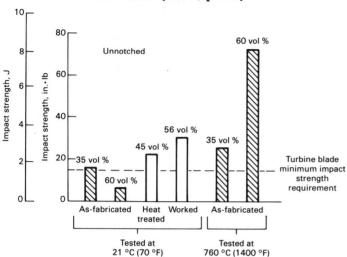

Ultimate Tensile Strength of Nickel and of Some Mo-, Thornel-, W-, and B-Reinforced Superalloy Composites (Ref 8, p 337)

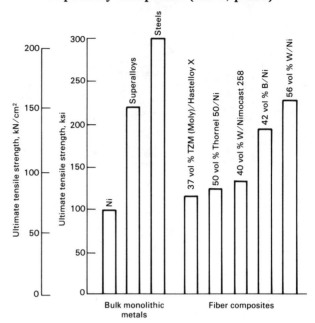

Larson-Miller Plot of Stress Rupture Strengths of W-1ThO₂/Fe-Cr-Al-Y, D.S. Eutectics, and Superalloys (Ref 117, p 130)

Larson-Miller Plot of Stress Rupture Strengths of W-1ThO$_2$/Fe-Cr-Al-Y, D.S. Eutectics, and Superalloys (Ref 117, p 130)

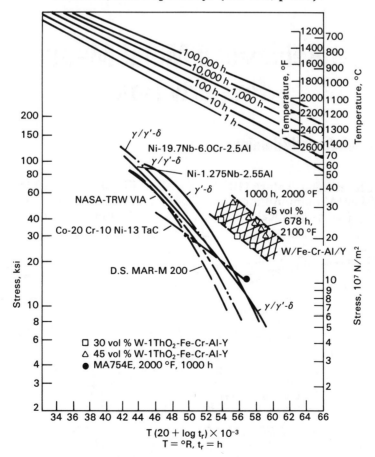

Ratio of Stress (To Cause Rupture) to
Density at 1093 °C (2000 °F) for Tungsten/
Nickel-Base Superalloy Composites
(Ref 92)

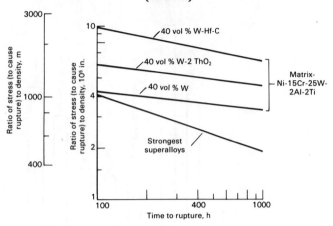

Specific Tensile Strengths of Mo-, W-, B-,
and Thornel-Reinforced Nickel
Composites (Ref 8, p 337)

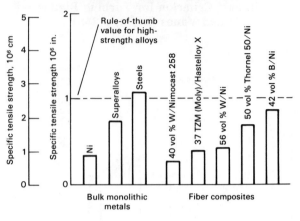

7.7. Whisker Reinforced Metal Matrix Composites Property Data

7.7.1. SiC Whisker Reinforced Aluminum

Specific Tensile Strength at Temperature (100-h
Exposure) for SiC Whisker/Aluminum Composite
and Aluminum Alloy (Arco Chemical Co.)

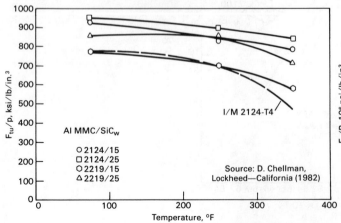

Specific Modulus of Elasticity (100-h Exposure)
for SiC Whisker/Aluminum Composite and
Aluminum Alloy (Arco Chemical Co.)

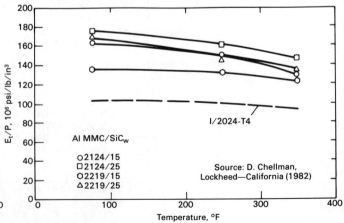

Coefficients of Thermal Expansion for SiC Whisker and Particulate Reinforced Aluminum Composites (Arco Chemical Co.)

Material	Coefficient of thermal expansion,[a] 10^{-6}/°F	Direction
6061-T7/25% SiC_w, VHP and extruded	6.7	Extruded
2024-T6/40% SiC_p, VHP	7.2	90° to pressing
2024-T6/25% SiC_w, CIP/HIP	8.3	Inapplicable
2024-T6/25% SiC_w, VHP	9.1	Pressing
2024-T6/25% SiC_w, VHP	8.3	90° to pressing

[a] At 50 °C (122° F).

Yield Strength and Elongation of SiC Whisker/Aluminum Composites (Arco Chemical Co.)

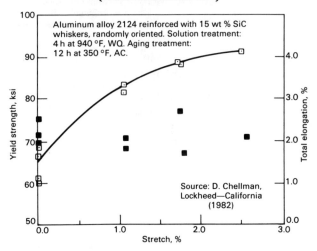

Stress-Strain Relation for Aluminum Alloy 2024 Reinforced With 20 Wt % Isotropically Dispersed SiC Whiskers (Ref 75, p 19)

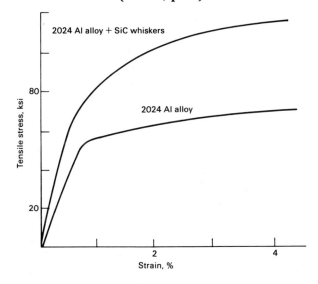

Stress-Rupture Data for SiC Whisker/Aluminum Composite and Aluminum Alloy at Different Temperatures (Arco Chemical Co.)

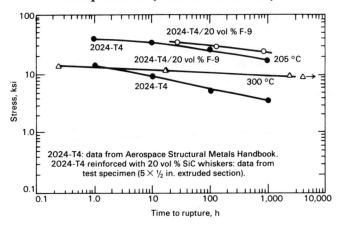

Tensile Strength at Temperature (100-h Exposure) for SiC Whisker/Aluminum Composite and Aluminum Alloy (Arco Chemical Co.)

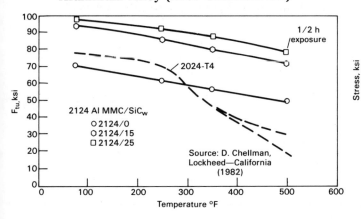

Axial Fatigue Comparison of Aluminum Alloy and SiC Whisker/Aluminum Composite (Arco Chemical Co.)

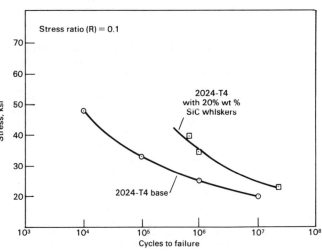

Thermal Conductivity of SiC Whisker/Aluminum Composite (Arco Chemical Co.)

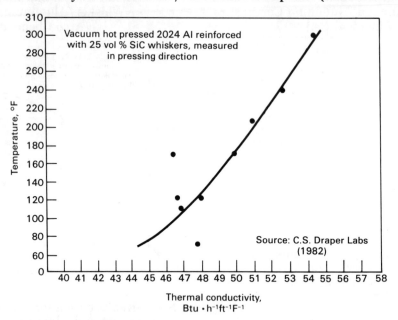

Vacuum hot pressed 2024 Al reinforced with 25 vol % SiC whiskers, measured in pressing direction

Source: C.S. Draper Labs (1982)

Temperature, °F

Thermal conductivity, Btu·h⁻¹ft⁻¹F⁻¹

7.7.2. Other Whisker Composites

Fatigue Behavior of 20 Vol % Al₂O₃/Al Composites at Room Temperature (Ref 113)

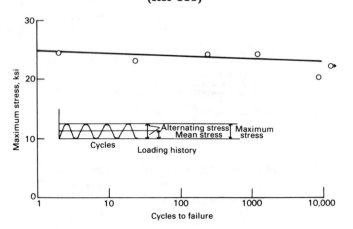

Specific Moduli of Elasticity of Some Al₂O₃ and SiC Whisker Composites and Bulk Monolithic Metals and Alloys (Ref 8, p 341)

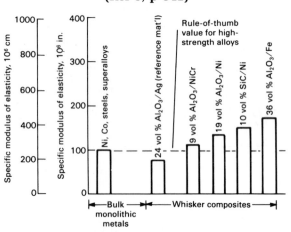

Tensile Strength and Effective Fiber Strength Factor, β, or Silver Specimens Reinforced With Al₂O₃ Whiskers (Ref 74, p 211)

Test temp, °C	d_f, μ	L/d_f	V_f	S_c, psi	$\bar{S}_f$, psi	$\bar{S}_f^*$, psi	β
Room temp	1–10	1300–2600	0.24	232,000	1,100,000	970,000	0.88
	>25	100–300	0.58	45,000[a]	850,000	73,000	0.86
	5–25	1000–2000	0.31	160,000	640,000	510,000	0.80
	5–25	1000–2000	0.20	142,000	730,000	710,000	0.97
	5–25	1000–2000	0.07	42,000	720,000	640,000	0.89
	1–10	1300	0.11	93,500[a]	1,740,000	830,000	0.48
870	>25	300	0.37	38,200[b]	280,000	100,000	0.35
870	15–25	400–800	0.46	82,500[a]	340,000	180,000	0.53
930	>25	100	0.39	40,600	160,000	100,000	0.63
940	>25	100	0.37	25,000	225,000	67,000	0.30

[a]Specimen fractured in the grips. [b]Specimen failed in gage length; whiskers pulled out of matrix.

Moduli of Elasticity of Some Al$_2$O$_3$ and SiC Whisker Composites and Bulk Monolithic Metals and Alloys
(Ref 8, p 341)

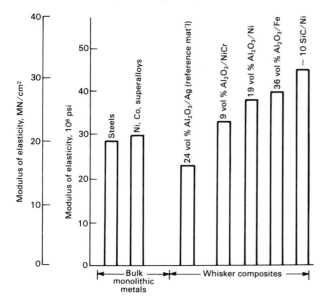

Stress-Strain Curve for Al$_2$O$_3$ Whisker/ Aluminum Composites

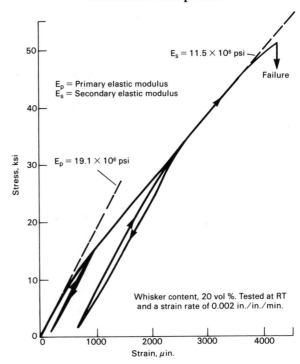

Specific Tensile Strengths of Some Al$_2$O$_3$ and SiC Whisker Composites and Bulk Monolithic Metals and Alloys
(Ref 8, p 341)

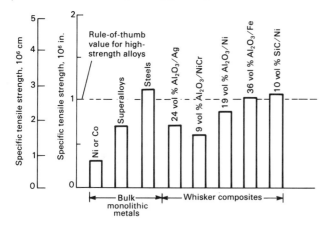

Tensile Strength of Al$_2$O$_3$ Whisker/Ni-Cr Composite Compared With Other Whisker-Reinforced Metals (Ref 75, p 19)

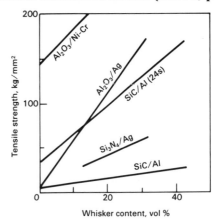

Room-Temperature Tensile Strengths of Some Al₂O₃ and SiC Whisker Composites
(Ref 8, p 341)

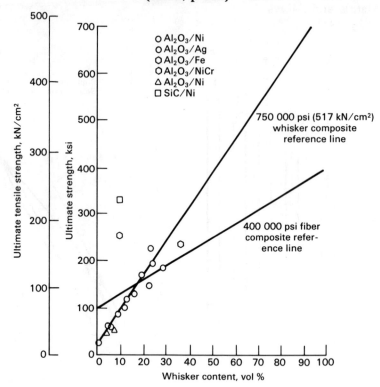

7.8. Eutectic Composites

7.8.1. Solidification Techniques for Various Matrices and Microstructures

Solidification Techniques for Various Matrices and Microstructures (Ref 43, p 127–131)

Microstructure	Matrix	Solidification technique	Properties or application field	Author(s)
Ti_5Si_3 fibers	Ti-base solution	Zone melting	Mechanical strength	Prud'homme
NiBe fibers	Ni-Cr solution	Normal solidification	Mechanical strength	Shen
$CuCd_3$ rods	Cd			
Mg_2Si rods	Mg	Normal solidification	...	Haour
ZrCuSi fibers	Cu	Zone melting	Mechanical strength	Sprenger
Mo fibers	NiAlTa alloy	Normal solidification	Mechanical strength	Pearson
γ (Ni, Al, Ta) composition fibers (structure based on Ni_3Al)	γ-Ni-base solution	Normal solidification	Mechanical strength	Jackson
β (Ni, Fe, Al) composition lamellae	γ-Ni-base solution	Normal solidification	Mechanical strength	Jackson
β (Ni, Co) Al lamellae	γ-Ni-base solution	Normal solidification	Mechanical strength	Jackson
Mo fibers	Ni_3Al	Normal solidification	Mechanical strength	Lemkey
Co_2Si fibers or lamellae	CoSiX alloy X = Al, Ga or both	Normal solidification	Mechanical strength	Livingston
Co_2Si irregular fibers or lamellae	CoSiX alloy X = Ta, Nb, V	Normal solidification	Mechanical strength	Livingston
Cr-rich lamellae	Ni-rich alloy	Different techniques	Mechanical strength	Shaw
Ni_3Al fibers	Ni alloy	Normal solidification	Mechanical strength	United Technologies
Mo alloy fibers	γ-Ni-base alloy	Normal solidification	Mechanical strength	United Technologies
Mo_2NiB_2 fibers	Ni-base alloy	Normal solidification	High-temperature strength	Sprenger
Al_6Fe fibers	Al-base alloy	Normal solidification	Mechanical strength	Comalco Aluminum
Co fibers	CoAl alloy	Normal solidification	Mechanical strength	Hubert
Co fibers (with fiber to lamella transition)	CoAlNi alloy	Normal solidification	Mechanical strength	Hubert
Cr fibers	NiAl alloy			
Mo fibers	NiAl alloy	Normal solidification	...	Walter
Ni base lamellae	ε-Ni_2In			
ε-Ni_2In rods	β-NiIn	Normal solidification	...	Livingston
$CrSi_2$ fibers	Si	...	...	Levinson
Al_3Ni fibers	Al	Normal solidification	Mechanical strength	Grabel
Cr fibers	NiAl	Normal solidification	Mechanical strength	Walter
ε-FeSi fibers or lamellae	x-Fe_2Si_5	Normal solidification	...	Nishada
Cr(Mo) plates	NiAl	Normal solidification	High temperature strength	Walter
Ni_3Al fibers	Ni_3Ta	Normal solidification	High temperature strength	Etat Français
Al_3Ni fibers	Al	EFG	...	Tyco
			...	U.S. Department of Energy
M_7C_3 fibers M = Cr, or Co	Co, Ni or Fe-base alloy	Normal solidification	Mechanical strength	United Aircraft
$(Cr,Co)_{23}C_6$	CoCr-base alloy	Normal solidification	Mechanical strength	United Aircraft
TaC fibers	CoCr-base alloy	Normal solidification	Mechanical strength	Walter
TaC fibers	CoCr or NiCr-base alloy	Normal solidification	Mechanical strength	Bibring
NbC, TiC, HfC fibers (or solid solutions thereof)	Co or Ni or Fe alloy	Normal solidification	Mechanical strength	Bibring
Cr_2C_2 fibers	Ni alloy	Normal solidification	Mechanical strength	Spiller
TaC fibers	Co, Cr, Ni solution	Normal solidification	Mechanical strength	Dunlevey
TaC fibers	Co-Cr	Normal solidification	Mechanical strength	Dunlevey
TaC fibers	two-phase γ-γ' Ni-Cr-Al-Ti alloy;	Normal solidification	Mechanical strength	Buchanan
TaC fibers	γ phase Ni-Cr alloy			
TaC, VC fibers (or mixture thereof)	Ni-base or Co-base	Normal solidification	Mechanical strength	Moore
TaC	CoNiWCrTa alloy	Normal solidification	Mechanical strength	Walter
$(Cr,Fe)_7Cr_3$ needles	α and/or γ Fe-base alloy	Different techniques	Mechanical strength	Van den Boomgaard

(continued)

Solidification Techniques for Various Matrices and Microstructures (continued) (Ref 43, p 127–131)

Microstructure	Matrix	Solidification technique	Properties or application field	Author(s)
(W,X)C lamellae				
C = IVa metal	W-base alloy	Casting technique	Wear resistance	Ruay
TaC needles	Ta	Zone melting	Mechanical strength	Lemkey
Carbides of Ti, V, Cb, Hf, Zr				
or Ta or mixtures thereof:				
fibers	Ni-base alloy	Normal solidification	Mechanical strength	Gigliotti
TaC	Re containing Ni-base alloy	Normal solidification	Mechanical strength	Henry
Cr_3C_2	Ni_3Al	Normal solidification	Mechanical strength	Chadwick
Nb_2C	Nb	Different techniques	Mechanical strength	National Research Development
Ta_2C	Ta			
NbC	NiCoAl-base alloy	Normal solidification	Mechanical strength	O.N.E.R.A.
$(Cr,Co,Ni)_7C_3$ fibers	CoCr-base alloy	Normal solidification	Mechanical strength	Brown, Boveri
Fibers of transition metal carbides	Co and Ni-base alloy	Normal solidification	Mechanical strength	O.N.E.R.A.
Fibers of TaC, VC, WC or mixtures thereof	Ni or Co-base alloy	Normal solidification	Mechanical strength	General Electric
M_7C_3 fibers				
M = Fe, Co or Cr	FeCrCo alloy	Normal solidification	Mechanical strength	Philips
TaC fibers	NiCrTa alloy	Normal solidification	Mechanical strength	General Electric
TaC fibers	CoCrTa alloy	Normal solidification	Mechanical strength	General Electric
Interlamellar $\alpha Mg/\beta Li$		Normal solidification	⋯	Prud'homme
Interlamellar Cd/Zn		Pulling a silica slide from the melt (film growth)	⋯	Albers
Interlamellar		⋯	Mechanical strength after aging	Rhodes
α-Al base phase/$CuAl_2$				
α-Al base phase/$CuMgAl_2$				
Interlamellar $Ag_3Mg/AgMg$		Normal solidification	Mechanical strength after aging	Kim
Interlamellar NiAl/V		Normal solidification	⋯	Pellegrini
Interlamellar In_2Bi/In		Normal solidification	⋯	Favier
Interlamellar Zn/Al		Normal solidification	⋯	Singh
Interlamellar Ag/Cu Cd/Zn Al/AlAg		Normal solidification	⋯	Cantor
Interlamellar Pb/Sn		Normal solidification	⋯	Labulle
Interlamellar Pb/Sn		Normal solidification	⋯	FDO
Interlamellar Al/Al_2Cu		Normal solidification	⋯	Riquet
Interlamellar $Co_2Si/CoSiX$ alloy				
X = W, Mo or both		Normal solidification	Mechanical strength	Livingston
Interlamellar Ni_3Al/Ni_3Cb		Normal solidification	Oxidation resistance	Tarshis
Interlamellar $CoFe_2O_4/BaTiO_3$		Normal solidification	Piezoelectric, magnetic	Philips
Interlamellar WXC/W alloy				
X = IVa metal		Normal solidification	Wear resistant	Aerojet General
Interlamellar $SnSe_2/SnSe$		Normal solidification	Semiconductor	Philips
Interlamellar Ni alloy/Cr alloy		Normal solidification	Mechanical strength	Doner
Interlamellar $CuAl_2/Al$		Normal solidification	Mechanical strength	Dean
MgO fibers	MgF_2	Normal solidification	Optical fibers	Parsons
NaF fibers	NaCl	⋯	Optical fibers	Yue
LiF fibers	NaCl			
Sb lamellae	InSb	Zone melting	Optical infrared polarizer	Clawson
Various halides, nitrates, complex oxides, etc. aligned crystallites	Various halides, nitrates, complex oxides, etc.	Normal solidification	Optical and electromagnetic	Lasko
PbS fibers	NaCl	Normal solidification	Optical	Philips
LiF lamellae	CaF_2	EFG	⋯	Tyco
NbC fibers	(Fe, Co, Ni) solid solution	Bridgman technique	Magnetic	Batt et al.
Co fibers	Sm_2Co_{17}	Normal solidification	Magnetic	Sahm et al.
Co_3Nb lamellae	lamellar Co	Normal solidification	Magnetic	Arnson
Ag platelets	Bi	Normal solidification	Galvano-thermomagnetic	Digges
MnBi whiskers	Bi	Solidification in magnetic field	Magnetic	Savitsky
CeAl alloy fibers	Al	Normal solidification	Nuclear reactor material	Euratom
Interlamellar Pb/Na		Normal solidification	Superconductor	Gupta
MgO rods	ZrO_2	Normal solidification	⋯	Kennard
MnO rods	Mn_2SiO_4	EFG	⋯	Finch

(continued)

Solidification Techniques for Various Matrices and Microstructures (continued) (Ref 43, p 127–131)

Microstructure	Matrix	Solidification technique	Properties or application field	Author(s)
Interlamellar NiO/Y_2O_3		Floating zone melting	...	Barailler
Fibers of Ta_2Cr; Cr; Ni; or mixtures of Ni and Cr	Al_2O_3, Cr_2O_3; Fe_2O_3, Fe_3O_4; MgO, Cr_2O_3; MgO, Al_2O_3; CoO. Al_2O_3; mixtures of Al_2O_3, Cr_2O_3, $FeCr_2O_4$ and $FeAl_2O_4$	Floating zone melting	Oxidation resistant (turbine airfoil)	Hulse
W fibers	UO_2, ThO_2, ZrO_2, MgO, or Cr_2O_3	Zone melting	Metal conductor	Clark
Re fibers	Cr_2O_3		In insulator	
Mo fibers	Cr_2O_3			
ZrO_2 lamellae	CaO-ZrO_2 phase	Floating zone melting	High temperature strength	Hulse

7.8.2. Eutectic Composite Property Data

Summary of Room-Temperature Mechanical and Certain Physical Properties of Aligned Nickel and Cobalt Alloys (Ref 14, p 106, 107)

System A-B	Vol % B	Melting point, °C	Density, g/cm^3	E, GPa	Tensile strength, MPa	Elongation, %
Ni-NiBe	38–40	1157	...	215	.918	9.0
Ni-Ni_3Nb	26	1270	8.8	...	745	12.4
Ni-Cr	23	1345	8.0	...	718	29.8
Ni-NiMo	50	1315	9.5	...	1250	<1
Ni-Ni_3Ti	29	~1300	8.2	...	650	<1
Ni-W	6	1500	...	...	830	45
Ni-TiC	5.5	1307	...	...	...	...
Ni-HfC	15–28	1260	...	...	...	...
Ni-NbC	11	1328	8.8	...	890	9.5
Ni-TaC	~10	...	...	...	...	...
Ni-Cr-NbC	11	1320 range	...	...	...	...
Ni, Co, Cr, Al-TaC	~9	...	8.8	...	1650	~5
Ni_3Al-Ni_3Nb	44	1280	8.44	242	1240	0.8
Ni_3Al-Ni_3Nb	32	1280 range	...	...	1230	2.0
Ni_3Al-Ni_3Nb	~32	1270–1285 range	8.5	...	1130	29
Ni-Ni_3Al-Ni_3Nb	...	1270	...	...	1140	2.3
Ni_3Al-Ni_3Ta	~65	~1360	10.8	...	930	<1
Ni-Ni_3Al-Ni_3Ta	...	~1360	...	...	1060	5
Ni_3Al-Ni_7Zr_2	42	1192	...	...	...	...
Ni_3Al-Mo	26	1306	8.18	138	1120	21
NiAl-Cr (rod)	34	1450–1455 range	6.4	182	1240	<1
(Ni,Cr)-(Cr,Ni)$_7C_3$	30	1305 range	...	200–290	685–960	2–11
Co-CoAl	35	1400	...	172	500–585	~6
Co-CoBe	23	1120	...	...	...	...
Co-Co_3Nb	~50	1235	...	...	...	...
Co-Co_2Ta	35	1276	...	...	...	...
Co-Co_7W_6	23	1480	...	...	750	<1
Co-TiC	16	1360	...	...	...	...
Co-HfC	15	...	...	...	...	...
Co-VC	20	...	...	...	...	...
Co-NbC	12	1365	8.8	...	1030	2
Co-TaC	16	1402	9.1	222	1035	11.8
Co-Cr-NbC	12	1340 range	...	...	1280	<2
Co-Cr-TaC	~9	1360 range	9.0	210	1035–1160	16–20
(Co,Cr)-(Cr,Co)$_7C_3$	30	1300 range	8.0	296	1280	1.5
(Co,Cr,Al)-(Cr,Co)$_7C_3$	28	1295 range	7.8	283	1730–2011	2.5–1.0
(Co,Cr)-(Cr,Co)$_{23}C_6$	~40	1340 range	7.91	~276	1200	0.96

Ultimate Tensile Strengths of Directionally Solidified Refractory Metal Matrix Eutectics (Ref 8, p 354)

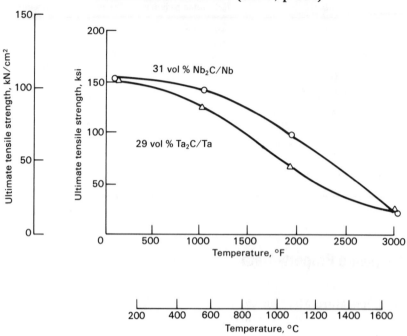

Stress-Rupture Strengths of Directionally Solidified Laminar Eutectics and High-Strength Superalloys (for Comparison) Represented by Larson-Miller Plots (Ref 8, p 357)

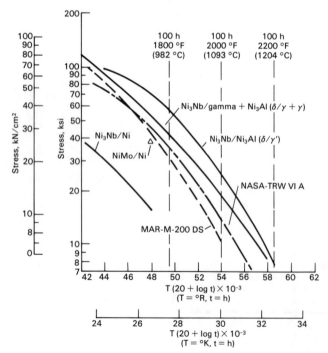

Stress-Rupture Strengths of Directionally Solidified Fibrous Eutectics and High-Strength Superalloys (for Comparison) Represented by Larson-Miller Plots (Ref 8, p 355)

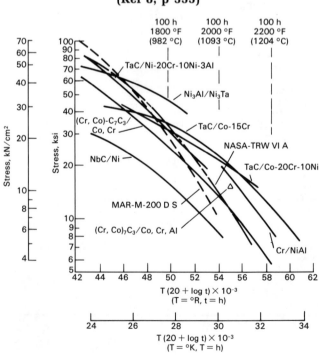

Properties of Some DS Eutectics[a] (Ref 7, p 2.101)

Eutectic	Density kg/m³	lb/in³	mp °C	°F	Potential service temperature[b] °C	°F	Strength[c] MPa	ksi	Transverse tensile elongation,[d] %	Resistance to oxidation and heat corrosion
γγ'δ	8580	0.310	1271	2320	1000	1832	172	25	0.5	Poor
NiTaC-13	8719	0.315	1349	2640	1005	1841	...	...	3	Poor
γγ'α	8498	0.307	1310	2390	1015	1859	220	31.9	5	Good
NiTaC3-116A	8580	0.310	...	...	1030	1886	...	...	...	Good
γβ	8027	0.29	1371	2500	...	...	170	24.6	<1	Excellent
COTAC-74	8580	0.310	1329	2425	...	...	340	49.3	2	Good

[a]Based on *Mater. Eng.*, December, 1977, p. 22. [b]Baseline service temperature is 960 °C (1760 °F) for DS MAR-M-200 + HF. [c]Shear-rupture strength at 760 °C (1400 °F) for 100 h. [d]At 760 °C (1400 °F).

Compositions of Some DS Eutectics* (Ref 7, p 2.100)

	γγ'δ	NiTaC-13	γγ'α	NiTaC3-116A	γβ	COTAC-74
Ni	Rem	Rem	Rem	Rem	Rem	Rem
Co	...	3.3	...	3.7	10.0	20.0
Cr	6.0	4.4	...	1.9	...	10.0
Al	2.5	5.4	6.0	6.5	11.0	4.0
Cb	20.1	...	...	...	...	4.9
Ta	...	8.1	...	8.2	...	...
W	...	3.1	...	...	8.0	10.0
Re	...	6.2	...	6.3	...	...
V	...	5.6	...	4.2	...	...
Mo	...	...	32.0	...	...	...
C	0.06	0.48	...	0.24	...	0.6

*α = alpha, β = beta, γ = gamma, δ = delta. Table based on *Mater. Eng.*, December, 1977, p. 22.

Ultimate Tensile Strengths of Directionally Solidified Fibrous Eutectics and High-Strength Superalloys (for Comparison) as a Function of Test Temperature (Ref 8, p 355)

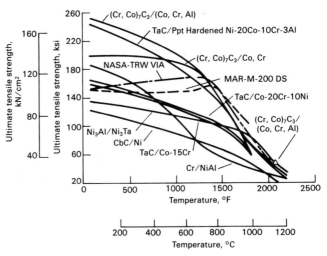

Ultimate Tensile Strengths of Directionally Solidified Laminar Eutectics and High-Strength Superalloys (for Comparison) as a Function of Test Temperature (Ref 8, p 357)

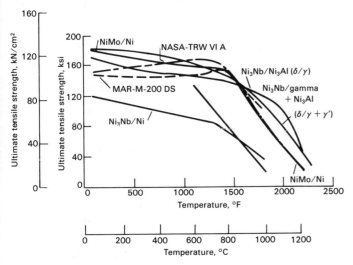

Toughness of Some *In Situ* Composites Vs Boron-Fiber-Reinforced Aluminum (Ref 13, p 202)

Systems:	B-Al alloy[a] (6061)	W-Ni[a] alloy	NbC-Ni[a]	(CoCr)-[a] (CrCo)₇C₃
V_f	0.5	0.6	0.11	0.3
Toughness:[b]	6[c]	9[d]	340	94[e]
	1.5[f]	...	...	17[e]
	1.5[f]	...	...	14[e]
Matrix	130	240	630	...

[a]*In situ* composites. [b]Toughness values, in kJ · m⁻², taken from impact tests at 20 °C. [c]Toughness increased with increasing V_f in range 0.3 to 0.5. [d]Toughness increased to 100 kJ · m⁻² at 370 °C, and to 500 kJ · m⁻² at 1100 °C. Hot working increased the toughness at 20 °C to 44 kJ · m⁻². [e]Slow bending tests gave much lower values (approximately 1/10). [f]Toughness independent of V_f in range 0.3 to 0.5.

Fatigue Properties of High-Temperature Lamellar and Fibrous Eutectic Materials
(Ref 14, p 140, 141)

Material	Temperature, °C	Endurance limit at 10^7 cycles, MPa	Tensile strength, MPa	Type of test
Lamellar Eutectics				
Ni-Cr	24	585	~830	Fluctuating tension
	760	310	~380	
Ni-Ni$_3$Nb	24	415	745	Notched specimen, fluctuating tension, 13.8 MPa minimum
Ni$_3$Al-Ni$_3$Nb	24	690	1170	Fluctuating tension, $R = 0.1$
Ni$_3$Al-Ni$_3$Nb	871	430	860	Reversed bending
Fibrous Eutectics				
Co-15Cr-TaC	24	600	1150	Rotating bending
	800	400	680	Fluctuating tension, 20 MPa minimum
Co-20Cr-10Ni-NbC	24	650	1090	Rotating bending
	800	500	700	Fluctuating tension, 20 MPa minimum
Co-20Cr-10Ni-NbC	24	650	1030	Rotating bending
	800	500	...	Fluctuating tension, 20 MPa minimum
Ni-20Co-10Cr-3Al-TaC	24	450	1480	Rotating bending
	800	600	~900	Fluctuating tension, 20 MPa minimum
(Co,Cr,Al)-(Cr,Co)$_7$C$_3$	24	620	1730	Fluctuating tension, $R = 0.1$

7.9. In-Situ Composites

Step-Load Creep Data for Tungsten/Refractory Compound Composites at 1650 °C (3000 °F) (Ref 83, p 22)

Billet	Composition	Second-stage creep elongation mm	mils	Load MPa	ksi	Time, h	Creep rate µm/h	mils/h	s⁻¹	Increase in creep rate, %	Stress-rupture life, h Estimated from graph below	Calculated
18	Tungsten/8 vol % hafnium boride	1.37	54	55	8	100.0	2.64	0.104	2.88×10^{-8}	...	545.0	538.0
				69	10	124.0	3.58	0.141	3.91×10^{-8}	35.6	375.0	383.0
				83	12	148.0	3.81	0.150	4.15×10^{-8}	6.3	330.0	359.0
				97	14	172.0	6.43	0.253	7.01×10^{-8}	69.6	207.0	213.5
				110	16	196.0	11.4	0.450	1.25×10^{-7}	77.9	135.0	120.0
				124	18	220.0	18.4	0.725	2.01×10^{-7}	61.2	77.0	74.6
				138	20	223.3ª	...	...	...	...	...	...
19	Tungsten/8 vol % hafnium nitride	7.87	310	55	8	239.1	23.4	0.921	2.55×10^{-7}	...	388.0	337.0
				62	9	264.1	35.8	1.41	3.91×10^{-7}	53.1	205.0	220.0
				69	10	292.2ª	49.5	1.95	5.40×10^{-7}	38.3	109.0	159.0
20	Tungsten/8 vol % hafnium carbide	4.32	170	55	8	44.0	3.89	0.153	4.28×10^{-8}	...	1145.0	1110.0
				69	10	68.0	6.35	0.250	7.00×10^{-8}	63.3	638.0	679.0
				83	12	92.0	13.0	0.510	1.42×10^{-7}	104.0	330.0	333.0
				97	14	116.4	22.5	0.884	2.47×10^{-7}	73.3	182.0	192.0
				110	16	140.4	35.8	1.41	3.91×10^{-7}	59.5	124.0	120.0
				124	18	164.9ª	...	...	...	...	...	...
21	Tungsten/8 vol % tantalum carbide (nose of bar)	2.54	100	55	8	310.5	2.82	0.111	3.08×10^{-8}	...	920.0	900.0
				69	10	359.5	6.73	0.265	7.36×10^{-8}	138.8	370.0	378.0
				83	12	407.8	16.6	0.655	1.82×10^{-7}	147.0	150.0	152.0
				97	14	114.6ª	52.1	2.05	5.70×10^{-7}	221.5	...	...

ªSpecimen failed.

Room-Temperature Ultimate Tensile Strength of *In Situ* Formed Cu-Nb Composites as a Function of True Strain (Ref 20, p 106)

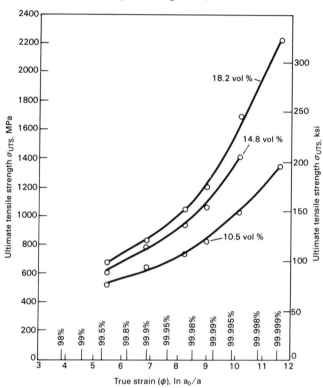

Stress to Rupture in 10 h for Fibered Oxide/Niobium Composites and Niobium Alloys at 2200 and 2500 °F (Ref 82, p 103)

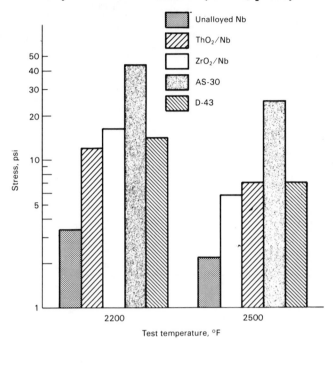

Step-Load Creep Curves for Tungsten/Refractory Compound Composites at 3000 °F (Ref 83, p21)

Composition

- ● Unreinforced tungsten
- ■ Tungsten/8 vol % hafnium carbide
- ▲ Tungsten/8 vol % tantalum carbide (nose of bar)
- ▲ Tungsten/8 vol % hafnium boride
- ◆ Tungsten/8 vol % hafnium nitride

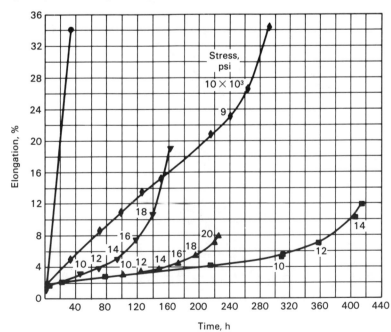

Ultimate Tensile Strength of *In Situ* Formed Cu-Nb Multifilamentary Composites as a Function of Temperature (Ref 20, p 107)

Temperature, K	Composite diameter (d), mm	True strain (φ), ln a₀/a	Ultimate tensile strenth (σ_UTS)			
			Cu–14.8 vol % Nb GPA	ksi	Cu–10.5 vol % Nb GPa	ksi
297	0.256	6.78	0.78	113	0.64	93
	0.129	8.16	0.93	135	0.73	106
	0.089	8.90	1.06	154	0.82	119
	0.050	10.05	1.42	206	1.03	149
	0.024	11.52	...	...	1.34	195
423	0.129	8.16	0.84	122	...	...
	0.085	8.99	0.98	142	...	...
	0.050	10.05	1.21	176	...	...
568	0.254	6.80	0.52	76	0.36	52
	0.192	7.36	0.60	87	0.41	60
	0.129	8.16	0.70	101	0.52	76
	0.085	8.99	0.86	125	0.59	85
	0.050	10.05	1.03	150	0.72	105
768	0.254	6.80	0.30	43	0.16	23
	0.192	7.36	0.35	50	0.19	28
	0.129	8.16	0.36	52	0.22	32
	0.085	8.99	0.42	61	0.26	38
	0.050	10.05	0.52	75	0.31	41

Effect of Extrusion Temperature on Length-to-Diameter Ratio of Fibered Oxides in Niobium (Ref 82, p 96)

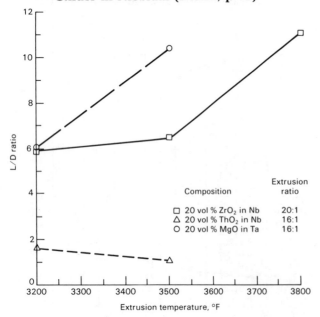

Tensile Strength/Density Ratio of Oxide Fiber/Niobium Composites and Niobium Alloys at Various Temperatures (Ref 82, p 102)

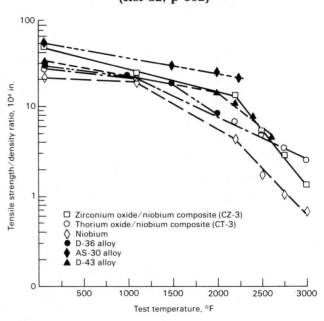

Properties of *In Situ* Tungsten Composites (Ref 83, p 16)

Billet	Composition	Fiber length/ diameter ratio	Theor- etical density, %	Diamond pyramid hardness		Average grain diameter of matrix, cm		Stress-rupture properties				Stress- rupture life, h
				Trans- verse	Longi- tudinal	Trans- verse	Longi- tudinal	Temperature		Stress		
								°C	°F	MPa	ksi	
1	Tungsten	...	99.32	376	360	0.0040	0.0040	1650	3000	55	8	1.9
2	Tungsten	...	99.51	383	371	0.0028	0.0045	1650	3000	55	8	1.5
												2.6
										34	5	53.8
										21	3	348.7
3	Tungsten/8 vol % zirconia	19.1	99.37	371	396	0.0016	0.0023	1650	3000	55	8	8.0
												12.1
										34	5	74.5
								2095	3800	21	3	4.3
4	Tungsten/8 vol % yttria	12.7	99.86	401	377	0.0019	0.0025	1650	3000	55	8	7.2
												9.7
										48	7	23.5
										41	6	44.9
										28	4	161.0
								1870	3400	21	3	25.5
								2095	3800	21	3	1.6
5	Tungsten/10 vol % hafnia	23.3	100.16	410	420	0.0022	0.0030	1650	3000	55	8	10.1
												10.6
										34	5	261.4
								1870	3400	21	3	34.7
6	Tungsten/5 vol % zirconia	12.5	99.04	...	399	...	0.0020	1650	3000	55	8	4.4
												3.1
7	Tungsten/5 vol % zirconia	29.3	99.71	...	384	...	0.0018	1650	3000	55	8	3.6
												4.3
												5.4
8	Tungsten	...	99.60	369	369	0.0047	0.0045	1650	3000	55	8	21.6
	Tungsten (swaged 50%)	...	...	429	446	0.0017	0.0042	1650	3000	55	8	11.8
												14.7
9	Tungsten	...	99.68	383	380	0.0023	0.0026	1650	3000	55	8	20.1
												29.1
10	Tungsten	...	...	...	...	...	...	1650	3000	55	8	18.9
11	Tungsten/14.5 vol % zirconia	23.8	99.69	441	423	0.0009	0.0014	1650	3000	55	8	1.9
												1.7
												1.0
												1.6
12	Tungsten/20.5 vol % zirconia	10.4	101.10[a]	459	440	0.0008	0.0011	1650	3000	55	8	2.0
												1.6
13	Tungsten/20.5 vol % yttria	6.8	100.83[a]	413	419	0.0009	0.0011	1650	3000	55	8	1.7
												1.3
												2.4
14	Tungsten/25.5 vol % hafnia	14.1	97.20[a]	466	458	0.0010	0.0020	1650	3000	55	8	8.8
												6.2
15	Tungsten/23.5 vol % hafnia[b]	16.7	101.10[a]	578	515	0.0005	0.0007	1650	3000	55	8	4.9
												4.7
												4.3
												5.4
16	Tungsten/8 vol % thoria	7.8	99.22	381	389	0.0026	0.0036	1650	3000	55	8	34.2

(continued)

Properties of *In Situ* Tungsten Composites (Ref 83, p 16) (continued)

Billet	Composition	Fiber length/ diameter ratio	Theoretical density, %	Diamond pyramid hardness Transverse	Diamond pyramid hardness Longitudinal	Average grain diameter of matrix, cm Transverse	Average grain diameter of matrix, cm Longitudinal	Temperature °C	Temperature °F	Stress MPa	Stress ksi	Stress-rupture life, h
17	Tungsten/16 vol % thoria	13.2	98.66	420	418	0.0012	0.0013	1650	3000	55	8	36.9
18	Tungsten/8 vol % hafnium boride	1.0	95.63[a]	385	425	0.0011	0.0024	1650	3000	55–138	8–20	223.3[c]
										55	8	545.0[d]
										124	18	83.5
										193	28	3.2
								1925	3500	55	8	4.0
								1370	2500	193	28	323.9
19	Tungsten/8 vol % hafnium nitride	18.4	98.50	543	471	0.0006	0.0007	1650	3000	55–69	8–10	292.2[c]
										55	8	388.0[d]
										83	12	101.1
										124	18	21.5
								1925	3500	55	8	4.2
								1370	2500	165	24	>45.4[e]
20	Tungsten/8 vol % hafnium carbide	6.6	98.19	418	406	0.0015	0.0014	1650	3000	55–124	8–18	164.9[c]
										55	8	1145.0[d]
										124	18	77.9
21	Tungsten/8 vol % tantalum carbide (nose of bar)	...	...	412	406	0.0012	0.0013	1650	3000	55–97	8–14	414.6[c]
										55	8	920.0[d]
	Tungsten/8 vol % tantalum carbide (middle of bar)	5.1	99.44	315	372	0.0011	0.0033	1650	3000	55	8	788.6
								1370	2500	83	12	487.9

[a]Questionable because of changes in theoretical density. [b]18.5 vol % hafnia was added to Curtiss-Wright powder already containing 5 vol % hafnia. [c]Total time with increasing load (see previous table). [d]Estimated time from creep data in previous table. [e]Test incomplete: specimen-support rod on stress-rupture machine failed.

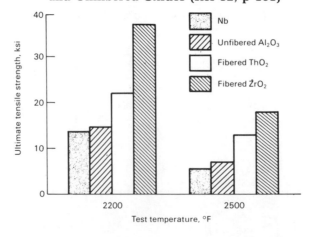

Comparison of Tensile Strengths of Niobium Composites Containing Fibered and Unifibered Oxides (Ref 82, p 101)

Effect of Extrusion Ratio on Length-to-Diameter Ratio of Fibered Oxides in Niobium (Ref 82, p 97)

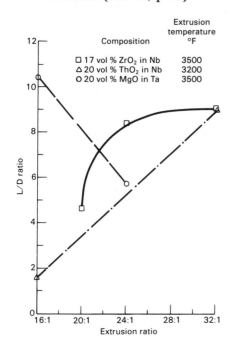

Comparison of Stress-Rupture Strengths of Zirconium Oxide/Niobium and Thorium Oxide/Niobium Composites at 2200 °F (Ref 82, p 102)

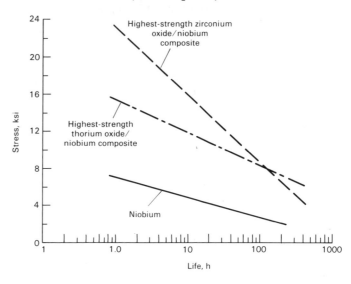

Stress for 100-h Stress-Rupture Life at 3000 °F for Tungsten Composites (Ref 83, p 24)

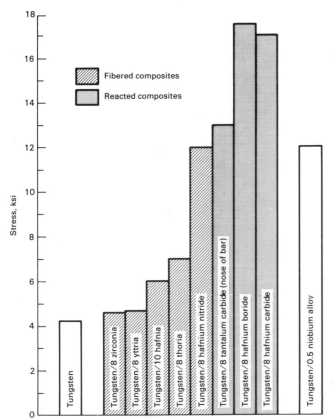

Stress-Rupture Life of Tungsten and of Tungsten/Oxide and Tungsten/Refractory Compound Composites at 3000 °F and 8 ksi (Ref 83, p 20)

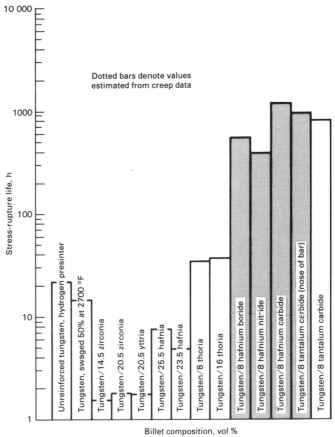

7.10. Particulate Reinforced Metals

Mechanical Properties of SiC Particulate Reinforced Aluminum Alloys[a] (DWA Composite Specialties, Inc.)

Matrix alloy	SiC$_p$, vol %	Modulus		Yield strength		Tensile strength		Ductility, %
		GPa	10^6 psi	MPa	ksi	MPa	ksi	
6061	15	97	14	400	58	455	66	7.5
	20	103	15	414	60	496	72	5.5
	25	114	16.5	427	62	517	75	4.5
	30	121	17.5	434	63	552	80	3.0
	35	134	19.5	455	66	552	80	2.7
	40	145	21	448	65	586	85	2.00
2124	20	103	15	400	58	552	80	7.0
	25	114	16.5	414	60	565	82	5.6
	30	121	17.5	441	64	593	86	4.5
	40	152	22	517	75	689	100	1.1
7090	20	103	15	655	95	724	105	2.5
	25	115	16.7	676	98	793	115	2.0
	30	128	18.5	703	102	772	112	1.2
	35	131	19	710	103	724	105	0.9
	40	145	21	689	100	710	103	0.9
7091	15	97	14	579	84	689	100	5
	20	103	15	621	90	724	105	4.5
	25	114	16.5	621	90	724	105	3.0
	30	128	18.5	676	98	765	111	2.0
	40	139	20.2	621	90	655	95	1.2

[a]DWAl 20.

Combined Plot of Strain Hardening Exponent Vs Volume Fraction for Several Copper Dispersion Alloys (Ref 111, p 239)

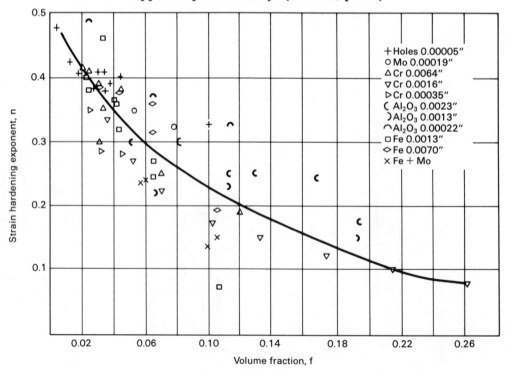

Properties of Al/SiC$_p$ Composites (Arco Chemical Co.)

Property	Al 2124-T6/ 30 vol % SiC$_p$	Al 2124-T6/ 40 vol % SiC$_p$	Property	Al 2124-T6/ 30 vol % SiC$_p$	Al 2124-T6/ 40 vol % SiC$_p$
Microyield strength:			Density:		
MPa	117	...	g/cm^3	2.91	2.96
ksi	17	...	lb/in.^{3}	0.105	0.107
Thermal expansion:			Specific modulus:		
10^6 m/m·K	12.4	10.8	10^6 in.	162	196
10^6 in./in.·°F	6.9	6.0	10^6 m	4.11	4.98
Modulus:			Thermal conductivity:		
GPa	117	145	W/m·K	125	116
10^6 psi	17	21	Btu/ft·h·°F	72	67
			Specific heat:		
			kJ/kg·K	0.80	0.67
			Btu/lb·°F	0.19	0.16

True Stress-Strain Curves Plotted on Logarithmic Paper for a Series of Copper-Chromium Alloys Containing Various Volume Fractions of Particles (Ref 111, p 238)

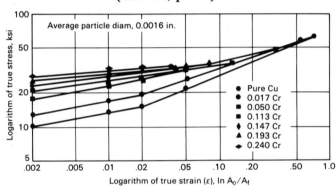

Tensile Stress-Strain Curves of Continuous Skeleton Tungsten-Copper Composites Vs Volume Fraction Dispersoid (Ref 10, p 478)

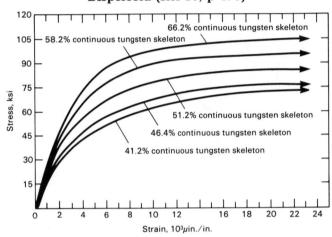

Relation of Yield Stress to Log Reciprocal Mean Free Path Between Particles in Copper-Chromium and Copper-Iron Alloys (Gensamer Plot) (Ref 111, p 244)

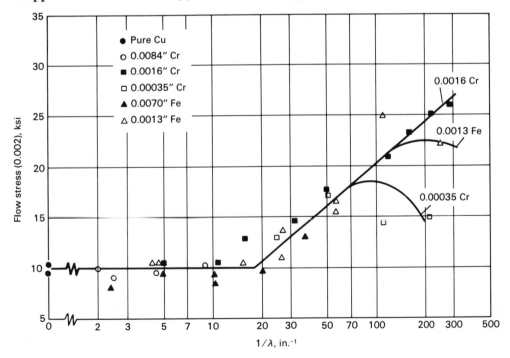

Elastic Modulus of Liquid-Phase Sintered Iron-Copper Composite Vs Volume Fraction Dispersoid (Ref 10, p 464)

Note that points fall above the lower-bound isostress curve, but between the rule-of-mixtures upper and lower bounds.

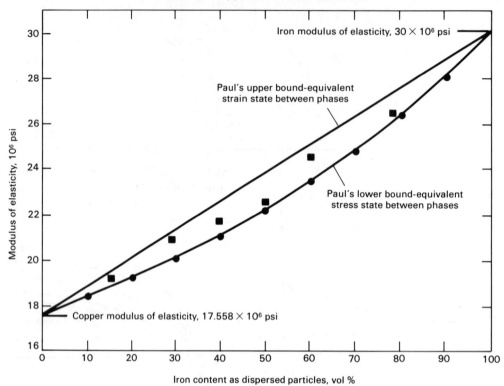

Effect of Temperature on Flow Curves of Iron-Copper Composites (Ref 10, p 470)

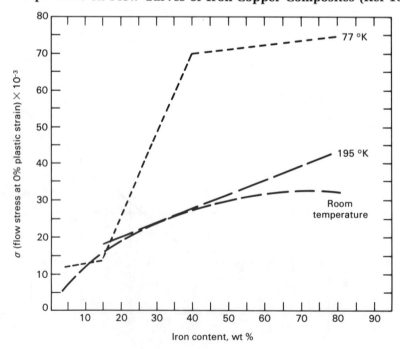

Proportional Limit and Flow Stress at 5% Plastic Strain for Four Tungsten-Nickel-Iron Composites of Different Compositions (Ref 10, p 466)

Both parameters are constant with volume fraction at the several strain rates.

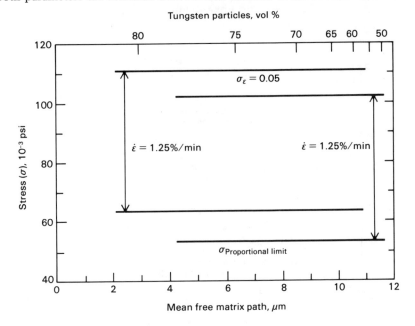

Elastic Modulus of Liquid-Phase Sintered Tungsten-Nickel-Iron Composite Vs Volume Fraction Dispersoid (Ref 10, p 461)

Note that points fall above the rule-of-mixtures upper bound.

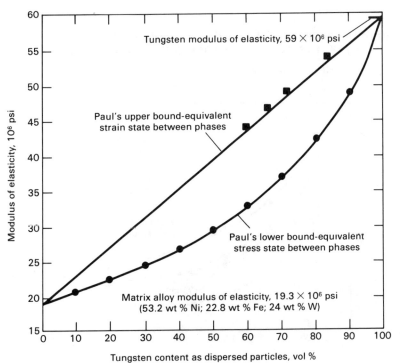

Elastic Modulus of Liquid-Phase Sintered Tungsten-Nickel-Copper Composite Vs Volume Fraction Dispersoid (Ref 10, p 462)

Note that the modulus of only the highest-volume-fraction specimen exceeds the rule-of-mixtures upper-bound prediction.

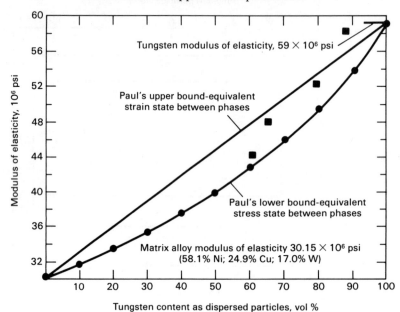

Temperature Dependence of Proportional Limit and Flow Stress ($\epsilon = 1\%$) of an 85 Wt % W–10.5 Wt % Ni–3.5 Wt % Fe Composite (Ref 10, p 468)

Note the similarity of the composite temperature dependence with that of single-crystal tungsten.

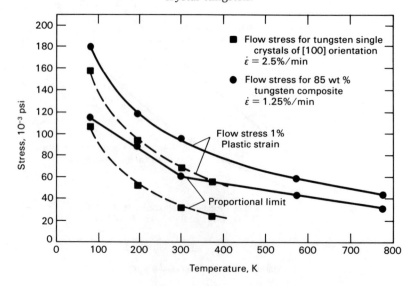

Stress-Strain Curves of Four Tungsten-Nickel-Iron Composites at Room Temperature (Ref 10, p 466)

In composites where the hard phase deforms, the flow curves are independent of the volume fraction dispersoid.

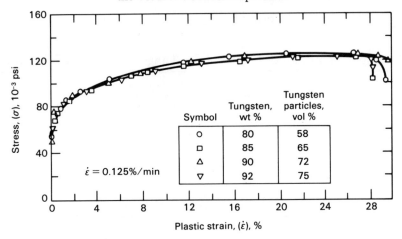

Symbol	Tungsten, wt %	Tungsten particles, vol %
○	80	58
□	85	65
△	90	72
▽	92	75

$\dot{\varepsilon} = 0.125\%/\text{min}$

Preliminary Wear Comparison of SiC Particulate Reinforced Aluminum (DWAl 20), Steel, and Aluminum (DWA Composite Specialties, Inc.)

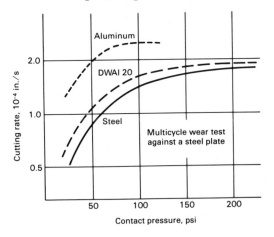

Coefficient of Thermal Expansion as a Function of SiC$_p$ Reinforcement (DWA Composite Specialties, Inc.)

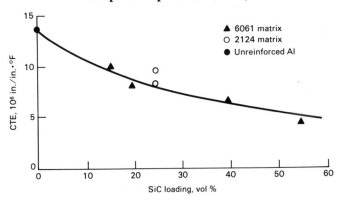

▲ 6061 matrix
○ 2124 matrix
● Unreinforced Al

Comparison of Microcreep of SiC Particulate Reinforced Aluminum Vs Beryllium (Arco Chemical Co.)

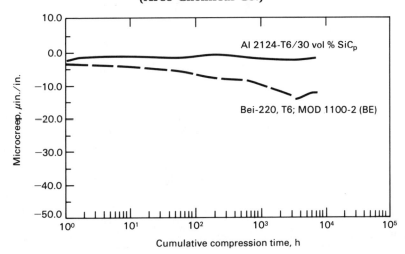

7.11. Metal Laminate Composite Property Data

7.11.1. Tungsten-Copper Laminate Composites

Thickness and Composition of Materials Used for Tungsten-Copper Laminates (Ref 85, p 18)

Material	Thickness cm	Thickness in.	Al	Cd	Si	Cr	Fe	Ni	Mn	Mg	Sn	Mo	Co	Zr	O$_2$	C	N$_2$	H$_2$	Cu	W
Copper	0.00381	0.0015																	[c]99.92 wt %	
	0.01651	0.0065																		
	0.03302	0.013																		
	0.2413	0.095																		
Tungsten	0.00254	0.001	...	...	...	...	20	...	...	...	...	200	...	...	30	30	10	6	...	Rem
	0.0127	0.005	<6	3	<7	3	13	12	<6	4	<6	45	<3	<3	...	...	...	...	<3	Rem
	0.0254	0.010	<6	3	<7	3	13	12	<6	4	<6	45	<3	<3	...	...	...	...	<3	Rem

[a]Compositions furnished by vendor. [b]Values given in ppm unless otherwise indicated. [c]OFHC copper (nominal composition), trace elements not reported.

Constitution of Unnotched and Notched Tungsten-Copper Laminar Composite Test Specimens (Ref 85, p 19)

Specimen type[a]	Tungsten laminae thickness cm	Tungsten laminae thickness in.	Number	Copper laminae thickness cm	Copper laminae thickness in.	Number	Volume fraction of tungsten
1A	0.00254	0.001	10	0.04445	0.0175	11	0.05
1B			5	0.04445	0.0175	6	0.05
2A			40	0.00965	0.0038	41	0.20
2B			20	0.00965	0.0038	21	0.20
3A			80	0.00381	0.0015	81	0.39
3B			40	0.00381	0.0015	41	0.39
4A			120	0.00127	0.0005	121	0.58
4B			60	0.00127	0.0005	61	0.58
5B			80	0.00635	0.00025	81	0.78
6A			190	0.000254	0.0001	191	0.95
6B			95	0.000254	0.0001	96	0.91
7A	0.0127	0.005	2	0.16256	0.064	3	0.05
8A			8	0.04375	0.0175	9	0.20
8B			4	0.04375	0.0175	5	0.19
9A			16	0.01651	0.0065	17	0.42
9B			8	0.01651	0.0065	9	0.42
10A			24	0.007112	0.0028	25	0.63
10B			12	0.007112	0.0028	13	0.62
11B			16	0.00381	0.0015	17	0.76
12A			48	0.000635	0.00025	49	0.95
12B			24	0.000635	0.00025	25	0.95
13A	0.0254	0.010	1	0.2413	0.095	2	0.05
14A			4	0.08128	0.032	5	0.20
14B			2	0.08128	0.032	3	0.17
15A			8	0.03302	0.013	9	0.41
15B			4	0.03302	0.013	5	0.38
16A			12	0.01651	0.0065	13	0.59
16B			6	0.01651	0.0065	7	0.57
17A			16	0.007112	0.0028	17	0.79
17B			8	0.007112	0.0028	9	0.80
18A			19	0.00127	0.0005	20	0.95
18B			10	0.00127	0.0005	11	0.94

[a]Type A specimens were nominally 0.508 cm (0.20 in.) thick; type B specimens were 0.254 cm (0.10 in.) thick.

Room-Temperature Tensile Strength Plotted Against Volume Fraction of Reinforcement for Unnotched Tungsten/ Copper Laminar Composites (Ref 85, p 26)

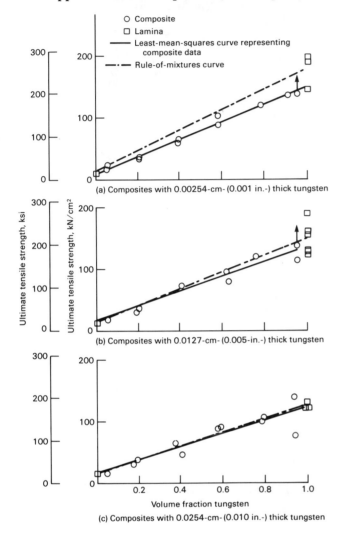

(a) Composites with 0.00254-cm- (0.001 in.-) thick tungsten

(b) Composites with 0.0127-cm- (0.005-in.-) thick tungsten

(c) Composites with 0.0254-cm- (0.010 in.-) thick tungsten

Room-Temperature Tensile Strength Plotted Against Volume Fraction of Reinforcement for Notched Tungsten/ Copper Laminar Composites (K_t = 5.8) (Ref 85, p 27)

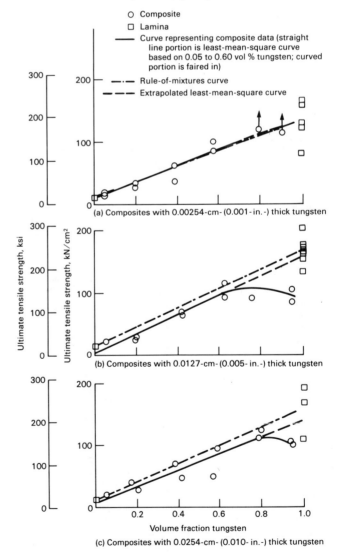

(a) Composites with 0.00254-cm- (0.001- in.-) thick tungsten

(b) Composites with 0.0127-cm- (0.005- in.-) thick tungsten

(c) Composites with 0.0254-cm- (0.010- in.-) thick tungsten

Examples of Load-Strain Curves for Unnotched Tungsten/Copper Laminar Composites (Ref 85, p 28)

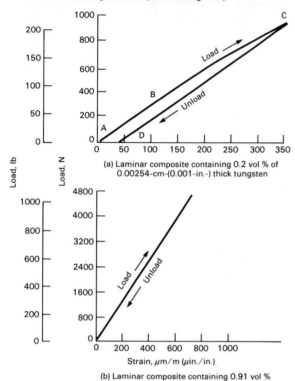

(a) Laminar composite containing 0.2 vol % of 0.00254-cm-(0.001-in.-) thick tungsten

(b) Laminar composite containing 0.91 vol % 0.00254-cm-(0.001-in.-) thick tungsten

Elastic Modulus Plotted Against Volume Fraction of Tungsten Reinforcement for Unnotched Tungsten/Copper Laminar Composites at Room Temperature (Ref 85, p 28)

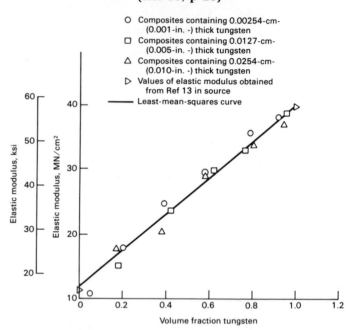

- ○ Composites containing 0.00254-cm-(0.001-in. -) thick tungsten
- □ Composites containing 0.0127-cm-(0.005-in. -) thick tungsten
- △ Composites containing 0.0254-cm-(0.010-in. -) thick tungsten
- ▷ Values of elastic modulus obtained from Ref 13 in source
- —— Least-mean-squares curve

7.11.2. Tungsten-Nickel Alloy Laminate Composites

Tensile and Stress-Rupture Results Obtained for Tungsten/Nickel Alloy Sheet and Foil Specimens at 871 °C (1600 °F) (Ref 84, p 7)

Material	Sheet or foil Thickness cm	in.	Condition	Ultimate tensile strength MPa	ksi	Stress MPa	ksi	Life, h
Tungsten	0.0025	0.001	As-received	...	...	759	110	10.0[a]
Tungsten	0.0025	0.001	As-received	...	...	690	100	56.0[a]
Tungsten	0.0025	0.001	As-received	...	...	621	90	9.0[a]
Tungsten	0.0025	0.001	As-received	...	...	552	80	500.0[a,b]
Tungsten	0.0125	0.005	As-received	1079	156	...	...	...
Tungsten	0.0125	0.005	Annealed for 4 h at 982 °C (1800 °F)	669	97	...	...	...
Tungsten	0.0125	0.005	As-received	...	...	689	100	2.3
Tungsten	0.0125	0.005	As-received	...	...	621	90	14.7
Tungsten	0.0125	0.005	As-received	...	...	517	75	45.0
Tungsten	0.0125	0.005	As-received	...	...	430	62.5	570.8
Tungsten	0.0125	0.005	Annealed for 4 h at 982 °C (1800 °F)	...	...	655	95	3.0
Tungsten	0.050	0.020	As-received	877	127	...	...	...
Tungsten	0.050	0.020	Annealed for 4 h at 982 °C (1800 °F)	669	97	...	...	...
Tungsten	0.050	0.020	As-received	...	...	621	90	6.6[a]
Tungsten	0.050	0.202	As-received	...	...	517	75	143.2[a]
Tungsten	0.050	0.020	As-received	...	...	517	75	65.8
Tungsten	0.050	0.020	As-received	...	...	476	69	268.3[a]
Nichrome V	0.0125	0.005	As-received	14.5	2	...	...	...
Nichrome V	0.0125	0.005	As-received	...	...	101.0	158	369.6

[a] Tested at commercial laboratory. [b] Test discontinued.

Composition and Thickness of Materials Used for Specimens for Tungsten/Nickel Alloy Laminates
(Ref 84, p 4)

Material	Nominal composition	Thickness cm	Thickness in.	Source
Tungsten	99.9+ W	0.0025	0.001	National Research Corp.
	99.9+ W	0.0125	0.005	General Electric Co.
	99.9+ W	0.050	0.020	General Electric Co.
	Unknown	0.25	0.10	Unknown
Nichrome V	80Ni-20Cr	0.0025	0.001	Driver Harris
		0.0125	0.005	Driver Harris
Tungsten alloy	W-0.4Re-0.35Hf-0.02C	0.0875	0.035	NASA-Lewis Research Center
Inconel 600	Ni-7Fe-15Cr	0.0125	0.005	International Nickel Co.

Constitution of Tungsten/Nickel Alloy Laminar Composite Tensile and Stress-Rupture Test Specimens
(Ref 84, p 6)

Specimen type	Constituents	Reinforcing constituent (tungsten or tungsten alloy) No. of laminae	Thickness cm	Thickness in.	Matrix constituent No. of laminae	Thickness cm	Thickness in.	Tungsten or tungsten alloy, vol %	Remarks
1a	Tungsten/Nichrome V	24	0.0025	0.001	25	0.0025	0.001	49	...
1b	Tungsten/Nichrome V	20	0.0025	0.001	21	0.0025	0.001	49	...
2	Tungsten/Nichrome V	10	0.0125	0.005	11	0.0125	0.005	48	...
3	Tungsten/Nichrome V	2	0.05	0.020	1[a]	0.05	0.020	50	...
					2[b]	0.025	0.010		
4	W-Re HfC/Inconel 600	1	0.0875	0.035	2	0.0125	0.005	77	Sandwich specimen
5	Tungsten arc sprayed with Nichrome V	1	0.0125[c]	0.005[c]	...	...	...	91	Diffusional effect study
			0.0139[d]	0.0055[d]	...	...	...		

[a]The inner lamina was 0.05 cm (0.020 in.). [b]The outer laminae were 0.025 cm (0.010 in.); in both cases (a and b), the matrix laminae were diffusion bonded 0.0125 cm (0.005 in.) Nichrome V foil. [c]Original dimension of tungsten. [d]After diffusion heat treatment.

Tensile and Stress-Rupture Strengths of Tungsten/Nickel Alloy Laminar Composites, Fiber Composites, and Superalloys at 1093 °C (2000 °F)
(Ref 84, p 17)

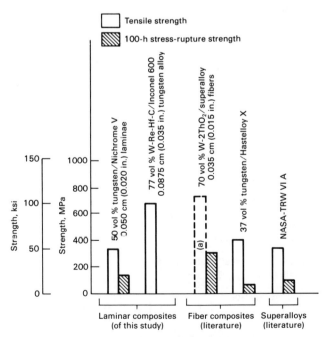

Specific Tensile and Stress-Rupture Strengths of Tungsten/Nickel Alloy Laminar Composites, Fiber Composites, and Superalloys at 1093 °C (2000 °F)
(Ref 84, p 18)

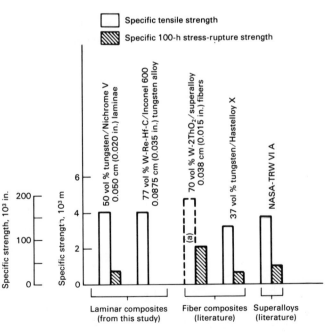

Comparison of Stress-Rupture Strengths of Laminar Composites at 1093 °C (2000 °F) Based on Percentage of Rule-of-Mixtures Values (Ref 84, Fig. 20)

Abbreviations and chemical symbols used in this figure: W = tungsten; SA = superalloy; NiCr V = Nichrome V; ThO$_2$ = thoria.

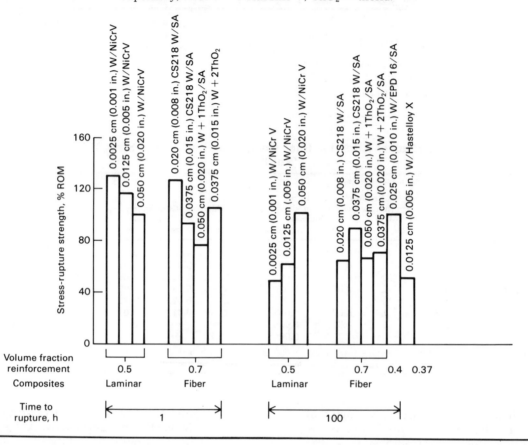

7.11.3. Other Laminate Composites

Thermal Conductivity of Stainless Steel/Copper Laminates (D. E. Makepeace Div., Engelhard Industries)

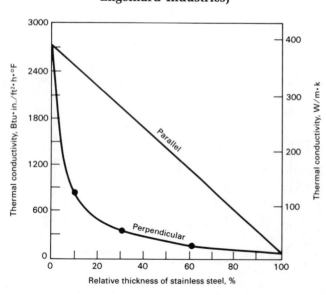

Comparison of Copper and Copper-Clad Stainless Steel (TiGuard)[a] (Ref 14, p 76)

Property	Copper	Copper-clad stainless steel
Nominal temper	Soft	Soft
Standard thickness:		
mm	0.51	0.38
in.	0.020	0.015
Relative cost	100	75
Yield strength:		
MPa	76	241
ksi	11	35
Tensile strength:		
MPa	241	434
ksi	35	63
Elongation in 50 mm or 2 in., %	36	30
Weight:		
kg/m^2	4.882	3.066
lb/ft^2	1.000	0.628
Coefficient of thermal expansion:		
10^{-6} m/m·K	17.6	11.0
10^{-6} in./in.·°F	9.8	6.1
Thermal conductivity:		
W/m·K	391.1	64
Btu·ft/ft^2·h·°F	226	37

[a]By permission of Texas Instruments, Inc.

Mechanical Properties of Stainless Steel/Aluminum Laminates[a,b] (Ref 14, p 54)

Thickness mm	in.	Al	Yield strength MPa	10³ psi	Tensile strength MPa	10³ psi	Elongation %	Young's modulus GPa	10⁶ psi
1.27	0.050	40	330	47.6	470	68.1	34	150	21.8
1.52	0.060	50	282	40.7	410	59.4	34	137	19.8
1.78	0.070	57	247	35.7	368	53.2	33	127	18.4
2.03	0.080	63	222	32.1	336	48.6	33	120	17.4
2.29	0.090	67	201	29.1	310	44.9	32	104	16.5
2.79	0.110	73	172	24.9	274	39.7	32	106	15.4
3.18	0.125	76	156	22.6	254	36.8	32	102	14.7
Aluminum			41	6.0	110	16.0	30	69	10.0
Stainless steel			505	73.0	690	100.0	35	200	29.0

[a]3003 aluminum core, 0.38-mm (0.015-in.) 304 stainless steel faces. [b]Data supplied by Charles Pfizer and Co., Inc. (1971).

Silver Brazed Steel Laminate — Homogeneous Steel (Ref 18)

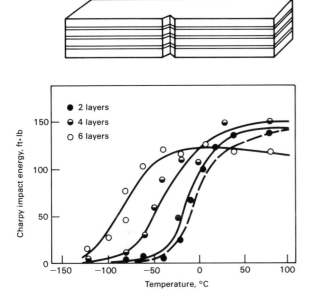

Electrical Conductivity of Stainless Steel/Copper Laminates (D. E. Makepeace Div., Engelhard Industries)

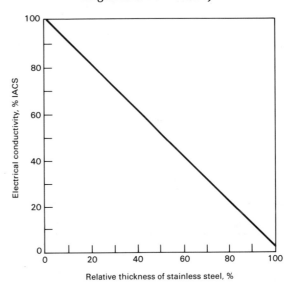

Comparison of Measured and Predicted Properties of Titanium-Beryllium Laminates (Ref 5, p 630)

	Laminate and direction					
	40Ti-58Be		55Ti-36Be		63Ti-31Be	
Property	RD	TD	RD	TD	RD	TD
Modulus, GPa:[a]						
Measured	203.5	206.9	165.5	165.5	175.9	169.0
Predicted	211.0	212.4	163.5	165.5	157.9	160.0
Difference, %	3.7	2.7	−1.2	0	−10.2	−5.1
Poisson ratio:						
Measured	0.20	0.25	0.26	0.27	0.26	0.28
Predicted	0.27	0.27	0.28	0.29	0.29	0.29
Difference, %	35.0	8.0	7.7	7.4	11.5	3.6
Fracture stress, MPa:[b]						
Measured	646.9	466.2	723.5	713.8	787.0	716.6
Predicted	677.3	642.8	713.8	694.5	767.0	749.0
Difference, %	4.7	37.9	−1.4	−2.7	−2.5	−4.7
Density, g/cm³:[c]						
Measured	2.85	···	3.24	···	3.40	···
Predicted	2.82	···	3.21	···	3.46	···
Difference, %	0.9	···	−0.8	···	1.6	···

[a]To convert GPa to psi, multiply by 145 000. [b]To convert MPa to psi, multiply by 145. [c]To convert g/m³ to lb/in.³, divide by 27.68.

Stress-Strain Curves for Mild Steel, 60:40 Pb-Sn Solder, and a 50:50 Steel-Solder Laminate (Ref 14, p 53)

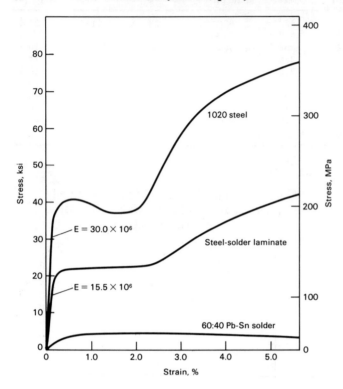

Aluminum Clad With Stainless Steel To Achieve Improved Laminate Bending Stiffness (Ref 18)

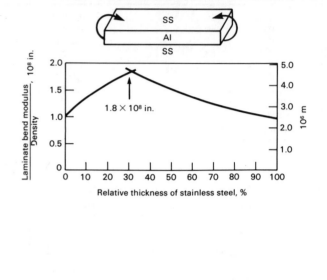

Young's Modulus and Yield Strength of Mild Steel/Solder Laminates (Ref 14, p 54)

Composition, %			Predicted				Observed			
			Young's modulus		Yield strength		Young's modulus		Yield strength	
Steel	Solder		GPa	10^6 psi	MPa	ksi	GPa	10^6 psi	MPa	ksi
100	0	....	...	...	...	...	207	30.0	276	40.0
90	10	....	190	27.2	250	36.3	187	27.1	255	37.0
75	25	....	160	23.0	210	30.7	158	22.8	215	31.2
50	50	....	110	16.0	147	21.3	107	15.5	144	20.8
0	100	....	...	...	...	...	14	2.0	128	4.0

SECTION 8
Property Data: Ceramic and Glass Matrix Composites

8.1. General Data on Ceramic Composites

Fabrication Techniques for Ceramic Composites

I Architectures	II Matrix densification
Filament winding	Infiltration
Chopped fiber	Glass
Braiding	Polymer precursor
Fabric lay-up	Sol gel
1D, 2D, 3D	Si
Whiskers	CVD
Particle dispersion	Hot pressing
	Sintering
	Reaction sintering
	Plasma spraying

Some Potential Matrix, Fiber, and Dispersion Material Options for Ceramic Composites

I Matrix Materials

- Si$_3$N$_4$
- ZrO$_2$, HfO$_2$
- Glass
- Mullite
- SiC
- Al$_2$O$_3$
- Glass ceramic
- Cordierite

II Fiber Materials

- SiC
- Si$_3$N$_4$
- Graphite
- Mullite
- BN
- Al$_2$O$_3$·B$_2$O$_3$·SiO$_2$
- Coatings
- Al$_2$O$_3$

III Dispersion Materials

- SiC
- ZrO$_2$
- TiC
- BN

Some Ceramic/Metal Combinations

Oxide matrix	Metal	Oxide matrix	Metal
Binary			
Cr$_2$O$_3$	Cr, Mo, W, Re	Nd$_2$O$_3$(CeO)	Nb
(Cr,Al)$_2$O$_3$	Cr, Mo, W	TiO$_2$	Cr, Nb, Ta
Gd$_2$O$_3$	Mo, W	UO$_2$	Mo, Ta, W
Gd$_2$O$_3$(CeO$_2$)	Mo	UO$_2$(ThO$_2$)	W
HfO$_2$(CaO)	Mo, W	Y$_2$O$_3$(CeO$_2$)	Mo, W
HfO$_2$(Y$_2$O$_3$)	W	ZrO$_2$	Ta
La$_2$O$_3$	Mo, W	ZrO$_2$(CaO)	W
Nd$_2$O$_3$	Mo, W	ZrO$_2$(Y$_2$O$_3$)	W
LaCrO$_3$	Cr, Mo, W	SiO$_2$	Cr
YCrO$_3$	Cr, Mo, W	TiC	Mo, Fe, Ni, Co
SiC	Ag, Co, Cr	Al$_2$O$_3$	Al, Co, Fe, Cr
WC	Co		

Oxide matrix	Metal	Oxide matrix	Metal
Ternary			
UO$_2$-MgO	W	La$_2$O$_3$-LaCrO$_3$	W
MgO-ZrO$_2$	W	CaO-CaCr$_2$O$_4$	W
MgO(Cr$_2$O$_3$)-ZrO$_2$	Mo	Cr$_2$O$_3$-ZrO$_2$	W, Mo
Cr$_2$O$_3$-LaCrO$_3$	W	Cr$_2$O$_3$-HfO$_2$	Mo

Ceramic Matrix Composite Toughening Concepts (Ref 109, p 4)

Concept	Basic requirements	Status of verification and modeling
1. Modulus transfer of load from matrix to fibers	$E_f > E_m$, preferably by a factor of >2	Verified, reasonable modeling
2. Prestressing of fibers and matrix	$\alpha_f > \alpha_m$ so axial tensile stresses in fibers < their fracture stress to give reasonable compressive axial stress in matrix	Not verified; basic modeling not expected to be difficult
3. Crack-impeding second phases	Fracture toughness of fibers (or particles) > local matrix so crack is either arrested or bows out, i.e., gives line tension effects between fibers or particles	Arrest impractical; line tension modeling, but uncertain verification
4. Fiber pull-out	Fiber (or elongated particles) have high enough transverse fracture toughness so failure occurs along fiber matrix interface	Limited verification and modeling
5. Crack deflection or multiplication	Sufficiently weak fiber (or particle) matrix interfaces, or appropriate mismatch of properties, especially thermal expansions between matrix and particles (fibers) and use of appropriate particles (fiber sizes)	Limited verification: no modeling. Some verification: possible modeling developing
6. Phase-transformation toughening	Second-phase particles (fibers) increase one or more dimensions.	Verified with ZrO_2 particles; modeling developing

Dependence of Composite Strength and Initial Elastic Modulus on Fiber Content for Cases B and C Where Matrix Failure Occurs First (Ref 18)

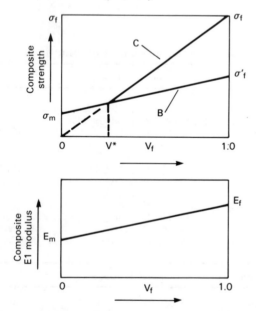

Note: V* = critical vol % fibers present, above which the matrix is immediately overloaded when the fibers fail. At less than V*, sufficient matrix is present to carry additional load imposed by the broken fibers.

Model Stress-Strain Behavior for Typical Ceramic-Matrix Composite and Its Constituents (Ref 18)

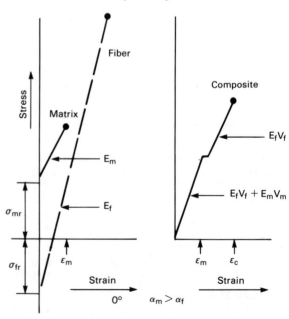

Fracture Types for Fiber-Reinforced Composites (Ref 18)

(A) Fiber failure occurs first in a metal-matrix composite. (B) Matrix and fiber fail simultaneously in a strongly bonded ceramic-matrix composite. (C) Matrix failure occurs first in a ceramic-matrix composite with low fiber/matrix interfacial strength.

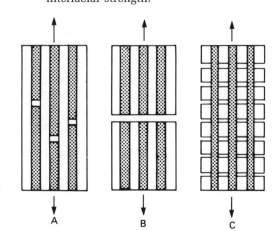

Schematic of Typical Stress-Strain Curves for (a) Monolithic Ceramic and (b) Ceramic-Matrix Composite With Continuous Fibers Aligned in the Loading Direction (Ref 68, p 11)

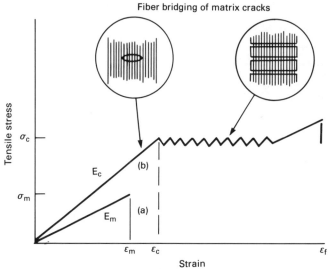

Fabrication Methods for Metal Fiber/Ceramic Matrix Composites

Fibers	Matrix	Fabrication method
Cr	Al_2O_3-Cr_2O_3	Hot pressing of grains of previously directionally solidified eutectic composites
Mo	CeO_2-doped Gd_2O_3	
V, Nb, Ta	Cr_2O_3	Directional solidification
Cr, Nb, Ta	TiO_2	Hot pressing of grains of previously grown eutectic
Ta	ZrO_2	
Cr	Fe_3O_4, Al_2O_3, Cr_2O_3, and mixtures	Directional solidification
Ta	Unstabilized HfO_2	Directional solidification
W, Mo	Stabilized HfO_2	Hot pressing
Ni, Fe, Co	MgO	Hot pressing
W	Fused SiO_2	Hot pressing
Ta, Mo, Nb	UO_2	Directional solidification
Stainless steel	Wustite	Hot pressing
Cu, Cu-Be, Be	Be_4B, Be_2B	Hot pressing, plasma spraying, or vapor deposition
Ti, Cr	SiC	Whisker formation *in situ*
Ta, W	Si_3N_4	Hot pressing
W, Mo	Si_3N_4	Flame spraying of silicon and heating in nitriding atmosphere
Mo, Ta, W	Sialon, Si_3N_4, Si_3N_4-C, TaC	Hot pressing
Ta	TaC	Hot pressing
W, W-Re	TaC	Hot pressing
Nb	$MoSi_2$	Hot pressing
Nb	Borosilicate glass	Hot pressing
Ni	Glass-ceramic	Hot pressing
W, Mo, stainless steel, or carbon steel	Glass, glass-ceramic	Fusing of glass-coated fibers together using pressure
Stainless steel	PbO glass	Hot pressing, vacuum injection, or pultrusion

Fabrication Methods for Ceramic Fiber/Ceramic Matrix Composites

Fibers	Matrix	Fabrication method
Al_2O_3	Al_2O_3	Sintering
AlN	Al_2O_3	Tape casting, aligning of AlN needles, and sintering
AlN, Si_3N_4	AlN, Si_3N_4	Hot pressing
Al_2O_3, C, ZrO_2	$Mg_3(PO_4)_2$	Hot pressing
Al_2O_3, C, B, SiC, SiO_2	Al_2O_3, $3Al_2O_3 \cdot 2SiO_2$	Hot pressing
Al_2O_3, C, B, BN, SiC	Si_3N_4	Reaction sintering
$3Al_2O_3 \cdot 2SiO_2$	$3Al_2O_3 \cdot 2SiO_2$-Al_2O_3	Slip casting and firing
BN	BN	Hot pressing
BN	BN	Chemical vapor deposition
BN	BN	Firing of B_2O_3-containing composite in nitriding atmosphere
C	Al_2O_3, $3Al_2O_3 \cdot 2SiO_2$	Coating of fibers with LiC; sintering
C	Al_2O_3	Hot pressing
C	Carbides, borides, silicides, oxides	Hot pressing
C	Pyrolytic materials	Chemical vapor deposition
C	C-SiC, TiC	Chemical vapor deposition
C	Si_3N_4	Hot pressing, reaction sintering; coating of fibers — e.g., with SiC — to improve compatibility
C	Sialon, Si_3N_4, Si_3N_4-C, TaC	Hot pressing
C	TaC	Vacuum impregnation with precursor solution and pyrolyzing
C	C-TaC	Hot pressing of Ta-coated fibers
C	ZrB_2-Si-C	Hot pressing
C, fused SiO_2	Powdered ceramic	Application of aqueous slurry and drying
MgO	Cubic ZrO_2	Directional solidification
SiC	Si	Impregnation of carbon fiber preform with molten silicon
SiC	Si	Heating of a mixture of carbon fibers and silicon powder
SiC	Si	Infiltration of SiC fibers with molten silicon
SiC	SiC	Chemical vapor deposition
SiC	SiC, Si_3N_4, AlN, BN	Hot pressing or sintering
SiC	Si_3N_4	Reaction sintering
Si_3N_4	Si_3N_4	Hot pressing
ZrO_2	Al_2O_3	Directional solidification
ZrO_2	CaO-ZrO_2	Directional solidification
ZrO_2	MgO	Hot pressing
ZrO_2	ZrO_2	Hot pressing
ZrO_2	ZrO_2	Impregnation

Fabrication Methods for Some Ceramic Whisker/Ceramic Matrix Composites

Whiskers	Matrix	Fabrication method
$3Al_2O_3 \cdot 2SiO_2$, α-Al_2O_3, ZrO_2	Oxides and nitrides	Hot pressing
$3Al_2O_3 \cdot 2SiO_2$, α-Al_2O_3, SiC, Si_3N_4, ZnO	TiO_2	Hot pressing
$3Al_2O_3 \cdot SiO_2$	Al_2O_3, Al_2O_3-Mo, Cr_2O_3, ZrO_2, Al_2O_3-Cr, AlN, BN, Si_3N_4, V_2O_3, TiN, SiO_2	Hot pressing
α-Al_2O_3, AlN, SiC	$3Al_2O_3 \cdot 2SiO_2$-Al_2O_3	Hot pressing
Si_3N_4	Si_3N_4	Sintering
SiC, BN, C	Si_3N_4, AlN	Sintering or hot pressing
ZrO_2	Stabilized ZrO_2	Hot pressing
ZrO_2	MgO	Hot pressing
Ground whiskers	Several oxides	Powder metallurgy techniques

8.2. Fiber Reinforced Ceramic Matrix Composites Property Data

8.2.1. Comparative Fiber Reinforced Ceramic Composite Data

Comparison Between Experimental Strength and Theoretical Strength of Various Discontinuous Randomly Oriented Fiber Systems (Ref 107, p 964)

System	V_f	$\Delta\alpha$, $10^{-6}\,°C^{-1}$	Experimental flexural strength, $MN \cdot m^{-2}$		Theoretical[a] tensile strength, $MN \cdot m^{-2}$	
			Absolute	Normalized	Upper-bound	Lower-bound
M_gO–carbon fiber	0.05	+13.5	30	0.15	7.43	6.50
	0.20	+13.5	20	0.10	14.50	8.90
Borosilicate glass–carbon fiber	0.20	+3.5	58	0.61	3.55	2.05
Al_2O_3–carbon fiber	0.20	+8.8	81	0.27	4.57	3.23
M_gO–ZrO_2 fiber	0.10	+5.9	76	0.38	2.89	2.42
	0.20	+5.9	40	0.20	6.00	4.20
Al_2O_3–Mo fiber	0.05	+2.8	92	0.37	2.99	2.63
	0.10	+2.8	67	0.27	4.46	3.48
	0.20	+2.8	113	0.45	3.07	1.90
Al_2O_3–Mo fiber	0.06	+2.7	135	~1.0	1.27	0.96
	0.12	+2.7	135	~1.0	1.53	0.91
Al_2O_3–Mo fiber	0.20	+2.8	158	0.67	2.13	1.28
HfO_2–Mo fiber	0.10	+0.2	72	1.03	1.89	0.88
	0.20	+0.2	98	1.36	2.06	0.58
ThO_2–Mo fiber	0.10	+3.6	35	0.33	4.83	2.94
	0.20	+3.6	76	0.71	3.05	1.33
Mullite–Mo fiber	0.20	−0.4	161	1.95	1.32	0.49
SiO_2–W fiber	0.20	−4.3	186	3.86	1.41	0.30
	0.30	−4.3	276	5.71	1.33	0.21
Mullite–W fiber	0.20	+0.6	157	1.89	1.84	0.52
Glass–W fiber	0.10	+3.5	68	1.02	2.53	1.12
	0.15	+3.5	83	1.24	2.70	0.98
	0.20	+3.5	101	1.50	2.73	0.85
	0.25	+3.5	57	0.86	5.77	1.60
Glass–W fiber	0.05	−0.9	66	1.15	1.67	0.92
	0.10	−0.9	131	2.30	1.24	0.50
	0.15	−0.9	157	2.75	1.38	0.44
	0.20	−0.9	142	2.50	1.89	0.51
Glass–Ni fiber	0.10	+3.5	47	0.66	1.77	1.55
	0.20	+3.5	64	0.91	1.47	1.19
	0.30	+3.5	73	1.02	1.45	1.07
	0.40	+3.5	92	1.44	1.27	0.88
Glass–Ni fiber	0.05	−6.3	41	0.61	1.80	1.68
	0.10	−6.3	47	0.70	1.70	1.49
	0.20	−6.3	66	0.98	1.39	1.11
	0.30	−6.3	73	1.08	1.41	1.03
	0.40	−6.3	92	1.37	1.25	0.84
	0.50	−6.3	112	1.67	1.13	0.71
Glass-ceramic–Ni fiber	0.02	−8.3	78	0.35	2.85	2.85
	0.08	−8.3	76	0.34	2.89	2.89
	0.12	−8.3	85	0.38	2.58	2.58
	0.08	+1.7	131	0.63	1.57	1.57

[a]Based on rule-of-mixtures calculations.

Variation in Work of Fracture With Fiber Volume Fraction for Short Random (sr) and Continuously Aligned (ca) Fiber Ceramics (Ref 107, p 963)

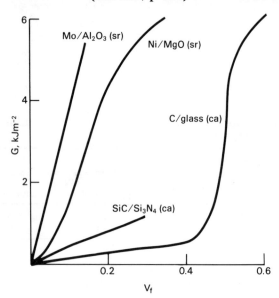

Normalized Flexural Strength of Randomly Oriented Short Brittle Fiber Ceramics (Ref 107, p 961)

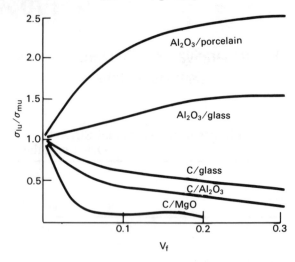

Physical Properties of Various Ceramic and Graphite Reinforced CVI SiC Composites[a]

Material system	Density g/cm³	lb/in.³	MOR[b] MPa	ksi	MOE[c] GPa	10⁶ psi	Strain, %
PAN 8HS/SWB	1.74	0.063	229	33.2	40.7	5.9	3.0
PAN knit/KFB	1.95	0.070	215	31.2	43.4	6.3	3.2
Nicalon/8HS ..	2.09	0.076	262[d]	38.0[d]	53.1	7.7	0.64
Saffil/Al₂O₃ paper	⋯	⋯	296	43.0	208.3	30.2	1.0
Saffil/Al₂O₃ paper	⋯	⋯	125	18.2	94.5	13.7	1.0
Saffil/Al₂O₃ paper	⋯	⋯	101	14.6	17.9	2.6	0.8

[a]Data from Refractory Composites, Inc. [b] Modulus of rupture. [c]Modulus of elasticity. [d]Ceramic-grade fiber; results for nonceramic-grade fiber were 108 MPa (15.6 ksi) high and 37 MPa (5.4 ksi) low.

Physical Properties of Fiber-Grain CVI SiC Matrix Composites[a]

Material system	Density g/cm³	lb/in.³	MOR[b] MPa	ksi	MOE[c] GPa	10⁶ psi	Strain, %
Carbon fibers/ SiC grain	2.06	0.074	40.0	5.8	20	2.9	0.57
Mullite fibers/ SiC grain	1.80	0.065	35.9	5.2	15.9	2.3	0.56
Silica fibers/ SiC grain	1.95	0.070	33.1	4.8	22.1	3.2	0.46

[a]Data from Refractory Composites, Inc. [b]Modulus of rupture. [c]Modulus of elasticity.

Theoretical Maximum Temperatures for Some Fiber-Reinforced Ceramics (Based on Melting or Softening Points)[a]

System	Maximum temperature, °C
Carbon–Pyrex glass	700–800
Carbon–glass ceramic	1300
Silicon carbide–glass	>650
FP alumina–glass	>650
Silicon carbide–silicon	1410
Silicon carbide–silicon nitride	1900
Carbon–carbon......................	3550

[a]Carbon oxidizes in air above 400 °C, and SiC oxidizes rapidly between 980 and 1150 °C but is stable between 1150 and 1400 °C.

Variation in Normalized Composite Strength With Fiber Volume Fraction for Randomly Oriented Ductile Fiber Ceramics (Ref 107, p 963)

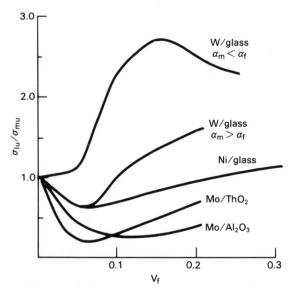

Flaw Size Particle or Fiber Spacings, and Strengths of Ceramic Composites (Ref 108, p 311)

Material	Flaw size[a], μm	Particle or fiber Material	Diameter, μm	Spacing, μm, for volume fraction of: 10%	20%	30%
Al_2O_3	65	Random carbon fibers	8	32	21	14
Glass	23	Al_2O_3 particles	60	45	23	12
			15	11	5.7	3.1
Glass	30	Aligned carbon fibers	8	14	8	5
Glass	30	Random carbon fibers	8	32	21	14
Glass	72	Al_2O_3 particles	3.5	2.6	1.4	0.7
			11	8	4.2	2.2
			44	33	17	9
MgO	82[b]	Random Ni fibers	89	360	240	160
	G[c]	Co, Fe, or Ni particles	~40	30	15	8
Si_3N_4	62	SiC	5	3.7	1.9	1.0
			9	6.7	3.4	1.8
			32	24	12	6.6

[a]Flaw size calculated from S, E, and γ of matrix. [b]Assuming uniform spacing of equal spherical or cylindrical particles, for random fibers, only one-third were assumed to be oriented to significantly affect crack propagation. [c]Grain size (G), 10 μm or less and $0.03 < P < 0.07$.

Physical Characteristics and Microsturcture Parameters of FP Alumina and SiC Fiber Reinforced Alumina Matrix Composites (Ref 58, p 576, 581)

Parameters[a]	FP alumina AKP50 system[b] Sintering temperature (°C) 1025	1145	1240	SiC AKP50 system[b] Sintering temperature (°C) 1025	1145	1240
Vf (%)	38.3	40.4	40.7	19.1	19.5	19.8
E_mTh. (GPa)	32	36.5	55.2	55.8	64.8	74.3
γ_mTh. (J/m²)	2.4	2.8	4.3	4.2	5.1	5.8
ϵ_mTh. (10^{-3})	0.58	0.58	0.58	0.58	0.58	0.58
σ_mTh. (MPa)	18.7	21.4	32.3	32.6	37.9	43.5
$\emptyset_f$ (μm)	20	20	20	12	12	12
E_f (GPa)	380	380	380	206	206	206
σ_f (MPa)	1900	1900	1900	2060	2060	2060
ρ_c	2.72	2.81	3.0	2.42	2.51	2.59
ρ_m	1.94	2.04	2.35	2.35	2.46	2.56
P_m (%)	50.9	48.4	40.6	40.4	37.6	35.1
σ_{cnl}Ex. (MPa)	335	186	187	195	120	80
$\epsilon_m E_{co}$Ex. (MPa)	92	109	130	42.5	50.6	54.7
σ_{cmax}Ex. (MPa)	423	215	137	217	243	365
E_{co}Th. (GPa)	165.3	175.3	187.4	84.5	92.3	100.4
E_{co}Ex. (GPa)	158.8	187.4	224	73.2	87.2	94.3
ϵ_{cnl}Ex. (10^{-3})	2.1	1.0	0.6	2.7	1.4	0.85

[a]Vf = volume fraction of fiber. ρ = density. P = porosity. E = Young's modulus. τ_0 = interfacial shear strength. γ = fracture surface energy. m = matrix. c = composite. f = fiber. ε = ultimate strength. σ = strength. $\emptyset_f$ = fiber diameter. Th. = theoretical. Ex. = experimental. E_{co}Th. = initial Young's modulus, theoretical. E_{co}Ex. = initial Young's modulus, experimental. [b]AKP50 = Sumitomo alumina.

The strain at which the matrix cracks in the composite is given by the following expression: $\epsilon_{mu}^* = \left[\dfrac{24\tau_0\gamma_m E_f V_f^2}{E_c E_m^2 \emptyset_f V_m} \right]^{1/3}$

The critical tensile stress is the corresponding stress at which the nonlinear behavior begins: $\sigma_{cnl}^* = \epsilon_{mu}^* E_c$

Comparison Between Strengths of Fiber/Matrix Interface and Matrix Strengths in the SiC/AKP50 Composite System[a] (Ref 58, p 584)

Sintering temperature (°C)	τ_0 (MPa)	σ_m Th. (MPa)	σ_{max} (MPa)
1025	40	32.6	217
1145	9	37.9	243
1240	3	43.5	365

[a]AKP50 = Sumitomo alumina. See table above for identification of τ_0, σ_mTh., and σ_{max}.

8.2.2. Silicon Carbide Fiber Reinforced and Silicon–Silicon Carbide Composites

Density and Porosity Data for SiC/RBSN Composites (Ref 67, p 9)

| Volume fraction of fibers, % | Before nitridation | | After nitridation | |
	Density, g/cm³	Matrix porosity[a], %	Density, g/cm³	Matrix porosity[a], %
0	1.56	35	1.98	37
23 ± 3	1.70	54[a]	2.19	39[a]
40 ± 2	1.90	51[a]	2.36	40[a]

[a]Matrix porosity calculated from composite density and from theoretical density for CVD SiC fiber (3.0 g/cm³) and from density for silicon (2.4 g/cm³) or for Si₃N₄ (3.2 g/cm³).

Room-Temperature Strengths of RBSN and SiC/RBSN Composites (Ref 67, p 10)

| Test | Axial strength, MPa | | |
	0% fiber	23 ± 3% fiber	40 ± 2% fiber
4-point bend (L/h[a] = 15)	107 ± 26[b]	539 ± 48[b]	616 ± 36[b]
4-point bend (L/h[a] = 45)	⋯	675 ± 42	868 ± 32
3-point bend (L/h[a] = 35)	⋯	717 ± 80	958 ± 45
Tensile[c]	⋯	352 ± 73	536 ± 20

[a]L/h refers to span-to-height ratio of test specimen (h ≈ 1.2 mm). [b]Tested at 50-mm gauge length. [c]Standard deviation for five tests.

Mean Matrix Crack Spacing and First Cracking Stress for SiC/RBSN Composites (Ref 67, p 10)

Fiber fraction, %	Matrix crack spacing, mm	Composite stress at which matrix first cracked, MPa
23 ± 3	2.0 ± 0.3[a]	237 ± 25[b]
40 ± 2	0.9 ± 0.2	293 + 15

[a]Standard deviation for 30 cracks on five bend specimens. [b]Standard deviation for 5 specimens measured in 3-point bend.

Load-Deflection Behavior in Three-Point Bending for 20 Vol % SiC Fiber/RBSN Composite (Ref 67, p 11)

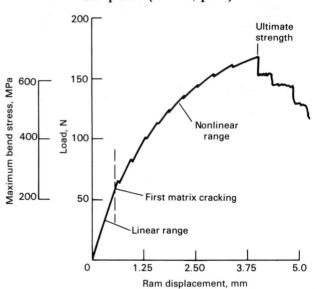

Tensile Stress-Strain Behavior for 20 Vol % SiC Fiber/RBSN Composite Showing Linear and Nonlinear Ranges (Ref 67, p 12)

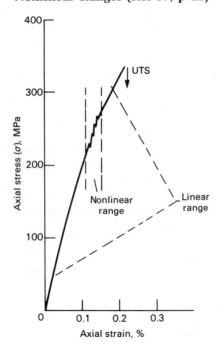

Flexural Strength, Determined by Three-Point Bending in Air, as a Function of Test Temperature for SiC-Monofilament-Reinforced Borosilicate 7740 Glass Composites (Ref 61, p 467)

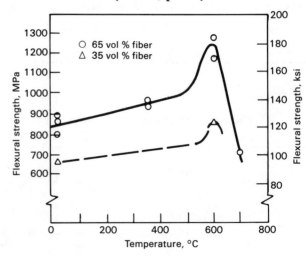

Three-Point Flexural Strength, σ, and Elastic Modulus, E, of SiC-Yarn-Reinforced Borosilicate 7740 Glass Composites (Ref 63, p 1203)

Test temperature, °C	0° specimens				0°/90° specimens			
	35 vol% SiC		50 vol% SiC		35 vol% SiC[a]		50 vol% SiC	
	σ, MPa	E, GPa	σ, MPa	E, GPa	σ, MPa	E, GPa	σ, MPa	E, GPa
22	414	97	814	116	347	53	619	98
	476	102	840	120	299	59	629	101
	520	99	763	118	339	71	...	...
	...	...	786	119	320	78	...	...
	...	...	...	...	301	72	...	...
600	702	136	940	101	272	33	685	81
	584	95	...	...	303	52	...	...
	569	104	...	...	...	...	...	...
700	698	80	545	76	192	21	200	45
	593	82	...	...	192	22	...	...
	663	88	...	...	145	16	...	...
750	556	60	...	...	...	...	...	...
	476	62	...	...	...	...	...	...
	561	69	...	...	...	...	...	...

[a]Span:depth ratio, 12:1 (all others tested at 20:1).

Properties of SiC-Fiber-Reinforced Borosilicate 7740 Glass (Ref 61, p 467)

Property	Monofilament		Yarn
Fiber content, vol %	35	65	40
Density, g/cm^3	2.6	2.9	2.4
Axial flexural strength, MPa:			
At 22 °C.............................650		830	290
At 350 °C	...	930	360
At 600 °C 825		1240	520
Axial elastic modulus, GPa, at 22 °C 185		290	120
Axial fracture toughness, MN·m$^{-3/2}$:			
At 22 °C............................. 18.8		...	11.5
At 600 °C 14.3		...	7.0

Coefficients of Thermal Expansion for Borosilicate 7740 Glass Matrix Composites (Ref 61, p 467)

Filament	Orientation	Coefficient of thermal expansion[a], 10^{-6}/°C
35 vol % SiC monofilament	0°	4.20
	90°	4.60
40 vol % SiC yarn	0°	3.25
	90°	2.70

[a]Average value between 22 and 500 °C.

Flexural Strength, Determined by Three-Point Bending in Air, as a Function of Test Temperature for 40 Vol % SiC-Yarn-Reinforced Borosilicate 7740 Glass Composites (Ref 61, p 465)

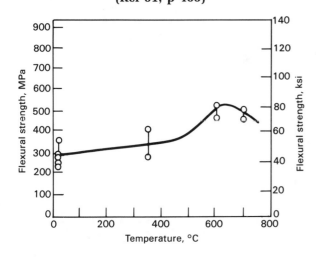

Three-Point Flexural Strength of SiC-Yarn-Reinforced Borosilicate Glass Composites as a Function of Test Temperature (Ref 63, p 1203)

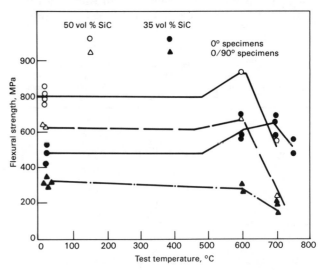

Residual Flexural Strength (at 22 °C) of SiC/Borosilicate 7740 Glass Composites After Exposure to Air at 540 °C (Ref 63, p 1204)

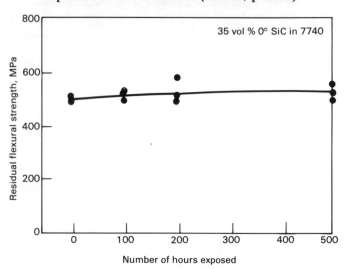

Three-Point-Bending Flexural Strength of SiC-Yarn-Reinforced High-Silica-Glass-Matrix Composites as a Function of Test Temperature (Ref 63, p 1206)

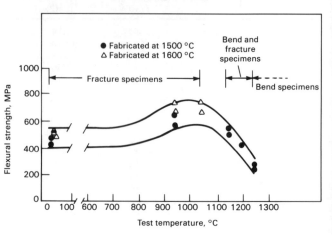

Three-Point Flexural Strength of SiC-Yarn-Reinforced LAS Glass-Ceramic Composites Tested in Argon (Ref 18)

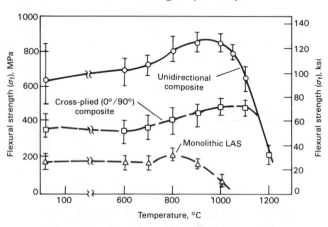

Three-Point Bending Strength of SiC-Yarn-Reinforced Blended-Glass-Matrix Composites (Ref 63, p 1204)

Test temperature, °C	Flexural strength, MPa, for 7930 Vycor/7740 borosilicate blend, wt %/wt %, of:			
	0/100	25/75	50/50	75/25
22	470	550	608	535
750	531[a]	560	670	433
850	···	160[a]	470	382

[a]Specimen bent extensively during test.

Three-Point Flexural Strengths and Elastic Moduli of SiC-Yarn-Reinforced 7930 Glass Composites (Ref 63, p 1205)

Test temperature, °C	Specimens fabricated at 1500 °C		Specimens fabricated at 1600 °C	
	Flexural strength, MPa	Flexural modulus, GPa	Flexural strength, MPa	Flexural modulus, GPa
22	467	111	509	109
	413	107	482	97.1
	447	112	527	101
950	643	108	724	101
	565	88	657	102
1050	···	···	742	83
	···	···	668	87
1150	541	64	···	···
	498	65	···	···
1200	419	60	···	···
1250	243	48	···	···
	280	52	···	···

Types and Properties of Si/Sic (Ref 6, p 246)

Designation	SiC content, vol %	Properties at 25 °C (77 °F)[a]					
		Bond strength		Elastic modulus		Density	
		MPa	ksi	GPa	10⁶ psi	g/cm³	lb/in.³

G.E.-reported data

Type TH[a]	80–85	483	70	393	57	3.30	0.012
Type THL	38–40	276	40	303	44	2.70	0.098
Type F[b]	20–25	172	25	200	29	2.60	0.094

Fansteel-reported data

Tensile strength at room temperature:
MPa 827
ksi 120
Charpy impact strength:
J 13.6
ft·lb 10
Oxidation resistance .... Excellent
Thermal shock resistance .... Excellent

[a]Measured on 0.1 × 0.1 × 0.625 in. specimens tested in three-point bending.
[b]Unidirectional orientation. [c]Omnidirectional orientation.

Effect of Fiber Orientation and Temperature on Compressive Strength, Yield Strength, and Acoustic-Emission Threshold Stress of Continuous SiC-Fiber-Reinforced Lithium Aluminosilicate (LAS) Glass-Ceramic Composites (Ref 58, p 591)

LAS-II contains 5 wt % Nb_2O_5, which enhances high-temperature oxidation resistance by forming an NbC oxidation barrier layer on the SiC fibers.

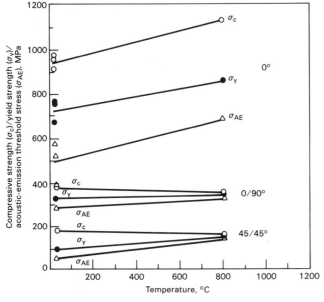

Effect of Loading Mode, Environment, and Temperature on Strength of Continuous SiC-Fiber-Reinforced Lithium Aluminosilicate (LAS) Glass-Ceramic Composites (Ref 58, p 593)

LAS-II contains 5 wt % Nb_2O_5, which enhances high-temperature oxidation resistance by forming an NbC oxidation barrier layer on the SiC fibers.

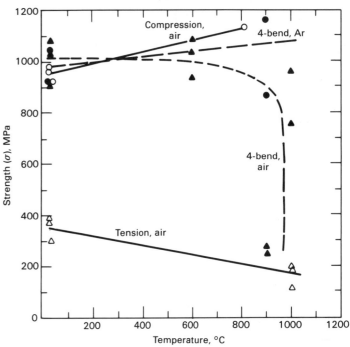

Effect of Fiber Orientation and Temperature on Compressive Strength of Continuous SiC-Fiber-Reinforced Lithium Aluminosilicate (LAS) Glass-Ceramic Composites (Ref 58, p 592)

LAS-II contains 5 wt % Nb_2O_5, which enhances high-temperature oxidation resistance by forming an NbC oxidation barrier layer on the SiC fibers.

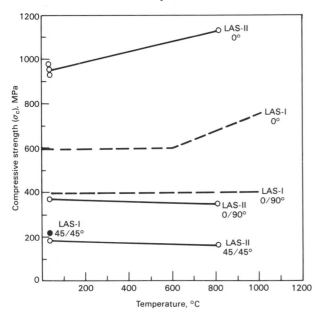

Elastic Modulus as a Function of Temperature for Various Grades of Silcomp Si/SiC

TH = unidirectional, 82 vol % silicon carbide. CD = woven, 70 vol % silicon carbide. F = omnidirectional, 25 vol % silicon carbide.

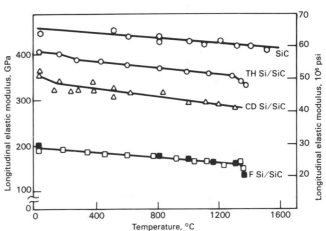

Strength as a Function of Temperature for Three Grades of Silcomp Si/SiC

TH = unidirectional, 82 vol % silicon carbide. CD = woven, 70 vol % silicon carbide. F = omnidirectional, 25 vol % silicon carbide.

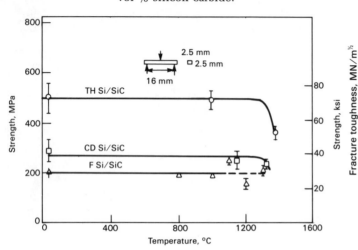

Fracture Toughness as a Function of SiC Volume Fraction for Three Si/SiC Structures and Machined Notches (Open Points) and a Sharp Crack (Ref 105, p 362)

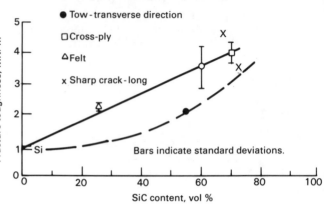

Strength as a Function of SiC Volume Fraction for Three Si/SiC Structures (Ref 105, p 362)

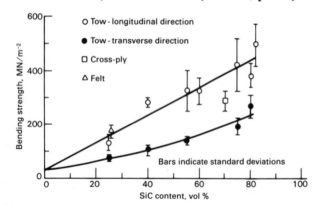

Bending Strength of Hot Pressed Si₃N₄ as a Function of SiC Whisker Content (Ref 75, p 24)

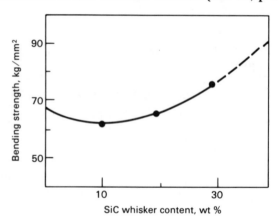

Shock Strength of SiC-Whisker-Reinforced Ceramics (Ref 75, p 24)

Matrix	Whisker content, vol %	Aid/wt %	Temperature °C	°F	Pressure MPa	ksi	Time, min	Density g/cm³	lb/in.³	Charpy shock strength, J (in.·lb) RT	1095 °C (2000 °F)	1315 °C (2400 °F)
Si₃N₄	5	MgO/1	1700	3092	28	4	65	3.15	0.114	0.150 (1.33)	0.160 (1.40)	0.15 (1.31)
	10	MgO/4	1700	3092	28	4	65	3.12	0.113	0.195 (1.73)	0.165 (1.46)	0.11 (0.99)
	38	MgO/1	1700	3092	28	4	75	3.00	0.108	0.094 (0.83)	0.080 (0.73)	0.21 (1.85)
	10	MgO/1	1700	3092	28	4	95	2.79	0.101	0.069 (0.61)	0.230 (2.03)	0.146 (1.29)
SiC	38	Al₂O₃/3; C/2	2140	3884	28	4	180	3.12	0.113	0.094 (0.83)	0.076 (0.67)	0.069 (0.61)
	50	Al₂O₃/3	2170	3938	28	4	180	3.11	0.112	0.072 (0.64)	0.079 (0.70)	0.069 (0.61)
	65	Al₂O₃/3	2140	3884	28	4	180	2.93	0.106	0.067 (0.59)	0.060 (0.54)	0.11 (0.96)

8.2.3. Carbon/Graphite Fiber Reinforced Composites

8.2.3.1. Various Carbon/Graphite Reinforced Composites

Modulus and Strength Fractions of Rule-of-Mixtures Values for Some Carbon-Fiber-Reinforced Ceramics[a] (Ref 13, p 218)

Matrix	Modulus fraction	Strength fraction
Pyrex glass	0.86	0.81
Glass-ceramic	0.71–0.80	0.81
Carbon	1.00	0.50

[a]Fibers are continuous and aligned, with V_f between 0.4 and 0.5.

Thermal Strain Parameters (at 20 °C) and Measured Works of Fracture for 20 Vol % Carbon-Fiber Composites (Ref 106, p 673)

Material	Work of fracture, $J \cdot m^{-2}$	$(\bar{\alpha}_m - \alpha_r)\Delta T$
MgO	10	...
CM	110	6.6×10^{-3}
Al_2O_3	38–66	...
CA	40	0.4×10^{-3}
Pyrex	4	...
CP	344	-2.3×10^{-3}
Glass-ceramic	4	...
CGC	100	-6.0×10^{-3}

Mechanical Properties of Graphite Fiber (MODMOR II) Reinforced Ceramic Composites[a]

Property[b]	RT	650 °C (1200 °F)
Flexural strength:		
MPa	153	194
ksi	22.2	28.2
Flexural modulus:		
GPa	86.2	80.7
10^6 psi	12.5	11.7
Compressive strength:		
MPa	205	247
ksi	29.8	35.8
Compressive modulus:		
GPa	106	...
10^6 psi	15.4	...
Tensile strength[c]:		
MPa	303	...
ksi	44.0	...
Tensile modulus:		
GPa	68.7	...
10^6 psi	9.96	...
Interlaminar shear strength:		
MPa	13.0	13.2
ksi	1.89	1.92

[a]Data from Acurex Corp. Ceramic matrix is Chemceram, a modified aluminum phosphate. [b]Flexural specimens failed in shear. Specific gravity, 1.77 g/cm³ (0.064 lb/in.³). Load applied once temperature achieved on specimen. [c]228 MPa (33.0 ksi) at 1260 °C (2300 °F); 210 MPa (30.4 ksi) at 1650 °C (3000 °F).

Typical Fabrication Process for Carbide-Fiber-Reinforced Glass-Matrix Composites

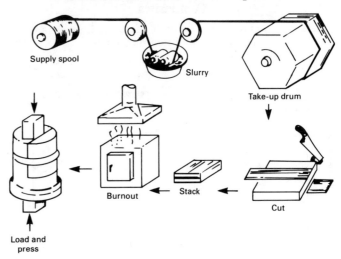

- Impregnate graphite fiber tows with a slurry of finely divided glass powder particles and a binder.
- Collimate the impregnated fibers to form a tape, and dry tape to form a prepreg.
- Cut and stack tape plies in a shaped die.
- Heat to remove binder.
- Heat die in an inert atmosphere and apply pressure to densify glass powder.
- Cool and remove fully dense part.
- Omit winding process for discontinuous fiber composites.

Specific Tensile Strength Vs Temperature for Graphite-Reinforced Ceramic Composites

Data from Acurex Corp. Ceramic matrix is Chemceram, a modified aluminum phosphate.

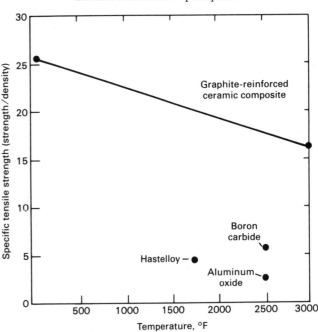

Flexural Properties Vs Temperature for Graphite (MODMOR II) Reinforced Ceramic Composites

Data from Acurex Corp. Ceramic matrix is Chemceram, a modified aluminum phosphate.

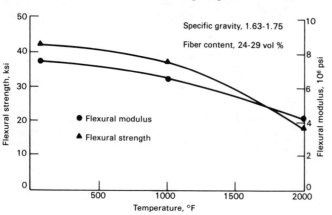

Physical Properties of Carbon-Fiber-Reinforced CVI SiC Matrix Composites (Refractory Composites, Inc.)

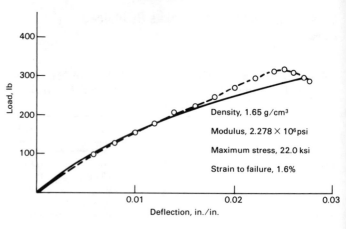

Physical Properties of Carbon-Fiber-Reinforced CVI SiC Matrix Composites at 1880 °C (Refractory Composites, Inc.)

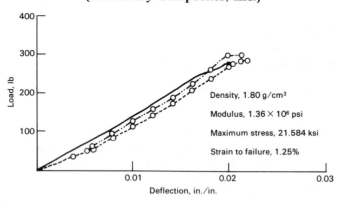

Physical Properties of Carbon-Fiber-Reinforced Si₃N₄ Matrix Composites (Refractory Composites, Inc.)

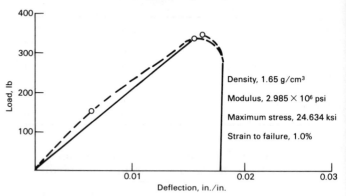

8.2.3.2. Carbon/Graphite Fiber Reinforced Borosilicate Glass

Fracture Toughness of HM Fiber Reinforced Borosilicate Glass Matrix Composites (Ref 17, p 149)

Test temperature, °C	Test speed, cm·s⁻¹	Fracture toughness (K_c), MPa·m$^{1/2}$
22	330.0	21.4
22	0.002	22.4
600	330.0	15.8
650	330.0	19.0

Coefficients of Thermal Expansion at 295 to 423 K for Some Graphite-Reinforced Borosilicate Glass Composites

Composite type	Coefficient of thermal expansion, 10^{-6} K^{-1}	
	Longitudinal	Transverse
60 vol % GY-70/7740	−0.29	7.6
60 vol % HMS/7740	−1.0	3.6
60 vol % T-300/7740.......	0.38	4.3

Tensile Stress-Strain Curve for Unidirectional 54 Vol % P-100 Graphite-Reinforced Borosilicate Glass Matrix Composite (Ref 62)

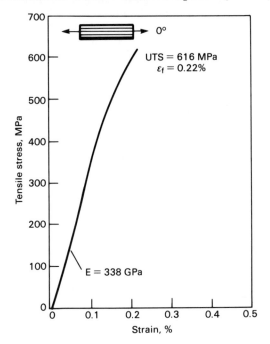

UTS = 616 MPa
ϵ_f = 0.22%

E = 338 GPa

Properties of Graphite-Reinforced Borosilicate Composites[a] (Ref 62)

Unidirectional		0/90° cross-ply

Thornel 300 Graphite-Reinforced Borosilicate (With Scrim)

E_{11}^T = 168 GPa	E_{22}^T = 16.8 GPa	E_{11}^T = 72.1 GPa
σ_{11}^T = 580 MPa	σ_{22}^T = 29.0 MPa	σ_{11}^T = 246 MPa
$\epsilon_{f\,11}^T$ = 0.36%	$\epsilon_{f\,22}^T$ = 0.45%	$\epsilon_{f\,11}^T$ = 0.32%
υ_{12}^T = 0.21	υ_{21}^T = 0.012	υ_{12}^T = 0.04
E_{11}^F = 124 GPa	E_{22}^F = 25.4 GPa	Vol % fiber = 60
σ_{11}^F = 905 MPa	σ_{22}^F = 94 MPa	
Vol % fiber = 54	ρ = 2.0 g/cm³	

HM Graphite-Reinforced Borosilicate (With Scrim)

E_{11}^T = 215 GPa	E_{22}^T = 10.2 GPa	E_{11}^T = 92.0 GPa
σ_{11}^T = 352 MPa	σ_{22}^T = 12.2 MPa	σ_{11}^T = 205 MPa
ϵ_{f11}^T = 0.17%	ϵ_{f22}^T = 0.26%	ϵ_{f11}^T = 0.23
υ_{12}^T = 0.22	υ_{21}^T = 0.022	υ_{12}^T = 0.023
E_{11}^F = 180 GPa	ρ = 2.0 g/cm³	Vol % fiber = 50
σ_{11}^F = 527 MPa		
Vol % fiber = 67		

[a]Superscript T denotes property measured in tension. Superscript F denotes property measured in three-point flexure.

Coefficient of Thermal Expansion at 22 °C for Graphite Fiber Unidirectionally Reinforced Borosilicate Matrix Composites (Ref 62)

Fiber type	Fiber elastic modulus, GPa	Fiber content, vol %	Coefficient of thermal expansion, 10⁻⁶ °C⁻¹ 0°	Coefficient of thermal expansion, 10⁻⁶ °C⁻¹ 90°
Thornel 300	234	54	−0.10	+4.6
HM	350	70	−0.50	+6.5
P-100	654	50	−1.0	+4.4
Chopped Cel 6000	234	30	1.7	1.7(4.2)
Borosilicate glass	...	0	3.25	3.25

Tensile Stress-Strain Curves for Discontinuous Graphite-Fiber-Reinforced Borosilicate Glass and Epoxy Matrix Composites (Ref 62)

The fibers are in a 2-D random array. The epoxy matrix composite contains 20 vol % fibers, and the borosilicate glass matrix composite contains 30 vol % fibers.

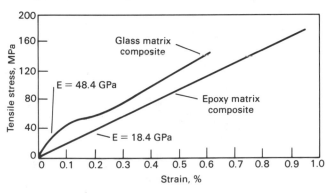

Glass matrix composite

E = 48.4 GPa

Epoxy matrix composite

E = 18.4 GPa

Flexural Strengths of HM Graphite Unidirectional Fiber Reinforced Composites Tested at Temperature in Inert Argon (Ref 62)

○ Borosilicate glass matrix composite
□ Polyether sulfone resin matrix composite
△ Polysulfone resin matrix composite

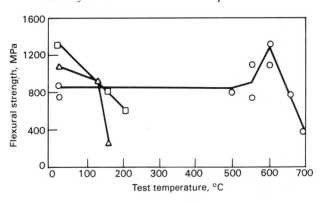

Flexural Strengths of Thornel-300 Graphite Unidirectional Fiber Reinforced Borosilicate Glass Composites Tested at Temperature in Air With a 30-Min Hold at Temperature Prior to Test (Ref 62)

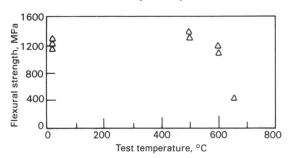

Flexural Elastic Moduli of Thornel-300 Graphite Unidirectional Fiber Reinforced Borosilicate Glass Composites Tested at Temperature in Air With a 30-Min Hold at Temperature Prior to Test (Ref 62)

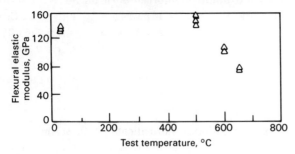

Three-Point Longitudinal Flexural Fatigue of HMS/7740 With R = 0.10 (Ref 6, p 87)

HMS/7740 is borosilicate glass reinforced with 60 vol % graphite fiber.

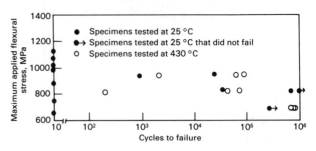

Three-Point Longitudinal Flexural Strength of HMS/7740 as a Function of Test Temperature

HMS/7740 is borosilicate glass reinforced with 60 vol % graphite fiber. Note: Specimen tested at 700 °C deformed extensively and did not fracture.

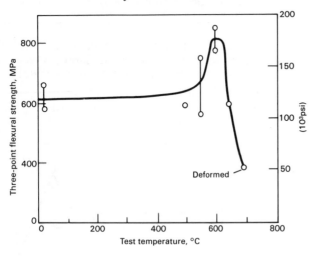

Three-Point Longitudinal Flexural Strength of 7740 Glass Matrix Composites as a Function of Time of Exposure to Air at Temperature (Ref 6, p 88)

Borosilicate glass matrix reinforced with 60 vol % graphite fiber.

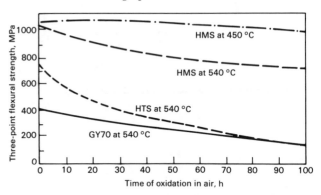

Change in Three-Point Longitudinal Flexural Strength of HMS/7740 as a Function of Thermal Cycles in Air (Ref 6, p 88)

HMS/7740 is borosilicate glass reinforced with 60 vol % graphite fiber.

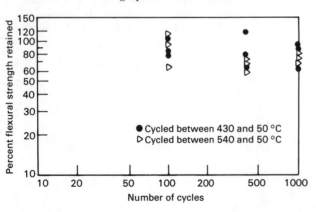

Transverse (90°) Thermal Strain Vs Temperature for a Unidirectional 70 Vol % HM Graphite Fiber Reinforced Borosilicate Glass Matrix Composite (Ref 62)

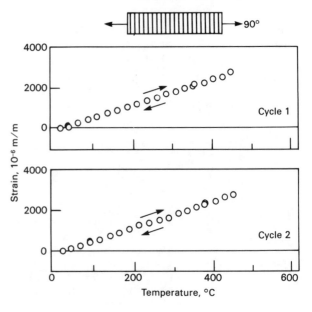

Axial (0°) Thermal Strain Vs Temperature for a Unidirectional 70 Vol % HM Graphite Fiber Reinforced Borosilicate Glass Matrix Composite (Ref 62)

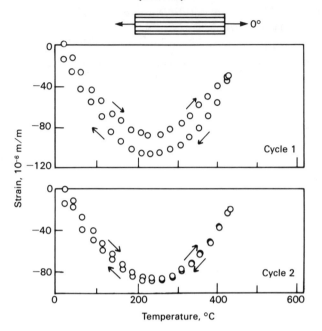

Axial (0°) Thermal Strain Vs Temperature for a Unidirectional 54 Vol % Thornel-300 Graphite Fiber Reinforced Borosilicate Glass Matrix Composite (Ref 62)

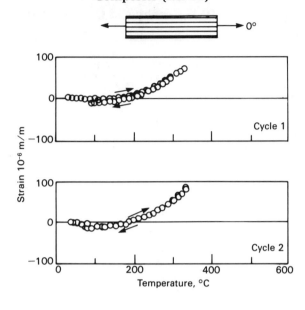

Longitudinal Thermal Expansion of 7740 Borosilicate Glass Matrix Composites Containing Approximately 60 Vol % Fiber for Three Different Graphite Fiber Types (Ref 6, p 91)

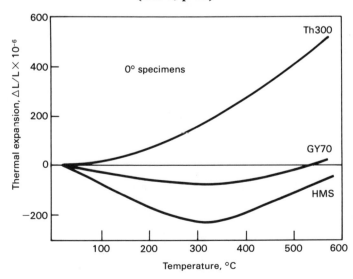

Thermal Expansion of 0°/90° HMS Graphite Reinforced 7740 Borosilicate Glass Composites (Ref 124)

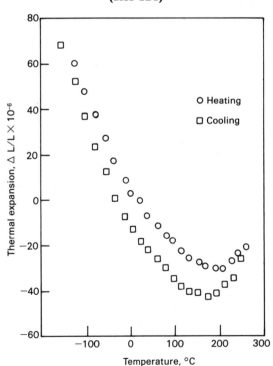

8.2.3.3. Carbon/Graphite Fiber Reinforced Lithium Aluminosilicate Glass

Properties of HMS Graphite/$Li_2O \cdot Al_2O_3 \cdot 8SiO_2$ Composites[a] Tested Perpendicular and Parallel to Hot Pressing Direction (Ref 104, p 799)

Test direction	Modulus of rupture MPa	ksi	Average modulus of rupture MPa	ksi	Standard deviation MPa	ksi	Coefficient of variation	Modulus of elasticity GPa	10^6 psi	Average modulus of elasticity GPa	10^6 psi	Standard deviation GPa	10^6 psi	Coefficient of variation
Perpendicular to hot pressing direction	804.6	116.7	793.2	115.0	57	8.2	7.1	146	21.2	144	20.9	11.7	1.7	8.1
	756.4	109.7						149	21.6					
	777.7	112.8						143	20.7					
	766.7	111.2						123	17.9					
	721.9	104.7						134	19.5					
	892.9	129.5						153	22.2					
	832.2	120.7						158	22.9					
Parallel to hot pressing direction	837.7	121.5	860.7	124.8	41	5.9	4.7	159	23.0	159	23.0	4.8	0.7	3.1
	897.7	130.2						155	22.5					
	815.0	118.2						152	22.0					
	877.0	127.2						160	23.2					
	912.2	132.3						166	24.1					
	824.6	119.6						159	23.0					

[a]Fiber content, 39.6 vol %. Bulk density, 2.06 g/cm³.

Modulus of Rupture Vs Thermal Shock Cycles Between Room Temperature and 1200 °C (2190 °F) for Graphite Fiber/$Li_2O \cdot Al_2O_3 \cdot 8SiO_2$ Composites (Ref 104, p 800)

Thermal shock cycles	Modulus of rupture MPa	ksi	Average modulus of rupture MPa	ksi
0	951.5	138.0	857.7	124.4
	839.8	121.8		
	781.9	113.4		
1	897.7	130.2	893.6	129.6
	889.4	129.0		
5	838.4	121.6	861.8	125.0
	885.3	128.4		

Short-Beam Shear Strength of HMS Graphite Fiber/$Li_2O \cdot Al_2O_3 \cdot 8SiO_2$ Composites[a] (Ref 104, p 800)

Short-beam shear stress MPa	ksi	Span/depth ratio	Failure mode
50.3	7.3	4/1	Shear, tension, compression
46.9	6.8	5/1	Shear, tension, compression
42.7	6.2	6/1	Shear
40.0	5.8	6/1	Shear
40.0	5.8	6/1	Shear
44.5	6.45	6/1	Shear
37.2	5.4	6/1	Shear

[a]Hot pressed for 5 min at 1375 °C (2510 °F) and 6.9 MPa (1.0 ksi).

Izod Impact Energy of HMS Graphite Fiber/$Li_2O \cdot Al_2O_3 \cdot 8SiO_2$ Composites, Udimet 700, MAR-M200, and Alumina (Ref 104, p 801)

Composition	Test bar No.	Density, g/cm³	Fiber content, vol %	Notched	Izod impact energy J/cm²	ft·lb/in.²
HMS graphite fiber/ $Li_2O \cdot Al_2O_3 \cdot 8SiO_2$ composites	2	2.19	31.6	No	15.9	75.5
	1	2.19	31.6	Yes	5.94	28.3
	3	2.18	34.2	Yes	4.20	20.0
	3	2.18	34.4	Yes	4.68	22.3
Udimet 700	2	⋯	⋯	Yes	34.31	163.4
MAR-M200	3	⋯	⋯	Yes	18.9	90.0
Al_2O_3 AP 35	3	⋯	⋯	Yes	0.315	1.50
	2	⋯	⋯	No	0.701	3.34

Physical Properties of HMS Graphite/$Li_2O \cdot Al_2O_3 \cdot nSiO_2$ Composites Where n = 3, 4, and 8 (Ref 104, p 804)

Composition	Modulus of rupture MPa	ksi	Average modulus of rupture MPa	ksi	Standard deviation MPa	ksi	Coefficient of variation	Modulus of elasticity GPa	10^6 psi	Average modulus of elasticity GPa	10^6 psi	Standard deviation GPa	10^6 psi	Coefficient of variation
$Li_2O \cdot Al_2O_3 \cdot 3SiO_2$[a]	452	65.5	410	59.4	45	6.5	10.9	152	22.0	148	21.5	6.9	1.0	4.5
	433	62.8						140	20.3					
	459	66.5						148	21.5					
	371	53.8						141	20.5					
	334	48.4						160	23.2					
	410	59.4						150	21.7					
	408	59.2						146	21.2					
$Li_2O \cdot Al_2O_3 \cdot 4SiO_2$[b]	403	58.5	506	73.3	57	8.3	11.3	142	20.6	149	21.6	6.2	0.9	4.3
	547	79.4						154	22.3					
	462	67.0						150	21.7					
	536	77.8						141	20.5					
	571	82.8						158	22.9					
	498	72.2						145	21.0					
	522	75.7						154	22.3					
$Li_2O \cdot Al_2O_3 \cdot 8SiO_2$[c]	906.7	131.5	866.0	125.6	46	6.7	5.4	150	21.7	147	21.3	4.1	0.6	2.9
	820.5	119.0						142	20.6					
	934.2	135.5						148	21.4					
	892.2	129.4						152	22.0					
	830.8	120.5						143	20.7					
	815.6	118.3						142	20.6					
	861.8	125.0						150	21.8					

[a]Hot pressing temperature, 1400 °C (2500 °F); fiber content, 37.5 vol %; density, 2.11 g/cm³. [b]Hot pressing temperature, 1425 °C (2600 °F); fiber content, 41.0 vol %; density, 2.15 g/cm³. [c]Hot pressing temperature, 1375 °C (2510 °F); fiber content, 32.3 vol %; density, 2.18 g/cm³.

Properties of HMS Graphite Fiber/$Li_2O \cdot Al_2O_3 \cdot 8SiO_2$ Composites[a] (Ref 104, p 797)

Fiber content, vol %	Bulk density, g/cm³	Modulus of rupture MPa	ksi	Average modulus of rupture MPa	ksi	Standard deviation MPa	ksi	Coefficient of variation
29.0	2.19	637	92.4	633	91.8	37	5.4	5.9
		641	93.0					
		689	99.9					
		601	87.1					
		597	86.6					
33.1	2.16	698.4	101.3	782.1	113.4	48	7.0	6.2
		799.1	115.9					
		821.2	119.1					
		806.7	117.0					
		785.3	113.9					
35.1	2.17	845.3	122.6	834.4	121.0	58	8.4	7.0
		878.4	127.4					
		830.8	120.5					
		737.7	107.0					
		879.8	127.6					
36.3	2.15	804.6	116.7	878.4	127.4	51	7.4	5.8
		839.1	121.7					
		896.3	130.0					
		951.5	138.0					
		892.9	129.5					
		886.0	128.5					

[a]Hot pressed for 5 min at 1375 °C (2510 °F) and 6.9 MPa (1.0 ksi).

Modulus of Rupture Vs Fiber Vol % for HMS Graphite/Li₂O · Al₂O₃ · 8SiO₂ Composites (Ref 104, p 797)

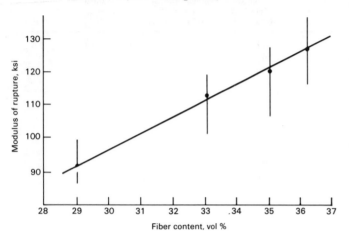

Cyclic Flexural Test Results for HMS Graphite/Li₂O · Al₂O₃ · 8SiO₂ Composites[a] (Ref 104, p 801)

Cycle No.	Average stress MPa	ksi	Average modulus of elasticity GPa	10⁶ psi
1 606		87.9	144	20.9
2–10 606		87.9	136	19.7
11 ....873.6		126.7	136	19.7

[a]Hot pressed for 5 min at 1375 °C (2510 °F) and 6.9 MPa (1.0 ksi). Fiber content, 32.2 vol %. Bulk density, 2.17 g/cm³.

8.2.4. Other Fiber Reinforced Composites

Thermal and Physical Properties of Glass-Reinforced Ceramic Composites[a]

Specific gravity .... 1.8 to 1.9
Specific heat:
 kJ/kg · K 0.84
 Btu/lb · °F 0.2
Coefficient of thermal expansion:
 Parallel to reinforcement
  10⁻⁶ m/m · K 3.6
  10⁻⁶ in./in. · °F 2.0
 Perpendicular to reinforcement
  10⁻⁶ m/m · K 1.1
  10⁻⁶ in./in. · °F 0.6
Thermal conductivity:
 W/m · K 0.58
 Btu · in./ft² · h · °F 4.0
Total normal emissivity:
 At 425 °C (800 °F) 0.73
 At 650 °C (1200 °F) 0.76

[a]S-994, style 181 fiberglass fabric in matrix of Chemceram, a modified aluminum phosphate. Data from Acurex Corp.

Electrical Properties Vs Temperature for Silica-Fiber-Reinforced Ceramic Composite[a]

Temperature °C	°F	Dielectric constant	Loss tangent
25	77	2.70	0.0050
106	223	2.70	0.0050
225	437	2.71	0.0053
410	770	2.71	0.0080
495	923	2.71	0.0105
580	1076	2.72	0.0140
673	1243	2.72	0.0177
760	1400	2.72	0.0228
827	1520	2.73	0.0265
916	1680	2.74	0.0315
967	1772	2.75	0.0380
25	77	2.71	0.0047

[a]Chopped Astroquartz fiber reinforcement in matrix of Chemceram, a modified aluminum phosphate. Frequency, 8.52 GHz. Composite density, 1.54 g/cm³. Data from Acurex Corp. (test performed by MIT).

Mechanical Properties of Glass-Reinforced Ceramic Composite[a] at Various Temperatures

Test temperature °C	°F	Flexural strength MPa	ksi	Tensile strength MPa	ksi	Compressive strength MPa	ksi	Flexural modulus GPa	10⁶ psi
RT	RT	149.96	21.750	246.66	35.775	80.09	11.616	21.7	3.15
290	550	89.63	13.000	194.16	28.160	35.65	5.170	16.1	2.34
425	800	81.01	11.750	223.80	32.460	57.92	8.400	15.2	2.21
540	1000	155.82	22.600	77.91	11.300	106.18	15.400	28.6	4.15
595	1100	187.54	27.200	21.07	3.056	80.88	11.730	31.2	4.52
650	1200	114.11	16.550	13.14	1.906	88.25	12.800	29.8	4.32

[a]S-994, style 143 fiberglass fabric in matrix of Chemceram, a modified aluminum phosphate. Data from Acurex Corp.

Results of Oxyacetylene Torch Test of Glass-Reinforced Ceramic Composite[a]

Surface temperature[b] °C	°F	Initial thickness mm	in.	Ablation mm	in.	Ablation index μm/s	mil/s	Time to 205 °C (400 °F) backside, min:s	Time to 425 °C (800 °F) backside, min:s	Insulation index 205 °C (400 °F) s/mm	s/in.	425 °C (800 °F) s/mm	s/in.
1260–1370	2300–2500	9.55	0.376	0	0	0	0	1:19	2:55	8.26	210	18.3	466
1260–1370	2300–2500	9.60	0.378	0	0	0	0	1:30	2:40	9.37	238	16.7	424
1425–1540	2600–2800	9.60	0.378	2.79	0.110	17	0.67	1:15	2:45	7.80	198	17.2	437
1425–1540	2600–2800	9.55	0.376	2.31	0.091	16	0.63	1:00	2:24	6.30	160	15.1	383
>1705	>3100	9.60	0.378	(c)	(c)	110	4.35	0:55	1:20	5.71	145	8.35[d]	212[d]

[a]S-994, style 181 fiberglass fabric in Chemceram, a modified aluminum phosphate. Data from Acurex Corp. [b]Optical pyrometer measurement. [c]Burn-through (1:27). [d] Data inaccurate due to rapid erosion.

Izod Impact Strength After Elevated-Temperature Exposure for Glass-Reinforced Ceramic Composite[a]

Temperature (2-h exposure) °C	°F	Izod impact strength J/cm	ft·lb/in.
315	600	17.1	32.0
425	800	9.0	16.9

[a]S-994, style 181 fiberglass fabric in matrix of Chemceram, a modified aluminum phosphate. Data from Acurex Corp.

Elastic Moduli of Some Mo-Reinforced Composites[a] (Ref 74, p 228)

Material	Elastic modulus GPa	10⁶ psi
$6ZrO_2 \cdot CeO_2$	56.5	8.2
Mo-reinforced composite	135.1	19.6
$ZrSiO_4$	96.5	14.0
Mo-reinforced composite	142.0	20.6
$3Al_2O_3 \cdot 2SiO_2$	129.6	18.8
Mo-reinforced composite	129.6	18.8
ThO_2	251.0	36.4
Mo-reinforced composite	185.5	26.9
MgO	239.2	34.7
Mo-reinforced composite	112.4	16.3

[a]All composites contain 10 vol % randomly dispersed Mo fibers.

Flexural Properties of Sapphire-Filament-Reinforced Ceramic Composites

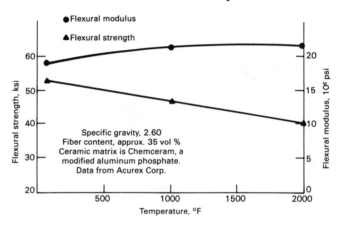

Specific gravity, 2.60
Fiber content, approx. 35 vol %
Ceramic matrix is Chemceram, a modified aluminum phosphate.
Data from Acurex Corp.

Room-Temperature Flexural Properties After Thermal Aging for Sapphire-Filament-Reinforced Ceramic Composite[a]

Specimen conditioning	Flexural strength MPa	ksi	Flexural modulus GPa	10⁶ psi
None	360	52.2	129	18.7
24 h at 1095 °C (2000 °F)	406	58.9	188.5	27.34

[a]Specific gravity, 2.60. Fiber content, 35 vol %. Ceramic matrix is Chemceram, a modified aluminum phosphate. Data from Acurex Corp.

Flexural Properties Vs Temperature Before and After Thermal Aging for Silica-Fiber-Reinforced Ceramic Composites[a]

Test temperature °C	°F	Flexural strength Unaged MPa	ksi	Aged 5 h at 870 °C (1600 °F) MPa	ksi	Elastic modulus Unaged GPa	10⁶ psi	Aged 5 h at 870 °C (1600 °F) GPa	10⁶ psi
RT	RT	144.8	21.00	22.4	3.25	27.2	3.94	17.8	2.58
595	1100	118.3	17.16	34.6	5.02	21.0	3.05	23.6	3.42
870	1600	62.9	9.13	46.3	6.72	18.8	2.73	23.4	3.39

[a]Style 181 Astroquartz fabric in matrix of Chemceram, a modified aluminum phosphate. Data from Acurex Corp.

Strength Data for an Alumina-Mullite Ceramic Reinforced With Mo Fiber (Ref 74, p 227)

Fiber dimensions		Fiber content, vol %		Modulus of rupture prior to thermal shock		Modulus of rupture following four thermal cycles[a]	
mm	in.			MPa	ksi	MPa	ksi
...	...	0	....	149	21.6	0	0
0.05 × 3.175	0.002 × 1/8	4.2	....	165	24.0	9.7	1.4
		10	....	157	22.8	137	19.8
		20	....	163	23.7	143	20.8
		30	....	179	25.9	212	30.7
0.05 × 12.7	0.002 × 1/2	10	....	101	14.6	81.4	11.8
		20	....	182	26.4	219	31.7
0.25 × 3.175	0.010 × 1/8	10	....	141	20.4	78.6	11.4
		20	....	108	15.6	86.9	12.6
0.25 × 12.7	0.010 × 1/2	10	....	148	21.4	60.7	8.8
		20	....	131	19.0	105	15.3

[a]Air quenched from 1205 °C (2200 °F).

8.3. Non-Fiber Reinforced Ceramic Composites, Cermets, and Other Metal-Ceramic Composites

8.3.1. Particulate and Other Non-Fiber Ceramic Composite Property Data

8.3.1.1. Al₂O₃ Composites

Properties of Simultaneously Pressed and Sintered Si₃N₄/Hardened Al₂O₃ Composite[a]

Mechanical Properties

Transverse rupture strength[b], MPa (ksi) 350 (50)
Fracture toughness, MPa·m$^{1/2}$ (ksi·in.$^{1/2}$) 3.1 (2.8)
Elastic constants:
　Elastic modulus, GPa (10⁶ psi)
　　At 25 °C (77 °F) 355 (51.5)
　　At 500 °C (930 °F) 336 (48.8)
　　At 1000 °C (1830 °F) 324 (47.0)
　Shear modulus, GPa (10⁶ psi)
　　At 25 °C (77 °F) 145 (21.1)
　　At 500 °C (930 °F) 141 (20.5)
　　At 1000 °C (1830 °F) 134 (19.4)
　Poisson's ratio
　　At 25 °C (77 °F) 0.22
　　At 500 °C (930 °F) 0.19
　　At 1000 °C (1830 °F) 0.21
Microhardness, Vickers DPH, GPa (kg/mm²)
　100-g load 26.0 (2650)
　500-g load 21.6 (2200)

Thermal Properties

Linear thermal expansion, 10⁻⁶/°C
　From 25 to 300 °C (77 to 570 °F) 5.5
　From 25 to 500 °C (77 to 930 °F) 5.9
　From 25 to 800 °C (77 to 1470 °F) 6.5
Thermal conductivity, W/cm·K (cal/cm·s·°C)
　At 25 °C (77 °F) 0.19 (0.046)
　At 500 °C (930 °F) 0.10 (0.024)
　At 1000 °C (1830 °F) 0.071 (0.017)
Thermal diffusivity, cm²/s
　At 25 °C (77 °F) 0.067
　At 500 °C (930 °F) 0.024
　At 1000 °C (1830 °F) 0.015
Specific heat (C_P), J/g·K (cal/g·°C)
　At 25 °C (77 °F) 0.76 (0.18)
　At 500 °C (930 °F) 1.14 (0.27)
　At 1000 °C (1830 °F) 1.26 (0.30)

Electrical Property

Volume resistivity, dc, Ω·cm
　At 25 °C (77 °F) ....>10¹⁶
　At 100 °C (212 °F) ....1.4 × 10¹³
　At 300 °C (570 °F) ....1.6 × 10⁹

Dielectric Properties

Dielectric constant
　At 1 kHz 9.9
　At 1 MHz 9.8
Dissipation factor
　At 1 kHz ....8.4 × 10⁻⁴
　At 1 MHz ....8.6 × 10⁻⁴
Loss factor
　At 1 kHz ....8.3 × 10⁻³
　At 1 MHz ....8.5 × 10⁻³

[a]Density, 3.78 g/cm³. Porosity, vacuum tight. Water absorption, none. Grain size, <1 μm. Color, gray, Data from Greenleaf Corp. [b]In four-point bending.

Properties of Simultaneously Pressed and Sintered $Al_2O_3/40B_4C$ Composite[a]

Mechanical Properties

Transverse rupture strength,[b] MPa (ksi) 620 (90)
Fracture toughness, MPa·m$^{1/2}$ (ksi·in.$^{1/2}$) 5.1 (4.6)
Elastic constants:
  Elastic modulus, GPa (10^6 psi)
    At 25 °C (77 °F) 353 (51.2)
    At 500 °C (930 °F) 341 (49.5)
    At 1000 °C (1830 °F) 292 (42.3)
  Shear modulus, GPa (10^6 psi)
    At 25 °C (77 °F) 147 (21.3)
    At 500 °C (930 °F) 142 (20.6)
    At 1000 °C (1830 °F) 122 (17.7)
  Poisson's ratio
    At 25 °C (77 °F) 0.20
    At 500 °C (930 °F) 0.20
    At 1000 °C (1830 °F) 0.19
Microhardness, Vickers DPH, GPa (kg/mm^2)
  100-g load 23.0 (2350)
  500-g load 19.6 (2000)

Thermal Properties

Linear thermal expansion, 10^{-6}/°C
  From 25 to 300 °C (77 to 570 °F) 5.7
  From 25 to 500 °C (77 to 930 °F) 6.3
  From 25 to 800 °C (77 to 1470 °F) 7.0
Thermal conductivity, W/cm·K (cal/cm·s·°C)
  At 25 °C (77 °F) 0.20 (0.049)
  At 500 °C (930 °F) 0.11 (0.025)
  At 1000 °C (1830 °F) 0.077 (0.018)
Thermal diffusivity, cm^2/s
  At 25 °C (77 °F) 0.074
  At 500 °C (930 °F) 0.024
  At 1000 °C (1830 °F) 0.015
Specific heat (C$_P$), J/g·K (cal/g·°C)
  At 25 °C (77 °F) 0.82 (0.20)
  At 500 °C (930 °F) 1.30 (0.31)
  At 1000 °C (1830 °F) 1.50 (0.36)

Electrical Property

Volume resistivity, dc, Ω·cm
  At 25 °C (77 °F) 90
  At 100 °C (212 °F) 30
  At 300 °C (570 °F) 5

[a]Composite is electrically conductive. Density, 3.39 g/cm^3. Porosity, vacuum tight. Water absorption, none. Grain size, 1–3 μm. Color, black. Data from Greenleaf Corp. [b]In four-point bending.

Properties of Simultaneously Pressed and Sintered $Al_2O_3/50B_4C$ Composite[a]

Mechanical Properties

Transverse rupture strength,[b] MPa (ksi) 620 (90)
Fracture toughness, MPa·m$^{1/2}$ (ksi·in.$^{1/2}$) 4.5 (4.1)
Elastic constants:
  Elastic modulus, GPa (10^6 psi)
    At 25 °C (77 °F) 377 (54.7)
    At 500 °C (930 °F) 363 (52.7)
    At 1000 °C (1830 °F) 326 (47.3)
  Shear modulus, GPa (10^6 psi)
    At 25 °C (77 °F) 157 (22.8)
    At 500 °C (930 °F) 150 (21.7)
    At 1000 °C (1830 °F) 132 (19.2)
  Poisson's ratio
    At 25 °C (77 °F) 0.20
    At 500 °C (930 °F) 0.21
    At 1000 °C (1830 °F) 0.23
Microhardness, Vickers DPH, GPa (kg/mm^2)
  100-g load 23.5 (2400)
  500-g load 20.1 (2050)

Thermal Properties

Linear thermal expansion, 10^{-6}/°C
  From 25 to 300 °C (77 to 570 °F) 5.3
  From 25 to 500 °C (77 to 930 °F) 6.0
  From 25 to 800 °C (77 to 1470 °F) 6.8
Thermal conductivity, W/cm·K (cal/cm·s·°C)
  At 25 °C (77 °F) 0.23 (0.054)
  At 500 °C (930 °F) 0.12 (0.029)
  At 1000 °C (1830 °F) 0.085 (0.020)
Thermal diffusivity, cm^2/s
  At 25 °C (77 °F) 0.083
  At 500 °C (930 °F) 0.028
  At 1000 °C (1830 °F) 0.017
Specific heat (C$_P$), J/g·K (cal/g·°C)
  At 25 °C (77 °F) 0.84 (0.20)
  At 500 °C (930 °F) 1.32 (0.32)
  At 1000 °C (1830 °F) 1.57 (0.38)

Electrical Property

Volume resistivity, dc, Ω·cm
  At 25 °C (77 °F) 90
  At 100 °C (212 °F) 25
  At 300 °C (570 °F) 4

[a]Composite is electrically conductive. Density, 3.28 g/cm^3. Porosity, vacuum tight. Water absorption, none. Grain size, 2–3 μm. Color, black. Data from Greenleaf Corp. [b]In four-point bending.

Temperature-Dependent Properties of Pressed and Sintered $Al_2O_3/40B_4C$ Composite
(Data From Greenleaf Corp.)

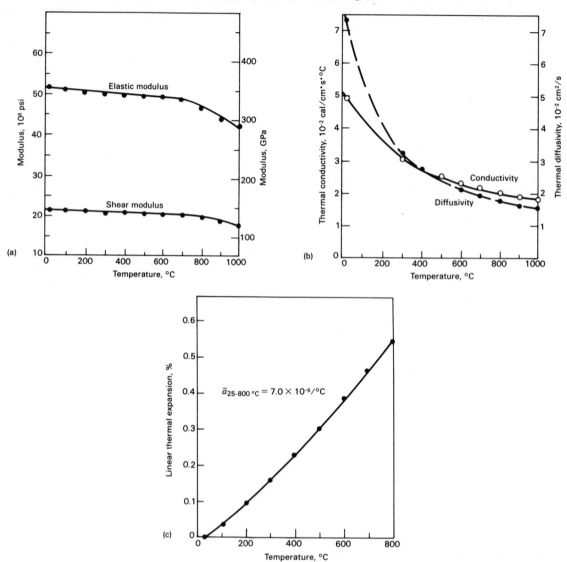

$$\bar{\alpha}_{25\text{-}800\,°C} = 7.0 \times 10^{-6}/°C$$

Selected Properties of BN/Al_2O_3 and ZrO_2/Al_2O_3 Composites

Material[a]	Bend strength, MPa	Young's modulus (E), GPa	γ_{Ic}, J/m²	ΔT_c, °C	K_{Ic}, MPa·m$^{1/2}$	σ_f, MPa	σ_T/σ_i
30 vol % BN-1/Al_2O_3, $\perp$ HPA	170	120	60	700–850[b]	3.8	...	...
30 vol % BN-2/Al_2O_3, $\perp$ HPA	400	190	40–>100	450[b]	3.9–9.0	...	...
30 vol % BN-2/Al_2O_3, $\parallel$ HPA	160	140	25	...	2.6	...	...
30 vol % BN/Al_2O_3, $\perp$ HPA[c]	200	200	3	400–800[b]	1.1	...	...
0.5 vol % ZrO_2/Al_2O_3	...	400	20	600	...	410	0.35
4.0 vol % ZrO_2/Al_2O_3	...	410	25	600	...	420	0.4
9.0 vol % ZrO_2/Al_2O_3	...	380	55	950	...	700	...
11.5 vol % ZrO_2/Al_2O_3	...	380	60	1150	...	780	0.6–0.75
14.0 vol % ZrO_2/Al_2O_3	...	375	35	800	...	680	0.65
19.0 vol % ZrO_2/Al_2O_3	...	350	35	800	...	350	0.65

[a]$\perp$ HPA–stress direction perpendicular to hot pressing axis. $\parallel$ HPA–stress direction parallel to hot pressing axis. [b]Determined by 22 °C water quench test using 3-mm-square bars. [c]Calculated treating BN as pseudoporosity.

Temperature-Dependent Properties of Pressed and Sintered Al₂O₃/50B₄C Composite
(Data From Greenleaf Corp.)

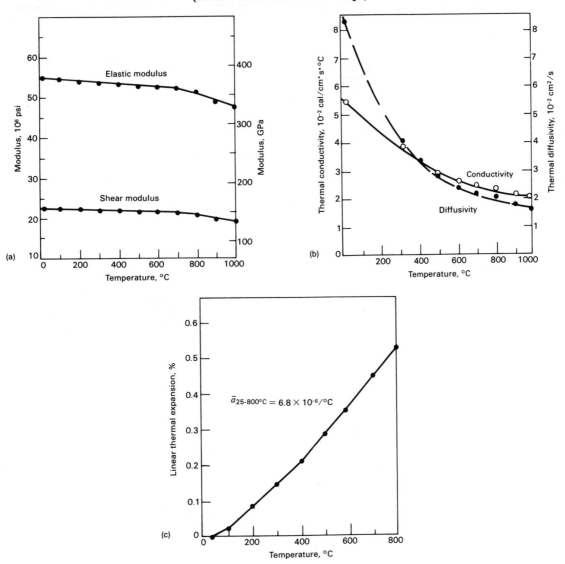

Comparative Thermal Shock Behavior of NRL-Developed Ceramic Composites
(Ref 23, p 624)

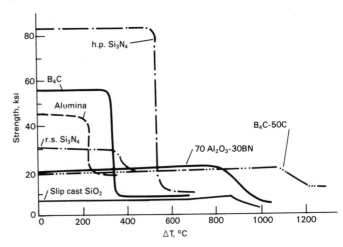

Results are shown for Al₂O₃ containing fine BN particles and B₄C containing fine graphite particles, along with Al₂O₃ alone, HPSN, RSSN, and fused SiO₂ for comparison.

8.3.1.2. Zirconium Diboride Composites

Compositions of Zirconium Diboride Composites
(Ref 25, p 5.1.8-1)

Material	Content, wt %	Reported final composition[a]
ZrB_2	93 to 96	ZrB_2 particles bonded with
B	4 to 7	boron
ZrB_2	81 to 87	Solid solution of $MoSi_2$ in
$MoSi_2$	~13	ZrB_2; SiC occurs as a minor phase when B_4C is added[b]

[a]After densification by hot pressing or sintering. [b]Small additions of either boron nitride (1 wt %) or boron carbide (about 6 wt %) have been used as additions to the basic composition to improve strength at elevated temperatures.

Physical Properties of ZrB_2 Composites
(Ref 25, p 5.1.8-2)

Composite material	Density, g/cm³	Melting temperature °C	°F
ZrB_2/B	5.2–5.4	~2980	~5400
$ZrB_2/MoSi_2$	4.87–5.50[a]	>2370	>4300

[a]Higher value is typical for hot pressed specimens; lower value is typical for sintered specimens.

Average Coefficients of Linear Thermal Expansion of ZrB_2 Composites (Ref 25, p 5.1.8-2)

Temperature range °C	°F	Coefficient of thermal expansion ZrB_2/B $10^{-6}/°C$	ZrB_2/B $10^{-6}/°F$	$ZrB_2/MoSi_2$ $10^{-6}/°C$	$ZrB_2/MoSi_2$ $10^{-6}/°F$
21–260	70–500	4.3	2.4	7.0	3.9
21–540	70–1000	4.9	2.7	6.5	3.6
21–705	70–1300	5.2	2.9	6.7	3.7
21–980	70–1800	5.6	3.1	6.8	3.8
21–1205	70–2200	5.8	3.2	7.2	4.0
21–1345	70–2450	···	···	8.3	4.6

Mechanical Properties of ZrB_2/B Composites
(Ref 25, p 5.1.8-4)

Ultimate tensile strength at 980 °C (1800 °F):	
MPa	241–276
ksi	35–40
Hardness, Rockwell A	88–91
Stress to rupture at 980 °C (1800 °F):	
In 5 h	
MPa	207
ksi	30
In 10 h	
MPa	172
ksi	25
In 100 h	
MPa	128
ksi	18.5
In 1000 h	
MPa	97
ksi	14

Oxidation Resistance of ZrB_2 Composites in Static Air (Ref 25, p 5.1.8–6)

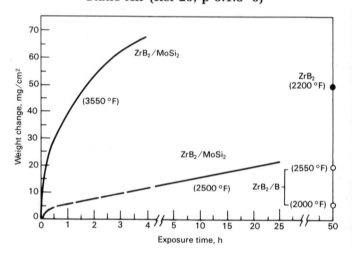

Average Creep Rates for $ZrB_2/MoSi_2$ Composites

Specimens[a]	Temperature °C	°F	Test conditions Initial stress MPa	ksi	Duration, h	Maximum equivalent strain rate[b], 10^{-4} m/m·h (10^{-4} in./in.·h)	Total strain[c], %
Sintered[d]	1500	2730	68.9	10.0	5	7	0.64
			41.4	6.0	5	2.5	0.18
	1850	3360	3.4	0.5	2	35	1.3
			0.8	0.12	5	4	0.27
Hot pressed[e]	1500	2730	44.8	6.5	2	38	1.4
			20.7	3.0	4	3	0.4
	2000	3630	5.2	0.75	2	47	1.28
			1.2	0.18	5	15	0.91

[a]Bar specimens 100 × 12.7 × 6.35 mm (4 × ¹/₂ × ¹/₄ in.), with polished surfaces. [b]Calculated from reported average deflection values obtained in flexure test, four-point loading, 25.4-mm (1-in.) span, dead load; strain rate determined after 1 h of testing. [c]Percent maximum equivalent strain = 0.284 × deflection at centerline × 100. [d]Average density of 4.87 g/cm³, ±0.03 standard deviation. [e]Average density of 5.50 g/cm³, ±0.03 standard deviation.

Specific Heat, Thermal Conductivity, and Linear Thermal Expansion of ZrB₂ Composites (Ref 25, p 5.1.8–3)

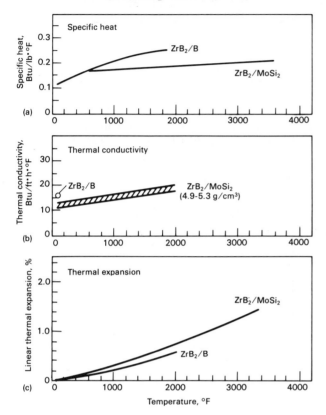

Bend Strength and Young's Modulus of ZrB₂ Composites (Ref 25, p 5.1.8–5)

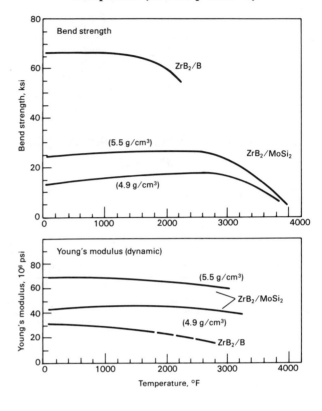

8.3.1.3. Other Ceramic Composites

Thermal Diffusivity of SiC/MgO Composites (Ref 38, p 328, 329)

(a) Thermal diffusivity of 20 wt % SiC/MgO thermally cycled to 1300 °C. (b) Thermal diffusivity of 20 wt % SiC/MgO thermally cycled to 1400 °C. (c) Effect of thermal cycling to 1400 °C on room-temperature thermal diffusivity of SiC/MgO composites.

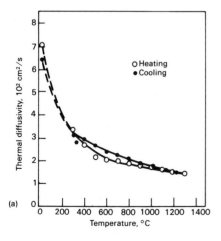
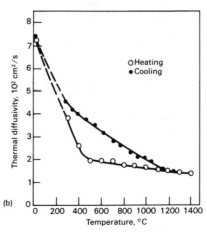
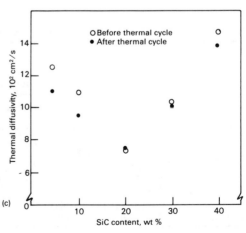

Cyclic Fatigue Results for Some Ceramic Composites

Material	Environment	Mean strength, MPa	Maximum stress, MPa	Number of cycles to failure
30 vol % BN/mullite	Liquid N_2	...	320–360	2–100
30 vol % BN/alumina	Liquid N_2	...	410	3
30 vol % BN/Si_3N_4	Silicone oil	330	275	10–100
Woven carbon/carbon	Silicone oil	97	80–110	2–70

8.3.2. Cermets and Metal-Ceramic Composite Property Data

8.3.2.1. Alumina Systems

Adhesion of Metals to Polycrystalline Al_2O_3

Metal	Sintering temperature, °C	Sintering atmosphere	Adhesion, erg/cm^2
Cu	450	H_2	815
Ni	1000	H_2	435
Au	1000	Air	530
Fe	1000	H_2	810
Ni	1400	Ar	518
Ag	700	H_2	435

Resistance of Al_2O_3/Cr and Sintered Al_2O_3 Cermets to Rapid Changes in Temperature (Ref 36, p 491)

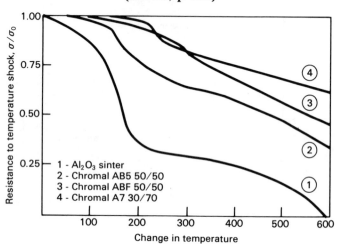

1 - Al_2O_3 sinter
2 - Chromal AB5 50/50
3 - Chromal ABF 50/50
4 - Chromal A7 30/70

Coefficients of Linear Thermal Expansion of Alumina/Chromium Composites Compared With Those of Polycrystalline Alumina (Ref 25, p 5.4.12-2)

Material	Method	Temperature °C	Temperature °F	Coefficient of linear thermal expansion 10^{-6} m/m·K	Coefficient of linear thermal expansion 10^{-6} in./in.·°F
Al_2O_3	(a)	27–799	80–1470	7.90	4.39
		27–1315	80–2400	9.40	5.22
70Al_2O_3/30Cr	Telemicroscope	25–800	77–1472	8.6	4.8
		22–1315	72–2400	9.45	5.25
34Al_2O_3/66(Cr,Mo)	Telemicroscope	24–802	75–1475	7.96	4.42
		24–1315	75–2400	10.5	5.82
28Al_2O_3/72Cr	Telemicroscope	25–800	77–1472	8.62	4.79
		22–1315	72–2400	10.4	5.75
23Al_2O_3/77Cr (LT-1)	...	25–1000	77–1832	8.91	4.95

[a]Average of many reported values obtained by a variety of methods.

Typical Compositions and Physical Properties of Alumina/Chromium Composites (Ref 25, p 5.4.12-1)

Typical composition wt %	Typical composition Approximate vol %	Density, g/cm^3	Melting point °C	Melting point °F
70Al_2O_3/30Cr	81Al_2O_3/19Cr	4.60–4.65	>1705[a]	>3100[a]
34Al_2O_3/66(Cr,Mo)[b]	49.9Al_2O_3/50.1(Cr,Mo)	5.82	>1730[a]	>3150[a]
28Al_2O_3/72Cr	42Al_2O_3/58Cr	5.9	>1730[a]	>3150[a]
23Al_2O_3/77Cr	35Al_2O_3/65Cr	5.9–6.0	1850[c]	3360[c]
Al_2O_3		3.98	2050	3720

[a]Temperature given is sintering temperature at which composite was prepared. [b]Second phase added as 80Cr-20Mo alloy.
[c]Maximum service temperature: 1315 °C (2400 °F) for long term; 1650 °C (3000 °F) for short term.

Creep Characteristics of 23Al₂O₃/77Cr LT-1 Composite Compared With Those of Alumina
(Ref 25, p 5.4.12-4)

| Parameter | Reported value to cause 0.125% equivalent tensile strain in outer fiber | | Test conditions | | | | | |
	LT-1[a]	Al₂O₃[b]	Temperature °C	°F	Stress MPa	ksi	Time, h
Time, h	344	650	1000	1830	28	4.0	···
	16	88	1100	2010	28	4.0	···
Temperature, °C (°F)	1040 (1905)	1095 (2005)	···	···	28	4.0	100
	1000 (1830)	1040 (1905)	···	···	28	4.0	300
Stress, MPa (ksi)	46 (6.6)	···	1000	1830	···	···	10
	28 (4.0)	55 (8.0)	1000	1830	···	···	300

[a]Slip cast and sintered. Density, 5.8–5.9 g/cm³ (~5.9 g/cm³ theoretical). Specimens, 82.6 × 2.8 × 2.8 mm (3.25 × 0.11 × 0.11 in.). Values determined in four-point-load bending: 76-mm (3-in.) span, 41.28-mm (1.625-in.) gage length. [b]Pressed and sintered. 94.5% pure (major impurity, SiO₂). 93% dense. Specimen size and test details same as for LT-1.

Temperature-Dependent Properties of Pressed and Sintered Al₂O₃/TiC Cermet
(Data From Greenleaf Corp.)

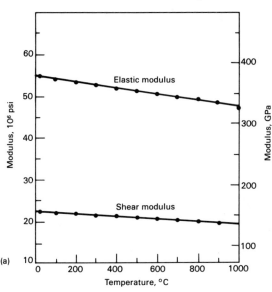

(a)

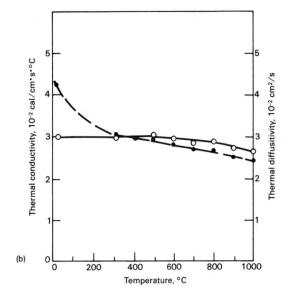

(b)

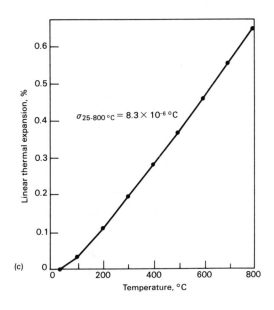

(c)

$$\sigma_{25\text{-}800\,°C} = 8.3 \times 10^{-6}\,°C$$

Temperature-Dependent Properties of Pressed and Sintered $Al_2O_3/30TiC$ Composite
(Data from Greenleaf Corp.)

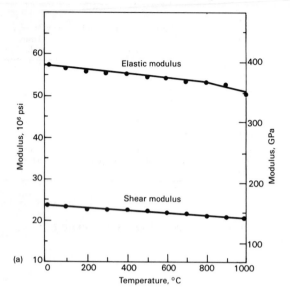

(a)

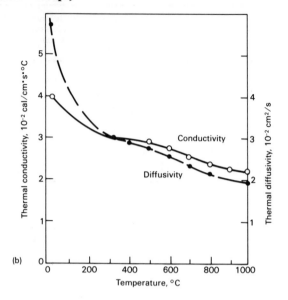

(b)

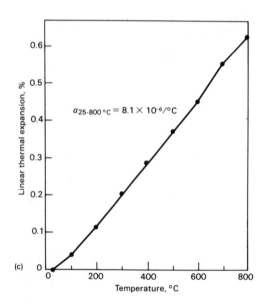

(c)

$\alpha_{25\text{-}800\,°C} = 8.1 \times 10^{-6}/°C$

Wetting Angles in Various Alumina/Liquid Metal Systems

System	Wetting angle	Temperature, °C, atmosphere
Al_2O_3/Ni	>90°	1450, air
Al_2O_3/Cu	138°	1200, vacuum
$Al_2O_3/Fe\text{-}Cr$	~40°	1650, reducing
Al_2O_3/Cr	1–10°	1950, reducing
Al_2O_3/Fe	139°	1550, N_2

Hardness and Fracture Toughness of Al_2O_3 Ceramic Cutting-Tool Materials Compared With Those of Sialon and $Si_3N_4 + Y_2O_3$ (Ref 36, p 459)

Material	Hardness (KHN), GN/m^2	Fracture toughness (K_{Ic}), $MN/m^{3/2}$
Al_2O_3	15.6	2.3
$Al_2O_3 + TiC$	17.2	3.2
$Al_2O_3 + ZrO_2$	15.2	3.2
Sialon	12.2	4.0
$Si_3N_4 + Y_2O_3$	13.4	4.8

Strength Properties of Alumina/Chromium Composites Compared With Those of Polycrystalline Alumina
(Ref 25, p 5.4.12-3, 5.4.12-4)

Property	Al₂O₃	70Al₂O₃/ 30Cr[a]	34Al₂O₃/ 66(Cr,Mo)[b]	28Al₂O₃/ 72Cr[c]	23Al₂O₃/ 77Cr (LT-1)[d]
Bend strength, MPa (ksi):					
At 24 °C (75 °F)	448 (65)	379 (55.0)	605 (87.8)	552 (80.0)	310 (45)
At 870 °C (1600 °F)	345 (50)	297 (43.1)	433 (62.8)	427 (62.0)	186 (27)[e]
At 1095 °C (2000 °F)	317 (46)	226 (32.8)	353 (51.2)	403 (58.4)	124 (18)[f]
At 1315 °C (2400 °F)	276 (40)	168 (24.4)	266 (38.6)	241 (35.0)	32 (4.6)
Compressive strength at 24 °C (75 °F), MPa (ksi)	2415 (350)	2205 (320)	···	···	758 (110)
Tensile strength, MPa (ksi):					
At 24 °C (75 °F)	262 (38)	241 (35)	369 (53.5)	269 (39.0)	145 (21)
At 870 °C (1600 °F)	241 (35)	149 (21.6)	276 (40.0)	175 (25.4)	141 (20.5)[g]
At 1095 °C (2000 °F)	234 (34)	128 (18.5)	188 (27.2)	150 (21.7)	80.7 (11.7)
At 1315 °C (2400 °F)	48 (7)	97.2 (14.1)	68.9 (10.0)	131 (19.0)	···
Impact strength at 24 °C (75 °F), J (in. · lb)	0.130 (1.15)	1.42 (12.6)	···	1.1–1.49 (9.6–13.2)	···

[a]Bend strength: pressed, hydrostatically re-pressed, and sintered; prepared from 99.5% pure alumina, >99% pure aluminum hydrate, and >99% pure Cr; density, 4.60-4.65 g/cm³ (~4.7 g/cm³ theoretical); bar specimens; no test details. Compressive strength: same as for bend strength, except specimens 9.52 mm (0.375 in.) diam by 25.4 mm (1 in.) long. Tensile strength: same as for bend strength, except rod specimens. Impact strength: same as for bend strength, except unnotched rectangular specimen, 41-mm (1.6-in.) span; Charpy impact test. [b]Bend strength: pressed, hydrostatically re-pressed, and sintered; prepared from >99.5% pure alumina, >99% pure Cr, and >99.75% pure Mo; metal added as 80Cr-20Mo (wt %) alloy; density, 5.82 g/cm³ (5.85 g/cm³ theoretical); specimens 114 × 12.7 × 3.8 mm (4.5 × 0.5 × 0.15 in.); three-point loading, 57-mm (2.25-in.) span. Tensile strength: same as for bend strength, except specimens 9.52 mm (0.375 in.) diam by 241 mm (9.5 in.) long, necked to 6.35-mm (0.25-in.) diam over 38-mm (1.5-in.) gage length; tested in tension. [c]Bend strength: pressed, hydrostatically re-pressed, and sintered; prepared from >99.5% pure Cr; density, 5.92 g/cm³ (5.85 g/cm³ theoretical); specimen size and test details same as for 34Al₂O₃/66(Cr,Mo) (footnote b), except 76-mm (3-in.) span. Tensile strength: material details same as for bend strength; specimen dimensions and test details same as for 34Al₂O₃/66(Cr,Mo) (footnote b). Impact strength: material details same as for bend strength; specimen dimensions and test details same as for 70Al₂O₃/30Cr (footnote a). [d]All data are typical property values; no material or test details. [e]At 980 °C (1800 °F). [f]At 1150 °C (2100 °F). [g]At 425 °C (800 °F).

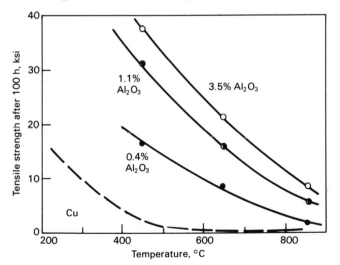

Dependence of Tensile Strength on Temperature (at Internal Oxidation Temperature of 950 °C) for Al₂O₃/Cu Composites and Pure Copper (Ref 36, p 493)

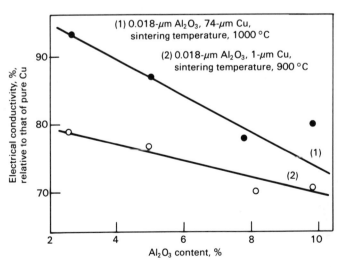

Relative Electrical Conductivity as a Function of Al₂O₃ Content and Grain Size for Al₂O₃/Cu Composites (Ref 36, p 493)

Properties of Simultaneously Pressed and Sintered Al$_2$O$_3$/TiC Cermet[a]

Mechanical Properties

Transverse rupture strength[b], MPa (ksi) 760 (110)
Fracture toughness, MPa · m$^{1/2}$ (ksi · in.$^{1/2}$) 5.9 (5.3)
Elastic constants:
  Elastic modulus, GPa (10^6 psi)
    At 25 °C (77 °F) 376 (54.5)
    At 500 °C (930 °F) 353 (51.2)
    At 1000 °C (1830 °F) 324 (47.1)
  Shear modulus, GPa (10^6 psi)
    At 25 °C (77 °F) 154 (22.4)
    At 500 °C (930 °F) 144 (21.0)
    At 1000 °C (1830 °F) 134 (19.4)
  Poisson's ratio
    At 25 °C (77 °F) 0.22
    At 500 °C (930 °F) 0.22
    At 1000 °C (1830 °F) 0.21
Microhardness, Vickers DPH, GPa (kg/mm^2)
  100-g load 23.0 (2350)
  500-g load 18.6 (1900)

Thermal Properties

Linear thermal expansion, 10^{-6}/°C
  From 25 to 300 °C (77 to 570 °F) 7.0
  From 25 to 500 °C (77 to 930 °F) 7.6
  From 25 to 800 °C (77 to 1470 °F) 8.3
Thermal conductivity, W/cm · K (cal/cm · s · °C)
  At 25 °C (77 °F) 0.13 (0.030)
  At 500 °C (930 °F) 0.13 (0.030)
  At 1000 °C (1830 °F) 0.11 (0.027)
Thermal diffusivity, cm^2/s
  At 25 °C (77 °F) 0.042
  At 500 °C (930 °F) 0.029
  At 1000 °C (1830 °F) 0.025
Specific heat (C$_P$), J/g · K (cal/g · °C)
  At 25 °C (77 °F) 0.58 (0.14)
  At 500 °C (930 °F) 0.84 (0.20)
  At 1000 °C (1830 °F) 0.88 (0.21)

Electrical Property

Volume resistivity, dc, Ω · cm
  At 25 °C (77 °F) .... 2 × 10^{-2}
  At 100 °C (212 °F) .... 2 × 10^{-2}
  At 300 °C (570 °F) .... 3 × 10^{-2}

[a]Composite is electrically conductive. Density, 5.16 g/cm^3. Porosity, vacuum tight. Water absorption, none. Grain size, ~2 μm. Color, black. Data from Greenleaf Corp. [b]In four-point bending.

Properties of Simultaneously Pressed and Sintered Al$_2$O$_3$/30TiC Composite[a]

Mechanical Properties

Transverse rupture strength[b], MPa (ksi) 760 (110)
Fracture toughness, MPa · m$^{1/2}$ (ksi · in.$^{1/2}$) 4.0 (3.6)
Elastic constants:
  Elastic modulus, GPa (10^6 psi)
    At 25 °C (77 °F) 395 (57.3)
    At 500 °C (930 °F) 367 (53.3)
    At 1000 °C (1830 °F) 338 (49.0)
  Shear modulus, GPa (10^6 psi)
    At 25 °C (77 °F) 162 (23.5)
    At 500 °C (930 °F) 150 (21.7)
    At 1000 °C (1830 °F) 137 (19.9)
  Poisson's ratio
    At 25 °C (77 °F) 0.22
    At 500 °C (930 °F) 0.22
    At 1000 °C (1830 °F) 0.23
Microhardness, Vickers DPH, GPa (kg/mm^2)
  100-g load 24.0 (2450)
  500-g load 21.1 (2150)

Thermal Properties

Linear thermal expansion, 10^{-6}/°C
  From 25 to 300 °C (77 to 570 °F) 7.3
  From 25 to 500 °C (77 to 930 °F) 7.7
  From 25 to 800 °C (77 to 1470 °F) 8.1
Thermal conductivity, W/cm · K (cal/cm · s · °C)
  At 25 °C (77 °F) 0.17 (0.040)
  At 500 °C (930 °F) 0.12 (0.029)
  At 1000 °C (1830 °F) 0.092 (0.022)
Thermal diffusivity, cm^2/s
  At 25 °C (77 °F) 0.057
  At 500 °C (930 °F) 0.028
  At 1000 °C (1830 °F) 0.020
Specific heat (C$_P$), J/g · K (cal/g · °C)
  At 25 °C (77 °F) 0.69 (0.16)
  At 500 °C (930 °F) 1.03 (0.25)
  At 1000 °C (1830 °F) 1.12 (0.27)

Electrical Property

Volume resistivity, dc, Ω · cm
  At 25 °C (77 °F) 0.1
  At 100 °C (212 °F) 0.1
  At 300 °C (570 °F) 0.1

[a]Composite is electrically conductive. Density, 4.26 g/cm^3. Porosity, vacuum tight. Water absorption, none. Grain size, 1–2 μm. Color, black. Data from Greenleaf Corp. [b]In four-point bending.

8.3.2.2. Boride Systems

Wetting Angles of Refractory Borides With Various Metal Melts (Ref 122, 123)

Boride	Cu 1130 °C	Al 900 °C	Ga 800 °C	In 250 °C	Si 1500 °C	Ge 1000 °C	Sn 300 °C	Pb 400 °C	Bi 320 °C
TiB$_2$	143	98	115	124	15	⋯	114	106	141
ZrB$_2$	135	106	117	114	⋯	102	110	⋯	⋯
HfB$_2$	⋯	134	⋯	114	⋯	140	⋯	⋯	⋯
NbB$_2$	109	125	101	133	0	60	102	125	110
TaB$_2$	⋯	138	⋯	117	⋯	⋯	⋯	⋯	⋯
CrB$_2$	26	107	123	97	⋯	126	100	124	128
Mo$_2$B$_5$	0	134	⋯	⋯	⋯	28	100	⋯	⋯
W$_2$B$_5$	104	⋯	⋯	130	22	128	100	⋯	⋯

Wetting Angles of Borides by Iron-Group Melts (Ref 122, 123)

Boride	Iron Wetting angle Vacuum	Iron Wetting angle Argon	Iron Work of adhesion, erg/cm²	Nickel Wetting angle Vacuum	Nickel Wetting angle Argon	Nickel Work of adhesion, erg/cm²	Cobalt Wetting angle Vacuum	Cobalt Wetting angle Argon	Cobalt Work of adhesion, erg/cm²
TiB$_2$	39	94	1650	25	72	2220	⋯	64	2590
ZrB$_2$	97	102	1410	72	78	2020	⋯	81	2100
HfB$_2$	⋯	98	1530	⋯	99	1430	⋯	⋯	⋯
NbB$_2$	⋯	0	⋯	⋯	24	⋯	⋯	22	3460
TaB$_2$	⋯	0	⋯	⋯	⋯	⋯	⋯	⋯	⋯
CrB$_2$	⋯	0	⋯	20	21	3310	⋯	0	⋯
Mo$_2$B$_5$	⋯	0	⋯	⋯	0	⋯	⋯	22	3460
W$_2$B$_5$	⋯	⋯	⋯	⋯	⋯	⋯	⋯	94	1680

Elevated-Temperature Strength of TiB$_2$/Ni and TiB$_2$/Fe Composites Compared With That of Pure TiB$_2$ (Ref 20, p 324)

The numbers in parentheses indicate the content (in vol %) of the metal phase on the polished surface.

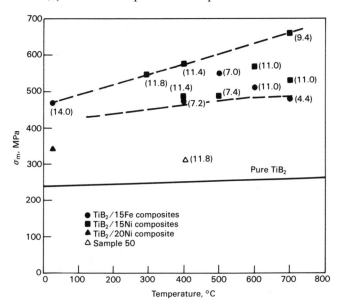

Empirical Relationship Between Strength and Porosity for TiB$_2$/Fe Composites and Pure TiB$_2$ (Ref 20, p 323)

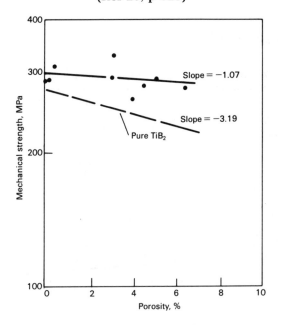

Illustration of Linear Dependence of Fracture Strength on Iron Content on Polished Surface for TiB₂/Fe Composites (Ref 20, p 321)

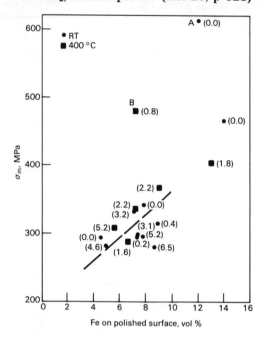

Parenthetical numbers are vol % of porosity on the sample surface as determined by point counting. Data points A and B represent deviation from the linear behavior of the other data points. Specimen A showed an unusual microstructure with micron-size regions of eutectic-like structure.

8.3.2.3. Carbides

Influence of Cobalt Content on Room- and High-Temperature Hardness of Sintered WC-Co Specimens (Ref 10, p 476)

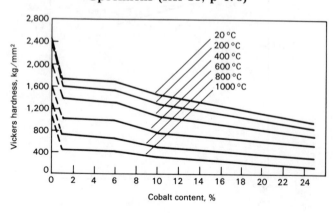

Dependence of Young's Modulus on Temperature for Cemented Carbides (Ref 10, p 476)

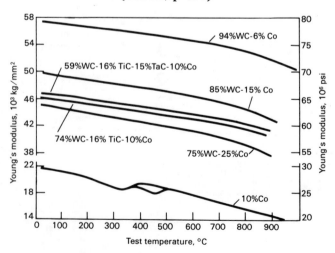

Wetting of Some Refractory Carbides by Liquid Metals

Carbide	Wetting melt	Atmosphere	Temperature, °C (°F)	Contact angle, degrees
WC	Cu	Argon	1100 (2010)	30
	Ni	Vacuum	1500 (2730)	0
	Co	Hydrogen	1500 (2730)	0
	Fe	Vacuum	1500 (2730)	0
TiC	Ni	Hydrogen	1450 (2640)	17
	Co	Hydrogen	1500 (2730)	36
	Fe	Hydrogen	1550 (2820)	39
TiC-WC	Ni	Vacuum	1477 (2690)	6
	Ni	Hydrogen	1477 (2690)	16
	Co	Vacuum	1477 (2690)	4
	Co	Hydrogen	1477 (2690)	26

General Property Data for Tungsten Carbide Cermets (Ref 25, p 5.2.10-6)

Property	Typical values reported at 24 °C (75 °F)[a] WC/Co composite	WC
Density, g/cm^3	11–15.2	~15.8
Melting point:		
°C	1455[b]	~2775
°F	2650[b]	~5030
Specific heat:		
kJ/kg·K	0.13–0.25	~0.13
Btu/lb·°F	0.03–0.06	~0.03
Thermal conductivity:		
W/m·K	28–52	...
Btu/ft·h·°F	16–30	...
Coefficient of linear thermal expansion (21–980 °C or 70–1800 °F):		
10^{-6}/°C	1.7–3.3	1.5
10^{-6}/°F	3.0–6.0	2.7
Bend strength:		
MPa	1380–2415	483–827
ksi	200–350	70–120
Compressive strength:		
MPa	2240–3790	2760–2965
ksi	325–550	400–430
Tensile strength:		
MPa	552–1240[c]	...
ksi	80–180[c]	...
Impact strength:		
J	2.7–20.3	...
in.·lb	24–180	...
Young's modulus:		
GPa	276–621	~690
10^6 psi	40–90	~100
Hardness, Rockwell A	80–92	~90
Oxidation resistance (20 h at 870–980 °C or 1600–1800 °F), % weight gain	>25	(d)

[a]Data are for composite materials containing up to 50 wt % metal; values were obtained from various sources. [b]Melting point for cobalt component. [c]Yield of 0.2%. [d]Catastrophic degradation.

General Property Data for Chromium Carbide Cermets (Ref 25, p 5.2.10-5)

Property	Typical values reported at 24 °C (75 °F)[a] Cr$_x$C$_y$[b]/Ni composite	Cr$_3$C$_2$
Density, g/cm^3	6.4–7.0	6.68[c]
Melting point:		
°C	1455[d]	~1895
°F	2650[d]	~3440
Specific heat:		
kJ/kg·K	~0.50	0.50
Btu/lb·°F	~0.12	0.12
Thermal conductivity:		
W/m·K	10–12	...
Btu/ft·h·°F	6–7	...
Coefficient of linear thermal expansion (21–980 °C or 70–1800 °F):		
10^{-6}/°C	3.3–3.9	~3.4
10^{-6}/°F	6.0–7.0	~6.1
Bend strength:		
MPa	483–793	241–324
ksi	70–115	35–47
Compressive strength:		
MPa	896[e], 3450[f]	1035
ksi	130[e], 500[f]	150
Tensile strength:		
MPa	>241[e]	...
ksi	>35[e]	...
Impact strength:		
J	0.113–0.903	~0.113
in.·lb	1–8	~1
Young's modulus:		
GPa	~345	386
10^6 psi	~50	56
Hardness, Rockwell A	~88	91
Oxidation resistance (20 h in air at 870–980 °C; 1600–1800 °F), % weight gain	<1	<1

[a]Data are for composite materials containing up to 50 wt % metal; values were obtained from various sources. [b]Cr$_3$C$_2$ or mixed carbides of chromium (Cr$_3$C$_2$, Cr$_7$C$_3$, and Cr$_4$C). [c]Theoretical density. [d]Melting point for nickel component. [e]No yield. [f]Yield of 1.1%.

Components of Typical Carbide/Metal Cermets
(Ref 25, p 5.2.10-4)

Carbide components[a]		Metal components[a]	
Major	Minor	Major	Minor
TiC	⋯	Ni or Co	⋯
TiC	⋯	Ni or Co	Cr, Mo, W, Al, or Fe
TiC (NbC, TaC)[b]	⋯	Ni or Co	(c)
TiC	Cr_3C_2	Ni or Co	⋯
TiC	(d)	Ni or Co	(c)
Cr_xC_y[e]	(d)	Ni	(c)
WC	(d)	Co	(c)

[a]Composite range of about 30 to 70 wt % of either major component. [b]A solid-solution component. [c]With or without other metal additions, such as Cr, Mo, W, Al, or Fe. [d]With or without other additions of transition-metal carbides. [e]As Cr_3C_2 or mixed carbides of chromium.

Creep Curves for Various WC Composites at 800 °C (1470 °F) Under a Compressive Stress of 490 MPa (71 ksi) (Ref 10, p 477)

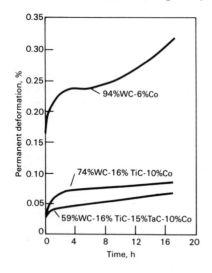

General Property Data for Titanium Carbide Cermets (Ref 25, p 5.2.10-5)

Property	Typical values reported at 24 °C (75 °F)[a]	
	TiC/metal composite	TiC
Density, g/cm³	5.5–6.8	4.65–4.92
Melting point:		
°C	1455[b]	3065
°F	2650[b]	5550
Specific heat:		
kJ/kg·K	0.46	0.50
Btu/lb·°F	0.11	0.12
Thermal conductivity:		
W/m·K	28–35	26–35
Btu/ft·h·°F	16–20	15–20
Coefficient of linear thermal expansion (21–980 °C or 70–1800 °F):		
10^{-6}/°C	2.8–3.6	2.4
10^{-6}/°F	5.0–6.5	4.3
Bend strength:		
MPa	689–1380	414–827
ksi	100–200	60–120
Compressive strength:		
MPa	2760–3450[c]	1380–2760
ksi	400–500[c]	200–400
Tensile strength:		
MPa	689–965[d]	241–276
ksi	100–140[d]	35–40
Impact strength:		
J	5.42–21.47	<1.36
in.·lb	48–190	<12
Young's modulus:		
GPa	310–379	276–448
10^6 psi	45–55	40–65
Hardness, Rockwell A	84–89	93
Oxidation resistance (20 h at 870–980 °C or 1600–1800 °F), % weight gain	<1	<1

[a]Data are for composite materials containing up to 50 wt % metal; values were obtained from various sources. [b]Melting point for nickel or cobalt metal component. [c]Compressive yield stress. [d]Yield of 0.2 to 0.3% for indicated stresses.

Physical, Thermal, and Mechanical Property Data for Silicon Carbide Composites (Ref 25, p 5.2.10-2)

Property	Reported property value[a]			
	Silicon nitride-bonded SiC[b]	Silicon oxynitride-bonded SiC[c]	SiC-bonded graphite aggregate[d]	Self-bonded SiC[e]
Density, g/cm³	2.5–2.8	~2.7	2.3–2.8	3.10
Porosity, %	13–15	~18	8–26	~5
Maximum recommended working temperature:		1650		
Oxidizing environment				
°C	1650		815–1540	1650
°F	3000	3000	1500–2800	3000
Neutral environment		2205		
°C	2205		2205	2315
°F	4000	4000	4000	4200
Specific heat (27–1370 °C; 80–2500 °F):		0.84–1.55		
kJ/kg·K	0.84–1.55		0.71–1.47	0.71–1.38
Btu/lb·°F	0.2–0.37	0.2–0.37	0.17–0.35	0.17–0.33
Thermal conductivity (980 °C; 1800 °F):		16		
W/m·K	16		48–59	42
Btu/ft·h·°F	9.5	9.5	28–34	24
Coefficient of linear thermal expansion (21–1370 °C; 70–2500 °F):		1.4		
10^{-6}/°C	1.4		1.5[f]	1.51
10^{-6}/°F	2.6	2.6	2.7[f]	2.72
Bend strength:		41–55		
24 °C (75 °F)				
MPa	41–69		34–69	165
ksi	6–10	6–8	5–10	24
1095 °C (2000 °F)		55–62		
MPa	48–69		34–69	172
ksi	7–10	8–9	5–10	25
1500 °C (2730 °F)		~21		
MPa	21		48–83	124
ksi	3	~3	7–12	18
Compressive strength (24 °C; 75 °F):		~138		
MPa	138		434[f]	1380
ksi	20	~20	63[f]	200
Impact strength (21 °C (70 °F):		...		
J	<0.113		<0.113	<0.113
in.·lb	<1	...	<1	<1
Young's modulus:		117		
24 °C (75 °F)				
GPa	117		~241[f]	379
10^6 psi	17	17	~35[f]	55
1000 °C (1830 °F)		69		
GPa	69		207[f]	365
10^6 psi	10	10	30[f]	53
Bend-creep resistance — conditions to promote a tensile creep strain of 0.125%:		...		
Time, h	>300 (1095 °C; 2000 °F)		>630 (1205 °C; 2200 °F)	>1000 (1205 °C; 2200 °F)
Temperature (300 h)		...		
°C	>1095		>1205	>1300
°F	>2000		>2200	>2370
Bend stress (100 h)		...		
MPa	28–41 (1000 °C)		28–41 (1000 °C)	159 (1205 °C)
ksi	4–6 (1830 °F)	...	4–6 (1830 °F)	23 (2200 °F)

[a]Data were obtained primarily from company literature. [b]Composition is approximately 80 wt % SiC and 20 wt % Si_3N_4. [c]Composition is approximately 80 wt % SiC and 20 wt % Si_2ON_2 (silicon oxynitride). [d]Variable graphite content; compositions reportedly range from 78 to 50 wt % SiC (dense bond phase), 20 to 46 wt % graphite aggregate, and 1 to 4 wt % silicon. [e]Fine-grain, high-density material. [f]Value for low-graphite-content material.

8.3.3. Si₃N₄ Ceramic Composites Property Data

Oxidation of Si₃N₄-Base Composites as a Function of Time in Air at 1200 °C (Ref 36, p 467)

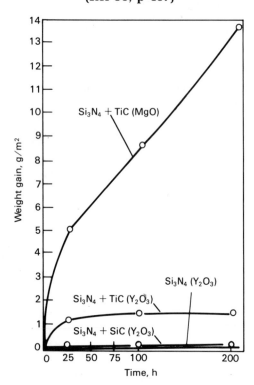

Calculated Abrasion Wear Resistance Parameters for Si₃N₄-Base Composites Compared With Some Al₂O₃ Systems (Ref 36, p 463)

Material	Abrasion wear resistance parameter, $H^{1/2} \cdot K_{Ic}^{3/4}$	Fracture toughness (K_{Ic}), MN/m³ᐟ²
Si₃N₄ + 6 wt % Y₂O₃	11.87	4.8
Si₃N₄ + 10 vol % TiC	11.91	4.8
Si₃N₄ + 30 vol % TiC	11.84	4.4
Si₃N₄ + 50 vol % TiC	8.48	2.7
Si₃N₄ + 30 vol % (W,Ti)C	9.57	3.5
Si₃N₄ + 30 vol % WC	13.06	5.2
Si₃N₄ + 30 vol % TaC	11.15	4.6
Si₃N₄ + 30 vol % HfC	9.81	3.6
Si₃N₄ + 30 vol % SiC	9.74	3.6
Al₂O₃	7.38	2.3
Al₂O₃ + ZrO₂	9.32	3.2
Al₂O₃ + TiC	9.92	3.2
Sialon	9.87	4.0

Room- and Elevated-Temperature Fracture Toughness of Si₃N₄ + 30 Vol % TiC Composite Materials (Ref 36, p 462)

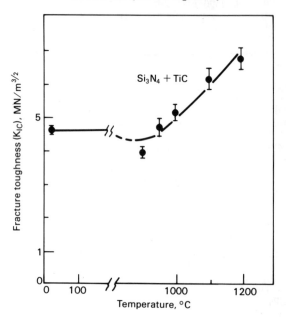

Mechanical Properties of Si₃N₄ Composites Containing 30 Vol % of Metal Carbide Dispersoid (2 µm Average Particle Diameter) (Ref 36, p 462)

Matrix	Dispersed phase	Density, g/cm³	Hardness (KHN), GPa	Fracture toughness (K_{Ic}), MN/m³ᐟ²	Modulus of rupture, MPa RT	1000 °C	1200 °C
Si₃N₄ + 6 wt % Y₂O₃	None	3.26	13.4 ± 0.3	4.8 ± 0.3	110.9 ± 1.6	88.3 ± 3.5	49.2 ± 5.0
Si₃N₄ + 6 wt % Y₂O₃	TiC	3.81	15.21 ± 0.3	4.4 ± 0.5	80.6 ± 5.9	120.4 ± 12.2	64.4 ± 2.9
	(Ti,W)C	4.55	14.06 ± 0.3	3.5 ± 0.3	75.5 ± 3.2	86 ± 0	52.9 ± 0.5
	WC	7.70	14.4 ± 0.4	5.2 ± 0.4	89.1 ± 31.8	136.4 ± 1.6	55.7 ± 0.5
	TaC	6.87	12.6 ± 0.2	4.6 ± 0.4	86.2 ± 7.3	124.5 ± 16.0	43.2 ± 2.0
	HfC	5.74	14.1 ± 0.4	3.6 ± 0.2	86 ± 0.8	···	68.6 ± 0.5
	SiC	3.24	13.6 ± 0.2	3.65 ± 0.5	97.6 ± 8.5	94.0 ± 4.9	52.3 ± 3.2
Al₂O₃	TiC	4.28	17.2 ± 0.2	3.2 ± 0.4	72.2 ± 13.0	69.4 ± 4.3	57.0 ± 4.1

Effect of Metal Carbide Content on Hardness of Si₃N₄ Composites (Ref 36, p 460)

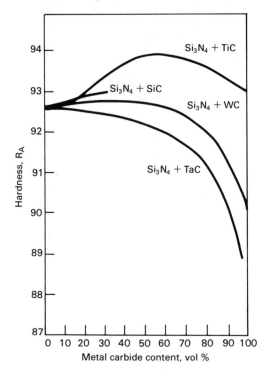

Influence of TiC Content on Room-Temperature Fracture Toughness of Si₃N₄ + TiC Composites (Ref 36, p 461)

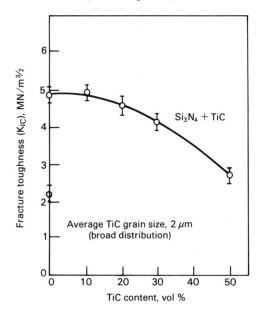

8.4. Carbon/Carbon Composites

8.4.1. Fabrication Methods and Processes

Methods for Developing Carbonaceous Matrix in Carbon/Carbon Composites (Ref 2, p 398)

1. Molten pitch from coal tar sources or petroleum sources is impregnated into the C/G fiber structure under heat and pressure. This may be followed by pyrolysis and subsequent re-impregnations.

2. Resin matrices which have been known to have high char strength after pyrolysis, such as those from certain grades of phenolic resins and furfuryl alcohol resins, will penetrate effectively into thick fibrous structures. Pyrolysis followed by re-impregnation cycles are necessary.

3. Chemical vapor deposition (CVD) from a gaseous phase (usually of methane and nitrogen, sometimes with a small amount of hydrogen) forms upon heated substrates such as the C/G fibers, high-strength pyrolytic graphite. This process may be intercalated into either of the first two processes to enhance physical properties of the composites of carbon/carbon.

4. Impregnation of a still-porous carbon/carbon composite prepared by the steps above with liquid monomers capable of forming heat-resistant structures represents an additional refinement. The choice of such monomers is very limited, although penetrating liquids of tetraethyl silicate coupled with strong mineral acid catalysts will yield a silicon-oxygen network with good heat resistance. Silicon resins may also serve this purpose.

Furnace Used in Constant-Temperature Infiltration Process for Carbon/Carbon Composites

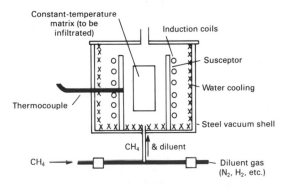

Infiltration by pyrolytic carbon or pyrolytic graphite of porous substrates, such as carbon felts, woven carbon/graphite products, and three-dimensional fabrics prepared by tufting or needling of woven carbon-fiber products, is done in furnaces such as the one shown here.

Schematic Illustration of Pore-Filling and Pore-Blocking Mechanisms During Vapor Deposition of Pyrolytic Carbon (Ref 94, p 375)

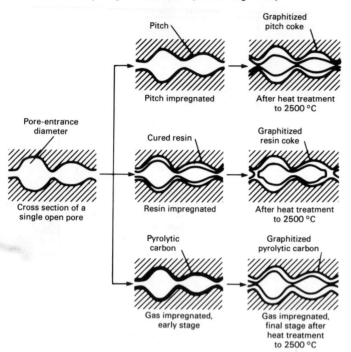

Two Fiber Arrangements for Carbon/Carbon Composites (Fiber Materials, Inc.)

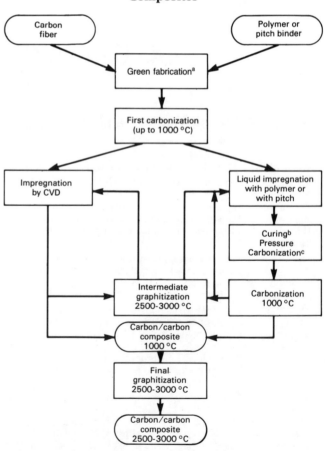

Cylindrical 3-D reinforcement employs tailored yarn placement to ensure uniform density

Fiber reinforcements in three orthogonal directions for stability and shear resistance

Standard HPIC (High-Pressure Impregnating Carbonization) Cycle at Approximately 100 MPa for Impregnation and Carbonization of Carbon/Graphite Fibrous Bases (Ref 2, p 399)

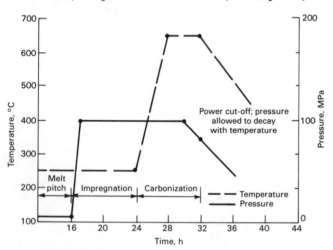

This process is used to densify carbon/carbon composites by impregnating with pitch and carbonizing in a hot autoclave. The two lines on the graph indicate temperature and pressure requirements of the process. Carbon is introduced in the form of melted pitch, goes through the impregnation cycle, and finally carbonizes at elevated temperature and pressure.

Flowchart for Production of Carbon/Carbon Composites

[a]Wet winding; prepreg technology; liquid impregnation of fiber preform [b]for polymer [c]for pitch.

8.4.2. Carbon/Carbon Composites Property Data

Typical Properties of Carbon/Carbon Composite Compared With Polycrystalline Bulk Graphite (Ref 102, p III-A-Three-5)

Property	Carbon/carbon composite[a]	Bulk graphite[b]
Density, g/cm³	1.65	1.83
Tensile strength, MPa (ksi):		
Room temperature	103 (15.0)	35.2 (5.1)
2480 °C (4500 °F)	68.9 (10.0)[c]	53.1 (7.7)
Tensile modulus, GPa (10⁶ psi):		
Room temperature	41.4 (6.0)	11.7 (1.7)
2480 °C (4500 °F)	10.3 (1.5)	11.0 (1.6)
Compressive strength, MPa (ksi):		
Room temperature	68.9 (10.0)	80.7 (11.7)
2480 °C (4500 °F)	159 (23.0)	···
Compressive modulus, GPa (10⁶ psi):		
Room temperature	22.8 (3.3)	···
2480 °C (4500 °F)	10.3 (1.5)	···
Flexural strength, MPa (ksi):		
Room temperature	96.5 (14.0)	42.0 (6.1)
1925 °C (3500 °F)	103 (15.0)	72.4 (10.5)
Flexural modulus, GPa (10⁶ psi):		
Room temperature	27.6 (4.0)	···
1925 °C (3500 °F)	17.2 (2.5)	···
Coefficient of thermal expansion from RT to 1095 °C (2000 °F), 10⁻⁶ m/m·K (10⁻⁶ in./in.·°F)	0.54 (0.3)	3.24 (1.8)
Thermal conductivity, W/m·K (Btu/ft·h·°F):		
260 °C (500 °F)	55.4 (32)	114 (66)
2480 °C (4500 °F)	24.2 (14)	46.7 (27)

[a]Properties measured in z direction. [b]Properties are with grain and were taken from Union Carbide Bulletin No. 463-201-H1. [c]Shear failure.

Physical Properties of Graphite-Fiber-Reinforced Graphite Composite[a]

Property	Standard product	Special grade
Density:		
g/cm³	1.6	2.0
lb/ft³	100	125
Tensile strength:		
MPa	11.7	24.1
ksi	1.7	3.5
Tensile modulus:		
GPa	8.3	17.9
10⁶ psi	1.2	2.6
Compressive strength:		
MPa	44.8	82.7
ksi	6.5	12.0
Compressive modulus:		
GPa	6.9	···
10⁶ psi	1.0	···
Shear modulus:		
GPa	3.4	···
10⁶ psi	0.5	···
Flexural strength:		
MPa	···	34.5
ksi	···	5.0
Thermal conductivity at 538 °C (1000 °F):		
W/m·K	86.4	144
Btu·in./ft²·h·°F	600	1000
Coefficient of thermal expansion at 538 °C (1000 °F):		
10⁻⁶ °C⁻¹	0.58	0.97
10⁻⁶ °F⁻¹	0.32	0.54
Open porosity, %	···	0.2

[a]All properties measured in the plane panel direction. Data from Fiber Materials, Inc.

Properties of Some Commercial Carbon/Carbon Composites (Ref 6, p 244)

Material	Bond	Fiber content, vol %	Fiber orientation	Density, g/cm³	Tensile strength MPa	ksi	Flexural strength MPa	ksi	Interlaminar strength MPa	ksi
Ag Carb 101	Resin char	66	Flat	1.45	75.8	11.0	96.5	14.0	5.2	0.75
PTE	Pitch char	>60	Flat	1.57	74.5	10.8	75.8	11.0	···	···
Carbitax 530	···	66	Flat	···	75.8	11.0	110	16.0	14	2.0
Carbitax 513	Resin char	66	Filament wound	1.45	103–138	15.0–20.0	···	···	···	···
Carbitax 515	···	66	Unidirectional	···	276	40.0	···	···	12	1.8
Pyrolarex	Resin char	···	Chopped	1.00	8.674	1.258	17.47	2.534	···	···
Pyrolarex 350	Resin char	>60	Flat	1.20	43	6.2	86.2	12.5	6.9	1.0
Pyrocarb 400	Char/CVD	>60	Flat	1.30	96.5	14.0	138	20.0	6.6	0.95
Haveg 513	Char/CVD	···	···	···	···	···	···	···	···	···
Haveg 41 G(L)	Char	>60	Flat	1.45	40.3	5.84	44.1	6.39	14.6	2.12
LTV–CG	Char/CVD	>60	Flat	1.38	64.0	9.28	65	9.4	15	2.2
RPG	CVD	5–7	Felt	1.75	43	6.2	82.7	12.0	41	6.0
Pyro-Bond	CVD	50	Filament wound; circular wrap	1.34	827	120	···	···	···	···
Pyco-Bond	CVD	34	Flat	1.65	···	···	169	24.5	31	4.5
Pyco-Bond	CVD	44	Filament wound; 75° wrap	1.45	To 391	To 56.7	···	···	···	···

Typical Strength Vs Temperature for Carbon/Carbon Composites and Various Other Materials

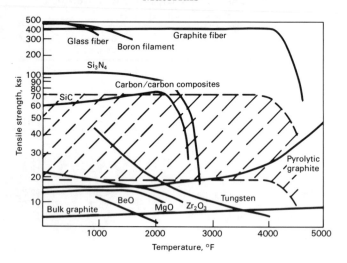

Thermal Conductivity Vs Temperature for Carbon/Carbon Composite and Polycrystalline Bulk Graphite (Ref 102, p III-A-Three-6)

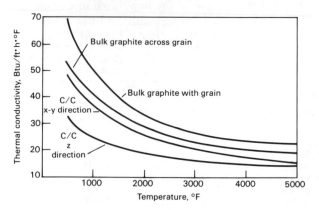

Weight Loss Vs Oxidation Time for Various Carbons

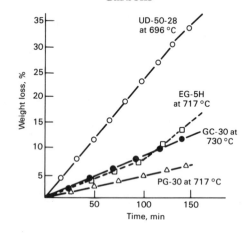

Influence of Oxidation (SiC Coating) Protection of Carbon/Carbon Composites: Surface Mass Loss, Δm, Vs Heat Treatment Time in Air, τ_{HT} (TEOS = Tetraethylorthosilicate)

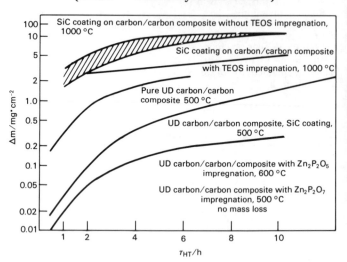

Effect of Heat Treatment Temperature on CVD/Felt Carbon/Carbon Composites With Various Fiber Contents (Ref 8, p 116)

Fiber content, vol %	Heat treatment temperature, °C		Strength		Modulus		d_{002}, Å	Macro-porosity[a], vol %
			MPa	ksi	GPa	10^6 psi		
9	1100	....	61.3	8.9	21.4	3.1	3.49	15
	2630	....	62.0	9.0	16.5	2.4	3.45	
	3000	....	52.4	7.6	11.0	1.6	3.41	
25	1100	....	95.2	13.8	22.8	3.3	3.49	10.7
	2630	....	83.5	12.1	21.4	3.1	3.46	
	3000	....	76.5	11.1	17.2	2.5	3.41	
36	1100	....	111.0	16.1	24.1	3.5	3.49	8.8
	2630	....	91.0	13.2	23.4	3.4	3.47	
	3000	....	89.6	13.0	17.9	2.6	3.42	
46	1100	....	130.2	18.9	26.2	3.8	3.49	5.5
	2630	....	108.9	15.8	23.4	3.4	3.47	
	3000	....	97.9	14.2	17.2	2.5	3.42	

[a]Determined by Quantimet image-analyzing microscope.

Thermal Expansion Vs Temperature for Carbon/Carbon Composite and Polycrystalline Bulk Graphite (Ref 102, p III-A-Three-5)

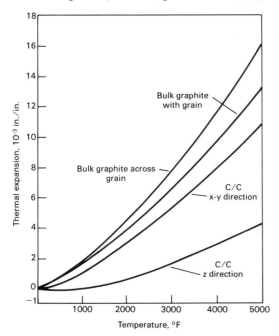

Effect of Heat Treating Temperature on Oxidation of CF/GC Composite

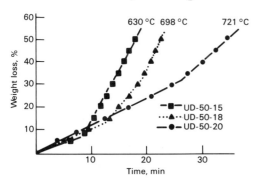

Impact Strength of Unidirectional Carbon/Carbon Composites Vs Strain to Failure of the Reinforcing Fibers (Ref 94, p 388)

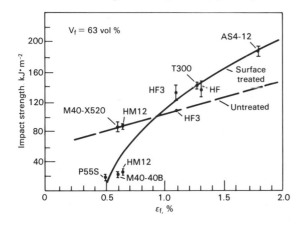

Inflammability of Various Carbon/Carbon Composites and Materials

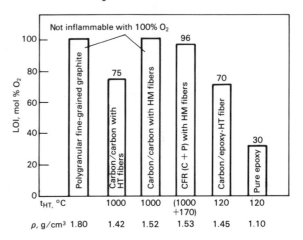

Rate of Oxidation Vs 1/T for Various Carbons

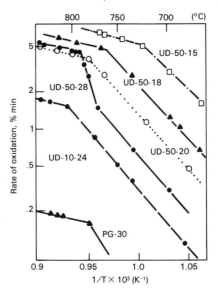

Short-Term Strength Vs Temperature for Carbon/Carbon Composites Compared With Other Materials

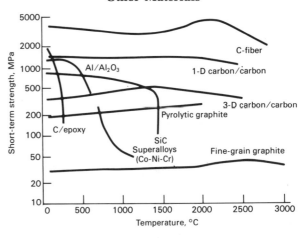

Oxidation of Carbon/Carbon Composites (Ref 93, p 496, 498)

No.	Symbol	Material	$2d_{002}$, Å	Fiber content, %	Heat treatment temperature, °C	Apparent energy, kcal/mol		
1	GC-30	Glassy carbon	...	...	3000	52 ± 2		
2	PG-30	Pyrolytic graphite	3.364	...	3000	54 ± 1		
3	EG-5H	Commercial graphite	3.367	...	...	55 ± 1 (0–8%), 57 ± 1 (>8%)		
4	UD-50-28	Carbon fiber/glassy carbon composite	3.360	50	2800	50 ± 2, 53 ± 2[a]	...	...
5	UD-10-24	Carbon fiber/glassy carbon composite	3.363	10	2400	49 ± 2[a]	50 ± 2[b]	48 ± 3[c]
6	UD-50-20	Carbon fiber/glassy carbon composite	...	50	2000	...	46 ± 2[b]	48 ± 2[c]
7	UD-50-18	Carbon fiber/glassy carbon composite	...	50	1800	...	44 ± 2[b]	46 ± 1[c]
8	UD-50-15	Carbon fiber/glassy carbon composite	...	50	1500	...	38 ± 1[b]	40 ± 2[c]

[a]Apparent activation energy of oxidation for graphite. [b]Apparent activation energy of oxidation for anisotropic area. [c]Apparent activation energy of oxidation for isotropic area.

SECTIONS 5a/8a
References for Sections 5, 6, 7 and 8

1. S. W. Tsai, *Composite Design–1985,* Think Composites, Dayton, Ohio, 1985

2. John Delmonte, *Technology of Carbon and Graphite Fiber Composites,* Van Nostrand Reinhold Co., New York, 1981

3. *Fabrications of Composite Materials,* edited by Mel M. Schwartz, American Society for Metals, Metals Park, Ohio, 1985

4. *Handbook of Composites,* edited by George Lubin, Van Nostrand Reinhold Co., New York, 1982

5. *Encyclopedia of Composite Materials and Components,* edited by Martin Grayson, John Wiley & Sons, New York, 1983

6. *Advanced Fibers and Composites for Elevated Temperatures,* edited by I. Ahmad and B. R. Noton, Proceedings of the 108th AIME Annual Meeting and Symposium, sponsored by the Metallurgical Society of AIME and ASM Joint Composite Materials Committee, New Orleans, Louisiana, February 20–21, 1979, The Society of AIME, Warrendale, Pennsylvania, 1980

7. Mel M. Schwartz, *Composite Materials Handbook,* McGraw-Hill, New York, 1984

8. *Composites: State of the Art,* edited by J. W. Weeton and E. Scala, Proceedings of the Metallurgical Society of AIME, Fall Meeting 1971, Detroit, Michigan, AIME, New York, 1974

9. *Fatigue of Fibrous Composite Materials,* ASTM STP 723, Proceedings of the ASTM Symposium on High Modulus Fibers and Their Composites, San Francisco, California, May 22–23, 1979, ASTM, Philadelphia, 1981

10. *Modern Composite Materials,* edited by Lawrence J. Broutman and Richard H. Krock, Addison-Wesley Publishing Co., Reading, Massachusetts, 1967

11. H. W. Rauch, W. H. Sutton, and L. R. McCreight, *Survey of Ceramic Fibers and Fibrous Composite Materials,* AFML-TR-66-365, October 1966

12. *Summary of the Eighth Refractory Composites Working Group Meeting,* prepared by Lt. D. R. James and L. N. Hjelm, Vol III, ML-TDR-64-233, January 1964

13. Michael R. Piggott, *Load-Bearing Fibre Composites,* Pergamon Press, Elmford, New York, 1980

14. *Metallic Matrix Composites,* Vol 4, edited by Kenneth G. Kreider, Academic Press, New York, 1974

15. Karl M. Prewo, "The Impact Tolerance of Fiber-and-Particulate-Reinforced Metal-Matrix Composites," *Mechanical Behavior of Metal-Matrix Composites,* edited by John E. Hack and Maurice F. Amateau, Proceedings of the 111th AIME Annual Meeting and Symposium, sponsored by the Composite Materials Committee of the Metallurgical Society of AIME and the Materials Science Division of the American Society for Metals, Dallas, Texas, February 16–18, 1982

16. Karl M. Prewo, "The Charpy Impact Energy of Boron Aluminum," *Journal of Composite Materials,* Vol 6, October 1972

17. Karl M. Prewo, "The Importance of Fibres in Achieving Impact Tolerant Composites," *Philosophical Transactions of the Royal Society of London A 294,* 1980

18. Karl M. Prewo, "Fiber Reinforced Metal and Glass Matrix Composites," presented at the Distinguished Lecture Series "Frontiers in Materials Science," Scandia National Labs, October 1983

19. Karl M. Prewo, "The Development of Impact Tolerant Metal Matrix Composites," *Proceedings of Failure Modes in Composites III,* edited by Chiao and Schuster, AIME, New York, 1976

20. *New Developments and Applications in Composites,* edited by Doris Kuhlmann-Wilsdorf and William C. Harrigan Jr., Proceedings of the TMS-AIME Fall Meeting and Symposium sponsored by the TMS-AIME Physical Metallurgy and Composites Committees, St. Louis, Missouri, October 16–17, 1978, The Metallurgical Society of AIME, Warrendale, Pennsylvania, 1979

21. *Commercial Opportunities for Advanced Composites,* ASTM STP 704, edited by A. A. Watts, ASTM, Philadelphia, 1980

22. *Composite Materials and Composite Structures,* Proceedings of the Sixth Sagamore Ordnance Materials Research Conference, Co-sponsored by the Ordnance Materials Research Office and the Office of Ordnance Research of U.S. Army Contract No. DA-30-069-ORD-2566, Racquette Lake, New York, August 18–21, 1959

23. *Proceedings of the DARPA/NAVSEA Ceramic Gas Turbine Demonstration Engine Program Review,* sponsored by Naval Ship Engineering Center-Washington and Naval Research Laboratory, Maine Maritime Academy, Castine, Maine, August 1–4, 1977, MCIC-78-36, Metals and Ceramics Information Center, Columbus, Ohio

24. *Mechanical Behavior of Materials—IV,* Vol 1 edited; J. Carleson and N. G. Ohlson, Proceedings of the Fourth International Conference, Stockholm, Sweden, August 15–19, 1983, Pergamon Press, Oxford, 1983

25. J. F. Lynch, C. G. Ruderer, and W. H. Duckworth, *Engineering Properties of Ceramics,* AFML-TR-66-52, June 1966

26. *Automotive Engineering,* Society of Automotive Engineers, May 1982

27. Jacques E. Schoutens, *Introduction to Metal Matrix Composite Materials,* MMC No. 272, DOD Metal Matrix Composites Information Analysis Center, Santa Barbara, California, June 1982

28. Leo Alting, *Manufacturing Engineering Processes,* English version edited by Geoffrey Boothroy, Marcel Dekker, Inc. New York, 1974

29. *Source Book on Wear Control Technology,* edited by David A. Rigney and W. A. Glaeser, American Society for Metals, Metals Park, Ohio, 1978

30. *Pyrolytic Reinforcements for Ablative Plastic Composites,* Technical Report No. ML-TDR-64-201, Contract No. AF 33(657)-11094, July 1964

31. *Tough Composite Materials,* compiled by Louis F. Vosteen, Norman J. Johnson, and Louis A. Teichman, Proceedings of a Workshop, sponsored by NASA Langley Research Center, Hampton, Virginia, May 24–26, 1983, NASA Conference Publication 2334, NASA, 1984

32. *Aerospace Engineering,* SAE, Vol 5, No. 5, May 1985

33. *Plastics Engineering,* SAE, Vol XLI, No. 4, April 1985

34. *Plastics Compounding,* Harcourt Brace Jovanovich Publications, November/December 1981

35. *Proceedings of the Seventh Annual Conference on Composites and Advanced Ceramic Materials,* sponsored by the Ceramic-Metal Systems Division, The American Ceramic Society, Cocoa Beach, Florida, January 16–19, 1983, The American Ceramic Society, Columbus, Ohio, 1983

36. *Sintered Metal-Ceramic Composites,* edited by G. S. Upadhyaya, Proceedings of the Third International School on Sintered Materials, New Delhi, December 6–9, 1983, Elsevier, Amsterdam, Holland, 1984

37. D. W. Richardson, *Modern Ceramic Engineering Structural Applications,* Short Course by National Institute of Ceramic Engineers, 1985

38. *Surfaces and Interfaces in Ceramic and Ceramic-Metal Systems,* edited by Joseph Pask and Anthony Evans, Plenum Press, New York, 1981

39. *Composite Materials: Testing and Design,* (Sixth Conference) ASTM STP 787, edited by I. M. Daniel, ASTM, Philadelphia, 1982

40. John W. Weeton, "Fiber-Metal Matrix Composites," *Machine Design,* February 20, 1969

41. *ICCM/2—Proceedings of the 1978 International Conference on Composite Materials,* Toronto, Canada, April 16–20, 1978, American Institute of Mining Metallurgical and Petroleum Engineers, Inc., New York, 1978

42. *Proceedings of the SAE Fatigue Conference P-109,* Dearborn, Michigan, April 14–16, 1982, Society of Automotive Engineers, Warrendale, Pennsylvania, 1982

43. P. Bracke, H. Schurmans, and J. Verhoest, *Inorganic Fibres and Composite Materials,* Pergamon Press, Oxford, 1984

44. *Automotive Engineering,* Society of Automotive Engineers, May 1985

45. H. L. Ewalds and R. J. H. Wanhill, *Fracture Mechanics,* Edward Arnold Ltd., London, 1984

46. *Aluminum, Vol 1, Properties, Physical Metallurgy and Phase Diagrams,* edited by Kent R. Van Horn, American Society for Metals, Metals Park, Ohio, 1967

47. *Materials and Processes,* SAMPE Vol 1, Papers No 1–13, Proceedings of the Fifth Technology Conference of the European Chapter, Montreux Convention Center, June 12–14, 1984

48. David L. McDanels, Robert A. Signorelli, and John W. Weeton, *Analysis of Stress-Rupture and Creep Properties of Tungsten-Fiber-Reinforced Copper Composites,* Report NASA TN D-4173, NASA, Cleveland, Ohio, September 1967

49. D. L. McDanels, R. W. Jech, and J. W. Weeton, *Stress-Strain Behavior of Tungsten-Fiber-Reinforced Copper Composites,* Report NASA TN D-1881, NASA, Cleveland, Ohio, October 1963

50. John V. Milewski, Harry S. Katz, and Kenneth W. Lee, *Evaluation of a New Submicron Ceramic Reinforcing Fiber,* Proceedings of the 40th Annual Conference, Reinforced Plastics/Composites Institute, January 28–February 1, 1985, The Society of the Plastics Industry, Inc.

51. *Short Fiber Reinforced Composite Materials,* ASTM STP 772, Proceedings of the Symposium of Short Fiber Reinforced Composite Materials, sponsored by ASTM Committee E-9 on Fatigue, Society of Automotive Engineers and the Society of Civil Engineers, Minneapolis, Minnesota, April 14–15, 1980 ASTM, Philadelphia, 1982

52. *Polymer Composites,* Vol 6, No. 1, Society of Plastics Engineers, January 1985

53. *Plastics Design Forum,* Harcourt Brace Jovanovich Publications, May/June 1985

54. *ACEE Composite Structures Technology,* Proceedings of NASA Conference, Seattle, Washington, August 13–16, 1984, NASA Conference Publication 2321, NASA, 1984

55. *Alumina as a Ceramic Material,* edited by Walter H. Gitzen, The American Ceramic Society, Columbus, Ohio, 1970

56. *Ceramic and Graphite Fibers and Whiskers,* Vol 1, edited by L. R. McCreight, H. W. Rauch, Sr., and W. H. Sutton, Academic Press, New York, 1965

57. Leroy W. Davis and Samuel W. Bradstreet, *Metal and Ceramic Matrix Composites,* Cahners Publishing Co., Boston, Massachusetts, 1970

58. *Fifth International Conference on Composite Materials ICCM-V,* edited by W. C. Harrigan, Jr., J. Strife, and A. K. Dhingra, Proceedings of the Fifth International Conference on Composite Materials, sponsored by the TMS Composite Committee, San Diego,

California, July 29, 30, August 1, 1985, The Metallurgical Society, Inc., Warrendale, Pennsylvania, 1985

59. *Proceedings of Colloquium/Workshop on Composite Materials and Structures,* edited by Stanley L. Channon, Washington, D.C., May 8–10, 1984, IDA Document D-70, Institute for Defense Analysis, Alexandria, Virginia, July 1984

60. *Plastics Engineering,* SAE, August 1985

61. Karl M. Prewo and John J. Brennan, "High-Strength Silicon Carbide Fibre-Reinforced Glass-Matrix Composites," *Journal of Materials Science,* No. 15, 1980

62. K. M. Prewo and E. J. Minford, "Graphite Fiber Reinforced Thermoplastic Glass Matrix Composites for Use at 1000 °F," *SAMPE Journal,* Vol 21-2, March/April 1985

63. K. M. Prewo and J. J. Brennan, "Silicon Carbide Yarn Reinforced Glass Matrix Composites," *Journal of Materials Science,* No. 17, 1982

64. *Textile Fibers for Industry,* Owens-Corning Fiberglas Corporation, 1979

65. *Textile World,* August 1970

66. *Fiberglas Reinforced Plastics,* Owens-Corning Fiberglas Corporation, 1962

67. Ramakrishna T. Bhatt, *Mechanical Properties of SiC Fiber-Reinforced Reaction Bonded Si_3N_4 Composites,* Report NASA TM-87085, USAAVSCOM Technical Report 85-C-14, NASA, Cleveland, Ohio, 1985

68. James A. DiCarlo, *High Performance Fibers for Structurally Reliable Metal and Ceramic Composites,* Report NASA TM-86878, NASA, Cleveland, Ohio, 1984

69. W. R. Bratschun, A. J. Mountvala, and A. G. Pincus, *Uses of Ceramics in Microelectronics,* NASA, Washington, D.C., 1971

70. David L. McDanels, *A Review of Carbon Fiber Reinforced Metal Matrix Composites—The Potential of Large-Diameter Carbon-Base Monofilaments,* Report NASA TMX-52922, NASA, Cleveland, Ohio, 1970

71. *Ceramic Industry,* Vol 125, No. 4, September 15, 1985

72. R. E. Gluyas and K. J. Bowles, *Improved Fiber Retention by the Use of Fillers in Graphite/Fiber Resin Matrix Composites,* Report NASA TM-79288, NASA, Cleveland, Ohio, 1980

73. R. G. Barrows, *A Weibull Characterization for Tensile Fracture of Multicomponent Brittle Fibers,* Report NASA TM-73790, NASA, Cleveland, Ohio, 1977

74. *Fiber Composite Materials,* Papers presented at the American Society for Metals, Metals Park, Ohio, October 17, 18, 1964, American Society for Metals, Metals Park, Ohio, 1965

75. "Silicon Nitride Whisker and Silicon Carbide Whisker of Tateho Chemical Industries Ltd.," Report by Tateho Chemical Industries Co., Ltd., Hyogo-ken, Japan

76. Donald W. Petrasek and Robert A. Signorelli, *Preliminary Evaluation of Tungsten Alloy Fiber-Nickel-Base Alloy Composites for Turbojet Engine Applications,* Report NASA TN D-5575, NASA, Cleveland, Ohio, February 1970

77. *Whisker Technology,* edited by Albert P. Levitt, John Wiley and Sons, New York, 1970

78. Anon, "Astroquartz, the Flexible Approach," J. P. Stevens Co., New York, 1983

79. Anon, "Kevlar 49 Data Manual," E. I. DuPont de Nemours and Co., Wilmington, Delaware

80. A. G. Metcalfe and G. K. Schmitz, "Effect of Length on the Strength of Glass Fibers," ASTM Preprint No. 87, presented at 67th Annual Meeting, June 1964

81. "Processing of Gafite," GAF Technical Bulletin, GAF Corporation, New York, 1978

82. R. W. Jech, J. W. Weeton, and R. A. Signorelli, "Fibering of Oxides in Refractory Metals," *Fiber-Strengthened Metallic Composites,* ASTM STP 427, ASTM, Philadelphia, 1967

83. Max Quatinetz, John W. Weeton, and Thomas P. Heibell, *Studies of Tungsten Composites Containing Fibered or Reactive*

Additives, Report NASA TN D-2757, NASA, Cleveland, Ohio, April 1965

84. C. A. Hoffman and J. W. Weeton, *Metal-Metal Laminar Composites for High Temperature Applications,* Report NASA TM X-68056, NASA, Cleveland, Ohio, May 1972

85. C. A. Hoffman and J. W. Weeton, *Tensile Behavior of Unnotched and Notched Tungsten-Copper Laminar Composites,* Report NASA TN D-8254, NASA, Cleveland, Ohio, 1976

86. E. J. Peters, and Dr. Paul Boymal, "Ceramic Fibers in Advanced Composites," Fiber Development Conference, Impact 1985, Sponsored by Marketing Technology Service, Inc., Fort Lauderdale, Florida, March 10–12, 1985

87. "Developmental Data, Fiberfrax Reinforcement of Plastics," Technical Bulletin, Sohio Engineered Materials Co., Fibers Division, Sohio Carborundum, Niagara, New York

88. J. W. Weeton and R. A. Signorelli, "Fiber-Metal Composites," *Strengthening Mechanisms-Metals and Ceramics,* edited by John J. Burke, Norman I. Reed, and Volker Weiss, Syracuse University Press, Syracuse, New York, 1966

89. John A. Alexander, Robert G. Shaver, and James C. Withers, *Critical Analysis of Accumulated Experimental Data on Filament-Reinforced Metal-Matrix Composites,* Contract No. NASW-1779, NASA, June 1969

90. "New Technology, New Horizons," Technical Bulletin, Allied Fibers Division, New York, 1985

91. Donald W. Petrasek, *High-Temperature Strength of Refractory-Metal Wires and Consideration for Composite Applications,* Report NASA TN D-6881, NASA, Cleveland, Ohio, August 1972

92. Donald W. Petrasek and Robert A. Signorelli, *Tungsten Fiber Reinforced Superalloys—A Status Review,* Report NASA TM-82590, NASA, Cleveland, Ohio, January 1981

93. *Carbon 76,* Conference proceedings of the Second International Carbon Conference, sponsored by the Working Group on Carbon of the German Ceramic Society, Baden-Baden, West Germany, 1976

94. Erich Fitzer, Antonios Gkogkidis, and Michael Heine, "Carbon Fibres and Their Composites," *High Temperatures—High Pressures,* Vol 16, 1984

95. C. T. Lynch, and J. P. Kershaw, *Metal Matrix Composites,* The Chemical Rubber Co., CRC Press, Cleveland, Ohio, 1972

96. *Plastics Engineering,* SAE, November 1985

97. *Plastics Compounding,* Harcourt Brace Jovanovich Publications, November/December 1985

98. *Material Engineering,* Penton/IPC, August 1978

99. *Modern Plastics Encyclopedia,* McGraw-Hill, 1978–1979

100. Rod Doerr, Ed Green, B. Lyon and S. Taha, *Development of Effective Machining and Tooling Techniques for Kevlar Composite Laminates,* Applied Technology Laboratories, U.S. Army Research & Technology Laboratories (AVRADCOM), June 1982

101. Anon, *A Guide to Cutting and Machining Kevlar Aramid Composites,* Du Pont Company, 1983

102. *Materials Review for '72,* SAMPE Vol. 17, Proceedings of National SAMPE Symposium and Exhibition, Los Angeles, California, April 11–13, 1972, SAMPE, Azusa, California, 1972

103. Leonard E. Samuels, *Metallographic Polishing by Mechanical Methods,* American Society for Metals, Metals Park, Ohio, 1982

104. S. R. Levitt, "High Strength Graphite Fibre/Lithium Aluminosilicate Composites," *Journal of Materials Science,* Vol 8, 1973

105. R. L. Mehan, "Effect of SiC Content and Orientation on the Properties of Si/SiC Ceramic Composites," *Journal of Materials Science,* Vol 13, 1978

106. J. Sambell, D. H. Bowen, and D. C. Phillips, "Carbon Fiber Composites with Ceramic Glass Matrices," *Journal of Materials Science,* Vol 7, 1972

107. I. W. Donald and P. W. McMillan, "Review Ceramic Matrix Composites," *Journal of Materials Science,* Vol 11, 1976

108. *Treatise on Material Science and Technology,* Vol 11, edited by R. K. MacCrone, Academic Press, Orlando, Florida, 1977

109. F. J. Jelinek, "Ceramic Composites," *Current Awareness Bulletin,* Metals and Ceramics Information Center, Columbus, Ohio, December 1985, Issue No. 154

110. *Composite Materials,* edited by L. J. Broutman and R. H. Krock, Vol. 1, *Interfaces in Metal Matrix Composites,* edited by Arthur G. Metcalfe, Academic Press, New York, 1974

111. B. I. Edelson and W. M. Baldwin, Jr., *The Effect of Second Phases on the Mechanical Properties of Alloys,* Transactions of the American Society for Metals, Volume LV, March, June, September, December, 1962, American Society for Metals, Metals Park, Ohio, 1962

112. G. E. Metzger, "Joining of Metal-Matrix Fiber-Reinforced Composite Materials," *WRC Bulletin No. 207,* Welding Research Council, 1975

113. R. L. Mehan, R. P. Jakas, and G. A. Bruch, *"Behavior Study of Sapphire Wool Aluminum Composites,"* AFML-TR-68-100, May 1968, Space Sciences Laboratory, General Electric Company, Air Force Contract No. F 33615-67-C-1308

114. A. W. H. Morris and A. Burwood-Smith, "Some Properties of a Fiber-Reinforced Nickel-Base Alloy," *Fibre Science Technology,* 3(1), 1970, pp 53–78

115. R. H. Baskey, *Fiber Reinforced Metallic Composite Materials,* Clevite Corporation (AFML-TR-67-196 AD-825364), September 1967

116. N. Nilsen and J. H. Sovik, "Fatigue of Tungsten Fiber Reinforced Nickel," *Practical Metallic Composites,* Proceedings of the Spring Meeting, Institution of Metallurgists, London, 1974, pp B51–B54

117. Diane M. Essock, "FeCrAlY Matrix Composites for Gas Turbine Applications," *Advanced Fibers and Composites for Elevated Temperatures,* edited by I. Ahmad and B. R. Noton, Proceedings of the 108th AIME Annual Meeting and Symposium, sponsored by the Metallurgical Society of AIME and ASM Joint Composite Materials Committee, New Orleans, Louisiana, February 20–21, 1979, The Society of AIME, Warrendale, Pennsylvania, 1980

118. R. W. Jech and R. A. Signorelli, *Evaluation of Silicon Carbide Fiber/Titanium Composites,* TM-79232, NASA Lewis Research Center, Cleveland, Ohio, July 1979

119. I. J. Toth, W. D. Brentnall, and G. D. Menke, "Making a Product from Composites," *Journal of Metals,* Vol. 24, October 1972, Metallurgical Society/AIME

120. D. Cratchley, "Factors Affecting the UTS of a Metal/Metal-Fibre Reinforced System," *Powder Metallurgy,* No. 11, 1963

121. K. M. Prewo and K. G. Kreider, "High-Strength Boron and Borsic Fiber Reinforced Aluminum Composites," *Journal of Composite Materials,* July 1972

122. G. V. Samsonov, A. D. Panasyk and H. S. Borivikova, "Wettability and Surface Properties of Melts and Solid Bodies," Naukova Dumka, Kiev, 1972

123. V. I. Tumanov, A. E. Goyburov and G. M. Kondratenko, ibid.

124. J. Theberge et al., 30th Annual Conference, Reinforced Plastics/Composites Institute, SPI, 1972, Section 14-C

125. K. M. Prewo, "New Dimensionally and Thermally Stable Composite," Conference on Advanced Composites, Technology Conference Associates, El Segundo, CA, December 1979, p. 1

SECTION 9
Bibliography of Reference Books
on Composites

Analysis and Performance of Fiber Composites

Bhagwan D. Agarwal and Lawrence J. Broutman
Wiley Interscience, New York
ISBN: 0-471-05928-5
355 pages, 1980

Society of Plastics Engineers Monographs. Individual study manual, requires knowledge of strengths of materials. Solutions manual available.

Advanced Fibers and Composites for Elevated Temperatures

I. Ahmad and B. R. Noton, editors
Metallurgical Society of AIME and American Society for Metals, Joint Composite Materials Committee at the 108th AIME Annual Meeting, New Orleans, LA, 20–21 February 1979
ISBN: 0-89520-366-9
252 pages, 1980

Topics restricted to metal- and ceramic-matrix composites. Subjects: high-temperature fiber reinforcements, composites for intermediate (<1,000 °F) applications, and composites for high-temperature (1,000–2500 °F) applications.

Inorganic Fibres and Composite Materials (A Survey of Recent Developments)

P. Bracke, H. Schurmans, and J. Verhoest
Pergamon Press, Elmsford, NY
ISBN: 0-08-031145-8
176 pages, 1984

Covers general methods for the manufacture of composite materials, including metal fiber-metal matrix systems, carbon-, boron-, carbide-, boride-fiber-metal matrix systems, and ceramic and glass.

Composite Materials (8 volumes)

Lawrence J. Broutman and Richard H. Krock, editors
Academic Press, New York
[Individual volumes are described below.]

Composite Materials, Volume 1 — Interfaces in Metal Matrix Composites

Arthur G. Metcalfe, editor
Academic Press, New York
ISBN: 0-12-136501-8
421 pages, 1974

Research on the study of interfaces, summarizing previous achievements in the field and discussing the effects of interfaces on longitudinal and transverse strength and on fracture.

Composite Materials, Volume 2 — Mechanics of Composite Materials

G. P. Sendeckyj, editor
Academic Press, New York
ISBN: 0-12-136502-6
503 pages, 1974

A summary of the development of composites, consisting of deformation behavior, strength, and experimental methods of characterizing these materials.

Composite Materials, Volume 3 — Engineering Applications of Composites

Bryan R. Noton, editor
Academic Press, New York
ISBN: 0-12-136503-4
515 pages, 1974

Emphasis on glass-reinforced plastics and their applications. Discusses the emergence of fiber-reinforced composites as substitutes for traditional materials.

Composite Materials, Volume 4 — Metallic Matrix Composites

Kenneth G. Kreider, editor
Academic Press, New York
ISBN: 0-12-136504-2
585 pages, 1974

Each chapter of this volume discusses a separate materials system, with the purpose of development of refined fabrication techniques and evaluation of the physical and mechanical properties of each composite material. Emphasis on aerospace applications.

Composite Materials, Volume 5 — Fracture and Fatigue

Lawrence J. Broutman, editor
Academic Press, New York
ISBN: 0-12-136505-0
465 pages, 1974

Properties of polymer-, metal-, and ceramic-matrix composites are discussed. Concepts, theories, and experiments with these materials are also considered.

Composite Materials, Volume 6 — Interfaces in Polymer Matrix Composites

Edwin P. Plueddemann, editor
Academic Press, New York
ISBN: 0-12-136506-9
294 pages, 1974

Examination of the role of intermediate materials to promote bonding across the interface with the ultimate goals of improved reproducibility of composite properties with familiar materials, combined high strength and toughness in composites, and improved reinforced thermoplastics.

Composite Materials, Volume 7 — Structural Design and Analysis — Part I

C. C. Chamis, editor
Academic Press, New York
ISBN: 0-12-136507-7
345 pages, 1974

Analysis and design principles are examined, including continuum mechanics principles dealing with anisotropic elasticity; composite failure mechanics; structural/stress analysis for struts, plates, and shells; wave propagation and impact mechanics; stress concentrations at discontinuities and joints; reliability; automated design; and various test methods.

Composite Materials, Volume 8 — Structural Design and Analysis — Part II

C. C. Chamis, editor
Academic Press, New York
ISBN: 0-12-136508-5
297 pages, 1974

A continuation of the analysis and design principles considered in Volume 7 — Part I.

Modern Composite Materials

Lawrence J. Broutman and Richard H. Krock, editors
Addison-Wesley Publishing Co., Reading, MA
581 pages, 1967

Describes the nature of composites, fibrous reinforcements, and the relations between structure and properties of particulate and fibrous composites. Topics include fundamentals, mechanics, chemical and time-dependent phenomena, properties, preparation of reinforcements, and technical applications.

Composite Materials: Quality Assurance and Processing

C. E. Browning, editor
From the Symposium on Producibility and Quality Assurance of Composite Materials, sponsored by ASTM Committee D-30 on High Modulus Fibers and Their Composites, St. Louis, MO, 20 October 1981
ASTM Special Technical Publication 797
PCN: 04-797000-30
173 pages

Includes developments in the applications of composites in the aerospace and automotive industries. Relevant methodologies, rationales, and experimental results are given.

Test Methods and Design Allowables for Fibrous Composites

C. Chamis, editor
ASTM Special Technical Publication 734
PCN: 04-734000-33
429 pages, 1981

Discussion of quantifying material properties for design by selecting and setting design allowables for composites.

Fiberglass-Reinforced Plastics Deskbook

Nicholas P. and Paul N. Cheremisinoff, Ann Arbor Science
ISBN: 0-250-40245-9
327 pages, 1978

Covers engineering plastics and their properties, FRP resins and their properties, pipe products and applications, pipe system design and installation, fiberglass tanks, air-handling equipment, industrial applications and products, industrial linings, and product standards and codes.

Mechanics of Composite Materials

Richard M. Christensen
Wiley Interscience, New York
ISBN: 0-471-05167-5
348 pages, 1979

Covers spherical inclusions, cylindrical and lamellar systems, and bounds of effective moduli. Includes solid media behavior and some fluid suspensions along with analyses of elasticity, viscoelasticity, and plasticity.

Encyclopedia/Handbook of Materials, Parts, and Finishes

Henry R. Clauser, editor
Technomic Publishing Co., Lancaster, PA
ISBN: 0-87762-189-6
564 pages, 1976

A basic source for industrial materials and production methods. Covers all major forming and fabrication methods for metals and their alloys, plastics, coatings and finishes, rubber and elastomers, adhesives, ceramics, films, composites, woods, and textiles.

Composite Materials: Testing and Design (Sixth Conference)

I. M. Daniel, editor
ASTM Special Technical Publication 787
PCN: 04-787000-33
587 pages, 1982

Covers test methods, material characterization, fracture and failure analysis, fatigue, nondestructive evaluation, time-dependent and dynamic response, environmental effects, durability and reliability, and testing of composite structures.

Technology of Carbon and Graphite Fiber Composites

John Delmonte
Van Nostrand Reinhold, New York
ISBN: 0-442-22072-3
464 pages, 1981

Associates the chemistry of resin binders, processing techniques, test methods, and applications with analyses of the structures of composites. Examines the qualities, characteristics, advantages, and disadvantages of these materials.

Design, Fabrication and Mechanics of Composite Structures

Technomic Publishing Co., Lancaster, PA
ISBN: 87762-358-9
352 pages, 1984

Seminar notes covering mechanics, fatigue and fracture of composites (with emphasis on designing for safety and durability), fabrication, and design.

New Composite Materials and Technology

Ashok K. Dhingra and R. Byron Pipes, editors
AIChE Symposium Series, No. 217, Vol 78
Detroit, MI 16–19 August 1981
66 pages, 1982

Papers discussing high-performance continuous fiber reinforced composites, formable fiber reinforced thermoplastic composites, metal/plastic laminates, short fiber and filler reinforcements, hybrid composites, and advanced composites for industrial use.

Encyclopedia of Composite Materials and Components

Martin Grayson, editor
Wiley Interscience, New York
ISBN: 0-471-87357-8
1161 pages, 1983

Details methods of manufacturing, properties, uses, and components of composites by 75 top materials scientists. Alphabetical arrangement with cross-references and an extensive index.

Effects of Defects in Composite Materials

A symposium sponsored by ASTM Committees D-30 on High Modulus Fibers and Their Composites and E-9

on Fatigue, San Francisco, CA, 13–14 December 1982
ASTM Special Technical Publication 836
PCN: 04-836000-33
272 pages, 1984

A discussion of defects in carbon/epoxy laminates in manufacturing or service use. Areas of concentration: methods of observing and measuring defect location and size, consequences of defects, analytical models for predicting defect behavior, and evaluation of failure surfaces involving defects.

Primer on Composite Materials: Analysis

John C. Halpin
Technomic Publishing Co., Lancaster, PA
ISBN: 87762-349-X
187 pages, 1984

Contains text on the properties of an orthotropic lamina, laminated composites, strength of laminated composites, analysis of composite structures, structure/property relationships for composite materials, and characterization and behavior of structural composites.

Polymer Blends and Composites in Multiphase Systems

C. D. Han, editor
American Chemical Society
ISBN: 0-8412-0783-6 Advances in Chemistry Series 206
383 pages, 1984

Reports research efforts on polymer blends and composites. Compatibility and characterization of polymer blends, rheology, processing, and properties of heterogeneous polymer blends, and polymer composites are discussed.

Fibre Composite Hybrid Materials

N. L. Hancox, editor
Macmillan Publishing Co., Inc., New York
ISBN: 0-02-949950-X
290 pages, 1981

Examination of fiber hybrids made from continuous or woven carbon, glass, or aramid fibers in thermosetting matrices with core materials of foamed or filled polymer, metal, honeycomb, or wood.

Nonmetallic Materials and Composites at Low Temperatures

Günther Hartwig and David Evans, editors
Plenum Press, New York
ISBN: 0-306-40894-5
410 pages, 1982

An examination of the cryogenic behavior of epoxies, polyethylenes, polymers, composites, and glasses by their thermal and dielectric properties, elasticity, cohesive strength, resistance to strain and fracture, and application.

Mechanics of Composite Materials — Recent Advances

Zvi Hashin and Carl T. Herakovich, editors
Pergamon Press, Elmsford, NY
ISBN: 0-08-029384-0
499 pages, 1983

A study of the mechanical properties of particulate and fiber composite systems. Properties examined include elasticity, thermal expansion, viscoelasticity and vibration damping, plasticity, nonlinear behavior, temperature dependence of mechanical properties, conductivity, moisture absorption, static strength, fracture mechanics, and fatigue failure.

Ultrastructure Processing of Ceramics, Glasses, and Composites

Larry L. Hench and Ronald R. Ulrich, editors
Wiley Interscience, New York
ISBN: 0-471-89669-1
564 pages, 1984

Compilation of material from the International Conference on Ultrastructure Processing of Ceramics, Glasses, and Composites, including sol-gel processing, organometallic precursors, chemical micromorphology, and phase transformation based processing.

Developments in Composite Materials — 1

G. S. Holister, editor
Applied Science Publishers Ltd., London
ISBN: 0-85334-740-9
245 pages, 1977

Improvements in and applications of composites. Results of research into behavior of composites in corrosive environments and under vibrational stress and extreme conditions of loading. Examination of stress fields that occur because of shapes, structures, and loading conditions.

Glass Reinforced Plastics in Construction — Engineering Aspects

Leonard Hollaway
John Wiley & Sons, New York
ISBN: 0-470-99338-3
228 pages, 1978

Theory, use, and basic characteristics of glass reinforced plastics, including mechanical properties, strength, durability, and methods of fire testing are discussed. Examples of methods of sandwich design and construction and analysis of components and structures are also given.

Handbook of Fillers and Reinforcements for Plastics

Harry S. Katz and John V. Milewski, editors
Van Nostrand Reinhold, New York
ISBN: 0-442-25372-9
784 pages, 1978

Coordinates material type and supplier for polymers, fiberglass, plastics and reinforced plastics/composites.

Kevlar Composites

Los Angeles, CA
ISBN: 0-938648-02-0
119 pages, 1980

Machining, materials selection, design, performance and applications of Kevlar composites.

Fiber Technology, from Film to Fiber

H. A. Krassig, J. Lenz, and H. F. Mark
ISBN: 0-8247-7097-8
333 pages, 1984

Examines the relationships between process limitations, performance, and product properties in film-to-fiber technology for packaging fabrics, carpet-backing fabrics, fibrillated yarns, outdoor carpeting, and other products.

Fatigue of Fibrous Composite Materials

K. N. Lauraitis, editor
ASTM Special Technical Publication 723
PCN: 04-723000-33
311 pages, 1981

Fatigue mechanisms are examined theoretically and by prediction models and test methods.

In Situ Composites IV

F. D. Lemkey, H. E. Cline, and M. McLean, editors
Elsevier Science Publishing Co., Inc., New York
ISBN: 0-444-00726-1
343 pages, 1982

Proceedings of the Materials Research Society Annual Meeting. New findings and technologically significant developments from studies of directionally transformed materials.

Fibrous Composites in Structural Design

Edward M. Lenoe, Donald W. Oplinger, and John J.
Burke, editors
Plenum Press, New York
ISBN: 0-306-40354-5
873 pages, 1980

Papers from Fourth Conference on Fibrous Composites, jointly sponsored by the Army, Navy, Air Force, and NASA, concerning non-aerospace applications such as bridges, flywheel energy storage systems, ship and surface vessel components, and ground vehicle components for advanced composites.

Handbook of Composites

G. Lubin
Van Nostrand Reinhold Co., New York
ISBN: 0-442-24897-0
786 pages, 1982

Detailed information on materials, processes, and applications of composites. Unsaturated polyester resins are covered extensively.

Composite Construction Materials Handbook

Robert Nicholls
Prentice-Hall, Inc., Englewood Cliffs, NJ
ISBN: 0-13-164889-6
580 pages, 1976

Associates chemical structure, mechanical behavior, and design optimization of composites used for construction. Emphasis on the dependence of mechanical properties on chemical structure. Many examples and problems accompany text.

**ICCM/Z — Proceedings of the 1978 International
Conference on Composite Materials**

Bryan Noton, Robert A. Signorelli, Kenneth N. Street,
and Leslie N. Phillips, editors
The Metallurgical Society of AIME, Warrendale, PA
ISBN: 0-89520-142-9
1657 pages, 1978

Detailed descriptions of the science, engineering, and utilization of composite materials. Topics include commercial products, fabrication methods, tool design, and nondestructive evaluation.

Long-Term Behavior of Composites

T. Kevin O'Brien, editor
ASTM Special Technical Publication 813
PCN: 04-813000-33
300 pages, 1983

Covers time-dependent behavior, fatigue behavior, long-term environmental effects, and reliability and life prediction of composites.

Load-Bearing Fibre Composites

Michael R. Piggott
Pergamon Press, Elmsford, NY
ISBN: 0-08-024230-8 (hardcover); 0-08-024231-6
(flexicover)
277 pages, 1980

Provides data on load-bearing capabilities, strength, creep resistance, and fatigue resistance in a general overview. Also discusses mechanical and composite matrix properties.

**Nondestructive Evaluation and Flaw Criticality for
Composite Materials**

R. B. Pipes, editor
A symposium sponsored by ASTM Committee D-30 on
High Modulus Fibers and Their Composites,
Philadelphia, 10–11 October 1978
ASTM Special Technical Publication 696
PCN: 04-696000-33
358 pages, 1979

Nondestructive evaluation methodology, flaw criticality, and flaw characterization are assessed to promote an understanding among engineers, materials scientists and physicists regarding composite structures.

**Determination of Dynamic Properties of Polymers and
Composites**

Brian Eric Read and Gregory Donald Dean
Wiley Interscience, New York
ISBN: 0-470-26543-4
207 pages, 1979

Original testing methods for measuring dynamic moduli and damping factors of polymers and composites. Tests done over a broad frequency range using nonresonance, torsion pendulum, audiofrequency resonance, and ultrasonic deformation modes.

Handbook of Composite Construction Engineering

Gajanan M. Sabnis, editor
Van Nostrand Reinhold, New York
ISBN: 0-442-27735-0
400 pages, 1979

Covers in detail combinations of structural steel, concrete, and wood needed to support stress loads on buildings and bridges. Contains construction methods and code applications.

Short Fiber Reinforced Composite Materials

B. Sanders, editor
ASTM Special Technical Publication 772
PCN: 04-772000-30
258 pages, 1982

Contains application information on high-volume processes in the automotive, consumer appliance, and commercial business machine industries. Topics include new material developments, test methods, engineering properties, fracture behavior, and environmental effects.

Introduction to Metal Matrix Composite Materials

Jacques E. Schoutens
Prepared under sponsorship of the DOD Metal Matrix
Composites Information Analysis Center
MMC: 272
640 pages, 1982

Covers metal matrix composites, interfacial reactions between fibers and matrix, metal matrix composite fabrication methods, mechanical behavior of composites, properties, and test methods.

Composite Materials Handbook

Mel M. Schwartz, editor
McGraw-Hill Book Co., New York
ISBN: 0-07-055743-8
672 pages, 1984

Describes each material and its method of manufacture and processing, compares metal to resin matrix systems, and explains industrial applications. Covers molding, curing, machining, casting, cutting, trimming, fastening, and joining.

Fabrication of Composite Materials

Mel M. Schwartz, editor
American Society for Metals, Metals Park, OH
ISBN: 0-87170-198-7
406 pages, 1985

Joining, trimming and cutting, painting, drilling, and machining of composites are covered in this source book of outstanding articles from the technical literature.

Seventh Annual Conference on Composites and Advanced Ceramic Materials

A collection of papers presented at the seventh annual conference of the Ceramic-Metal Systems Division of The American Ceramic Society, Cocoa Beach, FL, 16–19 January 1983
ISSN: 0196-6219
1983

Papers from leading government officials, businessmen, and academicians concerning composites and advanced materials.

Advanced Composites: Design and Applications

T. Robert Shives and William A. Willard, editors
Proceedings of the 29th meeting of the Mechanical Failures Prevention Group (MF PG), Gaithersburg, MD, 23–25 May 1979
National Bureau of Standards Special Publication 563
304 pages, 1979

Symposium subjects: design, applications, and failure modes of advanced composites for aerospace, aircraft, automotive, marine, and industrial use.

Carbon Black-Polymer Composites

E. K. Sichel, editor
Van Nostrand Reinhold, New York
ISBN: 0-8247-1673-6
224 pages, 1982

Examines the physical properties of these composites, including dc conductivity, high-frequency ac conductivity, rheology, morphology, and triboelectricity. Electrical conductivity covered in depth. Provides state-of-the-art materials characterization techniques.

Cracks in Composite Materials

George C. Sih and E. P. Chen, editors
Martinus Nijhoff Publishers, Boston, MA
622 pages, 1981

An analysis of stress loads that create cracks in composites. Mixed-mode crack extension is examined by the strain energy density criterion.

Environmental Effects on Composite Materials — Volume 1

George S. Springer, editor
Technomic Publishing Co., Lancaster, PA
ISBN: 087762-300-7
203 pages, 1981

Technical reports on the performance of advanced composites in relation to moisture and temperature variables. State-of-the-art theory and data with recent experimental and analytical results.

Environmental Effects on Composite Materials — Volume 2

George S. Springer, editor
Technomic Publishing Co., Lancaster, PA
ISBN: 87762-348-1
438 pages, 1984

A continuation of the moisture/temperature effects presented in Volume 1.

Biocompatible Polymers, Metals and Composites

Michael Szycher, editor
Technomic Publishing Co., Lancaster, PA
ISBN: 87762-323-6
1071 pages, 1983

Discusses plastics for biochemical applications, particularly the technical challenge presented by polymer compatibility with aggressive biological systems.

Technology Vectors

29th National SAMPE Symposium and Exhibition, Vol. 29, Reno, NV, 3–5 April 1984
ISBN: 0-0938994-24-7
1612 pages, 1984

Features many papers dealing with design histories of structures using composite materials and fabrication methods. Also contains papers on robotics and computer science as they interrelate with material and process engineering.

Mechanics of Fibre Composites

V. K. Tewary
Wiley Interscience, New York
ISBN: 0-470-99240-9
288 pages, 1978

Basic theory of fiber composites with emphasis on applications, including mechanics of solids, elastic constants, fracture and failure, propagation of elastic waves, and effect of defective fibers.

Composite Materials: Testing and Design (Fifth Conference)

S. W. Tsai, editor
ASTM Special Technical Publication 674
PCN: 04-674000-33
697 pages, 1979

Applications, design and data, testing and evaluation, environmental effects, fatigue, physiochemical properties, and mechanisms of failure of composite materials.

Introduction to Composite Materials

S. W. Tsai and H. Thomas Hahn
Technomic Publishing Co., Lancaster, PA
ISBN: 0-87762-288-4
457 pages, 1980

Introduces the governing principles of strength and stiffness of uni- and multi-directional composites. Explains the versatility, uniqueness, and structural advantages of composites as compared with conventional materials.

Materials Technology Series

Stephen W. Tsai and H. Thomas Hahn, editors
Technomic Publishing Co., Lancaster, PA

Volume 1 — Carbon Reinforced Epoxy Systems
295 pages, 1974

Volume 2 — Glass Reinforced Epoxy Systems
254 pages, 1974

Volume 3 — Boron Reinforced Epoxy Systems
127 pages, 1974

Volume 4 — Aluminum, Steel and Organic Reinforced Epoxy Systems
97 pages, 1974

Volume 5 — Reinforced Phenolic, Polyester, Polyimide and Polystyrene
143 pages, 1974

Volume 6 — Boron Reinforced Aluminum Systems
124 pages, 1974

Volume 7 — Carbon Composite and Metal Composite Systems
118 pages, 1974

Volume 8 — Carbon Reinforced Epoxy Systems, Part 2
243 pages, 1982

Volume 9 — Carbon Reinforced Epoxy Systems, Part 3
217 pages, 1982

Volume 10 — Glass Reinforced Epoxy Systems, Part 2
203 pages, 1982

Volume 11 — Boron Reinforced Aluminum Systems, Part 2
137 pages, 1982

Volume 12 — Carbon Reinforced Epoxy Systems, Part 4
271 pages, 1984

Volume 13 — Carbon Reinforced Epoxy Systems, Part 5
280 pages, 1984

Volume 14 — Glass Reinforced Polyester Systems
182 pages, 1984

Sintered Metal — Ceramic Composites (Materials Science Monographs, 25)

G. S. Upadhyaya, editor
Elsevier, New York
ISBN: 0-444-42401-6 (Volume 25), 0-444-41685-4 (Series)
540 pages, 1984

Proceedings of the Third International School on Sintered Materials, covering fundamentals of powder production, compaction and sintering, metal, refractory metal, silicon nitride, and oxide ceramic matrix composites, and hard metals.

Recent Advances in Composites in the United States and Japan

Jack R. Vinson and Minoru Taya, editors
ASTM Special Technical Publication 864
PCN: 04-864000-33
738 pages, 1985

A collection of 44 peer-reviewed papers on composite materials from the U.S. and Japan. Subjects include fracture, fatigue, stress analysis, dynamic behavior, design, fabrication methods, testing methods, elevated-temperature effects, and thermomechanical properties of composite materials.

Tough Composite Materials

Louis F. Vostun, Norman J. Johnston, and Louis A. Teichman, editors
Proceedings of a workshop sponsored by NASA Langley Research Center, Hampton, VA, 24–26 May 1983
391 pages, 1983

Areas of concentration: composite fracture toughness and impact characterization, constituent properties and interrelationships, and matrix synthesis and characterization. Discusses NASA's goal to achieve weight reduction by using composites in commercial transports.

Composites: State of the Art

J. W. Weeton and E. Scala, editors
Proceedings of the 1971 Fall Meeting of The Metallurgical Society of the American Institute of Mining, Metallurgical and Petroleum Engineers (AIME) Detroit, MI
365 pages, 1974

A comparison of composites with conventional materials, encompassing a review of fabrication techniques and applications, with the goal of establishing better composite properties, solving technical use problems, and reducing material and fabrication costs.

Other Books of Interest

Advanced Composite Materials — Environmental Effects

ISBN: 0-686-51998-1
ASTM: 04-658000-33
1978

BCC Staff: Advanced Composites P-023R: Commercial Applications

ISBN: 0-89336-160-7
1978

Powdered and Particulate Rubber Technology

C. W. Evans
Elsevier Applied Science, England
ISBN: 0-85334-773-5
107 pages, 1978

Developments in Rubber and Rubber Composites

C. W. Evans
Elsevier Applied Science, England
Volume I
ISBN: 0-85334-892-8
184 pages, 1980
Volume II
ISBN: 0-85334-173-7
183 pages, 1983

Advances in Composite Materials

G. Piatti, editor
Applied Science
ISBN: 0-85334-770-0
1978

Determination of Dynamic Properties of Polymers and Composites

Brian E. Read and Gregory D. Dean
Halsted Press
ISBN: 0-470-26543-4
1979

Polymer Engineering Composites

M. O. Richardson, editor
ISBN: 0-85334-722-0
1977

Composite Materials and Their Use in Structures

J. R. Vinson and T. W. Chou
438 pages

Textile Reinforcement of Elastomers

W. C. Wake, editor
Elsevier Applied Science, England
ISBN: 0-85334-998-3
271 pages, 1982

New Fibres and Their Composites
W. Watt
Royal Society, London
ISBN: 0-85403-130-8
189 pages, 1980

Journals of Interest

Advanced Materials and Processes

John C. Bittence, editor
American Society for Metals
Metals Park, OH

Comparative properties, design requirements, and processing characteristics of alternative engineering materials are discussed. Focus is on high-performance applications but is aimed at readers from all industries, including aerospace, automotive, machine tool, and military.

Ceramic Industry

Cahners Publishing Co.
Des Plaines, IL

Provides news, new product literature, and features on the advanced ceramics market, along with glass, whiteware, and porcelain enamel markets. Published monthly.

Composite Science and Technology

Bryan Harris, editor
Elsevier Applied Science Publishers
Essex, England

Formerly *Fibre Science and Technology*. The journal publishes referred original articles, occasional review papers, and letters on all aspects of fundamental and applied science of engineering composites. Eight issues are published annually, in two volumes.

Composites Technology Review

ASTM
Philadelphia, PA

Composite materials and their response to environment, application technology, and structural behavior. Topics include new materials, test methods, uses, and processing and manufacturing techniques. Published quarterly.

Journal of Composite Materials

Stephen W. Tsai and H. Thomas Hahn, editors
Technomic Publishing Co., Lancaster, PA

Reviews theoretical and experimental studies of the physical and structural properties of multiphase materials. Technological orientation; phenomenological and mechanistic approaches. Published bimonthly.

Journal of Reinforced Plastics and Composites

Stephen W. Tsai and George S. Springer, editors
Technomic Publishing Co., Lancaster, PA

Results of research into material composition, manufacturing processes, material properties, analytical methods for design, and design criteria of plastics and composites by industry, government, and universities. Published quarterly.

National Glass Budget

Box 7138
Pittsburgh, PA 15213

A newspaper covering the worldwide glass industry. Annual directory lists glass manufacturing facilities in the U.S. and Canada, along with their management, capacity, and products. Data base. Published biweekly.

Plastics Design Forum

Mel Friedman, editor
Resin Publications, Inc.
Duluth, MN

Emphasis is on current engineering technology and design. Magazine also includes calendar of events for conferences, exhibitions, and seminars relating to plastics design and engineering. Published bimonthly.

Plastics Engineering

A. A. Schoengood, editor
The Society of Plastics Engineers
Brookfield Center, CT

Covers design and engineering technology for the plastics engineer. Provides industry and company news and government regulatory information. Regular features also include up-to-date product listings, available literature, and conference and seminar dates. Published monthly.

Plastics Technology

Malcolm W. Riley, editor
A Bill Publication
New York

Besides extensive technological coverage, the magazine also includes government regulatory, pricing and industry updates. A calendar of events is also featured. Published monthly; semimonthly in June.

Plastics World

Cahners Publishing Co.
Boston, MA

Covers the economical developments in the plastics industry. Features market forecasts, mergers and acquisitions, company expansion news and general technological articles. Published monthly.

Polymer Composites

R. S. Porter, editor
Society of Plastics Engineers
Brookfield Center, CN

Covers developments in reinforced plastics and polymer composites. Contains abstracts and references. Published quarterly.

SAMPE Journal

Dr. S. M. Lee, editor
SAMPE Publishing Co.
Covina, CA

Provides information for material process engineers. Other features include industry news, publications and abstracts. SAMPE membership events, directories and buyers' guide are also included. Published bimonthly.

SECTION 10
Directory of Consultants

Diran Apelian
Metallurgical International Inc.
904 Clover Hill Road
Wynnewood, PA 19096
(215) 649-6761

Powder technology; rapid solidification by plasma deposition; plasma spraying; metallic coating; controlled solidification processes; spray forming; materials selection.

Charles L. Beatty, Professor
University of Florida
Department of Materials Science & Engineering
Center for Macromolecular Science & Engineering
Gainesville, Florida 32611
(904) 392-1574

High-strain-rate deformation; fatigue and fracture of polymers and polymer composites; improved high-strength, low-density metal-matrix composites.

Walter L. Bradley, Professor of Mechanical Engineering
Texas A & M University
College Station, TX 77843
(409) 845-1259

Delamination fracture of graphite/epoxy or graphite-thermoplastic composite materials loaded under mode I, mode II, and mixed mode conditions; fracture mechanics characterization of composite materials; observation of fracture micromechanisms during fracture in the scanning electron microscope; retention of neat material toughness in the composite in the form of resistance to delamination fracture; fracture mechanics characterization of neat material and composite toughness (delamination), done using standard fracture mechanics approaches (LEFM or J-integral); the micromechanics of the fracture process in the neat material and the composite are noted in the scanning electron microscope where small specimens are broken; differences in fracture process due to constraint, etc. are noted; failure analysis of composite materials.

Lawrence J. Broutman
L. J. Broutman and Associates
3424 S. State
Chicago, IL 60616
(312) 842-4100

Fiber matrix interface; fracture, fatigue; environmental effects, design impact strength, and toughness.

A. T. Chapman, Professor
Georgia Institute of Technology
School of Materials Engineering (Ceramics)
Atlanta, GA 30332
(404) 894-2851

Oxide-metal composites composed of submicron-size metal fibers arrayed in a ceramic (insulating) matrix; major application is charge injection into nonconducting fluids and gases, i.e., electrodes for the electrostatic atomization of combustible liquids.

Tsu-Wei Chou, Professor
University of Delaware
Center for Composite Materials
Newark, DE 19716
(302) 451-2904

Analytical modeling and experimental characterization of mechanical and physical properties of composites; includes composites composed of fibers in short, continuous, hybrid, and woven forms, in polymer, metal, and ceramic materials.

Dr. Linda L. Clements, Professor
San Jose State University
Materials Engineering Department
San Jose, CA 95192
(408) 277-3599

High-performance organic-matrix composites, including graphite, aramid, and glass fibers in epoxy and thermoplastic matrices; test methodology including test program design, mechanical test technique development, characterization, generation of engineering design data, and specimen fabrication and quality assurance; optical, SEM, and fractographic analysis leading to defect characterization and to an understanding of mechanisms of degradation and failure in composite materials; systematic study of moisture and temperature effects in composites, including mechanisms of resulting property changes; education, specializing in composite materials, polymers, mechanical behavior, and structure-property relations.

Thomas H. Courtney, Professor
Michigan Technological University
Department of Metallurgical Engineering
Houghton, MI 49931
(906) 487-2327 or (906) 487-2036

Metal-matrix composites; powder metallurgical composites.

Charles B. Criner
101 W. Woodburn Drive
Taylors, SC 29687
(803) 292-3345

Metal-matrix composites of nonferrous metal powders with silicon carbide; silicon nitride, boron carbide; aluminum oxide-silica fiber reinforcement.

Dr. K. B. Das; Affiliate Associate Professor
University of Washington
Department of Materials Science & Engineering
9522 49th N.E.
Seattle, WA 98115
(206) 655-2774 or (206) 543-2600

Metal-matrix composites; joining methodology of MMC; corrosion behavior of certain MMC; patents pending on welding method for MMC and silicon nitride particulate and/or whisker-reinforced MMC systems; teaching senior/graduate level composite materials course dealing mainly with resin matrix systems; R & D for 20 years with Boeing Aerospace Company; currently Manager of Materials and Processes Organization at Boeing.

John Delmonte
Delsen Testing Laboratories, Inc.
1024 Grand Central Avenue
Glendale, CA 91201
(213) 245-8517

Author and consultant in the plastics and composites field.

Charles G. Dodd
Connecticut Technology Consultants, Inc.
P.O. Box 524
Stratford, CT 06497
(203) 375-5015; Telex: 499 6940

Metallic coating: surface modification by deposition of thin films by sputtering, ion plating, and physical vapor deposition methods, with ion implantation; surface analysis of metals and materials treated by surface modification.

Roman Dubrovsky
New Jersey Institute of Technology
Mechanical Engineering Department
32B High Street
Newark, NJ 07102
(201) 596-3353

Bimetals; wear resistance of different materials on composite base.

David W. Dwight, Associate Professor
Virginia Polytechnic Institute
Materials Engineering
Blacksburg, VA 24061
(703) 961-6346

Failure analysis: precise determination of micromechanisms of property degradation, especially under environmental conditions that may allow accelerated service-life predictions; surface and interphase modification: the use of approaches such as additives that preferentially migrate to surfaces, and silane coupling agents, as well as chemical and ion beam etching, to tailor new surface structures and properties; adhesion: studies of the ways to use analysis of surface chemistry and structure to predict the formation of durable adhesive bonds to composite materials, as well as between fibers and the resin matrix.

Dr. Ross F. Firestone
L. J. Broutman & Associates
3424 S. State Street
Chicago, IL 60616
(312) 842-4100

Metal-matrix and ceramic-matrix composites.

Amit K. Ghosh
Rockwell International Science Center
1049 Camino Dos Rios
Thousand Oaks, CA 91360
(805) 498-4545

Metal-matrix composites.

Dr. Claus G. Goetzel
Stanford University
Department of Materials Science & Engineering
250 Cervantes Road
Portola Valley, CA 94025
(415) 851-1369

Metal-matrix composites; microcomposites; cermets. Materials selection: high-temperature materials; structural ceramics; carbon and graphite; materials for power plants and chemical propulsion systems (inert).

David K. Hackett
Aztech Services
6500 Trousdale Road
Knoxville, TN 37921
(615) 691-7835

Testing and control: forensic investigations; failure analysis; fracture mechanics; corrosion failure and weld failure of metals, plastics, and composites.

Dr. John E. Hockett
John E. Hockett, Inc.
Materials Consulting
P.O. Box 278
Aquila, AZ 85320
(602) 685-2416

Design and fabrication of filament-wound graphite/Kevlar epoxy pressure vessels; several have already withstood long-term use at internal pressurized conditions.

Professor C. O. Horgan
Michigan State University
Department of Metallurgy, Mechanics and Materials Science
East Lansing, MI 48824
(517) 355-5112

Saint-Venant edge effects in composite materials and structures; vibrations of composite structures; fundamental research on mechanical behavior of composites; associated mathematical issues.

Dr. Jerald E. Jones
Associate Research Professor
Colorado School of Mines
Department of Metallurgical Engineering
Golden, CO 80401
(303) 273-3414

Humidity degradation of graphite/epoxy composites; welding of metal-matrix composites.

James C. Kenney
5846 Carvel Avenue
Indianapolis, IN 46220
(317) 253-0360

Ferrite composites.

William J. Knapp, Professor
University of California
Materials Science and Engineering Department
6532 Boelter Hall
Los Angeles, CA 90024
(213) 825-2758

Finely laminated ceramic-metal composites.

Michael J. Koczak
Professor of Materials Engineering
Drexel University
Department of Materials Engineering
Philadelphia, PA 19104
(215) 895-2329

Metal-matrix composite structure studies of interface reactions, alteration of mechanical properties with processing, and failure modes; the effects of temperature on strength and degradation of properties in SiC/Al and B/Al systems; the development of tough ceramics using 2-D and 3-D fiber weaves in a glass-ceramic matrix.

Alan Lawley, Professor
Drexel University
Department of Materials Engineering
Philadelphia, PA 19104
(215) 895-2326

General processing–microstructure–mechanical property relationships in metal matrix composites. Role of the interface
between matrix and reinforcement on composite behavior
under load. Shaping and forming metal matrix composites.
Powder metallurgy processing for composite fabrication.

Dr. Ray Y. Lin, Assistant Professor
University of Cincinnati
M.L. #12, Department of Materials Science and
    Engineering
Cincinnati, OH 45221
(513) 475-3096

Thermodynamics of metal-matrix composites; materials
compatibility between metal-matrix and reinforcements;
strength of metal-ceramics interface as a function of the interface chemistry and microstructure; effect of processing environments on the interfacial properties of metal-matrix
composites.

Karen P. Martin
Material and Design Concepts, Inc.
R.R. 15, Box 410
Hanover, NM 88041

Engineering components, structures, and composites.

Dr. Alan G. Miller
Boeing Materials Technology
P.O. Box 3707 Mail Stop 73-43
Seattle, WA 98124
(206) 342-3640 or (206) 237-9725

Composite material failure analysis; composites fractogra-
phy; environmental durability of composites; polymer viscoelasticity; electron microscopy; material specifications;
process specifications; fracture mechanics of composites.

John C. Monsees
1088 Cajon Greens Drive
El Cajon, CA 92021
(619) 579-1988

High-temperature composite polyimide systems and general
aerospace composite fabrication; metallic and nonmetallic
coatings; adhesives; general design consultation, particularly
in jet engine technology.

Dr. Martin Perl
64 Forest Road
Valley Stream, NY 11581

Metallic coating: electroplating; electroless plating (Cu, Ni on
plastics); flame spraying; metallizing; Materials selection:
semiconductors; composites; refractory.

W. Arthur Potter
5, Beechwood Avenue
Burnley, Lancashire, England, BB 11 2PL
0282-22294

Carbon and aramid fiber composites.

George C. Richardson
Senior Member Technical Staff
RCA — Astro Electronics
P.O. Box 800 M/S T2F
Princeton, NJ 08540
(609) 426-3702

Aerospace-related use of graphite, glass, and Kevlar in epoxy.

Dr. Donald Saylak
Texas A & M University
Civil Engineering Department
College Station, TX 77843
(409) 845-9962

Sand, asphalt, sulfur paving materials; sulfur-modified concrete; phosphogypsum/cement composites; phosphogyp-
sum/fly ash composites.

Mel Schwartz
Chief, Metals and Metals Processing
Sikorsky Aircraft
S312A-N. Main St. Stratford, CT 06443
Home: 80 Rolling Meadow Rd. Madison, CT 06443
(203) 386-4562, Home (203) 245-4268

Author of *Composite Materials Handbook,* and numerous
composite-related articles for trade publications; directed
composite activities and adhesive bonding technology for 7
1/2 years at Rohr-Riverside; organized composite manufacturing at Sikorsky Aircraft.

Norman S. Stoloff, Professor
Rensselaer Polytechnic Institute
Materials Engineering Department
Troy, NY 12180
(518) 266-6371

Fatigue and creep behavior of metal-matrix composites.

Mirle Surappa
Cavendish Laboratory
Cambridge, England
0223-66477, Ext. 419

Cast composites.

David A. Thomas, Professor
Lehigh University
Department of Metallurgy & Materials Engineering
Whitaker Laboratory #5
Bethlehem, PA 18015
(215) 861-4237

Processing, properties, and applications of polymers and fiber-reinforced polymer composites.

Roy Vipond
The City University
Department of Mechanical Engineering
Northampton Square
London, England, EC1V OHB
01-253-4399; Telex: 263896 CIT UNIV-G

Fatigue and adhesion of CFRP composites.

Curtis R. Watts, President/Consultant
C. R. Watts Associates
P.O. Box 3539
Redondo Beach, CA 90277
(213) 542-5210

Over 20 years of aerospace materials and processes engineering experience. Extensive involvement with organic matrix composites, metal matrix composites and carbon-carbon
materials. Technical roles in composites have involved research and development, design, support, specification review, program monitoring, and failure investigations. Other

(continued)

work has included the development and/or evaluation of composite materials for structural applications, antennas, corrosion protection, chemical compatibility, lightning-strike protection, and fire resistance. Work also includes research on surface-finishing methods.

Franklin E. Wawner, Jr., Research Professor
University of Virginia
Thornton Hall, Department of Materials Science
Charlottesville, VA 22903
(804) 924-6341

Structural, microstructural, mechanical property characterization, and fracture analysis by scanning electron and transmission electron microscopy of reinforcing fibers, whiskers, and composites thereof; emphasis on metal- and ceramic-matrix composites, interfacial phenomena, fiber and matrix contribution to observed composite properties and processing conditions; chemical vapor deposition of protective layers on fibers and composites and characterization of the resulting product.

John Weeton
22647 Peachtree Lane
Rocky River, OH 44116
(216) 333-0982

Involved in pioneer work at NASA in metal-matrix composite research.

Yechiel Weitsman, Professor
Texas A & M University
Mechanics & Materials Center
Civil Engineering Department
College Station, TX 77843
(409) 845-7512

Mechanical characterization; time-dependent behavior; residual stresses due to cure and cool-down; environmental effects; moisture absorption and moisture-induced damage; characterization and modeling of damage in composites.

George J. Weng, Professor
Rutgers University
Department of Mechanics & Materials Science
New Brunswick, NJ 08903
(201) 932-2223

Theories of inclusion; effects of fiber length; stress concentration; bonding techniques; prediction of effective moduli; two-phase plasticity; the influence of rigid particles and voids on the high-temperature behavior of metals.

Donald M. Yenni, Consulting Engineer
Route 2, Box 182
East Jordan, MI 49727
(616) 582-9092

Invention and development of the plasma plating process (along with the design of equipment) used in metal-matrix composites; design and construction of a 7-ft diam filament winder.

SECTION 11
Directory of Laboratories and Information Centers

Ames Laboratory
Iowa State University
Ames, IA 50011
(515) 294-4037

Director: R. S. Hansen
Contact: D. K. Finnemore, Associate Director
Founded: 1949

Research work includes the study of dendritic growth in metal-metal composites and synthesis of ceramic powders. R & D services are offered for industry, but they must have consent of the U.S. Department of Energy. The DOE also determines patent and licensing policies. Ninety-eight percent of the lab's funding is provided by the government.

Brookhaven National Laboratory
Upton, NY 11973
(516) 282-2123;

Associate Director: Dr. Martin Blume
Contacts: Dr. Victor Emery, Assoc. Chairman, Physics Dept.
Dr. Allen Goland, Assoc. Chairman, Dept. of Applied Science
Dr. William Marcuse, Head, Office of Research and Technology
    Applications
Founded: 1947

The laboratory provides basic and applied research on materials, including polymer/aggregate composites and polymer concrete development. Collaborative and proprietary research is conducted at BNL with engineers and scientists from the private sector. There are 40 senior researchers at the lab.

Carnegie-Mellon University
Polymer Science Program
Department of Chemistry
4400 Fifth Avenue
Pittsburgh, PA 15213
(412) 578-3131

Head: Professor G. C. Berry
Founded: 1960

Research is under way involving suspensions in polymeric matrices (e.g. magnetic particles) and interfacial polymerization, and major work has been accomplished in elucidation of interparticle organization in suspensions. Individual research contracts are awarded and faculty consulting is available. Patents and licenses are primarily state and federally owned. The library contains conference proceedings and key journals.

California Polytechnic State University
Metallurgical & Welding Engineering Department
San Luis Obispo, CA 93407
(805) 546-2568

Acting Head: George T. Murray

Composites are part of the current research program. Individual research contracts are available and patent and licensing policies vary. Faculty consulting is also available.

Colorado School of Mines
Advanced Materials Institute
Golden, CO 80401
(303) 273-3830

Acting Director: Jerome G. Morse, Ph.D.
Founded: 1984

Amorphous materials, ceramics, coatings, composites, particulates, polymers, magnetic materials, metals, and semiconductors are part of the current research areas. Graphite/epoxy composites are also currently under investigation. Industrial affiliates have access to long and short range problem-solving assistance, seminars and workshops, an expanding materials library, computer programming development, access to National Standard Reference Data and Information Systems in related fields, and faculty consulting. Individual research contracts are available. Although funded primarily by the state legislature, research orientation is also determined by industry.

Georgia Institute of Technology
School of Ceramic Engineering
Atlanta, GA 30332
(404) 894-2850

Director: Joseph L. Pentecost
Founded: 1924

Current research includes in-situ composites and crystal growth. Research involvement extends to industry on a formal contract basis and the facility is available on a fee basis. Seminars are offered periodically and other services include the use or purchase of some computer programs. Some proprietary research for ultimate patenting or licensing is available, but not preferred.

Jet Propulsion Laboratory
Applied Mechanics Technology Section (354)
California Institute of Technology
4800 Oak Grove Drive
Pasadena, CA 91109
(818) 354-6580;

Section Manager: Charles E. Lifer
Founded: Mid-1930's

Advanced composite technology has been developed and applied to the space program. Current space research includes work on durable materials and coatings and ceramic composites for armor and space shielding. JPL transfers and disseminates much of its research through published reports. Research contracts and faculty consulting are available to industry but the lab does not engage in proprietary R & D. Computer programs and space flight and science data are available through NASA.

Massachusetts Institute of Technology
Materials Processing Center
77 Massachusetts Ave., Bldg. 12-007
Cambridge, MA 02139
(617) 253-3217;

Director: Professor H. Kent Bowen
Contact: Dr. George B. Kenney, Assistant Director
Founded: 1980

Metal- and polymer-matrix composites are under current research. Workshops and seminars are sponsored each year, as well as continuing education courses. Research contracts to industry are single and multiclient. Although somewhat flexible, patents generated from sponsored research programs are the property of MIT. The sponsors are granted nonexclusive royalty-free licenses.

Metals and Ceramics Information Center (MCIC)
Battelle-Columbus Laboratories
505 King Avenue
Columbus, OH 43201
(614) 424-5000

Director: Dr. Harold Mindlin
Contact: Frank Jelinck, Associate Director

This center covers the following subjects. Materials: type or base. Metals: Titanium; aluminum; magnesium; beryllium; refractory metals; high-strength steels; superalloys. Ceramics: Borides; carbides; carbon/graphite; nitrides; oxides; sulfides; silicides; selection glass; glass ceramics. Coverage: Composites of these materials; coatings; environmental effects; mechanical and physical properties; materials applications; test methods; sources/suppliers; specifications; design characteristics. Processes: Basic materials production; primary fabrication (forging, casting, rolling, extrusion, etc.); joining; powder processes; surface treatment; quality control and inspection.

Metal Matrix Composites Information Analysis Center
(MMCIAC)
Kaman-TEMPO (formerly General Electric-TEMPO)
816 State Street
P.O. Drawer QQ
Santa Barbara, CA 93102
(805) 963-6497

Director: Louis Gonzales
Contact: Jacques E. Schoutens, Manager Data Analysis
William E. Rogers, Manager Information Services

Metal-matrix composite materials technology includes: continuous fibers; wires; discontinuous whiskers with L/D 10; directionally solidified eutectics. Fibers: boron; graphite; silicon carbide; borsic; nitride; alumina; boron carbide; titanium diboride. Wires: stainless steel; tungsten; molybdenum; beryllium; titanium; niobium alloys; compounds. Whiskers: alumina; silicon carbide; silicon nitride. MMC systems: alumina/magnesium; beryllium/titanium; boron/stainless steel/aluminum; boron/titanium/aluminum; borsic/aluminum; borsic/titanium; graphite/copper; graphite/aluminum; graphite/lead; tungsten/nickel. MMC properties: physical properties (thermal, chemical, optical, electromagnetic); specific modulus; strength; fatigue; environmental response; creep and wear resistance. Technical areas cover manufacturing; fabrication process development; defense system applications; performance computations; cost; testing and evaluation techniques and methods; properties data; operational; serviceability/repair; environmental protection; other MMC-related areas.

Michigan State University
Center for Composite Materials and Structures
Room 330, Engineering Building
East Lansing, MI 48824
(517) 353-5466;

Interim Director: David L. Sikarski
Founded: 1983

Research areas include stress field determination, mechanical characterization and testing, and mechanical fastening of joints in composites. Studies of fatigue in composites and polymers and polymer processing are also under way. The center works closely with the Michigan Molecular Institute of Midland, Michigan, which provides expertise in the area of molecular-oriented research. The center offers short courses, workshops, contract research, consultation, industrial intern programs, and referral services. Future plans include an advanced materials/composites database containing pertinent materials selection information. The database will be available to industry.

Mississippi State University
Raspet Flight Research Laboratory
Department of Aerospace Engineering
Drawer A
Mississippi State, MS 39762
(601) 325-3623

Director: Dr. George Bennett
Founded: 1948

The lab specializes in fabrication of aircraft composite structures. Facilities also include flight test evaluation. Services to industry range from consulting to research contracts, and patent and licensing arrangements are liberal. The lab also offers courses on applied design and fabrication of composite structures with hands-on experience in prepreg fabrication. A wide range of composite structural design computer programs are available.

North Carolina A & T State University
Composite Materials Research Laboratory
Department of Mechanical Engineering
108 Graham Hall
Greensboro, NC 27411
(919) 379-7620

Head: Dr. V. Sarma Avva

Composite studies include the frequency and stacking sequence effects on composites, the effect of thermal loads and fatigue on graphite/glass and SiC/glass, and the buckling of composite materials under hot and wet environments. SEM, LM, and X-radiographic studies are also being conducted on composite materials. Cooperative research and development projects, as well as individual contracts, are available. Faculty consulting is provided, and seminars and workshops can be arranged. Patent and licensing policies are negotiable.

North Carolina State University
Department of Materials Engineering
Box 7907
Raleigh, NC 27695
(919) 737-2377

Department Head: Dr. Hans Conrad

Composite materials research is part of the engineering department's program. Services offered to industry include testing, consulting, and individual and affiliated research contracts.

North Carolina State University
School of Textiles
Raleigh, NC 27650
(919) 737-3057

Head: William K. Walsh, Associate Dean

Besides the research and production of fibers, there is a complete polymer analytical lab. Polymer research areas include physical chemistry, synthesis, physics, electrical properties, and diffusion and radiation curing methods. The school awards research contracts, offers faculty consulting, service agreements, and rental equipment to industries. The patent and licensing policies may be negotiated, with the institution retaining the rights to use the invention free of royalty fees.

Ohio State University
Materials Research Laboratory
174 W. 18th Avenue
Columbus, OH 43210
(614) 422-5190

Director: James C. Garland
Contact: Richard L. McGeery, Associate Director
Founded: 1982

Particulate composites are currently being researched. Individual research contracts are awarded and faculty consulting and use of the facilities are also available. The lab is funded primarily by the National Science Foundation.

Plastics Technical Evaluation Center (PLASTEC)
U.S. Army Armament Research and Development
    Command
Dover, NJ 07801
(201) 724-4222

Director: Harry E. Pebly
Contact: Lee Ann Chervnisk, Publications

The center is responsible for the generation, evaluation, and exchange of technical information related to plastics, adhesives, and organic matrix composites. Covers technology from applied research through fabrication with emphasis on properties and performance. Subject areas include structural, electrical, electronic, and packaging applications. Includes molded, formed, foamed, and laminated materials. Maintains computerized data file on compatibility of polymers with propellants and explosives (COMPAT); also HAZARD-FAILURE file. Maintains complete file of standards, specifications, and handbooks in subject areas. Provides following services on a fee basis: technical inquiries, state-of-the-art studies, data compilations, handbooks, consultant, analysis and evaluations, background studies, bibliographic, and literature searches.

Polytechnic Institute of New York
Polymer Research Institute
333 Jay Street
Brooklyn, NY 11201
(212) 643-5235

Director: Professor Eli M. Pearce
Founded: 1942

The study of thermosetting and thermoplastic composites are part of the polymer research at the institute. All phases of the polymer disciplines are currently under investigation.

Rensselaer Polytechnic Institute
Center for Manufacturing Productivity
    and Technology Transfer
JEC 5009
Troy, NY 12181
(518) 266-6950

Director: Dr. Leo E. Hanifin
Contact: Dr. Robert W. Messler, Jr., Associate Director

Current research of advanced composites is being performed at the institute. The work includes the study of high-modulus and high-strength fibers and chemical vapor deposition (CVD). The center publishes reports on individual projects and technical articles. Use of computer programs is proprietary to project sponsors and member companies. The center also employs project managers from industry, who manage staff members working on R & D programs.

Rensselaer Polytechnic Institute
Materials Engineering Department
Troy, NY 12181
(518) 266-6372

Chairman: Martin E. Glicksman
Contact: Roger N. Wright, Executive Officer

Current research areas include polymer-matrix composites. Services to industry include contract research, use of student talent, and consulting services. Patent and licensing policies are available upon request.

Rutgers, The State University of New Jersey
Center for Ceramics Research
Box 909
Piscataway, NJ 08854
(201) 932-2724

Director: John B. Wachtman, Jr.
Founded: 1982

Composite bonding and joining, as well as high-strength and high-temperature ceramic composites, are under research. The center provides cooperative research with industry and maintains confidential results for 1 year. Patents and licenses are usually held by the university with royalty-free, non-exclusive licenses given to sponsors. A company may obtain ownership of a patent or license at additional cost under single-client contract agreements.

Southern Illinois University
Materials Technology Center
Carbondale, IL 62901
(618) 536-2129

Director: Kenneth E. Tempelmeyer
Founded: 1983

The technology center specializes in the research of carbon and graphite composites and fibers. Major accomplishments include the development of applications of composite materials (particularly graphite) in commercial and construction areas. The center holds annual meetings on materials technology, and working relations with industry include an industrial advisory board, group and single client research projects, and consulting. Under patent and licensing agreements the center, inventors, and sponsors share benefits. Quarterly newsletters and research reports are published.

Southwest Research Institute
Materials Science Department
6220 Culebra Road
P.O. Drawer 28510
San Antonio, TX 78284
(512) 684-5111, Ext. 2500

Director: Dr. U. S. Lindholm
Contact: Dr. G. R. Leverant, Assistant Director

The study of polymer composites is included in the department's current research. R & D in the areas of materials development, processing, characterization, application, and

(continued)

service performance are available to industry, and both laboratory and field studies are conducted. Industrial subscription programs, consulting, and single-client research contracts are offered. Clients retain all rights to patents developed.

Textile Research Institute
P.O. Box 625
Princeton, NJ 08542
(609) 924-3150

President and Director: Dr. Ludwig Rebenfeld
Founded: 1930

The institute is currently researching adhesion- and fiber-reinforced composites. Also, special emphasis is on fiber structure and dye transport processes. Services offered to industry include conferences, symposiums, and individual and multi-client research projects. Sponsored projects are given licensing and patent rights. Along with the *Textile Research Journal*, the institute also publishes many technical reports.

U.S. Army Material Development and Readiness
Command
Army Materials and Mechanics Research Center
Director, AMMRC, Attn: ORXMR-PP
Watertown, MA 02172
(617) 925-5527

Contact: David W. Seitz, Technology Transfer
Coordinator
Founded: 1966

Composites are currently under research at the AMMRC. Contact the Technology Transfer Coordinator for questions about research contracts. U.S. Government policies apply to all patents and licenses.

University of California
Lawrence Livermore National Laboratory
Box 808
Livermore, CA 94550
(415) 422-6416

Director: R. E. Batzel
Contact: Charles Miller, interim Technology Transfer &
Exchange Officer

Fundamental and applied composite research.

University of Cincinnati
Polymer Research Center
Mail Location 172
Cincinnati, OH 45221
(513) 475-2453

Director: Professor James E. Mark

Reinforced polymer composites are part of the center's current research.

University of Connecticut
Institute of Materials Science
Storrs, CT 06268
(203) 486-4623/4

Director: Leonid V. Azaroff
Founded: 1966

Current research areas include composite materials studies. The institute also offers lectures and workshops in metallurgy and polymer science. Working arrangements with industry include faculty consulting and single- or multi-client research contracts. The institute has liberal licensing policies, including exclusive licenses to sponsors.

University of Delaware
Center for Composite Materials
201 Spencer Laboratory
Newark, DE 19716
(302) 451-8149;

Director: R. Byron Pipes, Ph.D.
Contact: William A. Dick, Assistant Director
Founded: 1974

The center deals exclusively with the research of composite materials. An encyclopedia of over 3,000 pages has been authored and distributed by the center. R & D services to industry are directed towards the transfer of composite technology to the workplace. The center also conducts annual workshops and symposiums. Nonexclusive royalty bearing agreements are offered on patents and licenses, but this policy is negotiable. Center publications include *Composite Update,* a quarterly newsletter, and over 20 technical reports a year. Computer programs and a materials micromechanics database, which includes fiber and matrix properties, are other services available to industry.

University of Detroit
Polymer Institute
400 West McNichols Road
Detroit, MI 48221
(313) 927-1270;

Director: Dr. Kurt C. Frisch
Contact: Dr. Daniel Klempner, Associate Director
Founded: 1968

Fiber-reinforced and laminate composites are part of the current polymer research. The institute conducts applied as well as basic research and development, on new and improved materials. Polymer coatings are the subject of periodic workshops and conferences. Services offered to industry include faculty consulting, training programs, and individual research contracts. Grants and contracts are generally lengthy and run 1 to 2 years in duration. The institute requires no licensing and all patents are the property of the sponsors.

University of Illinois at Urbana-Champaign
Materials Engineering Research Laboratory
100 Talbot Laboratory
Urbana, IL 61801
(217) 333-3751

Head: Dr. Frederick V. Lawrence, Jr.
Founded: 1978

The study of the fatigue behavior of composite structures is part of the lab's current research. The lab sponsors The Fracture Control Program and The Materials Processing Consortium as part of their university-industry cooperative program. Along with the usual grants, the lab also offers engineering testing agreements and consulting arrangements. The patent and licensing policies vary. "Material Engineering — Mechanical Behavior," a report series, is distributed to sponsors of industrial programs.

University of Florida
Program in New Materials (Composites)
Department of Engineering Sciences
Gainesville, FL 32611
(904) 392-0961

Head: Dr. L. E. Malvern

Research on filament-reinforced laminated plates and other fiber reinforced composites includes the dynamic response of composite materials and structures, characterization of damping of composite materials, and the response of composite materials under high-strain-rate load. Individual research contracts are available, as well as consulting services. Patent and licensing policies are flexible.

University of Michigan
Materials and Metallurgical Engineering Department
Dow Building
Ann Arbor, MI 48109
(313) 764-7489

Head: Professor Robert D. Pehlke

Metal, ceramic, polymer, and elastomer composites are under research at the university. Along with individual research contracts, faculty consulting and fellowships are available.

University of Notre Dame
Metallurgy and Materials Science Department
P.O. Box E
Notre Dame, IN 46556
(219) 239-5330

Chairman: Dr. Gordon A. Sargent

Composites are part of the current research at the university. The department offers fundamental research support to industry, along with workshops and seminars. Research contracts and consulting are available and patent/licensing policies are negotiable.

University of Southern California
Polymer Program
Department of Chemical Engineering
University Park, MC 1211
Los Angeles, CA 90089-1211
(213) 743-7051

Departmental Chairman: Professor J. D. Goddard
Contact: Professor R. Salovey, Director Polymer Program
Founded: 1975

The department specializes in the study of polymers and rubberlike materials. Composites, filled copolymers, and blends are also under research. Included in the seminar and workshop programs are instructional television and material science courses. Research contracts are available for specific work, and contracts with other industrial groups are through advisory boards. Faculty consulting is also available. The university usually retains patent and license rights, although this is sometimes negotiated. Some computer programs are under development concerning the prediction of material properties. Industrial support and advisory groups are: Urethane Group; TLARGI (The Los Angeles Rubber Group Inc.) Foundation, 1975–; SCARAB (Sealants, Composites, Adhesives Research Advisory Board), 1984–; Participating University in Adhesive and Sealants Council Fellowship Program, 1984–.

University of Texas at Austin
Center for Materials Science and Engineering
ETC 9.102
Austin, TX 78712
(512) 471-1504

Director: Harris L. Marcus, Ph.D.
Founded: 1983

Basic research is conducted on metals, polymers, ceramics, and semiconductors. Polymer and metal composites are also under study. The center offers short courses, single-client research, faculty consulting, and cooperative teaching/research programs with industry. Patents are owned by the university, but exclusive licenses with or without a royalty to the university are granted, depending on the balance of equity.

SECTION 12
Societies, Trade Associations, and Institutes

American Carbon Society (ACS)
The Stackpole Corporation
St. Marys, PA 15857
(814) 781-8410

Members: 500
Founded: 1957
Publications: *Carbon* — Bimonthly
Conference registration list, program and abstracts of
    conference papers
Meeting/Convention: Carbon Conference — Biennially
Conferences and publications dealing with the
    chemistry, physics, and scientific aspects of
    materials ranging from carbons and graphites to
    organic crystals and polymers.

American Ceramic Society (ACS)
65 Ceramic Drive
Columbus, OH 43214
(614) 268-8645

Members: 10,100
Staff: 32
Founded: 1899
Publications: *Journal of the American Ceramic Society* —
    Monthly
American Ceramic Society Bulletin — Monthly
Ceramics Abstracts — Bimonthly
Communications of the American Ceramic Society —
    Monthly
Meeting/Convention: Annually
Local Groups: 31

Provides scientific and technical information to scientists and
engineers involved in the glass, cements, refractories, ceramic-metal systems, nuclear ceramics, electronics, white
wares, and structural clay products industries. Affiliated with
the Ceramic Educational Council and the National Institute
of Ceramic Engineers.

American Society for Metals (ASM)
Metals Park, OH 44073
(216) 338-5151

Members: 52,000
Staff: 135
Founded: 1913
Publications: *Metal Progress* — Monthly
Metal Progress Data Book — Annually
Metal Progress Heat Treating Buyers Guide &
    Directory — Annually
Metal Progress Testing & Inspection Buyers Guide &
    Directory — Annually
Metals Abstracts — Monthly
Metallurgical Transactions A — Monthly
Metallurgical Transactions B — Quarterly
Metals Abstract Index — Monthly
ASM News — Monthly
Bulletin of Alloy Phase Diagrams — Bimonthly
Journal of Materials for Energy Systems — Quarterly
Journal of Applied Metalworking — Semiannually
Journal of Heat Treating — Semiannually

Advanced Materials & Processes — Monthly
Alloys Index — Monthly
Directory of Metallurgical Consultants and Translators —
    Annually
Meeting/Convention: Annual
Local Groups: 236

Serves the metals and materials community through education (Metals Engineering Institute), career-development programs, computerized information, a 9,000-volume library,
conferences and expositions.

ASTM
1916 Race Street
Philadelphia, PA 19103
(215) 299-5400

Members: 30,500
Staff: 200
Founded: 1898
Publications: *Standardization News* — Monthly
Composites Technological Review — Quarterly
Journal of Testing and Evaluation — Bimonthly
Geotechnical Testing Journal — Quarterly
Cement, Concrete and Aggregates Journal —
    Semiannually
ASTM Standards — Annually
Other technical papers and reports
Meeting/Convention: Monthly

Establishes voluntary standards for materials, products, systems, and services. The society features 140 technical committees, sponsors research projects, issues various awards,
and has developed over 6,800 standard test methods.

Fiber Society (FS)
Box 625
Princeton, NJ 08540
(609) 924-3150

Members: 400
Founded: 1941
Meeting/Convention: Semiannually

Serves physicists, engineers, biologists, chemists, and math-
ematicians involved in the research of fibers, fiber products,
and fibrous materials.

Fiberglass Fabrication Association (FFA)
1010 Wisconsin Avenue N.W.
Suite 630
Washington, DC 20007
(202) 544-0262

Members: 320
Staff: 3
Founded: 1979
Publications: *Who's Who in Fiberglass Fabrication* —
    Annually
News — Monthly
Annual Buyers Guide
Meeting/Convention: Annually

(continued)

Serves companies engaged in fabrication of fiberglass products.

Materials Research Society (MRS)
110 Materials Research Laboratory
University Park, PA 16802
(814) 865-3424

Members: 1,500
Staff: 2
Founded: 1973
Publications: Bulletin — 6 issues per year
Proceedings — As held
Membership Directory — Annually
Meeting/Convention: Semiannually

Promotes interaction among researchers and provides a forum for university and industry interface. Provides short courses, conferences, and tutorial lectures.

National Institute of Ceramic Engineers (NICE)
65 Ceramic Drive
Columbus, OH 43214
(614) 268-8645

Members: 1,900
Founded: 1938
Meeting/Convention: Annually, in conjunction with
   American Ceramic Society

Professional organization promoting the improved status of ceramic engineering through high ethical engineering standards, engineer registration, and accreditation of education programs in ceramic engineering.

Plastics Institute of America (PIA)
Stevens Institute of Technology
Castle Point Station
Hoboken, NJ 07030
(201) 420-5553

Members: 100
Staff: 11
Founded: 1961
Publications: *Institute Report* — Semiannually
Pipeline — Quarterly
Annual Report
Meeting/Convention: Annually

A cooperative venture of companies in the plastics industry to support education and research in the plastic fields. Conducts a graduate-level program of education in plastics in cooperation with over 100 universities and colleges.

Reinforced Plastics/Composites Institute (RP/CI)
355 Lexington Avenue
New York, NY 10017
(212) 573-9424

Members: 400
Staff: 4
Founded: 1946
Publications: Reprint of *Annual Technical Conference*
Meeting/Convention: Annually

A division of the Society of Plastics Industry addressing the needs of molders, fabricators, and suppliers of fiber-reinforced plastics.

Society for the Advancement of Material and Process
   Engineering (SAMPE)
P.O. Box 2459
Covina, CA 91722
(818) 331-0616

Members: 5,000
Staff: 26

Founded: 1944
Publications: *SAMPE Quarterly*
SAMPE Journal — Bimonthly
Roster of National Officers — Annually
Symposium and Conference Proceedings
Meeting/Convention: Semiannually
Local Groups: 26

A society for materials and process engineers, professionals, and scientists. Consults with educational institutes to establish study programs in materials and process engineering. Also provides job placement service for members.

The Refractories Institute (TRI)
3760 One Oliver Plaza
Pittsburgh, PA 15222
(412) 281-6787

Members: 86
Staff: 6
Founded: 1951
Publications: *Product Directory of the U.S. Refractories*
   Industry
Meeting/Convention: Semiannually

Serves producers and suppliers in the refractory industry. Supports research at The Ohio State University and awards annual scholarships to undergraduates majoring in ceramics with refractories emphasis.

Society of Plastics Engineers (SPE)
14 Fairfield Drive
Brookfield Center, CT 06805
(203) 775-0471

Members: 24,000
Staff: 32
Founded: 1942
Publications: *Polymer Engineering and Science* — 18
   issues per year
Plastics Engineering — Monthly
Journal of Vinyl Technology — Quarterly
Polymer Composites — Quarterly
Meeting/Convention: Annually
1986 — Toronto, Canada, May 5–8
1987 — Los Angeles, CA, May 4–7
Sections: 84

Provides technical and scientific information and services for plastics scientists, engineers, educators, and students. Numerous conferences and seminars are conducted each year.

Society of the Plastics Industry (SPI)
355 Lexington Avenue
New York, NY 10017
(212) 573-9400

Members: 1,400
Staff: 81
Founded: 1937
Publications: *News Briefs* — Biweekly
Facts and Figures — Annually
Labor Survey — Annually
Financial and Operating Ratios — Annually
Membership Directory and Buyers Guide — Annually
Proceedings of Conferences — Annually
Meeting/Convention: 1986 — Atlanta, GA, January 27–31
1987 — Cincinnati, OH, February 2–6
1988 — Cincinnati, OH, February 1–5
Regional Groups: 5

Serves most phases of the plastics industry through many specialized divisions. Also sponsors the National Plastics Exposition.

SECTION 13
Directory of Manufacturers, Suppliers, and Services

Following is a selected listing of manufacturers, suppliers, and services pertaining to the various composites industries. Space limitations preclude representing the entire field of composites-related organizations, or companies' entire product or service lines.

Company Name	Address	Phone	Description
A & M Engineered Composites	3 Hayes Memorial Dr. Marlboro, MA 01752	(617) 485-8000	Manufactures Kevlar, fiberglass, and graphite/epoxy structures for military and aerospace applications
Accudyne Engineering & Equipment Co.	Bell Gardens, CA		Supplies molding (compression) presses for laminates and other composite materials
ACI Fibreglass	Box 327 Frankston Rd. Victoria 3175 Australia	Telex No.: AA32275	Manufacturer of fiberglass products, including continuous-filament yarns, rovings, woven rovings, muffler fiber, chopped strands, chopped strand materials
Acoustic Emission Technology Corporation	1812 J Tribute Rd. Sacramento, CA 95815	(916) 927-3861 Telex: 171356	Testing and inspection equipment for composite materials and structures. Instruments for testing composite strengths within ASTM-SPI/CARP-ASME requirements
Acurex Corp., Aerotherm Division	555 Clyde Ave. Mt. View, CA 94039	(415) 964-3200 Telex: 34-6391 CA: FHX TWX: 910-379-6593	Manufactures composite structures and materials for extreme environmental conditions. Produces Chemceram, a family of ceramic composites currently under research, which include graphite fiber, silica fiber, sapphire filament, and glass-reinforced composites. Also produces ceramic foams
Advanced Composite Products & Technology, Inc.	7415 Mount Jay Huntington Beach, CA 92648	(714) 848-3123	Complete composite fabrication capability: from hand lay-up through commercial rolling to filament winding. A team of aerospace experts supplies technological support. Additional services include design and analysis and marketing functions
Advanced Composite Products, Inc.	37 Washington Ave. East Haven, CT 06512	(203) 469-4647	Product development center for composites and composite structures. Manufactures components for the aerospace industry and provides technology and composite structures for governmental and OEM applications
Aerospace Technologies, Inc.	P.O. Box 50727-T Fort Worth, TX 76105	(817) 451-0620	Manufactures composite structures: honeycomb bonded panels, glass-reinforced plastics, and aircraft parts. Also supplies design engineering services
Airco Carbon, Division of the BOC Group, Inc.	800-0 Theresia St. St. Marys, PA 15857	(814) 781-2611 TWX: 510-693-4515 TELEX: 914513	Manufactures carbon and graphite tooling products. Features tooling used for fabrication of reinforced plastic composites for aerospace applications
Airtech International	Carson, CA		Features a line of layup materials for ovens and autoclaves used in the manufacture of composites. Operations include tooling and bonding shops

Company Name	Address	Phone	Description
Aluminum Company of America, Chemicals Division	1501 Alcoa Building Pittsburgh, PA 15219	(412) 553-4545	Produces sintered tubular alumina in crushed, graded, and ground sizes. Applications include use as base material in refractories. Also produces aluminum trihydrate, a filler used in plastic systems
Amalga Corporation	10600 W. Mitchell St. West Allis, WI 53214	(414) 453-9555 Telex: 26-9503	Produces Amalgon, a filament-wound, glass fiber-reinforced epoxy resin tubing for hydraulic cylinder applications. Black Amalgon tubing has self-lubricating features and is used for low-pressure air and water applications
Amercom, Inc.	8949 Fullbright Ave. Chatsworth, CA 91311	(213) 882-4821	Fabricates metal and ceramic matrix composites. Products include aluminum, titanium, magnesium, and copper matrix materials reinforced with boron, silicon carbide, and graphite fibers and filaments
American Cyanamid Co.	Engineered Materials Department One Cyanamid Plaza Wayne, NJ 07470	(201) 831-2000	Manufactures Dura-Core aluminum honeycomb and Cycom structural resins and prepregs with glass, quartz, ceramic, graphite, and aramid fibers
American Klegecell Corp.	204 N. Dooley Grapevine, TX 76051	(817) 481-3547 Telex: 73318	Produces high strength-to-weight ratio polyvinyl foam cores for composite uses
American Tempering, Inc.	33200 Western Ave. Union City, CA 94587	(415) 471-6811 Telex: 336302	Manufactures full range of glass products including laminates. Also produces polycarbonate laminates
Amoco Chemicals Corporation	200 East Randolph Dr. Chicago, IL 60601	(800) 621-4557 312/856-3414	Manufactures Torlon, poly (amide-imide) resins in glass, carbon, or graphite fiber-reinforced grades. Other grades are reinforced or filled with graphite powders and fluorocarbons for wear-resistant properties
The Andersons	P.O. Box 119 Maumee, OH 43537	(800) 472-3220 (800) 532-3370	Produces corn cob granule extenders and fillers for plastics as well as other industrial, chemical and agricultural applications
Andus Corporation	Canoga Park, CA	(818) 882-5744	Develops and manufactures thin-film coatings on polymer substrates. Products are used in hi-tech applications
Applied Plastics Company, Inc.	612 E. Franklin St. El Segundo, CA 90245	(213) 322-8050	Features epoxy resins and hardeners for use in composite construction
Applied Polymer Technology, Inc.	6078-B Corte Del Cedro Carlsbad, CA 92008	(619) 438-8977 Telex: 821426	Manufactures composite fabrication control systems
Arco Chemical Company — Advanced Materials	Route 6, Box A Greer, SC 29651	(803) 877-0123	Manufactures high-performance silicon carbide-reinforced aluminum composite products
Artech Corp.	2901 Telestar Ct. Falls Church, VA 22042	(703) 560-3292	Products include: ceramic-to-metal seals and composites, thin-film resistors, ceramic adhesives and coatings, radiation detectors, and consumer protection equipment
Arvin Industries, Inc., Arvinyl Division	Department T Columbus, IN 47201	(812) 379-3000	Manufactures vinyl-to-metal laminates and glass-reinforced nylon and polypropylene
Asahi Nippon Carbon Fiber Co. Ltd.	Chiyoda-Ku Tokyo, Japan		Manufactures chopped carbon fiber and high-strength, high-strain continuous carbon fiber. Also produces Nicalon, a silicon carbide fiber
Astro Met Associates, Inc.	9974 Springfield Pike Cincinnati, OH 45215	(513) 772-1242 Telex: 23295526	Supplies research, development, and production of ceramics, including cermets

Company Name	Address	Phone	Description
Atacs Products, Inc.	Renton, WA		Features composite repair systems
Atlantic Research Corporation	5390 Cherokee Ave. Alexandria, VA 22314	(703) 642-4530	Supplies engineering, application analysis, testing, and fabrication services for advanced composite materials
Automation/Sperry	20327 Nordhoff Street Chatsworth, CA 91311	(818) 882-2600	Manufactures state-of-the-art ultrasonic inspection systems for composites and bond testing
Avco, Specialty Materials Division	2 Industrial Ave. Lowell, MA 01851	(617) 452-8961	Produces boron, graphite and silicon carbide reinforced aluminum and titanium. Also manufactures epoxy prepregs, fire protection, and graphite materials
Bentley-Harris Mfg. Co.	241 Welsh Pool Rd. Lionville, PA 19353	(215) 363-2600	Manufactures braided and woven reinforcements and materials of graphite, Kevlar, glass, ceramic, thermoset prepregs, graphite/PEEK, graphite/PPS, glass/nylon, and metal strip. Also provides design technology and is active in R & D with major industries and universities
Bisco Products, Inc.	Park Ridge, IL	(312) 298-1200	Manufactures specialty silicone products, including a number of silicone coated fiberglass products
Boeing Technology Services	P.O. Box 3707 M/S 9R-28 Seattle, WA 98124	(206) 237-4490 Telex: 32-9430	Features advanced, automated manufacturing, and fabrication of composite structures. Also serves the sporting goods industry with analysis and consultation for use of composites
Boeing Aerospace Company	P.O. Box 3999 M/S 3E-79 Seattle, WA 98124	(206) 773-5889	Composite fabrication technology includes layup, filament winding, injection and compression molding, foams and bonding capabilities. Facilities also include testing and inspection equipment and technology
Bonded Technology, Inc.	One Alcap Ridge Rd. P.O. Box 160 Cromwell, CT 06416	(203) 635-1150	Custom designing, engineering, and manufacturing of bonded honeycomb and fiber composites to aerospace and governmental specifications. Production technology includes two autoclaves, two ovens, platen presses, FPL etch line, and phosphoric acid anodizing. Extensive use of Kevlar, graphite, and S-glass in composite applications
Boots Company PLC	Nottingham N62 3AA England	0602 506255 Telex: 377811	Manufactures thermosetting bismaleimide resins used primarily for fiber-reinforced composites. Also produces structural laminates and parts for aerospace, electrical, and thermal industries
Borg-Warner Chemicals, Inc.	International Center Parkersburg, WV 26102	(304) 424-5411	Manufactures phenylene ether copolymer alloys and Prevex in reinforced and nonreinforced grades
Bridon Composites Limited	Fairoak Lane Whitehouse Industrial Runcorn Estate Runcorn, Cheshire WA7 3DU England	Telex: 629765 701515 STD Code 0928	Manufactures and distributes high-performance woven, braided, and knitted fabrics, pultrusions, and thermoplastic products
Briskheat	Columbus, OH	(614) 294-3376	Features line of curing and hot debulking composite equipment
Bristol Composite Materials, Inc.	P.O. Box 789 1363 North Gaffey St. San Pedro, CA 90733-0789	(213) 514-3755 Telex: 691-549	Manufactures advanced composite components for the aerospace industry. Also provides designing and testing services

Company Name	Address	Phone	Description
L. J. Broutman & Associates Ltd.	3424 S. State St. Chicago, IL 60616	(312) 842-4100	Mechanical testing, chemical/thermal analysis, failure analysis, stress analysis, and design and R & D in polymers and composites
Burnham Products, Inc.	4203-T W. Harry St. P.O. Box 12950 Wichita, KS 67277	(316) 942-3208	Features Fiber-Lok composite tooling reinforcements. Also manufactures honeycomb and laminate composites of fiberglass and graphite
C-E Refractories, Division of Combustion Engineering, Inc.	P.O. Box 828 Valley Forge, PA 19482	(215) 337-1100	Manufactures ceramic fibers
Standard Oil Engineered Materials Co.	P.O. Box 156 Niagara Falls, NY 14302	(716) 278-2000	Advanced materials division produces high-performance ceramics, and composite materials and structures
Celion Carbon Fibers, Division of BASF Corporation	Cherry Road Station Rock Hill, SC 29730	(803) 366-5656	Manufactures ultrahigh-modulus and intermediate-modulus carbon and graphite fibers for applications in the aerospace, aircraft, recreation, marine, and industrial markets
Chem-Tronics Inc.	1150 W. Bradley P.O. Box 1604 El Cajon, CA 92022	(619) 448-2320	Produces polyimide foams, adhesives, and laminating resins. Also manufactures Unistructure, a thermoplastic, titanium-reinforced composite, and other lightweight metal structures of aluminum, stainless steel, and nickel-base alloys for aerospace industries
Ciba-Geigy Corporation	Composite Materials 10910 Talbert Ave. Fountain Valley, CA 92708	(714) 964-2731	Manufactures impact- and damage-tolerant resin systems. Also features custom Kevlar, and graphite weavings, hybrid constructions, woven and unidirectional materials of Kevlar, graphite, Nomex, and high-temperature fiberglass, honeycomb core materials, fabricated sandwich panels, laminates, and preimpregnated fiberglass
Cincinnati Milacron Industries Inc.	Electronic Circuit Materials Division Route 28 Blanchester, OH 45107	Telex: 214-472	Manufactures glass-reinforced polyester resins and copper-clad laminates. Also produces aircraft parts of composite materials with the aid of the Milacron 7-axis, a numerically controlled computer
Composition Materials Co. Inc.	26 Sixth St. Stamford, CT 06905	(203) 324-0000 Telex: 131-454	Features a broad range of fillers, extenders, and abrasives, including cotton and cellulose fiber fillers for plastics
Comp-Tite SPS Aerospace & Industrial Products Division	Highland Ave. Jenkintown, PA 19046	(215) 572-3000	Manufactures blind fastener systems for use with advanced composites. Compatible with graphite-fiber composites
Cosby Newsom Associates	Norwalk, CA	(213) 921-1972	Designers and manufacturers of composite debulking and curing equipment
Custom Coating and Laminating Corp.	77 Goddard Industrial Park 717 Plantation St. Worcester, MA 01605	(617) 852-3072 Telex: 955329	Provides custom coatings and laminations
Cyro Industries	P.O. Box 8588 Woodcliff Lake, NJ 07015	(201) 285-1544 (800) 631-5384	Manufactures polymethacrylimide foam core materials for sandwich construction

Company Name	Address	Phone	Description
DWA Composite Specialties, Inc.	21119 Superior St. Chatsworth, CA 91311	(818) 998-1504	Specialists in metal matrix composites and hardware. Manufactures DWG, the only thin-ply, zero CTE uniaxial graphite/aluminum products available
Delsen Testing Laboratories, Inc.	1024 Grand Central Ave. Glendale, CA 91201	(213) 245-8517	Mechanical testing of advanced composites. Other testing performed includes thermal analysis, flammability testing, HPLC-GPC, electrical, metallographic, and environmental testing
Diab-Barracuda, Inc.	Grand Prairie, TX		Manufactures polyvinyl foam core for use at the high-temperature ranges required in the aerospace industry
Dow Chemical Co.	2040 Dow Center Midland, MI 49674	(800) 258-2436	Through various subsidiaries, manufactures thermoset resin matrix systems and high-technology silicone materials for composite applications. Also manufactures ceramic specialty products for the ceramics industry
Dresser Industries	Dresser Manufacturing Division Nil Cor Operations P.O. Box 2058 Alliance, OH 44601	(216) 823-0500 Telex: 98-3461	Produces Nil-Cor ball and butterfly valves from composite materials. Valves are constructed from graphite fiber-reinforced vinyl ester, glass fiber-reinforced vinyl ester, and graphite fiber-reinforced Ryton polyphenylene sulfide
E.I. du Pont de Nemours & Co.	Textile Fibers Department Kevlar Special Products Centre Road Building Wilmington, DE 19898	(302) 999-3728	Manufactures Kevlar (aromatic polyamide) fibers, fabrics, and preimpregnated thermoplastic resins, and Nomex aramid honeycomb. Ceramic (fiber FP) and pitch carbon fibers are currently under development
Duramic Products Inc.	426 Commercial Ave. Palisades Park, NJ	(201) 947-8313 Telex: 710-991-9632	Custom fabricators of ceramic composites and other high-temperature ceramics and materials
Eurocarbon Tilburg Bu	Tilburg, Holland		Produces tubular and flat fiber braidings, woven unidirectional, and bidirectional tapes and woven cloth, of carbon fiber, aramid, E or S2 glass, silicon carbide, and ceramics. Also features hybrid constructions and high-temperature thermoplastic strips
Ed Fagan, Inc.	Component Materials Division 33 Whitney Rd. Mahwah, NJ 07430	(201) 891-4003	Manufactures metal-to-metal laminates. Also produces ceramic and glass sealing alloys
Emser Industries	P.O. Box 1717 Industrial Park & Corporate Way Sumter, SC 29151	(803) 481-3172 Telex: 805077	Specializes in general purpose and specialty nylon engineering thermoplastics. Produces Grilon(R), a nylon 6 resin and Grilamid(R), a nylon 12 resin in reinforced and unreinforced grades
Emerson & Cuming Inc.	Canton, MA 02021	(617) 828-3300	Features a line of ceramic microsphere products, Eccospheres, for plastic, ceramic, glass, plaster and concrete applications
Ferro Corporation, Composites Division	8790 National Blvd. Culver City, CA 90232	(213) 870-7873	Fabricates epoxy, polyimide, phenolic, and polyester resin prepregs with graphite, glass, silica, quartz, aramid, silicon carbide, alumina, and other ceramic reinforcements. Also manufactures broad goods, unidirectional tapes, rovings, molding compounds, and adhesive films

Company Name	Address	Phone	Description
Fiber Composites, Inc.	101-T S. Highland Greens Port Ludlow, WA 98365	(206) 437-2152 Telex: 215406	Provides filament winding services
Fiber Innovations, Inc.	Norwood, MA	(617) 769-2400	Manufactures triaxial braided fiber reinforcements. Designs and develops fibrous structures for aerospace, industrial, and sporting goods industries
Fiber Materials, Inc.	Biddeford Industrial Park Biddeford, ME 04005	(207) 282-5911 Telex: 944480	Produces advanced composite materials, including carbon/carbon and resin and metal matrix composites. Also available are carbon and graphite felts and Microfil high-modulus graphite fibers
Fiber Science, Inc./ Edo Corp.	506 N. Billy Mitchell Rd. Salt Lake City, UT 84116	(801) 539-0747	Manufactures composite structures
Fiberite Corporation	501 West 3rd St. Winona, MN 55987	(507) 454-3611	Supplies advanced composite materials for aerospace and recreational industries. Also provides design assistance, materials selelction, product development, analysis and testing, fabrication processes and production troubleshooting
Frekote, Inc., A subsidiary of The Dexter Corporation, Hysol Division	1701 Spanish River Blvd. West Boca Raton, FL 33431	(305) 395-3083 Telex: 51-8918	Produces wide range of mold-release additives for plastics, advanced aerospace polymer composites, and rubber/elastomers
GAF Corporation	1361 Alps Road Wayne, NJ 07470	(201) 628-3000	Extensive line of reinforced and unreinforced engineering plastics. Products include Gafite, a thermoplastic polyester, and Gaflex, a thermoplastic polyester elastomer
General Electric, Plastics Operations	One Plastics Ave. Pittsfield, MA 01201	(413) 494-1110	Manufactures Valox, Noryl, and Xenoy thermoplastic glass fiber-reinforced and nonreinforced resins. Other resin systems include thermoplastic foams, polycarbonates, and polyetherimides, marketed under the tradenames Lexan and Ultrem
GGT	Aerospace Division 55 Gerber Rd. South Windsor, CT 06074	(203) 644-2401 Telex: 643-771	Produces advanced composite manufacturing systems for the aerospace industry
Genstar Stone Products Co.	Executive Plaza IV Hunt Valley, MD 21031	(301) 628-4000	Produces various grades of calcium carbonate for plastics, paint and paper industries
D.A. Gordon Co.	Downy, CA	(213) 862-8993	Supplies composite and polymer testing equipment
Gordon Plastics, Inc.	2872 S. Santa Fe Vista, CA 92083	(619) 727-2008	Provides custom manufacturing and design of FRP laminates and structural products
Great Lakes Carbon Corp.	360 Rainbow Blvd. S. P.O. Box 727 Niagara Falls, NY 14302	(800) 828-6601 (716) 285-5200	Produces Fortafil carbon fibers and prepreg composites. Fibers available in continuous filament tows and chopped strands. Also manufactures graphite and coke particles and graphite structural rods and shapes

Company Name	Address	Phone	Description
Greene, Tweed & Co.	North Wales, PA 19454	(215) 256-9521 Telex: 6851164	Produces Arlon 1000, a high temperature semi-crystalline thermoplastic in reinforced & non-reinforced grades, useful for seal applications, high-temperature connectors and insulators, valve slats and compressor rings
Greenleaf Technical Ceramics	25019 Viking St. Hayward, CA 94545	(415) 783-0120	Offers a complete line of full-density, high-purity ceramic composite materials, pressed and ground to specifications
Grumman Aerospace Corp.	S. Oyster Bay Rd. Bethpage, NY 11714	(516) 575-0574 Telex: 961-440	Produces advanced composites for aerospace applications
Hartford Steam Boiler, AE International Division	One State St. Hartford, CT 06102	(203) 722-1866 Telex: 99354	Provides acoustic emission testing services and equipment for composites
Heatcon, Inc.	Seattle, WA		Supplies advanced composite repair equipment
Heath Tecna Precision Structures, Inc.	19819 84th Ave. S Kent, WA 98031	(206) 872-7500	Plastic and fiberglass fabricators
Hercules, Inc.	Hercules Aerospace Division Hercules Plaza Wilmington, DE 19899	(302) 594-5000	Manufactures Magnamine graphite fibers and products. Provides composite structures for aircraft, missiles, and satellites. Also produces resins and esters and epoxy prepreg tapes
Hexcel Trevarno	P.O. Box 888 Lancaster, OH 43130	(614) 653-1528 Telex: 24-5391	Produces bonded honeycomb sandwich constructed composite materials, preimpregnated fabrics, resin systems, and fiberglass textiles
Hi-Tech Composites, Inc.	5447 Equity Ave. Reno, NV 89502	(702) 786-8666	Manufactures fabrics of polyester, glass, carbon, and Kevlar for the automotive, aerospace, marine and recreational markets
HITCO Materials Division	P.O. Box 1097 Gardena, CA 90249	(800) 243-8160	Manufactures epoxy and polyimide Kevlar-reinforced resin materials. Also produces carbon-carbon, ablative, honeycomb, and bonded structures and materials
J.M. Huber Corporation	P.O. Box 2831 Borger, TX 79008	(806) 274-6331	Specializes in reinforcing additives. A new amorphous silica fiber/whisker product is under development and will be available in mid-1986
Hysol Grafil Co.	2850 Willow Pass Rd. Pittsburg, CA 94565	(415) 938-5533	Manufactures resins including SynCore, a group of syntactic thermosetting films and pastes used in advanced composite systems. Also produces carbon/graphite fibers through its Courtaulds' Carbon Fibers Division.
ICD Group Inc.	641 Lexington Ave. New York, NY 10022	(212) 644-1500	Specialists in raw materials supply to the ceramic industry including refractories, abrasives, composites, cutting tools, electronics and glass. Also distributes Tateho SiC and Si_3N_4 whiskers in U.S.
ICI Americas, Inc.	Concord Pike & New Murphy Rd. Wilmington, DE 19897	(302) 575-4466	Manufactures alumina fiber in bulk and mat forms for ceramic, glass, and metal reinforcement. Also produces polymer resins
Instron Corp.	100 TR Royall St. Canton, MA 02021	(617) 828-2500 Telex: 92-4434	Supplies instruments and systems for testing and evaluating advanced materials
KDI Composite Technology, Inc.	6881 Eighth St. Buena Park, CA 90620	(714) 739-8045	Manufactures fiberglass-reinforced plastics and composites demonstrating laminating, compression molding and metal bonding

Company Name	Address	Phone	Descriptions
Kaiser Aerotech	P.O. Box 1678 San Leandro, CA 94577	(415) 562-2456	Manufactures carbon/carbon composite structures
Koppers Company, Inc.	1250 Koppers Bldg. Pittsburgh, PA 15219	(412) 227-2000	Produces extensive line of reinforced and unreinforced polyester resin systems. Continued study of polyester resins is conducted at a centralized research facility
LNP Corporation	412 King St. Malvern, PA 19355	(215) 644-5200	Produces Thermocomp, carbon, and glass-reinforced thermoplastic resins, LNP internally lubricated reinforced-thermoplastics and filled fluoropolymer composites. Also manufactures statically conductive and EMI attenuating composites
Lamination Technology, Inc.	2720 S. Main St. Santa Ana, CA 92707	(714) 556-1460	Products include: mass lamination for multi-layers, laminated copper-clad sheets, and epoxy fiberglass prepregs
Laminations	P.O. Box 1874 Auburn, WA 98002	(206) 833-8200	Features custom fiber-resin laminating and metal bonding. Also manufactures parts and assemblies
Laserage Technological Group	4201 Grove Ave. Gurnee, IL 60031	(312) 249-5900	Provides laser machining of all materials including composites, plastics, most metals, ceramics, sapphire, and quartz
Leesona Corp.	Regional Center Hwy. 521 Fort Hill, SC 29715	(401) 739-7100 Telex: 927715	Manufactures fiber producing machinery for textile and plastics industries
Lewcott Chemicals & Plastics	P.O. Box 319 Millbury, MA 01527	(617) 865-4466	Produces composite prepreg materials and phenolic foam
Lipton Steel & Metal Products Inc.	454 South St. P.O. Box 1159 Pittsfield, MA 01202	(413) 499-1661 Telex: 955317	Supplies bonding and laminating autoclave systems
C.A. Litzler Co. Inc.	4802 W. 160 St. Cleveland, OH 44135	(216) 267-8020 Telex: 98-0234	Manufactures and designs continuous process ovens for composite materials production
Lockheed Missiles & Space Co. Inc.	P.O. Box 504 Sunnyvale, CA 94086	(804) 742-4321	Produces advanced structural composites for the aerospace industry
Lord Corporation	Chemical Products Group 2000 West Grandview Blvd. Box 10038 Erie, PA 16514	(814) 868-3611 Telex: 91-4445	Produces structural adhesives, coatings, binders, mold-release agents, and specialty chemicals for polymers and elastomers
Lunn Industries, Inc.	Lunn Bldg. Wyandanch, NY	(516) 643-8900	Features molding and fabrication of reinforced plastics
Lydall, Inc.	Composite Materials Division 615 Parker St. Manchester, CT 06040	(203) 646-1233	Manufacturers fiber board, fiber composites, and fiber/polymer alloys
MFG, Molded Fiber Glass Companies	P.O. Box 675 Ashtabula, OH 44004	(216) 997-5851	Over 80% of manufacturing is devoted to custom-molding of FRP. The remaining capacity is applied to proprietary products including flat sheet and ribbed panels
MTS Systems Corp.	P.O. Box 24012 TR Minneapolis, MN 55424	(612) 937-4000	Supplies systems for material testing

Company Name	Address	Phone	Description
Manville	P.O. Box 5108 Denver, CO 80217	(303) 978-4900	Manufactures fiberglass fiber reinforcements and products. Also produces plastic and plastic material fillers and refractory products
Marshall Consulting, Inc.	720 Appaloosa Dr. Walnut Creek, CA 94596	(415) 945-6051	Offers assistance in the construction, design, production, tooling, and selection of materials for sandwich construction. Also offers in-plant seminars
Martin-Marietta Corporation	6801 Rockledge Dr. Bethesda, MD 20817	(301) 897-6000	Aerospace division manufactures advanced composite structures. Other divisions produce ceramic and cement industry-related supplies and products
Material Concept, Inc.	666 N. Hague Ave. Columbus, OH 43204	(614) 272-5785	Provides research and development in the field of metal matrix composites. Also produces graphite machined parts, tool and dies
Materials Sciences Corp.	Gwynedd Plaza II Spring House, PA 19477		Offers research, design, and analysis services for materials, structures and composites
Metglass Products, Division of Allied Corporation	6 Eastmans Rd. Parsippany, NJ 07054	(201) 581-7700 Telex: 136044	Supplies Metglas amorphous alloy ribbons, which are produced through rapid solidification of molten metals at cooling rates of about a million degrees centigrade per second. Composite applications are being explored with promising results being reported
Mobay Chemical Corporation	Mobay Rd. Pittsburgh, PA 15205	(412) 777-2000	Produces a wide range of plastics and polymer additives. Features reinforced and nonreinforced grades of polycarbonates, ABS blends, and polyesters. Products are marketed under Pocan, Merlon, Petlon, Makroblend tradenames
Monsanto Polymer Products Co.	800 N. Lindbergh Blvd. St. Louis, MO 63167	(314) 694-1000	Produces a wide range of reinforced and nonreinforced thermoplastic resins. Also manufactures Santoprene, a thermoplastic rubber compound. Engineering resins include Nyrim, a rubber modified nylon copolymer, Lustran ABS resins, polystyrene, Lustrex resins, Cadon resins, and Vydyne glass and mineral-reinforced nylon and polyamide resins
Mossberg Industries, Inc.	160 Bear Hill Road Cumberland, RI 02864	(401) 333-3000	Manufactures the New England Butt composite maypole braider for continuous braiding of advanced composite fibers. Machines can be custom designed for specific requirements
Mutual Industries	1400 Goldmine Rd. Monroe, NC 28110	(704) 283-2147	Manufactures knit and woven graphite, Kevlar, carbon, and fiberglass tapes
NYCO	Box 368 Willsboro, NY 12996	(518) 963-4262	Produces mineral fillers including mica particulate, alumina trihydrate and wollastonite, calcium metasilicate for the plastics industry. Also produces glass fillers including hollow microspheres
Narmco Materials, Inc.	1440 N. Kraemer Blvd. Anaheim, CA 92806	(714) 630-9400 Telex: 18307	Features structural adhesives and advanced composites. High-modulus Rigidite resin systems include epoxies and bismaleimides reinforced with carbon, glass, and Kevlar fibers. Structural Narmco plastics include reinforced modified epoxies and phenolics. Also provides extensive R & D in the composite area

Company Name	Address	Phone	Description
National Beryllia Corporation	Greenwood Ave. Haskell, NJ 07420	(201) 839-1600	Manufactures ceramic composite structures including ceramic-to-metal and glass-to-metal assemblies. Also produces pure oxide body composites
National-Standard	Woven Products Division P.O. Box 1620 Department IT Corbin, KY 40701	(606) 528-2141 Telex: 218486	Produces fine nickel fibers in loose or in sinter bonded mat forms. Also manufactures conductive fiber fabrics
National Starch & Chemical Corporation	Finderne Ave. Bridgewater, NJ 08807	(201) 685-5122	Manufactures Thermid, polyimides, and polyisoimide resins for high-performance structural composites, adhesives, and coatings
Newport Plastics, Subsidiary of Kidde, Inc.	Box 466 Derby Rd. Newport, VT 05855	(802) 334-7941	Specializes in low- to medium-volume production of fiberglass-reinforced plastic housings. Offers complete engineering and manufacturing services
Ontario Die Company of America	2735 20th St. P.O. Box 376 Port Huron, MI 48060	(313) 987-5060	Supplies cutting dies for composites. Includes cutting dies for graphite, Kevlar, fiberglass, and honeycomb core materials
PPG Industries	Fiber Glass Products One Gateway Center Pittsburgh, PA 15222	(412) 434-3131	Manufactures fiberglass reinforcements for thermoset and thermoplastic resins. Fibers available in rovings, chopped strands, yarns, mat and filament winding systems. Also produces Azdel, a polypropylene glass-reinforced laminate moldable sheet
PQ Corporation	P.O. Box 840 Valley Forge, PA 19482	(215) 293-7200	Produces Q-Cel, siliceous and ceramic hollow microsphere fillers for plastics and cements. Features wide range of particle sizes and densities
Perkin-Elmer Corporation	Main Ave. Norwalk, CT 06856	(203) 762-1000 Telex: 965-954	Manufactures and designs scientific instruments, including testing equipment for composites, metals, and polymer applications
Pfizer, Inc.	235 E. 42nd St. New York, NY 10017	(800) 421-6330	Produces a line of talc and calcium carbonate fillers for use as extenders and reinforcements in plastics
Phillips Petroleum Co.	91-G Research Center Bartlesville, OK 74004	(918) 661-6600	Produces thermoplastic composite materials reinforced with carbon or glass fibers. Is the major manufacturer of glass- and mineral-reinforced polyphenylene sulfide, also known as Ryton
Physical Acoustics Corporation	819 Alexander Rd. P.O. Box 3135 Princeton, NJ 08540	(609) 452-2510 Telex: 642236	Designs and manufactures acoustic emission systems for testing composites and metals
Pierce & Stevens Chemical Corp.	710 Ohio Street Buffalo, NY 14240	(716) 631-8991 Telex: 91-202	Features Miralite, an ultra-low density PVDC microsphere in preexpanded form. Unexpanded microspheres are also available
Pollux Corporation	8280 Patuxent Range Road Jessup, MD 20794	(301) 953-2008	Manufactures aluminum honeycomb core for composite construction
Potters Industries, Inc.	377 Route 17 Hasbrouck Hts., NJ 07604	(201) 288-4700 Telex: 133447	Manufactures solid glass microsphere particles for additives and reinforcements of plastics

Company Name	Address	Phone	Description
Quantum Composites, Inc.	4702 James Savage Rd. Midland, MI 48640	(517) 496-2884	Manufactures Lytex, an epoxy SMC reinforced with chopped glass fibers. Also produces Moulage, an epoxy composite, single component extruded sheet tooling compound, which eliminates layup method of creating mold tooling
Refractory Composites, Inc.	12220-A Rivera Rd. Whittier, CA 90606	(213) 698-8061	Specialists in the design, engineering, and manufacturing of ceramic composite structures, from inception to finished product. All aspects of customer needs and product applications are reviewed. Resources include consultants throughout industry who can assist in any phase of ceramic composites development
Reichhold Chemicals Inc.	Resins & Binders Division P.O. Box 1433 Pensacola, FL 32596	(904) 433-7621 Telex: 702-424	Features epoxy and phenolic resins, as well as glass cloth and fiber materials
Resinoid Engineering Corp.	7557 N. St. Louis Ave. Skokie, Il 60076	(312) 673-1050	Manufactures materials for engineering applications, including asbestos, glass, fabric, and mineral reinforcements
Rheometrics, Inc.	2438 U.S. Highway No. 22 Union, NJ 07038	(201) 687-4838 Telex: 138-816	Manufactures rheological and impact testing instrumentation
Ribbon Technology Corp.	Box 30758 Gahanna, OH 43230	(614) 864-5444 Telex: 246-518	Markets carbon and stainless steel fibers as reinforcements for Portland Cement concretes and refractory castables. R & D in flakes, fibers, foils, and powder materials for composite applications
Rockwell International	North American Aircraft Operations 100 N. Sepulveda Bl. P.O. Box 92098 Los Angeles, CA 90009	(213) 647-1000 Telex: 664363	Manufactures composite and metallic structures for the aerospace industries
Rogers Corporation, Molding Materials Division	Box 550 Manchester, CT 06040	(203) 646-5500	Produces fiber-reinforced thermoset molding compounds. The University of Delaware Center for Composite Materials assists with R & D efforts
RTP Co.	580 East Front St. P.O. Box 439 Winona, MN 55987	(507) 454-6900 Telex: 910-565-2276	Manufactures reinforced thermoplastics from a variety of over 24 resins and a large selection of fillers and fibers
SEP	Tour Roussel Nobel Cedex No. 3 F-92080 Paris France		Develops and manufactures carbon-carbon and ceramic materials for aircraft structures
Shur-Tronics	2541 White Rd. Irvine, CA 92714	(714) 474-6000	Manufacturer of Compositest, a composite testing system with user friendly software
Stackpole Fibers Co., Inc.	Foundry Industrial Park Lowell, MA 01852	(617) 454-0409	Produces carbon fibers and carbon fiber products. Panex products include carbon fiber fabrics, spun yarns, chopped carbon fibers, continuous carbon fiber filaments and tow, and carbon fiber paper. Also produces carbon fiber-reinforced injection-molded parts for industrial applications

Company Name	Address	Phone	Description
A.E. Staley Mfg. Co., Polymerizable Products Department	2200 East Eldorado St. Decatur, IL 62525	(217) 423-4411	Features Stalink polymerizable cellulosics for FRP resin binders and tie coats for FRP fillers and fibers
Stevens Products, Inc.	128 North Park St. East Orange, NJ 07019	(201) 672-2140	Manufactures ready-to-use shapes of high-modulus graphite epoxy, glass epoxy, and fiberglass laminates
Tateho Chemical Industries Co., Ltd.	974 kariya, Ako-shi Hyogo-ken, Japan	07914-5-2041 Telex: 5778625	Manufactures SiC and Si_3N_4 single crystal ceramic whiskers. Also produces other ceramic and refractory materials
Thermal Equipment Corp.	1301 W. 228th St. Torrance, CA 90501	(213) 775-6745 Telex: 182577	Designers and manufacturers of autoclave systems for composites, as well as fabricators of hydroclaves and composite presses
3-M Ceramic Fiber Products	225-4N 3-M Center St. Paul, MN 55144	(612) 733-1558	Produces continuous polycrystalline metal oxide ceramic fibers, Nextel. Fibers are available in rovings, ply-twisted yarns, tapes, sleevings, fabrics, and special constructions
Tiodize Co., Inc.	15701 Industry Lane Huntington Beach, CA 92649	(714) 898-4377	Features graphite composites, lubricants, and coatings
Unicel Corporation	Department TR 1520 Industrial Ave. Escondido, CA 92025	(619) 741-3912	Manufactures aluminum, low-carbon steel, and stainless steel honeycomb cores
Union Carbide Corporation	Old Ridgebury Rd. Danbury, CT 06817	(203) 794-5300	Produces carbon fibers and carbon fiber composite products
Versar Manufacturing Inc.	Specialty Products Division 6850 Versar Center P.O. Box 1549 Springfield, VA 22151	(703) 750-3000	Manufactures conductive or nonconductive composite filler particles of carbon
WSF Industries, Inc.	Box 400, Kenwood Dr. Buffalo, NY 14217	(800) 874-8265 (716) 692-4930	Engineers and manufactures composite bonding equipment
Washington Penn Plastic Co., Inc.	2080 North Main St. P.O. Box 236 Washington, PA 15301	(412) 228-1260	Features calcium carbonate, talc, mica, glass, and combination filled polypropylene and polyethylene resins
Westinghouse Electric Corp., Insulating Materials Div.	West Mifflin, PA 15122	(412) 256-3975	Manufacturers fiberglass-reinforced polyester composite shapes. Polyglass structures are produced through the pultrusion process and are available in general and flame-retardant grades. Over 100 shapes are available as stock items
Westlake Plastics Co.	P.O. Box 127 161 W. Lenni Rd. Lenni, PA 19052	(215) 459-1000 Telex: 83-5406	Features extrusion of engineering thermoplastics and thermosets. Extrudes thermosetting plastics with high-performance fibers/fillers
Wilson-Fiberfil International	P.O. Box 3333 2267 W. Mill Rd. Evansville, IN 47732	(812) 424-3831 Telex: 752708	Supplies an extensive and complete line of electrically conductive thermoplastic compounds. Electrofil(R) resins are reinforced with carbon fibers, carbon black, aluminum flakes, nickel coated carbon fibers or stainless steel fibers
Zircar Products, Inc.	110 North Main St. Florida, NY 10921	(914) 651-4481 Telex: 996-608	Produces zirconia, alumina, and alumina-silica fiber products. Also supplies alumina and zirconia bulk fibers, refractory sheets, rigidizers and cements and textiles of yttrium oxide, hafnium oxide, tantalum oxide, titanium oxide, and cerium oxide

SECTION 14
Glossary of Terms
Relating to Composites

A

A-stage. An early stage in the reaction of certain thermosetting resins in which the material is fusible and still soluble in certain liquids. Sometimes referred to as resol. See also *B-stage* and *C-stage*.

ABL bottle. An internal pressure testing vessel, used to determine the quality and properties of a filament-wound material in the vessel, that is about 18 in. (457 mm) in diameter and 24 in. (610 mm) long.

abhesive. A material that resists adhesion. Abhesive coatings are applied to surfaces to prevent sticking, heat sealing, etc.

ablation. An orderly heat and mass transfer process in which a large amount of thermal energy is expended by sacrificial loss of surface region material. The heat input from the environment is absorbed, dissipated, blocked, and generated by numerous mechanisms. The energy absorption processes take place automatically and simultaneously, serve to control the surface temperature, and greatly restrict the flow of heat into the substrate interior.

ablative plastic. A term applied to any polymer or resin which decomposes layer-by-layer when its surface is heated, leaving a heat-resisting layer of charred material which eventually breaks away to expose virgin material.

abrasion resistance. Ability of a plastic to withstand mechanical action such as rubbing and scraping.

absorption. Amount of energy absorbed by a shield. Absorption depends on energy, frequency, thickness, conductivity, and permeability of the shield.

accelerator. A material which, when mixed with a catalyzed resin, will speed up the curing reaction. Also known as *promoter* or *curing agent*.

acceptance test. A test, or series of tests, conducted by the procuring agency, or an agent thereof, upon receipt of an individual lot of materials to determine whether the lot conforms to the purchase order or contract or to determine the degree of uniformity of the material supplied by the vendor, or both.

acetal. Very strong, stiff engineering plastic with exceptional dimensional stability and resistance to creep and vibration fatigue, low coefficient of friction, and high resistance to abrasion and chemicals. It retains most properties when immersed in hot water and has a low tendency to stress-crack.

acid. A chemical compound containing one or more hydrogen atoms available for reaction with active metals or alkaline solutions.

acid number. The number of milligrams of potassium hydroxide required to neutralize the free fatty acid in one gram of fat, oil, wax, or resin.

acrylic. Any of a family of synthetic resins made by polymerizing esters of acrylic acids. Properties include: high optical clarity; excellent resistance to outdoor weathering; hard, glossy surface; excellent electrical properties; fair chemical resistance; availability in brilliant, transparent colors.

acrylonitrile-butadiene-styrene. Very tough, yet hard and rigid, resins that have fair chemical resistance, low water absorption, and hence good dimensional stability. They also have high abrasion resistance and are easily electroplated.

activator. An additive used to promote and reduce the curing time of resins. See also *accelerator*.

adaptive control. A method by which input from sensors automatically and continuously adjusts in an attempt to provide near-optimum processing conditions.

additives. Substances compounded into resin to improve certain characteristics, such as plasticizers, initiators, light stabilizers, catalysts, and flame retardants.

adhere. To cause two surfaces to be held together by adhesion.

adherend. A body held to another body by an adhesive.

adherend preparation. See *surface preparation*.

adhesion. The state in which two surfaces are held together by interfacial forces which may consist of valence forces or interlocking action, or both. See also *mechanical adhesion* and *specific adhesion*.

adhesive. A substance capable of holding materials together by surface attachment.

adhesive bonding. A materials joining process in which an adhesive, placed between facing surfaces, solidifies to bond the surfaces together.

adhesive dispersion. A two-phase system in which one phase is suspended in a liquid.

adhesive failure. Rupture of an adhesive bond, such that the separation appears to be at the adhesive/adherend interface.

adhesiveness. The property defined by the adhesion stress $A = F/S$, where F = perpendicular force to glue line and S = surface.

advanced composites. Strong, tough materials created by combining one or more stiff, high-strength reinforcing fibers with a compatible resin system. Advanced composites can be substituted for metals in many structural applications.

after-bake. See *postcure*.

aggregate. A hard, fragmented material used with an epoxy binder (or other resin), common in plastic tooling.

aggressive tack. See *dry tack*.

aging. Loss of properties through time-exposure to elevated temperature, ultraviolet radiation, moisture, or other hostile environments.

aging time. See *joint conditioning time*.

air-bubble void. Noninterconnected spherical air entrapment within and between the plies of a reinforcement.

alkyd plastics. A generic name for a family of synthetic resins made from organic acids, alcohols, and drying oils. The acids include such compounds as phthalic anhydride, maleic anhydride, fumaric acid, sebacic acid, and adipic acid. The alcohols include glycerin, ethylene glycol, trimethylolethane, and pentaerythrytol. Drying oils include soya, dehydrated castor, and linseed. These resins have excellent electrical properties and heat resistance. They are easier and faster to mold than most thermosets, with no volatile by-products. See also *polyester resins*.

alligatoring. A laminate surface flaw resembling the texture of an alligator's skin.

alloy. In plastics, a blend of polymers with other polymers or copolymers. In metals, a substance having metallic properties and being composed of two or more chemical elements of which at least one is a metal.

allyl plastics. Plastics based on resins made by addition polymerization of monomers containing allyl groups: diallyl phthalate (DAP); diallyl isophthalate (DAIP). These resins have outstanding dimensional stability and electrical properties, are easy to mold, and have excellent resistance to moisture and chemicals at high temperatures.

ambient. Prevailing environmental conditions such as the surrounding temperature, pressure, and relative humidity.

amino resins (urea, melamine). A large class of thermosetting resins made by the reaction of an amine with an aldehyde. The aldehyde is usually formaldehyde and the most important amines are urea and melamine. These resins are abrasion and chip resistant, and have good solvent resistance. Urea molds faster and costs less than melamine. Melamine has a harder surface and higher heat and chemical resistance.

amorphous phase. A phase devoid of crystallinity (noncrystalline). Most plastics are amorphous at processing temperatures.

amylaceous. Pertaining to, or of the nature of, starch; starchy.

angle-ply laminate. Laminate possessing equal plies with positive and negative angles. This bi-directional laminate is simple because it is orthotropic, not anisotropic. A [±45] is a very common angle-ply laminate. A cross-ply laminate is another simple laminate.

anisotropic. Not isotropic; exhibiting different properties when tested along axes in different directions.

anisotropic laminate. A laminate that has strength properties which are different in different directions.

antioxidant. Additive to prevent degradation of plastics through exposure to atmosphere. Deterioration may be caused by heat, age, radiation, chemicals, stress, etc.

antistatic agent. An agent applied to a molding material or the surface of a molded object, which makes it less conducting, and thus hinders the fixation of dust.

aramid. A generic classification for a group of strong, high-temperature-resistant, man-made fibers. Lightweight aromatic polyamide fibers offer excellent high-temperature, flame, and electrical properties. These fibers are used in protective clothing and as high-strength reinforcements in plastic composites.

arc resistance. The total time in seconds that an intermittent arc can play across a plastic surface without rendering the surface conductive.

arc spray. See *flame spray.*

artificial weathering. The exposure of plastics to cyclic laboratory conditions comprising high and low temperatures, high and low relative humidities, and ultraviolet radiant energy, with or without direct water spray, in an attempt to produce changes in their properties similar to those observed in long-time continuous exposure conditions outdoors. The laboratory exposure conditions are usually intensified beyond those encountered in actual outdoor exposure, in an attempt to achieve an accelerated effect. See also *weathering.*

aspect ratio. Length-to-width ratio of a rectangular plate, the length-to-diameter ratio of a discontinuous fiber, or the length-to-bundle-diameter ratio of a bundle of parallel fibers.

assembly. A group of materials or parts, including adhesive, which have been placed together for bonding or which have been bonded together.

assembly adhesive. An adhesive that can be used for bonding parts together, such as in the manufacture of boats, airplanes, furniture, and the like.

assembly time. The time interval between the spreading of the adhesive on the adherend and the application of pressure or heat, or both, to the assembly.

attenuation. As applied to the formation of fibers from molten glass, the process of making slender and thin.

autoclave. Pressure vessel that can maintain temperature and pressure of a desired air or gas for the curing of organic-matrix composite materials.

autoclave molding. A molding method in which, after final lay-up, an entire assembly is put into a steam or electrically heated autoclave at elevated pressure. Additional pressure achieves higher reinforcement loadings and improved removal of air.

automatic mold. In injection or compression molding, a mold that goes repeatedly through an entire cycle, including injection, without human assistance.

axial winding. In filament-wound reinforced plastics, a winding with the filaments parallel to the axis.

B

B-stage. An intermediate stage in the reaction of certain thermosetting resins in which the material softens when heated and swells when in contact with certain liquids, but may not entirely fuse or dissolve. The resin in an uncored thermosetting adhesive is usually in this stage. Sometimes referred to as *resitol.* See also *A-stage* and *C-stage.*

BFRA. Boron fiber reinforced aluminum.

BFRP. Boron fiber reinforced plastic.

BMC. See *bulk molding compound.* See also *sheet molding compound* and *premix.*

back pressure. The resistance of plastic material to flow during processing, due to its viscosity.

bag molding. The technique of molding reinforced plastic composites using a flexible cover (bag) over a rigid mold. The composite material is placed in the mold and covered with the bag. Then pressure is applied by vacuum, autoclave, press, or by inflating the bag.

balanced design. A winding pattern in filament-wound reinforced plastics designed so that all stresses in all filaments are equal.

balanced laminate. A composite laminate whose lay-up is symmetrical with relation to the midplane of the laminate.

balanced twist. An arrangement of twist in a plied yarn or cord which will not cause twisting on itself when the yarn or cord is held in the form of an open loop.

balanced-in-plane contour. A head contour, in a filament-wound part, in which the filaments are oriented within a plane and the radii of curvature are adjusted to balance the stresses along the filaments with the pressure loading.

barcol hardness. A hardness value obtained by measuring the resistance to penetration of a sharp steel point under load. The apparatus used to obtain this reading is called a barcol impressor and the readings are sometimes associated with the degree of cure of a plastic.

bare glass. The glass as it flows from the bushing in fiber form, before binder or size is applied.

batch. The manufactured unit, or a blend of two or more units of the same formulation and processing. See also *lot.*

batt. A felted fabric that is built by the interlocking action of fibers without spinning, weaving, or knitting. See also *felt.*

beam. A spool, on which is wound a number of parallel ends of singles or plied yarns, for use in weaving or similar processing operations.

beaming. An operation in which the yarn from several section beams is combined on the final warp beam.

bearing strength. The bearing stress at that point on the stress-strain curve where the tangent is equal to the bearing stress divided by $n\%$ of the bearing-hole diameter.

bearing stress. The applied load in pounds divided by the bearing area of a specimen.

biaxial load. (1) A loading condition of a pressure vessel under internal pressure and with unrestrained ends. (2) A loading condition in which a laminate is stressed in at least two different directions in the plane of the laminate.

biaxial winding. A type of winding, in filament winding, in which the helical band is laid in sequence, side by side, with no crossover fibers.

bidirectional laminate. Fibers, in a reinforced plastic laminate, oriented in various directions in the plane of the laminate; a cross laminate. See also *unidirectional laminate.*

binder. A material applied in liquid form to fibers, yarn, or fabric, to retain structural integrity during further processing.

bisphenol A. A condensation product formed by reaction of two (bi) molecules of phenol with acetone (A). This polyhydric phenol is a standard resin intermediate along with epichlorohydrin in the production of epoxy resins.

blanket. Plies laid up in a complete assembly and placed on or in the mold all at one time (flexible-bag process); also, the form of bag in which the edges are sealed against the mold.

bleeder cloth. A nonstructural layer of material used in manufacture of composite parts to allow the escape of excess gas and resin during curing.

bleed. (1) To give up color when brought into contact with water or solvents. (2) An undesired movement of certain materials in adhesives (sometimes plasticizers) to the surface of the bonded article (especially plastics) or into adjacent material. (3) To evacuate air or gases from an assembly during the curing cycle.

blister. An elevation on the surface of an adherend containing air or water vapor, somewhat resembling in shape a blister on the human skin. Its boundaries may be indefinitely outlined, and it may have burst and become flattened.

blocked curing agent. A curing agent or hardener rendered unreactive, which can be reactivated as desired by physical or chemical means.

blocking. An undesired adhesion between touching layers of a material, such as occurs under moderate pressure during storage or use.

bloom. A visible surface exudation—e.g., irregular spotting or clouding on plastic—caused by migration to the surface of incompatible particles, or by faulty coloring or lubricating methods.

blowing agent. An additive for resins to be foamed. When heated to a specific temperature, it decomposes to yield a large volume of gas that creates cells in foamed plastics.

blush. See *chalking.*

bond. (1) The adhesion at the interface between two surfaces. (2) To attach materials together by means of adhesives.

bond line. See *glue line*.

bond strength. The unit load applied in tension, compression, flexure, peel, impact, cleavage, or shear, required to break an adhesive assembly with failure occurring in or near the plane of the bond.

bonded joint. Location where the two adherends are bonded together by a layer of adhesive between them. In a lap joint, the adherends are positioned with an overlap; in a scarf joint, with matched taper sections; in a stepped joint, through steps.

bonding agent. See *glass finish*.

bonding angle. A connecting angle of several plies of reinforcement and resin used to connect two parts of a laminate, usually at right angles to each other.

boron filament. A filament made by vapor deposition of B_4C onto a tungsten substrate. Two filament diameters, 0.1 and 0.2 mm (0.004 and 0.008 in.), are made. Two outstanding features of boron are the high longitudinal compressive strength and the relative ease of fabrication of boron/aluminum composite material.

borsic. Silicon carbide coated boron fiber.

boss. Small projection from a part's surface designed to add strength, facilitate alignment with another part during assembly, or permit attachment to another part.

bottom plate. A steel plate fixed to the lower section of a mold, often used to join the lower section of the mold to the platen of the press.

branched. Chemical term referring to side chains attached to original chain (in a direction different from that of the original chain) in molecular structure of polymers.

breather. Porous material, such as a fabric or mat, placed inside the vacuum bag to facilitate removal of air, moisture, and volatiles during curing.

breathing. (1) Opening and closing of a mold to allow gases to escape early in the molding cycle (also called *degassing*). (2) Permeability to air of plastic sheeting.

bridging. A region of a contoured part which has been cured without being properly compared against the mold.

bristle. See *fiber*.

broad goods. Woven glass or synthetic fiber or combination thereof, over 18 in. (457 mm) in width.

brushability. The adaptability of a paint to application by brushing.

bubble. A spherical interval void; a globule of air or other gas trapped in a plastic.

buckling. Unstable lateral displacement of a structural part such as a panel caused by excessive compression and shear. Microbuckling of fibers in a composite material can also occur under axial compression.

bulk factor. The ratio of the volume of a molding compound or powdered plastic to the volume of the solid piece produced therefrom; the ratio of the density of the solid plastic object to the apparent density of the loose molding powder.

bulk modulus (*B*). The ratio of the hydrostatic pressure P to the volume strain (see also *modulus of elasticity*):

$$B = \frac{P}{\Delta V / V_0} = \frac{P V_0}{\Delta V}$$

where V_0 is the original volume and ΔV is the volume change due to applied pressure.

bulk molding compound (BMC). Thermosetting resins mixed with stranded reinforcement, fillers, etc., into a viscous compound for injection or compression.

bundle strength. Strength obtained from a test of parallel filaments, with or without an organic matrix. The bundle test is often used in place of the tedious monofilament test.

burned. Showing evidence of thermal decomposition on the surface of a plastic through some discoloration, distortion, or destruction.

burst strength. The hydraulic pressure required to burst a vessel of given thickness. This is commonly used in testing filament-wound composite structures.

bushing. An electrically heated alloy container incased in insulating material, used for melting and feeding of glass in the forming of individual fibers or filaments; the outer ring of any type of a circular tubing or pipe die which forms the outer surface of the extruded tube or pipe.

butt joint. A type of edge joint in which the edge faces of the two adherends are at right angles to the other faces of the adherends.

butt wrap. Tape wrapped around an object in an edge-to-edge fashion.

C

C-C. Carbonaceous heat-shield composites (C-C stands for carbon-carbon).

C-stage. The final stage in the reaction of certain thermosetting resins in which the material is relatively insoluble and infusible. Certain thermosetting resins in a fully cured adhesive layer are in this stage. Sometimes referred to as *resite*. See also *A-stage* and *B-stage*.

CAD. Computer aided design. The use of a computer to design a product. The use of a computer to develop the design and produce the NC program used by automated manufacturing equipment.

CAM. Computer aided manufacturing. Automated manufacturing of a product defined by an NC program. The output of a CAD system is often in NC program form for CAM.

CFG iron. Compact-flake-graphite iron.

CFRP. Carbon or graphite fiber reinforced plastic.

CIM. Computer integrated manufacturing. The use of computers to design and manufacture a product. Formerly called *CAD/CAM*.

CP. (1) Cross-ply. (2) Resinous heat-shield composite.

CPI. Condensation-reaction polyimide.

carbon fiber. Fibers produced by pyrolysis of an organic precursor fiber in an inert atmosphere at temperatures higher than 1800 °F (980 °C). Reinforcement for lightweight, high-strength, and high-stiffness structures. The high stiffness and high strength of fibers depends on the degree of preferred orientation.

carbonization. The process of pyrolyzation in an inert atmosphere at temperatures ranging from 1000 to 1500 °C (1830 to 2730 °F). All noncarbon elements are driven off in the process.

cartridge heater. Cylindrical-bodied electrical heater for providing heat for injection, compression, and transfer molds; injection nozzles; runnerless mold systems; hot stamping dies; sealing; etc.

catalyst. A substance that initiates or accelerates a chemical reaction. See also *curing agent, hardener, inhibitor, promoter*.

catastrophic failures. Gross failures of an unpredictable nature.

catenary. A measure of the difference in length of the strands in a specified length of roving as a result of unequal tension; the tendency of some strands in a taut horizontal roving to sag lower than the others.

caul plate. A smooth metal plate used in contact with the lay-up during curing to transmit normal pressure and to provide a smooth surface to the finished laminate.

cavity. (1) The space inside a mold into which the resin is poured or injected. (2) The space between matched molds in which the laminate is formed. (3) Also a term for a female mold.

cellular adhesive. See *foamed adhesive*.

cellular plastics. Materials with cell structure throughout their mass.

cellulosics. Family of tough, hard materials: cellulose acetate, propionate, butyrate, and ethyl cellulose. Property ranges are broad because of compounding; available with various degrees of weather, moisture, and chemical resistance; fair to poor dimensional stability; brilliant colors.

centrifugal casting. A high-production technique for cylindrical composites, such as pipe, in which chopped strand mat is positioned inside a hollow mandrel designed to be heated and rotated as resin is added and cured.

ceramic(s). (1) Any of a class of inorganic, nonmetallic products which are subjected to a high temperature during manufacture or use. (2) Product containing ceramic materials.

cermet. A powder metallurgy product consisting of ceramic particles bonded with a metal.

chalking. (1) The presence of loose powder on the surface of a paint easily detected by wiping the film. Controlled chalking results in a self-cleaning paint. (2) Dry, chalklike appearance of or deposit on the surface of a plastic.

charge. A measured amount of material used to load a mold.

Charpy test. A pendulum-type, single-blow impact test in which the specimen, usually notched, is supported at both ends as a simple beam and broken by a falling pendulum. The energy absorbed, as determined by the subsequent rise of the pendulum, is a measure of impact strength or notch toughness. Contrast with *Izod test*.

chase. (1) The main body of the mold, which contains the molding cavity or cavities, or cores, the mold pins, the guide pins or bushings, etc. (2) An enclosure of any shape used to shrink-fit parts of a mold cavity in place to prevent spreading or distortion in hobbing or to enclose an assembly of two or more parts of a split-cavity block.

china clay. Aluminum silicate, used as a filler.

circuit. In filament winding, (1) one complete traverse of the fiber-feed mechanism of a winding machine, or (2) one complete traverse of a winding band from one arbitrary point along the winding path to another point on a plane through the starting point and perpendicular to the axis.

circumferential winding. In filament-wound reinforced plastics, a winding with the filaments essentially perpendicular to the axis.

clamp tonnage. Rated clamping capacity of an injection or transfer molding machine.

clamping area. Largest molding area an injection molding machine can hold closed under full pressure.

clamping pressure. Pressure which is applied to an injection or transfer mold to hold it closed.

closed assembly time. See *assembly time.*

closed cell foam. Cellular plastic in which individual cells are completely sealed off from adjacent cells.

co-injection. The technique of injecting two materials into a single mold from two plasticating cylinders, either simultaneously or in sequence.

coagulation. The precipitation of colloid into a single solid mass.

coalescence. The fusing together of a latex film upon evaporation of the water.

cocuring. Simultaneous curing and bonding of a composite laminate to another material, like honeycomb core, or to other parts, such as stiffeners. Cocuring can reduce fabrication cost.

coefficient of cubical expansion. See *coefficient of expansion.*

coefficient of elasticity. The reciprocal of Young's modulus in a tension test. Also called *tensile compliance.*

coefficient of expansion. The fractional change in the dimensions of a material pertinent to a change in temperature. Also called *coefficient of thermal expansion.*

coefficient of friction. A measure of the resistance to sliding of one surface in contact with another surface.

coefficient of linear expansion. The change in length per unit length produced by a unit rise in temperature. See *coefficient of expansion.*

coefficient of thermal expansion. The change in volume per unit volume produced by a 1° rise in temperature:

$$a = \frac{I}{V}\left(\frac{\alpha V}{\alpha T}\right) = \left(\frac{\alpha I_n V}{\alpha T}\right)P$$

where V is volume and T is temperature. See *coefficient of expansion.*

cohesion. (1) The state in which the particles of a single substance are held together by primary or secondary valence forces. (2) As used in the adhesive field, the state in which the particles of the adhesive (or the adherend) are held together.

cold flow. The distortion which takes place in materials under continuous load at temperatures within the working range. See *creep, strain relaxation.*

cold molding. Matched die press molding process using a catalyst system requiring no external heat.

cold pressing. A bonding operation in which an assembly is subjected to pressure without the application of heat.

cold setting adhesive. A synthetic resin adhesive capable of hardening at normal room temperature in the presence of a hardener.

collimated. Rendered parallel (applies to filaments).

colloid. A material composed of ultramicroscopic particles of a solid, a liquid, or a gas dispersed in a different medium which can be a solid, a liquid, or a gas.

colophony. See *rosin.*

color concentrate. Plastic resin which contains a high loading of pigment. Concentrates provide a dust-free method of handling colors.

color retention. When a coating is exposed to the elements and shows no signs of changing its color, it is said to have excellent color retention.

colorant. A dye or pigment used in plastics.

compliance. See *tensile compliance* and *shear compliance.*

composite. A homogeneous material created by the synthetic assembly of two or more materials (a selected filler or reinforcing elements and compatible matrix binder) to obtain specific characteristics and properties. Composites are subdivided into the following classes on the basis of the form of the structural constituents: fibrous – the dispersed phase consists of fibers; flake – the dispersed phase consists of flat flakes; laminar – composed of layer or laminar constituents; particulate – dispersed phase consists of small particles; skeletal – composed of a continuous skeletal matrix filled by a second material.

compounding. The process of mixing the polymer with all the materials necessary for the finished resin to be shipped to the processor.

compression mold. A mold which is open when the material is introduced and which shapes the material by heat and by the pressure of closing.

compression molding. A technique of thermoset molding in which the molding compound (generally preheated) is placed in the open mold cavity, the mold is closed, and heat and pressure are applied until the material has cured.

compression molding pressure. The unit pressure applied to the molding material in the mold.

compressive modulus (E_c). Ratio of compressive stress to compressive strain below the proportional limit. Theoretically equal to Young's modulus determined from tensile experiments.

compressive strength. The maximum compressive stress which a material is capable of developing, based on the original area of the cross section.

compressive stress. The compressive load per unit area of original cross section carried

by the specimen during the compression test.

condensation. A chemical reaction in which two or more molecules combine with the separation of water or some other simple substance. If a polymer is formed, the process is called polycondensation. See also *polymerization.*

condensation agent. A chemical compound which acts as a catalyst and also furnishes a complement of material necessary for a polycondensation reaction to proceed.

condensation resin. A resin formed by polycondensation: for example, the alkyd, phenol-aldehyde, and urea formaldehyde resins.

conditioning time. See *joint conditioning time.*

conductive paints. Silver-, nickel-, copper-, graphite-, and copper-graphite-base paints used to coat nonconductive substrates for EMI/RFI shielding. Spray equipment or paddle guns are normally used to apply conductive paints.

conductivity. (1) Reciprocal of volume resistivity. (2) The conductance of a unit cube of any material.

consistency. That property of a liquid adhesive by virtue of which it tends to resist deformation.

constituent materials. Individual materials that make up the composite material: e.g., graphite and epoxy are the constituent materials of a graphite/epoxy composite material.

contact adhesive. An adhesive that is apparently dry to the touch and which will adhere to itself instantaneously upon contact; the surfaces to be joined must be no farther apart than about 0.004 in. (0.1 mm). Also called *contact bond adhesive* or *dry bond adhesive.*

contact bond adhesive. See *contact adhesive.*

contact molding. A process for molding reinforced plastics in which reinforcement and resin are placed on a mold, curing is done either at room temperature or by oven heat, and no pressure is applied.

contact-pressure resins. Liquid resins which thicken or polymerize when bonding laminates, and require only enough pressure to ensure intimate contact.

continuous filament. A strand in which individual filament lengths approach the strand length.

continuous filament yarn. Yarn formed by twisting two or more continuous filaments into a single, continuous strand.

cooling fixture. A fixture used to maintain the shape or dimensional accuracy of a molding or casting after it is removed from the mold and until the material is cool enough to hold its shape.

copolymer. The product obtained when two different monomers are polymerized together to yield a polymer. They are linked chemically. See *polymerization.*

copolymerization. The building up of linear or nonlinear macromolecules (copolymers) in which many monomers, possessing molecules having one or many double bonds, have located in every macromolecule of different size which constitutes the copolymerizate, following alternations which may be regular or not. See also *polymerization*.

core. (1) A channel in a mold for circulation of heat-transfer media. (2) Male part of a mold which shapes the inside of the mold. (3) The central member of a sandwich construction to which the faces of the sandwich are attached.

count. Number of warp and fill yarns per unit length: e.g., a fabric count of 24 × 26 in the English unit means 24 yarns per inch in the warp, and 26 in the fill.

coupling. Linking a side effect to a principal effect. Poisson coupling links the lateral contraction to an axial extension. For composite materials, an anisotropic laminate couples the shear to normal components; an unsymmetric laminate couples curvature with extension. These couplings are unique with composites and provide opportunities to perform extraordinary functions.

crack. (1) An actual separation of molding material visible on opposite surfaces of the part and extending through the thickness. (2) A fracture.

crazing. Fine resin cracks at or under the surface of a plastic.

cream time. Length of time between pouring of mixed urethane foam and the point when the material turns creamy or the beginning of foaming.

creel. That part of a twisting, winding, or warping machine that holds packages of strands for further fabrication.

creep. The dimensional change with time of a material under load, following the initial instantaneous elastic or rapid deformation. The creep strain occurring at a diminishing rate is called primary creep; that occurring at a minimum and almost constant rate, secondary creep; and that occurring at an accelerating rate, tertiary creep. Creep at room temperature is sometimes called *cold flow*.

creep strength. (1) The constant nominal stress that will cause a specified quantity of creep in a given time at constant temperature. (2) The constant nominal stress that will cause a specified rate of secondary creep at constant temperature.

crimp. (1) To fold over and fasten under pressure; to indent the adherend surface in order to obtain more positive contact at the resin interface. (2) The waviness of a fiber; it determines the capacity of fibers to cohere under light pressure, and is measured either by the number of crimps or waves per unit length or by the percent increase in extent of the fiber on removal of the crimp.

critical concentration. The minimal percentage of an electrically conductive additive needed to change an insulating plastic to a conductive one. Also known as *threshold concentration*.

critical longitudinal stress. (Fibers) The longitudinal stress necessary to cause internal slippage and separation of a spun yarn; the stress necessary to overcome the interfiber friction developed as a result of twist.

critical strain. The strain at the yield point.

cross-ply laminate. Special laminate that contains only 0 and 90° plies. This bidirectional laminate is orthotropic, and has nearly zero Poisson's ratio. The other simple bidirectional laminate is the angle-ply, which possesses one pair of balanced off-axis plies.

crosslinking. The establishment of chemical links between the molecular chains in polymers. Crosslinking can be accomplished by chemical reaction, vulcanization, and electron bombardment.

crosswise direction. Refers to cutting of specimens and to application of load. For rods and tubes, crosswise is the direction perpendicular to the long axis. For other shapes or materials that are stronger in one direction than in another, crosswise is the direction that is weaker. For materials that are equally strong in both directions, crosswise is an arbitrarily designated direction at right angles to the length.

crystallite. A perfect portion of an ordinary crystal—that is, a portion with its atoms and molecules arranged in a perfect crystal lattice. Ordinary crystals are composed of a large number of crystallites, which may or may not be arranged in perfect alignment with one another.

cull. Excess material in the transfer chamber of a transfer molding machine after the mold has been filled.

cure. To irreversibly change the properties of a thermosetting resin by chemical reaction—i.e., condensation, ring closure, or addition. Curing may be accomplished by addition of curing (crosslinking) agents, with or without heat.

curing temperature. Temperature at which a material is subjected to curing.

curing agent. A catalytic or reactive agent which when added to a resin causes polymerization; synonymous with *hardener*.

curing time. The period of time during which an assembly is subjected to heat or pressure, or both, to cure the resin. The interval of time between the instant that relative movement between the moving parts of a mold ceases and the instant that pressure is released. Further curing may take place after removal of the assembly from the conditions of heat or pressure.

cut. In the fiber industry, including glass and asbestos, the number of 100-yard lengths of fiber per pound.

cut-off. In compression molding, the line where the two halves of a mold come together; also called *flash groove* or *pinch-off*.

cycle. One full sequence in a molding operation, from a point in the process to the same point in the next sequence.

D

D glass. Glass with a high boron content, used for fibers in laminates which require a precisely controlled dielectric constant.

DS. Directionally solidified.

dam. Ridge circumventing a mold to prevent resin runout during curing.

daylight. Clearance between two platens of a bonding press in the open position. For a multidaylight press, daylight is the distance between adjacent platens.

debond. Area of separation within or between plies in a laminate, or within a bonded joint, caused by contamination, improper adhesion during processing, or damaging interlaminar stresses.

debulk. To reduce laminate thickness by application of pressure. The compaction is achieved by removing trapped air, vapor, and volatiles between plies.

deep-draw mold. A mold having a core which is long in relation to the wall thickness.

deflection temperature under load. The dimensional change of a material under load for a specific time following the instantaneous elastic deformation caused by the initial application of the load; also called *cold flow* or *creep*.

degassing. See *breathing*.

degradation. Deleterious change in the chemical structure of a plastic reflected in the appearance or physical properties.

degree of polymerization. Number of structural units or mers in the average polymer molecule in a sample measure of molecular weight. Generally, the degree of polymerization is in the thousands.

delamination. Debonding process primarily resulting from unfavorable interlaminar stresses. Edge delamination, however, can be effectively prevented by a wraparound reinforcement.

denier. A yarn and filament numbering system in which the yarn number is equal numerically to the weight in grams of 3000 ft (914 m) (used for continuous filaments). The lower the denier, the finer the yarn.

desizing. The process of eliminating sizing, which is generally starch, from gray goods prior to applying special finishes or bleaches (for yarn such as glass or cotton).

die. See *cavity*.

dielectric. A material with electrical conductivity less than one millionth of a reciprocal ohm per centimeter, thus so weakly conductive that different parts of its surface can have a different electrical charge. In radio-frequency heating operations, the term dielectric is used for the material being heated. The term is also used for the nonconductive material separating the conductive elements of a condenser.

dielectric constant. Normally, the relative dielectric constant. For practical purposes, the ratio of the capacitance of an assembly of two electrodes separated solely by an adhesive insulating material to its capacitance when the electrodes are separated by air.

dielectric curing. The curing of a synthetic thermosetting resin by the passage of an electric charge produced from a high-frequency generator through the resin.

dielectric heating. The plastic to be heated forms the dielectric of a condenser to which is applied a high-frequency (20-to-80-MHz) voltage. *Dielectric loss* in the material is the basis. See *high-frequency heating*.

dielectric loss. A loss of energy eventually showing through the rise in heat of a dielectric placed in an alternating electrical field.

dielectric loss angle. Dielectric phase difference. The difference between an angle of 90° and the dielectric phase angle.

dielectric loss factor. See *loss factor.*

dielectric phase angle. The angular difference in phase between the sinusoidal voltage applied to the dielectric and the resulting current.

dielectric strength. The electric voltage gradient at which an insulating material is broken down or "arced through" in volts per mil of thickness.

differential scanning calorimeter. Determines thermal histories of polymers in the laboratory analogous to those imposed on the polymer in processing so that the effects can be studied directly.

dilatancy. The property of some materials by which the resistance to flow increases with agitation.

diluent. An ingredient usually added to an adhesive to reduce the concentration of bonding materials. See also *extender.*

dimensional stability. In molded, cast, or formed parts, the ability of the plastic part to retain its shape.

dispersion. (1) Any heterogeneous system of solids, gases, or liquids. (2) Finely divided particles of a resin held in suspension in another material.

displacement angle. In filament winding, the distance of advance of the winding ribbon at the equator after one complete circuit.

doctor-bar or blade. A scraper mechanism that regulates the amount of adhesive on the spreader rolls or on the surface being coated.

doctor-roll. A roller mechanism that revolves at a different surface speed, or in an opposite direction, resulting in a wiping action for regulating the adhesive supplied to the spreader roll.

doily. In filament winding, the planar reinforcement applied to a local area between windings to provide extra strength in an area where a cutout is to be made — e.g., port openings.

dome. In filament winding, the portion of a cylindrical container that forms the integral ends of the container.

double spread. See *spread.*

doubler. Extra plies of reinforcement added to a laminate in areas of high stress.

draft. The tapered design of a mold wall which facilitates removal of molded part.

draft angle. The angle between the tangent to the surface at that point and the direction of ejection.

drape. Textile conformity; the ability of preimpregnated broad goods to conform to an irregular shape.

driers. Metallic salts, usually cobalt, lead, manganese, iron, zinc, and calcium. Many paints dry through oxidation processes. The inclusion of driers increases the oxygen absorption, hastening the curing process.

drooling. Leakage of resin from a nozzle or around the nozzle area during the injection step in injection molding. Also, the escape of low-viscosity resins from the nozzle.

dry. To change the physical state of an adhesive on an adherend by the loss of solvent constituents by evaporation or absorption, or both. See also *cure* and *set.*

dry bond adhesive. See *contact adhesive.*

dry lay-up. Construction of a laminate by layering preimpregnated reinforcement (partly cured resin) in a female or male mold, usually followed by bag molding or autoclave molding.

dry spot. Area of incomplete surface film on laminated plastics; an area over which the interlayer and the glass have not become bonded. See also *resin-starved area.*

dry strength. The strength of a laminate after conditioning in standard laboratory atmosphere. See *wet strength.*

dry tack. The property of certain adhesives, particularly nonvulcanizing rubber adhesives, to adhere to themselves on contact at a stage in the evaporation of volatile constituents even though they seem dry to the touch. Sometimes called *aggressive tack.*

dry winding. Filament winding using preimpregnated roving, as differentiated from wet winding. See also *wet winding.*

drying oils. Vegetable or animal oils which, when exposed to air, will absorb oxygen and harden to a tough elastic material.

drying time. The period of time during which an adhesive on an adherend or an assembly is allowed to dry with or without the application of heat or pressure, or both. See also *curing time; joint conditioning time;* and *setting time.*

ductility. Amount of plastic strain a material can withstand before it fractures.

dwell. (1) A pause in the application of pressure to a mold, made just before the mold is completely closed, to allow gas to escape from the molding material. (2) In filament winding, the time the traverse mechanism is stationary while the mandrel continues to rotate to the appropriate point for the traverse to begin a new pass.

E

E glass. A borosilicate glass, most used for glass fibers in reinforced plastics.

EMC. Elastomeric-molding tooling compound.

E.S.C. See *environmental stress cracking.*

edge distance ratio. The distance from the center of the bearing hole to the edge of the specimen in the direction of the principal stress, divided by the diameter of the hole.

edge joint. A joint made by bonding the edge faces of two adherends.

edgewise. In cutting, the application of forces in directions parallel to and actually in the plane of a sheet or specimen. For compression-molded specimens of square cross section, the edge is the surface parallel to the direction of motion of the molding plunger. For injection-molded specimens of square cross section, this surface is selected arbitrarily. For laminates, the edge is the surface perpendicular to the laminate. See also *flatwise.*

efflorescence. A phenomenon whereby a whitish crust of fine crystals forms on a painted surface. These are usually sodium salts which diffuse through the paint film from the substrate.

ejection. Removal of the molded part from the mold by mechanical means or with compressed air.

ejector plate. A plate which backs up the ejector pins and holds the ejector assembly together.

ejection ram. A small hydraulic ram fitted to a press to operate the ejector pins.

elastic deformation. The portion of deformation of an object under load which can be recovered after the load is removed.

elastic limit. The maximum stress to which a material may be subjected without any permanent strain remaining upon complete release of the stress.

elastic recovery. The fraction of a given deformation that behaves elastically. Elastic recovery equals elastic extension divided by total extension. Elastic recovery is 1 for perfectly elastic material and 0 for perfectly plastic material.

elasticity. The property of plastic materials by which they tend to recover their original size and shape after deformation.

elastomer. A macromolecular material which, at room temperature, is capable of recovering substantially in size and shape after removal of a deforming force.

electric glass. See *E glass.*

electric surface resistance. The surface resistance between two electrodes in contact with a material is the ratio of the voltage applied to the electrodes to that portion of the current between them which flows through the surface layers.

electric surface resistivity. The ratio of the potential gradient parallel to the current along the surface of a material to the current per unit width of surface.

electrical dissipation factor (D_e). The ratio of the power loss in a dielectric material to the total power transmitted through the dielectric, or the imperfection of the dielectric. Equal to the tangent of the *dielectric loss angle:*

$$D_e = \frac{\epsilon''}{\epsilon'} = \tan \sigma = \frac{1}{2\pi f C_p R_p}$$

where f is frequency of applied voltage in hertz, C_p is equivalent parallel capacity, and R_p is equivalent parallel resistance.

electroformed mold. A mold made by electroplating metal on the reverse pattern on the cavity.

electromagnetic interference (EMI). An electromagnetic energy that causes interference in the operation of electronic equipment.

electrostatic discharge (ESD). A large electrical potential (4000 V or more) moving from one surface or substance to another. ESD is also an abbreviation for electrostatic dissipation.

elongation. The fractional increase in length of a material stressed in tension. When expressed as a percentage of the original gage length, it is called percentage elongation.

emissivity. The ratio of the total heat-radiating power of a surface to that of a black body of the same area and of the same temperature.

emulsifier. A material which, when added to a mixture of dissimilar materials, such as oil and water, will produce a stable homogeneous emulsion.

emulsion. A suspension of fine particles or globules of a liquid within a liquid.

emulsion polymerization. The process of polymerization taking place in the presence of water to form a latex.

encapsulating. Enclosing an article in an envelope of plastic by immersing the object in a casting resin and allowing the resin to polymerize or, if hot, to cool.

end. Individual warp yarn, thread, monofilament, or roving. For glass fibers, an end contains 206 filaments.

end count. An exact number of ends supplied on a ball or roving.

endothermic reaction. A reaction which is accompanied by the absorption of heat.

endurance limit. See *fatigue limit.*

environmental stress cracking (E.S.C). The susceptibility of the thermoplastic resin to cracking or crazing when in the presence of surface active agents or other environments.

epichlorohydrin. A basic chemical used in the production of epoxies. It contains an epoxy group and is highly reactive with polyhydric phenols such as bisphenol A.

epoxides. Compounds containing the oxirane structure, a three-member ring containing two carbon atoms and one oxygen atom. The most important members are ethylene oxide and propylene oxide.

epoxy equivalent. The weight of resin in grams which contains one gram equivalent of epoxy. If it is assumed that the resin chains are linear and that an epoxy group terminates each end, then the epoxy equivalent is one half the average molecular weight of a diepoxy resin, one-third the average molecular weight of a triepoxy, etc.

epoxy molding compound. Compounds are mineral-filled powders which can be molded on compression or transfer molding presses.

epoxy plastics. An important class of structural adhesives. Based on ethylene oxide, its derivatives or homologs, epoxy resins form straight-chain thermoplastics and thermosetting resins — e.g., by condensation of bisphenol and epichlorohydrin. Epoxies exhibit exceptional mechanical strength, electrical properties, and adhesion to most materials, and low mold shrinkage; some formulations can be cured with or without heat or pressure.

equator. In filament winding, the line in a pressure vessel described by the junction of the cylindrical portion and the end dome.

even tension. The process whereby each end of roving is kept in the same degree of tension as the other ends making up that ball of roving. See also *catenary.*

exotherm. (1) The temperature-vs-time curve of a chemical reaction and the amount of heat given off. Maximum temperature occurs at peak exotherm. (2) The liberation or evolution of heat during curing of a plastic product.

exothermic reaction. A reaction in which heat is given off.

expansion coefficient. Measurement of

swelling or expansion of a composite material due to temperature changes or moisture absorption.

extend. To add fillers or lower-cost materials in an economy-producing endeavor; to add inert materials to improve void-filling characteristics and to reduce crazing.

extender. A substance, generally having some adhesive action, added to an adhesive to reduce the amount of the primary binder required per unit area. See also *binder, diluent,* and *filler.*

extender pigment. A pigment which contributes very little hiding power to the system. However, inerts do contribute other desirable physical properties to the coating.

extensibility. The ability of a material to extend or elongate upon application of sufficient force. Expressed as a percentage of the original length.

$$\text{extensibility} = \frac{\Delta L_{\max}}{L_0} \times 100 = \epsilon_B \times 100$$

extrusion. Forcing a material to flow plastically through a die orifice. In direct extrusion (forward extrusion), the die and ram are at opposite ends of the extrusion stock, and the product and ram travel in the same direction. Also, there is relative motion between the extrusion stock and the container. In indirect extrusion (backward extrusion), the die is at the ram end of the stock and the product travels in the direction opposite that of the ram, either around the ram or up through the center of a hollow ram. Impact extrusion is the process (or resultant product) in which a punch strikes a slug (usually unheated) in a confining die. The material flow may be either between punch and die or through another opening. Impact extrusion of unheated slugs is often called cold extrusion.

eyelet-type insert. Insert having a section which protrudes from the material and is used for spinning over in assembly.

F

FP. Polycrystalline alumina fiber.

FRAT. Fiber-reinforced advanced titanium.

FRP. Fiber-reinforced plastic; a general term covering any type of fiber-reinforced plastic.

fabric. Planar, woven material constructed by interlacing yarns, fibers, or filaments. Nonwovens are sometimes included in this classification.

failure criterion. Empirical description of the failure of composite materials subject to complex states of stresses or strains. The most commonly used are the maximum stress, the maximum strain, and the quadratic criteria.

fan. In glass-fiber forming, the fan shape that is made by the filaments between the bushing and the shoe.

fan gate. Opening between the mold runner and the mold cavity which has the shape of a fan. This shape helps reduce stress concentrations in the gate area by spreading the opening over a wider area.

fatigue. The phenomenon leading to fracture under repeated or fluctuating stresses having a maximum value less than the tensile strength of the material. Fatigue fractures are progressive, beginning as minute cracks that grow under the action of the fluctuating stress.

fatigue life. The number of cycles of stress that can be sustained prior to failure under a stated test condition.

fatigue limit. The maximum stress that presumably leads to fatigue fracture in a specified number of stress cycles. If the stress is not completely reversed, the value of the mean stress, the minimum stress, or the stress ratio also should be stated.

fatigue notch factor (K_f). The ratio of the fatigue strength of an unnotched specimen to the fatigue strength of a notched specimen of the same material and condition; both strengths are determined at the same number of stress cycles. See also *stress concentration factor.*

fatigue ratio. The ratio of fatigue strength to tensile strength. Mean stress and alternating stress must be stated.

fatigue strength. The maximum stress that can be sustained for a specified number of cycles without failure, the stress being completely reversed within each cycle unless otherwise stated.

fatigue-strength reduction factor. See *fatigue notch factor.*

felt. A fibrous material made from interlocked fibers by mechanical or chemical action, moisture, or heat; made from asbestos, cotton, glass, etc. See also *batt.*

fiber. This term usually refers to relatively short lengths of very small cross-sections of various materials. Fibers can be made by chopping filaments (converting). Also called *filament, thread,* or *bristle.* See also *staple fibers.*

fiber content. Percent volume of fiber in a composite material. Percent weight or mass of fiber is also used.

fiber glass. (1) An individual filament made by attenuating molten glass. (2) Major material used to reinforce plastics. Available as mat, roving, fabric, etc., it is incorporated into both thermosets and thermoplastics. The glass increases mechanical strength, impact resistance, stiffness, and dimensional stability of the matrix.

fiber glass chopper. Chopper guns, long cutters, and roving cutters cut glass into strands and fibers to be used as reinforcements in plastics.

fiber orientation. Fiber alignment in a nonwoven or a mat laminate where the majority of fibers are in the same direction, resulting in a higher strength in that direction.

fiber pattern. (1) Visible fibers on the surfaces of laminates or moldings. (2) The thread size and weave of glass cloth.

fiber strain in flexure. The maximum strain in the outer fiber occurring at midspan.

fiber stress in flexure. When a beam of homogeneous, elastic material is tested in flexure as a simple beam supported at two points and loaded at the midpoint, the maximum stress in the outer fiber occurs at midspan.

fiber-composite material. A material consisting of two or more discrete physical phases, in which a fibrous phase is dispersed in a continuous matrix phase. The fibrous phase may be macro-, micro-, or submicroscopic, but it must retain its physical identity so that it could conceivably be removed from the matrix intact.

fiber-matrix interface. The region separating the fiber and matrix phases, which differs from them chemically, physically, and mechanically. In most composite materials, the interface has a finite thickness (nanometers to thousands of nanometers) because of diffusion or chemical reactions between the fiber and matrix. Thus, the interface can be more properly described by the terms interphase or interfacial zone. When coatings are applied to the fibers or several chemical phases have well-defined microscopic thicknesses, the interfacial zone may consist of several interfaces.

Fick's equation. Diffusion equation for moisture migration. This is analogous to the Fourier's equation of heat conduction.

filament. Any fiber whose aspect ratio (length to effective diameter) is for all practical purposes infinity — i.e., a continuous fiber. For a noncircular cross section, the effective diameter is that of a circle which has the same (numerical) area as the filament cross section.

filament weight ratio. In a composite material, the ratio of filament weight to the total weight of the composite.

filament winding. A composite fabrication process which consists of winding a continuous reinforcing fiber (impregnated with resin) around a rotating and removable form (mandrel).

filamentary composite. A laminate composed of laminae in which continuous filaments are in nonwoven, parallel, uniaxial arrays. Individual uniaxial laminae can be combined into specifically oriented multiaxial laminates.

fill. Yarn running from selvage to selvage at right angles to the warp in a woven fabric. See *warp*.

filler. Inert, inexpensive substance added to resins to extend volume, improve properties, and lower the cost of the laminate. See also *binder, extender,* and *reinforced plastic.*

filler sheet. A sheet of deformable or resilient material that when placed between the assembly to be bonded and the pressure applicator, or when distributed within a stack of assemblies, aids in providing uniform application of pressure over the area to be bonded.

filler yarn. See *weft.*

fillet. That portion of an adhesive which fills the corner or angle formed where two adherends are joined.

film adhesive. A synthetic resin adhesive usually of the thermosetting type that has been placed on a carrier or calendered into a thin film (0.05 to 0.4 mm, or 0.002 to 0.016 in.).

finish. A mixture of materials for treating glass fibers. It contains a coupling agent to improve the bond of resin to glass and usually includes a lubricant to prevent abrasion and a binder to promote strand integrity. With graphite or other filaments, it may perform either or all of the above functions.

first-ply-failure. First ply or ply group that fails in a multidirectional laminate. The load corresponding to this failure can be the design limit load.

fish-eye. A fault in transparent or translucent plastic materials such as films or sheets, appearing as a small globular mass and caused by incomplete blending of the mass with surrounding material.

flame resistance. Ability of a material to extinguish flame once the source of heat is removed. See also *self-extinguishing.*

flame retardants. Chemicals used to reduce or eliminate the tendency of a resin to burn. (For polyethylene and similar resins, chemicals such as antimony trioxide and chlorinated paraffins are useful.)

flame spray shielding. Flame spray or arc spray shielding is a process of depositing molten metal onto a nonconductive substrate with an arc spray or flame spray pistol.

flame spray coating. Method of applying a plastic coating in which finely powdered fragments of the plastic, together with suitable fluxes, are projected through a cone of flame onto a surface.

flame-retarded resin. A resin compounded with certain chemicals to reduce or eliminate its tendency to burn.

flammability. A measure of the extent to which a material will support combustion.

flash. Excess plastic material or glass which forms at the parting line of a mold or which is extruded from a closed mold.

flash groove. See *cut-off.*

flash mold. A specific type of mold which is designed to allow excess material to escape as flash.

flat lay. (1) The property of nonwarping in laminating adhesives. (2) An adhesive material with good noncurling and nondistention characteristics.

flatwise. Refers to cutting of specimens and the application of load. The load is applied flatwise when it is applied to the face of the original sheet or specimen.

flexible molds. Molds made of rubber, elastomers, or flexible thermoplastics, used for casting plastics. They can be stretched to remove cured pieces with undercut.

flexible resin. A type of polyester resin which cures to a flexible, tough solid rather than a rigid solid. It is sometimes added to the rigid resins to improve laminate resiliency.

flexural modulus. The ratio, within the elastic limit, of the applied stress on a test specimen in flexure to the corresponding strain in the outermost fibers of the specimen.

flexural strength. (1) The resistance of a material to breakage by bending stresses. (2) The strength of a material in bending expressed as the tensile stress of the outermost fibers of a bent test sample at the instant of failure. For plastics this value is usually higher than the straight tensile strength. (3) The unit resistance to the maximum load before failure by bending, usually in megapascals (SI system) and/or kips per square inch (English system).

flocculate. To form agglomeration of undispersed pigments, etc.

flow. During processing by injection, compression, or transfer molding, a measure of the fluidity of a plastic; movement of an adhesive during the bonding process, before the adhesive is set.

flow lines. Marks on a molded piece made by the meeting of two flow fronts during molding. Also called *striae, weld-marks* and *weld-lines.*

fluidized-bed coating. The process of applying plastic coatings to objects of other materials, often metals, wherein a powdered resin is placed in a container provided with a porous or perforated bottom through which a gas is discharged upward to keep the resin particles in a state of flotation. The part to be coated is preheated and lowered into the fluidized bed until a deposit of the desired thickness is formed, then withdrawn. Subsequent heating is usually necessary to fuse the resin particles into a smooth, homogeneous layer.

fluoroplastics. Polyolefin polymers in which fluorine, fluorinated alkyl groups, or other halogens replace hydrogen atoms in the carbon chain. This structure has outstanding electrical properties, excellent resistance to chemical attack, low coefficient of friction, excellent fire resistance, exceptionally good performance at high and low temperatures, low moisture absorption, and outstanding weatherability. Fluoroplastics include PTFE, FEP, PFA, CTFE, ECTFE, ETFE, and PVDF. Strength is low to moderate; cost is high.

fluted core. An integrally woven reinforcement material consisting of ribs between two skins in a unitized sandwich construction.

foam pouring. The first method of foam processing. Resin mixture contains chemical blowing agent, and it is poured into an open mold. Usually molds into a bun, but other shapes are possible.

foam spray. Used in the construction industry for on-site application of urethane foam insulation.

foam-in-place. The dispensing of foam material at the work site.

foamed adhesive. An adhesive, the apparent density of which has been decreased substantially by the presence of numerous gaseous cells dispersed throughout its mass.

foamed plastics. Resinous materials that have been expanded into a multicellular structure, characterized by light weight and relatively low strength. May be rigid or flexible, porous or nonporous.

foaming agent. A chemical additive for resins that decomposes on mixing with the resin, yielding gases which produce lightweight cellular products.

force. (1) In matched die molding, the male half of the mold, which enters the cavity, exerts pressure on the resin, and causes it to flow (also referred to as the "punch"). (2) In compression molding, the downward-acting mold half, usually the male half.

Fourier's equation. Diffusion equation commonly associated with the heat conduction in a body. Fick's equation is a special case, applied to moisture migration and accumulation.

fracture. Rupture of the surface without complete separation of laminate.

fracture test. Test in which a specimen is broken and its fractured surface is examined with the unaided eye or with a low-power microscope to determine such factors as composition, grain size, case depth, or soundness.

fungus resistance. The resistance of a material to attack by fungi in conditions promoting their growth.

furane plastics. Dark-colored thermosetting resins obtained primarily by the condensation polymerization of furfuryl alcohol in the presence of strong acids, sometimes in combination with formaldehyde or furfuryldehyde. The term also includes resins made by condensing phenol with furfuryl alcohol or furfuryl, and furfuryl-ketone polymers. The resins are available as liquids ranging from low-viscosity fluids to thick, heavy syrups, which cure to highly crosslinked, brittle substances. Also spelled "furan."

fused quartz. See *silica glass.*

G

GPD. Grams per denier. See *tenacity.*

gage length. The original length of that portion of a test specimen over which strain or change in length under load is to be measured.

gap. In filament winding, the space between successive windings, which are usually intended to be next to each other.

gate. The opening through which the plastic enters the mold cavity.

gel. A partial cure of plastic resins. Transition of resin from a liquid to a jellylike state.

gel coat. A quick-setting resin used in molding processes to provide an improved surface for composites; it is the first resin applied to the mold after the mold-release agent.

gel point. The stage at which a liquid begins to exhibit pseudo-elastic properties. Also conveniently observed from the inflection point on a viscosity-time plot.

gelatin time. For synthetic thermosetting resins, the interval of time between introduction of a catalyst into a liquid adhesive system and gel formation.

gelation. Formation of a gel.

gelation rate. See *gel point.*

geodesic. Pertaining to the shortest distance between two points on a surface.

geodesic isotensoid. A filamentary structure in which there exists a constant stress in any given filament at all points in its path.

geodesic ovaloid. A contour for end domes, the fibers forming a geodesic line on the surface of revolution. The forces exerted by the filaments are proportioned to meet hoop and meridional stresses at any point.

geodesic-isotensoid contour. In a filament-wound reinforced plastic pressure vessel, a dome contour in which the filaments are placed on geodesic paths so that the filaments will exhibit uniform tension throughout their length under pressure loading.

glass. An inorganic product of fusion which has cooled to a rigid condition without crystallizing. Glass is typically hard and relatively brittle and has a conchoidal fracture.

glass fiber. A glass filament that has been cut to a measurable length. Staple fibers of relatively short length are suitable for spinning into yarn.

glass filament. A form of glass that has been drawn to a small diameter and an extreme length. (Most filaments are less than 0.005 in., or 0.13 mm, in diameter.)

glass filament bushing. The unit through which molten glass is drawn in making glass filaments.

glass finish. A material applied to the surface of a glass reinforcement to enhance the bond between the glass and the plastic binder. Also called *bonding agent.*

glass flake. Thin, irregularly shaped flakes of glass typically made by shattering a continuous thin-wall tube of glass.

glass former. An oxide which forms a glass easily; also one that contributes to the network of silica glass when added to it.

glass stress. In a filament-wound part, usually a pressure vessel, the stress calculated using only the load and the cross-sectional area of the reinforcement.

glass temper. To introduce compressive surface stresses into the glass by rapid cooling from above the annealing point.

glass transformation. The transition from the supercooled liquid to a true glass.

glassy temperature. The temperature at which the stiffness and strength of an organic resin undergo drastic reduction. This is also known as the glass transition temperature and can be the maximum use temperature.

gloss. The shine, sheen, or lustre of a dried film.

glue. Originally, a hard gelatin obtained from hides, tendons, cartilage, bones, etc., of animals. Also, an adhesive prepared from this substance by heating with water. Through general use the term is now synonymous with the term "adhesive." See also *adhesive, mucilage, paste,* and *sizing.*

glue line. The layer of adhesive which attaches two adherends. Also called *bond line.*

graphitization. The process of pyrolyzation in an inert atmosphere at temperatures in excess of 1800 °C (3270 °F), and usually as high as 2700 °C (4890 °F). This produces a turbostratic "graphite" crystal structure.

gravity closing. The closing motion in a down-stroke press, which is actuated by the weight of the ram and associated parts only.

gray goods. See *greige.*

Green strength. The ability of a material, while not completely cured, set or sintered, to undergo removal from the mold and handling without distortion.

greige. Fabric before finishing; yarn or fiber before bleaching or dyeing. Also called *gray goods, greige goods, greige gray.*

greige goods. See *greige.*

greige gray. See *greige.*

grex. The weight in grams of 10 kilometers of a yarn or fiber.

guide pins (dowel pins). In compression, transfer, and injection molding, hardened steel pins that maintain proper alignment of the mold halves as they open and close.

guide-pin bushing. The bushing through which the guide pin moves when the mold is closed.

gum. Any of a class of colloidal substances, exuded by or prepared from plants, sticky when moist, composed of complex carbohydrates and organic acids, which are soluble or swell in water. See also *adhesive, glue, resin.*

gunk. A viscous premix of a resin and filler or milled fibers used as a filler, as in bosses for sandwich cores.

gusset. A piece used to give added size or strength in a particular location of an object; the folded-in portion of a flattened tubular film.

H

HM. High-modulus.

hand. The softness of a piece of fabric, as determined by touch (individual judgment).

hand lay-up. Method of positioning successive layers of reinforcement mat or web (which may or may not be preimpregnated with resin) on a mold by hand. Resin is used to impregnate or coat the reinforcement, followed by curing of the resin to permanently fix the formed shape.

hardener. A substance or mixture of substances added to an adhesive to promote or control the curing reaction by taking part in it. The term is also used to designate a substance added to control the degree of hardness of the cured film. See also *catalyst.*

hardness. The resistance to surface indentation usually measured by the depth of penetration of a blunt point under a given load using a particular instrument according to a prescribed procedure.

heat-activated adhesive. A dry adhesive film that is rendered tacky or fluid by application of heat, or heat and pressure, to the assembly.

heat build-up. The temperature rise in a part resulting from the dissipation of applied strain energy as heat.

heat cleaned. Fiber glass reinforcement which has been exposed to elevated temperatures to remove preliminary sizings and binders which are not compatible with the resin to be applied.

heat distortion point. The temperature at which a standard test bar deflects a specified amount under a stated load.

heat endurance. The time of heat aging that a material can withstand before failing a specific load (physical test).

heat-convertible resin. A thermosetting resin that can be converted by heat to a solid.

heteropolymerization. See *polymerization.*

hexa. Short for hexamethylenetetramine, a source of reactive methylene for curing Novolaks.

hiding power. The ability of a paint to obscure the background over which it is applied.

high-pressure laminates. Laminates molded and cured at pressures not lower than 1 ksi (7 MPa), and more commonly in the range of 1.2 to 2.0 ksi (8.3 to 13.8 MPa) (typical of decorative laminates for surfacing applications).

high-pressure spot. An area containing very little resin; usually due to an excess of reinforcing materials.

high-frequency heating. The heating of materials by dielectric loss in a high-frequency electrostatic field. The material is exposed between electrodes and is heated quickly and uniformly by absorption of energy from the electrical field.

high-pressure molding. A molding process in which the pressure used is greater than 7 MPa (1 ksi).

homopolymer. The result of the polymerization of a single monomer, a homopolymer consists of a single type of repeating unit.

honeycomb. A low-density cellular material resembling natural honeycomb, used as a core in sandwich construction.

hoop stress. The circumferential stress in a material of cylindrical form subjected to internal or external pressure.

horizontal shear strength. Estimated from a short-beam-shear test. This test is approximate because the stresses calculated from the simple beam theory are not exact.

hot melt adhesive. An adhesive that is applied in a molten state and forms a bond on cooling to a solid state.

hot setting adhesive. An adhesive the setting of which requires a temperature at or above 100 °C (212 °F). See also *cold setting adhesive, intermediate temperature setting adhesive,* and *room temperature curing adhesive.*

hybrid. Composite with more than two constituents — e.g., a graphite/glass/epoxy hybrid. An intralaminar hybrid has hybrid plies made from graphite and glass filaments. An interlaminar hybrid has laminates made from two or more different ply materials.

hydraulic press. A press in which the platens are pressured by pressure exerted on a fluid.

hydromechanical press. A press in which the molding forces are created partly by a mechanical system and partly by a hydraulic system.

hydrophilic. Capable of absorbing water.

hydrophobic. Capable of repelling water.

hygroscopic. Capable of adsorbing and retaining atmospheric moisture.

hygrothermal effect. Change in properties due to moisture absorption and temperature change.

I

ILC. Integrated laminating center.

IM. Intermediate-modulus.

ignition loss. The difference in weight before and after burning; as with glass, the burning off of the binder or size.

immiscible. Incapable of mixing. Oil and water are immiscible.

impact shock. A stress transmitted to an adhesive interface which results from the sudden jarring or vibrating of the bonded assembly.

impact strength. Ability to withstand shock loading.

impact test. The impact test measures the energy necessary to fracture a standard notched bar by an impulse load. The notched test specimen is struck and fractured by a heavy pendulum, released from a known height, at the nadir of its swing. From a knowledge of the mass of the pendulum and the difference in the initial and final heights, the energy absorbed in fracture is calculated. See also *Izod test, Charpy test.*

impedence. A measure of the total resistance of an object (through its width, length, and depth) to the flow of ac electricity. A higher impedence reflects a higher resistance.

impingement mixing. Intermixing of two liquid chemical streams upon high-velocity injection into a small chamber in the mixing head.

impregnate. To provide liquid penetration into a porous or fibrous material; the dipping or immersion of a fibrous substrate into a liquid resin. Generally, the porous material serves as a reinforcement for the plastic binder after curing.

impregnated fabric. A fabric impregnated with a synthetic resin. See also *prepreg.*

impregnator. A mechanical device for wetting, or impregnating, glass fabrics with resin. Generally consists of a trough through which the fabric is drawn and a set of adjustable scraper bars to remove excess resin.

industrial robot. A reprogrammable, multifunction manipulator designed to move material, parts, tools or specialized devices through variable programmed motions in the performance of a variety of tasks.

inert filler. A material added to an adhesive to alter properties by physical rather than chemical means.

infrared. Part of the electromagnetic spectrum lying outside the visible light range at its red end. Radiant heat is in this range, and infrared heaters are used as a source of heat in sheet thermoforming.

inhibitor. A material added to a resin to slow down curing. It also retards polymerization, thereby increasing shelf life of a monomer.

initial modulus. See *modulus of elasticity.*

initiators. Peroxides used as sources of free radicals. They are used in free radical polymerizations, curing thermosetting resins, as crosslinking agents for elastomers and polyethylene, and for polymer modification.

injection molding. A process of shaping a material using heat and the pressure of injection of the resin into a closed mold.

inorganic. Applies to the chemistry of all elements and compounds not classified as organic; matter other than vegetable, such as earthy or mineral matter.

inorganic pigments. Natural or synthetic metallic oxides, sulfides, and other salts, calcified during processing at 1200 to 2100 °F (650 to 1150 °C). They impart heat- and light-stability, weathering resistance, and migration resistance to plastics.

insert. An integral part of a plastic molding consisting of plastic, metal, or other material that has been preformed and inserted into the mold so it becomes an integral part of the finished molding.

insert molding. Process by which components, such as terminals, pins, studs, and fasteners, may be molded into a part.

insert pin. A pin which keeps an inserted part (insert) inside the mold by screwing or friction; it is removed when the object is being withdrawn from the mold.

instron. An instrument used to determine the tensile and compressive properties of materials.

insulating resistance. The electrical resistance between two conductors or systems of conductors separated only by insulating material.

insulator. (1) A material of such low electric conductivity that the flow of current through it can usually be neglected. (2) A material of low thermal conductivity.

integral skin foam. Urethane foam with a cellular core structure and a relatively nonporous skin.

intelligent robot. A category of robots that have sensory perception, making them capable of performing complex tasks which vary from cycle to cycle. Intelligent robots are capable of making decisions and modifications to each cycle.

intermediate temperature setting adhesive. An adhesive that sets in the temperature range 31 to 99 °C (87 to 211 °F). See also *cold setting adhesive, hot setting adhesive,* and *room temperature curing adhesive.*

interaction. Same as coupling. For example, longitudinal tensile strength is affected by the presence of transverse stress. Similar interaction exists between the longitudinal buckling stress and the transverse or shear stress. As a rule, the interaction effects for composite materials are greater than for the conventional isotropic material. All anticipated stresses should be considered simultaneously.

interface. The boundary between the individual, physically distinguishable constituents of a composite.

interlaminar shear. The shear strength at rupture in which the plane of fracture is located between the layers of reinforcement of a laminate.

interlaminar stresses. Three stress components associated with the thickness direction of a plate. The remaining three are the in-plane components of the plate. Interlaminar stresses are significant only if the thickness is greater than 10% of the length or width of the plate. These stresses can also be significant in areas of concentrated loads, and abrupt changes in material and geometry. The effects of these stresses are not easy to assess because three-dimensional stress analyses and failure criteria are not well understood.

intermesh. The positioning of adjacent blocks of honeycomb so that the outermost edge of one block falls within the outermost edge of the adjacent block.

intumesce. To foam and swell up as the result of heat. Some fire-retardant paints exhibit this property, forming a heat insulating surface.

irradiation. As applied to plastics, the bombardment with a variety of subatomic particles, generally alpha-, beta-, or gamma-rays. Used to initiate polymerization and copolymerization of plastics and in some cases to bring about changes in the physical properties of a plastic.

irreversible. Not capable of redissolving or remelting. Refers to thermosetting synthetic resins; refers to chemical reactions which proceed in a single direction and are not capable of reversal.

isocyanate resin. Resins containing organic isocyanate radicals. They are generally reacted with polyols such as polyester or polyether and the reactants are joined through the formation of the urethane linkage. See also *urethane plastics.*

isomeric. Composed of the same elements united in the same proportion by weight, but differing in one or more properties because of differences in structure.

isotactic. Pertaining to a type of polymeric molecular structure containing a sequence of regularly spaced asymmetric atoms arranged in like configurations in a polymer chain.

isotropic. Having uniform properties in all directions. The measured properties of an isotropic material are independent of the axis of testing.

isotropic laminate. A laminate in which the strength properties are equal in all directions.

Izod test. A pendulum-type single-blow impact test for which the specimen, usually notched, is fixed at one end and broken by a falling pendulum. The energy absorbed, as measured by the subsequent rise of the pendulum, is a measure of impact strength or notch toughness. Contrast with *Charpy test.*

J

joint aging time. See *joint conditioning time.*

joint conditioning time. The time interval between the removal of the joint from the conditions of heat or pressure, or both, used to accomplish bonding and the attainment of approximately maximum bond strength. Sometimes called *joint aging time.*

K

K factor. The coefficient of thermal conductivity. The amount of heat that passes through a unit cube of material in a given time when the difference in temperature of two faces is 1°.

Kevlar fiber. DuPont company trade name for an aramid fiber.

knockout pin. A pin that ejects a molded piece from the mold.

L

LIM. Liquid injection molding.

LMC. Low-pressure molding compound.

lack of fillout. An area of reinforcement that has not been wet with resin; mostly seen at the edge of a laminate.

lacquer. A solution of a film-forming natural or synthetic resin in a volatile solvent, with or without a color pigment, which when applied to a surface forms an adherent film that hardens solely by evaporation of the solvent.

laminate. A product made by bonding together two or more layers of material or materials (normally used with reference to flat sheets). See also *bidirectional laminate, unidirectional laminate.*)

laminate ply (lamina). One layer of a product which is evolved by bonding together two or more layers of materials.

laminated plastics. A class of standard structural shapes, plates, sheets, angles, channels, rods, tubes, and zees that are produced by combining layers of resin-impregnated materials in a press under heat and pressure. Also called laminated molding.

laminated plate theory. The most common method for the analysis and design of composite laminates. Each ply or ply group is treated as a quasi-homogeneous material. Linear strain across the thickness is assumed. This is also called the lamination theory.

lamination. The process of preparing a laminate. Also, any layer in a laminate.

land. (1) In filament winding, the amount of overlay between successive windings, usually intended to minimize gapping. (2) The portion of the die which limits the closure.

land area. The area of surfaces of a mold which contact each other when the mold is closed, measured in a direction perpendicular to the direction of application of closing pressure.

landed force. A force with a shoulder which seats one land in a landed positive mold. Also called *landed plunger.*

landed plunger. See *landed force.*

lap joint. A joint made by placing one piece partly over another and bonding the overlapped portions. See *scarf joint.*

latex. A generic term describing any of many stable dispersions of insoluble resin particles in a water system.

lay. In the packaging of glass fibers, the spacing of the roving bands on the package expressed as the number of bands per inch; in filament winding, the orientation of the ribbon with some reference, usually the axis of rotation.

lay-up. A laminate that has been assembled, but not cured; or a description of the component materials and geometry of a laminate.

legging. The drawing of filaments or strings when adhesive-bonded substrates are separated.

lengthwise direction. Refers to cutting of specimens and the application of loads. For rods and tubes, lengthwise is the direction of the long axis. For other shapes of materials that are stronger in one direction than in the other, lengthwise is the direction that is stronger. For materials that are equally strong in both directions, lengthwise is the arbitrarily designated direction that may be with the grain, direction of flow in manufacture, longer direction, etc. See also *crosswise direction.*

level winding. See *circumferential winding.*

line pressure. The pressure under which an air or hydraulic system operates.

linear expansion. The increase of a planar dimension, measured by the linear elongation of a sample in the form of a beam which is exposed to two given temperatures.

liner. In a filament-wound pressure vessel, the continuous, usually flexible, coating on the inside surface of the vessel used to protect the laminate from chemical attack or to prevent leakage under stress.

liquid resin. An organic polymeric liquid which becomes a solid when converted into its final state for use.

liquidus. The maximum temperature at which equilibrium exists between the molten glass and its primary crystalline phase.

load-deflection curve. A curve in which the increasing flexural loads are plotted on the ordinate and the deflections caused by those loads are plotted on the abscissa.

loaded. Pertains to *roving* or *mat.*

log. Roughly cylindrical-shape charge of bulk molding compound which is cut into small sections and placed in the compression mold, or fed to the stuffer plunger of an injection molding machine.

longitudinal modulus. Elastic constant along the fiber direction in a unidirectional composite — e.g., longitudinal Young's and shear moduli.

longos. Low-angle helical or longitudinal windings.

loop strength. See *loop tenacity.*

loop tenacity. The strength value obtained by pulling two loops, like two links in a chain, against each other in order to test whether a fibrous material will cut or crush itself; also called *loop strength.*

loss angle. The antitangent of the electrical dissipation factor. See *dielectric loss angle.*

loss factor. The product of the power factor and the dielectric constant of a dielectric material. Also called *dielectric loss factor.*

loss index. A measure of a dielectric loss defined by the product of the power factor and the permittivity or *dielectric constant.*

loss modulus. A damping term describing the dissipation of energy into heat when a material is deformed.

loss on ignition. The loss of weight after burning off of an organic resin from a glass fiber laminate or an organic sizing from glass fibers. It is usually expressed as a percent of the total.

loss tangent. See *electrical dissipation factor.*

lot. A specific amount of material produced at one time and offered for sale as a unit quantity.

low-pressure mold. Mold manufactured of low-strength materials and limited to about 200 psi (1.38 MPa) maximum molding pressure.

low-pressure laminates. In general, laminates molded and cured at pressures ranging from 0.4 ksi (2.8 MPa) to contact pressure.

lyophilic. Oil-loving.

M

M-glass. A high-beryllia-content glass designed especially for high modulus of elasticity.

MEKP. Methyl ethyl ketone peroxide, a catalyst for polyester resins.

MMC. Metal-matrix composite.

MVT. Moisture vapor transmission.

mandrel. A form around which filament-wound and pultruded composite structures are shaped.

manufactured unit. A quantity of finished adhesive or finished adhesive component, processed at one time.

mass stress (fibers). Force per unit mass per unit length; grams per denier, etc. (Used the same way as force per unit area.)

mat. A randomly distributed felt of fibers used in reinforced plastics. Available in blankets of various widths, weights, and lengths.

mat binder. Resin applied to fibers and cured during the manufacture of mat, used to hold the fibers in place and maintain the shape of the mat.

matched metal molding. A reinforced-plastic manufacturing process in which matching male and female metal molds are used (similar to compression molding) to form the part as opposed to low-pressure laminating or spray-up.

matrix. The principal phase or aggregate in which another constituent (e.g., fibers, particles, or fillers) are embedded or surrounded.

mean strain. Analogous to *mean stress*.

mean stress. A dynamic fatigue parameter. The algebraic mean of the maximum and minimum stress in one cycle: $\sigma = 1/2 \ (\sigma_1 + \sigma_2)$ where σ_1 is maximum stress and σ_2 is minimum stress.

mechanical adhesion. Adhesion between surfaces in which the adhesive holds the parts together by interlocking action.

mechanical damping. The amount of energy dissipated as heat during the deformation of a material. Perfectly elastic materials have no mechanical damping.

melamine formaldehyde resin. Classified as a synthetic resin derived from the reaction of melamine (2,4,6-triamino-1,3,5 triazine) with formaldehyde or its polymers.

mer. The repeating structural unit of any high polymer.

metallic fiber. Manufactured fiber composed of metal, plastic-coated metal, metal-coated plastic, or a core completely covered by metal.

microcomputer. (1) A type of computer which utilizes a single chip microprocessor as its basic operating element. (2) A system comprised of a microprocessor plus other necessary electronic elements to provide the input, processing, memory, and output necessary for computing.

micron. A unit of length replaced by the micrometer (μm); 1 μm = 10^{-6} m = 10^{-3} mm

= 0.00003937 in. = 39.4 μin.

microprocessor. The basic element of a central processing unit developed on a single integrated circuit chip. A single integrated chip provides the basic core of a central processing unit, even though it may require additional components to operate as a central processing unit.

microstructure. A structure with heterogeneities that can be seen through a microscope.

migration. The extraction of an ingredient from a material by another material, such as the migration of a plasticizer from one material into an adjacent material with a lower plasticizer contact.

mil. A unit of measure often used in measuring the diameter of fibers, strands, wire, etc. (1 mil = 0.001 in.).

milled fibers. Continuous strands hammer-milled into small nodules or filamentized glass. Useful as anticrazing reinforcing fillers for adhesives.

mixing head. The mechanism in which polyol and isocyanate streams are combined by impingement mixing.

modifier. Any chemically inert ingredient added to an adhesive formulation that changes its properties. See also *filler, plasticizer, extender.*

modulus. A number which expresses a measure of some property of a material (e.g., modulus of elasticity, shear modulus, etc.); a coefficient of numerical measurement of a property. Using "modulus" alone without modifying terms is confusing and should be discouraged.

modulus of elasticity. The ratio of stress to corresponding strain within the range in which these are proportional. If a tensile strength of 2 ksi (14 MPa) results in an elongation of 1%, the modulus of elasticity is 2/0.01 = 200 ksi (1380 MPa); also called *Young's modulus*.

modulus of resilience. The energy that can be absorbed per unit volume without creating a permanent distortion. Calculated by integrating the stress-strain curve from zero to the elastic limit and dividing by the original volume of the specimen.

modulus of rupture. Calculated maximum stress (outer fiber) in test-beam loading. Applies only to nonductile materials.

Mohs hardness. A measure of the scratch resistance of a material on a scale from 1 (talc) to 10 (diamond).

moisture absorption. The pickup of water vapor from air by a material; it relates only to vapor withdrawn from the air by a material and must be distinguished from water absorption. See also *water absorption*.

moisture vapor transmission. The rate at which water vapor permeates through a plastic film or wall at a specified temperature and relative humidity.

mold. (1) Cavity in which a composite part is placed, and from which it takes its shape after curing. (2) To shape plastic parts or finished articles by heat and pressure. (3) The assembly of all the parts that function collectively in the molding process.

mold release. A substance added to the mixture or used to coat the mold to prevent

the molded laminate from sticking to the mold, thus facilitating removal of a part from the mold.

mold seam. Line on a molded or laminated piece, differing in color or appearance from the general surface, caused by the parting line of the mold.

mold shrinkage. (1) The immediate shrinkage which a molded part undergoes when it is removed from a mold and cooled to room temperature. (2) The difference in dimensions, expressed in inches per inch (millimeters per millimeter) between a molding and the mold cavity in which it was molded (at normal temperature measurement). (3) The incremental difference between the dimensions of the molding and the mold from which it was made, expressed as a percentage of the dimensions of the mold.

molding. The shaping of a plastic composition in or on a mold, normally accomplished under heat and pressure; sometimes used to denote the finished part.

molding compounds. Plastic in a wide range of forms (especially granules or pellets) to meet specific processing requirements.

molding cycle. The period of time required to complete the molding of a part. In injection molding, the cycle begins when the mold closes and ends with the opening of the mold and ejection of the molded part.

molding powder. Plastic material in varying stages of granulation, and comprising resin, filler, pigments, plasticizers, and other ingredients, ready for use in the molding operation.

molding pressure. The pressure applied to the ram of an injection machine or press to force the softened plastic to fill the mold cavities completely. See also *compression molding pressure*.

molecular weight. The sum of the atomic weights of all atoms in a molecule.

monofil. See *monofilament*.

monofilament. (1) A single fiber or filament of indefinite length generally produced by extrusion. (2) A continuous fiber sufficiently large to serve as yarn in normal textile operations; also called *monofil*.

monolayer. The basic laminate unit from which crossplies or other laminate types are constructed.

monomer. A single molecule which can join with another monomer or molecule to form a polymer or molecular chain.

mucilage. An adhesive prepared from a gum and water. Also in a more general sense, a liquid adhesive which has a low order of bonding strength. See also *adhesive, glue, paste, sizing*.

multidirectional. Having multiple ply orientations in a laminate.

multifilament yarn. A quantity of fine, continuous filaments (often 5 to 100), usually with some twist in the yarn to facilitate handling. Sizes range from 5 to 10 denier up to a few hundred denier. Individual filaments in a multifilament yarn are usually about 1 to 5 denier.

multiple-layer adhesive. A film adhesive, usually supported, with a different adhesive composition on each side; designed to bond dissimilar materials such as the core-to-face bond of a sandwich composite.

N

nested laminate. In reinforced plastics, a laminate in which the plies are placed so that the yarns of one ply lie in the valleys between the yarns in the adjacent ply.

netting analysis. The analysis of filament-wound structures which assumes that the stresses induced in the structure are carried entirely by the filaments, the strength of the resin being neglected, and that the filaments possess no bending or shearing stiffness, carrying only the axial tensile loads.

nol ring. A parallel filament-wound test specimen used for measuring various mechanical-strength properties of the material by testing the entire ring or segments of it.

nonhygroscopic. Not absorbing or retaining an appreciable quantity of moisture from the air (water vapor).

nonpolar. Having no concentration of electrical charges on a molecular scale; thus, incapable of significant dielectric loss. Examples among resins are polystyrene and polyethylene.

nonrigid plastic. A plastic which has a stiffness or apparent modulus of elasticity not over 10 ksi (69 MPa) at 73.4 °F (23 °C).

nonwoven fabric. A fabric, usually resin-impregnated, in which the reinforcements are continuous and unidirectional; layers may be crossplied.

notch factor. Ratio of the resilience determined on a plain specimen, to the resilience determined on a notched specimen.

notch rupture strength. The ratio of applied load to original area of the minimum cross section in a stress-rupture test of a notched specimen.

notch sensitivity. Extent to which the sensitivity of a material to fracture is increased by the presence of a surface inhomogeneity, such as a notch.

notch strength. The maximum load on a notched tensile test specimen divided by the minimum cross-sectional area (the area at the root of the notch). Also called *notch tensile strength.*

notch tensile strength. See *notch strength.*

Novolak. Trade name for a phenolic-aldehydic resin that, unless a source of methylene groups is added, remains permanently thermoplastic. See also *thermoplastic.*

nylon (polyamide). Family of engineering resins having outstanding toughness and wear resistance, low coefficient of friction, and excellent electrical properties and chemical resistance. Resins are hydroscopic; dimensional stability is poorer than that of most other engineering plastics.

O

offset yield strength. The stress at which the strain exceeds by a specific amount (the offset) an extension of the initial proportional portion of the stress-strain curve. It is expressed in force per unit area.

oil resistance. The ability to withstand contact with an oil without deterioration of physical properties or geometric change to a degree which would impair part performance.

open cell foam. Foamed or cellular material with cells which are generally interconnected. Closed cell refers to cells which are not interconnected.

open-cell-foamed plastic. A cellular plastic in which most of the cells are interconnected in a manner such that gases can travel freely from one cell to another.

orange peel. Surface roughness somewhat resembling the surface of an orange; describes injection moldings with unintentionally rugged surfaces.

organic. Designating or composed of matter originating in plant or animal life or composed of chemicals of hydrocarbon origin, natural or synthetic.

oriented materials. Materials, particularly amorphous polymers and composites, whose molecules and/or macroconstituents are aligned in a specific way. Oriented materials are anisotropic. Orientation is generally uniaxial or biaxial.

orthotropic. Having three mutually perpendicular planes of symmetry. Unidirectional plies, fabric, cross-ply, and angle-ply laminates are all orthotropic.

overcuring. The beginning of thermal decomposition resulting from too high a temperature or too long a molding time.

overflow groove. A small groove used in molds to allow material to flow freely to prevent weld lines and low density, and to dispose of excess material.

overlay sheet. A nonwoven fibrous mat (in glass, synthetic fiber, etc.) used as the top layer in a cloth or mat lay-up, to provide a smoother finish or minimize the appearance of the fibrous pattern. See also *surfacing mat.*

P

PAN. Polyacrylonitrile.

PVA. Polyvinyl alcohol, a parting agent used in film or liquid form.

package. The method of supplying the roving or yarn.

packing. Filling of the mold cavity or cavities as full as possible without causing undue stress on the molds or causing flash to appear on the molding.

paint mask. Stencil designed to conform to the shape of the part with the areas to be decorated cut out.

pan. Abbreviation for polyacrylonitrile, which is used as a base material in the manufacture of certain carbon fibers.

parallel laminate. A laminate in which all the layers of reinforcement are oriented approximately parallel to each other.

parameter. An arbitrary constant, as distinguished from a fixed or absolute constant. Any desired numerical value may be given as a parameter.

parting agent. Also, *release agent.* See *mold release.*

parting line. A mark on a molded piece where the sections of a mold have met in closing.

paste. An adhesive composition having a characteristic plastic-type consistency — that is, a high order or yield value, such as that of a paste prepared by heating a mixture of starch and water and subsequently cooling the hydrolyzed product. See also *adhesive, glue, mucilage, sizing.*

peel. A common type of bond failure characterized by one adherend "peeling" away from the other.

peel ply. A layer of resin-free material used to protect a laminate for later secondary bonding.

peel strength. Bond strength, in pounds per inch of width, obtained in a peel test. See also *bond strength.*

penetration. The entering of an adhesive into an adherend.

permanence. The resistance of a plastic to deteriorating influences; its resistance to appreciable change in characteristics with time and environment.

permanent set. The increase in length, expressed as a percentage of the original length, by which an elastic material fails to return to its original length after being stressed for a standard period of time.

permeability. (1) The passage or diffusion of a gas, vapor, liquid, or solid through a barrier without physically or chemically affecting it; (2) the rate of such passage.

phase angle. See *dielectric phase angle.*

phenolic, phenolic resin. A thermosetting synthetic resin for elevated-temperature service produced by the condensation of an aromatic alcohol with an aldehyde, particularly of phenol with formaldehyde. See also *A-stage, B-stage, C-stage, Novolak.*

phenylene oxide. Excellent dimensional stability (very low moisture absorption); superior mechanical and electrical properties over a wide temperature range. Resists most chemicals but is attacked by some hydrocarbons.

phenylsilane resins. Thermosetting copolymers of silicone and phenolic resins; furnished in solution form.

photoelasticity. Changes in the optical properties of isotropic, transparent dielectrics when subjected to stress.

pick. (1) To experience tack. (2) To transfer unevenly from an adhesive applicator mechanism due to high surface tack. (3) An individual filling yarn running the width of a woven fabric at right angles to the warp, also called *fill, woof, weft.*

pick-up roll. A spreading device wherein the roll for picking up the adhesive runs in a reservoir of adhesive.

pigment. A finely divided insoluble substance which imparts color or black or white to the material to which it is added.

pinch-off. In blow molding, a raised edge around the cavity in the mold which seals off the part and separates the excess material as the mold closes around the parison.

pit. Small craterlike defect in the surface of a plastic part.

pitch. A residual petroleum product which

is used in the manufacture of certain carbon fibers.

planar helix winding. A winding in which the filament path on each dome lies on a plane which intersects the dome while a helical path over the cylindrical section is connected to the dome paths.

planar winding. A winding in which the filament path lies on a plane intersecting the winding surface.

plane strain. The stress condition in linear elastic fracture mechanics in which there is zero strain in a direction normal to the axis of applied tensile stress.

plane stress. The stress condition in linear elastic fracture mechanics in which the stress in the thickness direction is zero; most nearly achieved in loading very thin sheet along a direction parallel to the surface of the sheet. Under plane-stress conditions, the plane of fracture instability is inclined 45° to the axis of the principal tensile stress.

plastic. A material that contains as an essential ingredient an organic substance of high molecular weight, is solid in its finished state, and at some stage in its manufacture or processing into finished articles can be shaped by flow; made of plastic.

plastic deformation. Deformation of an object after which the object does not return to its original shape or size (upon removal of pressure, stress, or load). See also *elastic recovery*.

plastic flow. Deformation under the action of a sustained force; flow of semisolids in molding of plastics.

plastic memory. The tendency of a thermoplastic material which has been stretched while hot to return to its unstretched shape upon being reheated.

plastic tooling. A term employed for structures composed of plastics, usually reinforced thermosets, which are used as tools in the fabrication of metals or other materials, including plastics.

plasticate. To soften by heating or kneading.

plasticity. A property of adhesives that allows the material to be deformed continuously and permanently without rupture upon the application of a force that exceeds the yield value of the material.

plasticize. To render a material softer, more flexible, and/or more moldable by the addition of a plasticizer.

plasticizer. A material incorporated in a resin to increase its flexibility, workability, or distensibility. The addition of the plasticizer may cause a reduction in melt viscosity, lower the temperature of the second-order transition, or lower the elastic modulus of the solidified resin.

platens. Mounting plates of a press to which the mold assembly is fastened.

plied yarn. An assembly of two or more previously twisted yarns.

plug. A male form identical in shape to the finished object, over which a female mold is fabricated.

ply group. Group formed by contiguous plies with the same angle.

ply strain. Those components in a ply which, by the laminated plate theory, are the same as those of the laminate.

ply stress. Those components in a ply which vary from ply to ply depending on the materials and angles in the laminate.

poise. A unit of measure used for the specific viscosity of a fluid (1 centipoise = 100 poises) See *viscosity*.

Poisson's ratio, ν. A constant relating change in cross-sectional area to change in length when a material is stretched:

$$\nu \approx \begin{pmatrix} 1/2 & \text{for rubbery materials} \\ 1/4 \text{ to } 1/2 & \text{for crystals and glasses} \end{pmatrix}$$

polar winding. A winding in which the filament path passes tangent to the polar opening at one end of the chamber and tangent to the opposite side of the polar opening at the other end. A one-circuit pattern is inherent in the system.

polyamide. See *nylon*.

polycarbonate. Highest impact resistance of any rigid, transparent plastic; excellent outdoor stability and resistance to creep under load; fair chemical resistance; some aromatic solvents cause stress cracking.

polycondensation. See *condensation*.

polyester resins. Family of resins produced by reaction of dibasic acids with dihydric alcohols. Polyethylene terephthalate (PET) is a thermoplastic which may be extruded, injection molded, or blow molded. Unsaturated polyesters are thermoset and used in the reinforced plastics industry for applications such as boats, auto components, etc.

polyethylene. Wide variety of grades: low-, medium-, and high-density formulations. LD types are flexible and tough. MD and HD types are stronger, harder, and more rigid. All are lightweight, easy-to-process, low-cost materials; poor dimensional stability and heat resistance; excellent chemical resistance and electrical properties. Also available in ultrahigh-molecular-weight grades.

polyimide. A polymer produced by heating polyamic acid. Polyimides exhibit outstanding resistance to heat (500 °F or 260 °C continuous, 900 °F or 480 °C intermittent) and to heat aging. High impact strength and wear resistance; low coefficient of thermal expansion; excellent electrical properties; difficult to process by conventional methods; high cost.

polymer. A compound formed by the reaction of simple molecules having functional groups which permit their combination to proceed to high molecular weights under suitable conditions. Polymers may be formed by polymerization (addition polymer) or polycondensation (condensation polymer). When two or more monomers are involved, the product is called a *copolymer*.

polymerization. A chemical reaction in which the molecules of a monomer link together to form a polymer. Curing of polyester resins is a polymerization process. When two or more monomers are involved, the process is called *copolymerization* or *heteropolymerization*.

polymerize. To unite molecules of the same kind into a compound having the elements in the same proportion but possessing much higher molecular weight and different physical properties.

polyol. Based on ethylene oxide/propylene oxide or propylene oxide and having molecular weights ranging from 700 to 6000. Produced to exacting specifications for the manufacture of foams.

polyphenylene sulfide. Outstanding chemical and heat resistance (450 °F, or 230 °C, continuous); excellent low-temperature strength; inert to most chemicals over a wide temperature range; inherently flame-retardant; requires high processing temperature.

polypropylene. Outstanding resistance to flex and stress cracking; excellent chemical resistance and electrical properties; good impact strength above 15 °F (-9 °C); good thermal stability; light weight, low cost, can be electroplated.

polystyrene. Low-cost, easy-to-process, rigid, crystal-clear, brittle material; low moisture absorption, low heat resistance, poor outdoor stability; often modified to improve heat or impact resistance.

polysulfone. Highest heat-deflection temperature of melt-processable thermoplastics; requires high processing temperature; tough (but notch-sensitive), strong, and stiff; excellent electrical properties and dimensional stability, even at high temperature; can be electroplated; high cost.

polyurethane resin. A thermosetting resin produced by reacting diisocyanate with organic compounds containing two or more active hydrogens to form polymers with free isocyanate groups. Polyurethanes are tough, extremely abrasion- and impact-resistant materials; they exhibit good electrical properties and chemical resistance; they can be made into films, solid moldings, or flexible foams; UV exposure produces brittleness, lower properties, and yellowing; also made in thermoset formulations.

polyvinyl acetate. Commonly called PVA compounds; (PVA consists primarily of a homopolymer or copolymer of vinyl acetate with more flexible monomers.

polyvinyl alcohol resin. A thermoplastic material composed of polymers of the hypothetical vinyl alcohol.

polyvinyl chloride. Many formulations available; rigid grades are hard, tough, and have excellent electrical properties, outdoor stability, and resistance to moisture and chemicals; flexible grades are easier to process but have lower properties and chemicals; flexible grades are easier to process but have lower properties; heat resistance is low to moderate for most types of PVC; low cost.

porosity. (1) The formation of undesirable clusters of air bubbles in the surface or body of the laminate. (2) The relationship between nonsolid materials (air or void) and total volume.

positive mold. A compression mold designed to prevent the escape of molding material during the molding cycle.

postcure. To expose a plastic assembly to an additional cure, following the initial cure, for the purpose of modifying specific properties. Complete curing and ultimate mechanical properties of certain resins are attained only by exposure of the cured resin to higher temperatures than those of curing.

postforming. A term used in the reinforced plastics industry to denote the heating and reshaping of a fully cured laminate. On cooling, the formed laminate retains the contours and shape of the mold over which it has been formed.

pot life. The time period during which a compound remains suitable for the intended use, after compounding ingredients such as solvent or catalyst have been added. Also called *working life.*

power factor (P.F.). The cosine of the phase angle. Ratio of the dielectric constant ϵ' to the absolute value of the complex dielectric constant ϵ^*, related to the dissipation factor D as follows:

$$P.F. = \frac{D}{\sqrt{(1 + D^2)}} = \frac{\epsilon''/\epsilon'}{\sqrt{1 + \left(\frac{\epsilon''}{\epsilon'}\right)^2}} = \frac{\epsilon'}{|\epsilon^*|}$$

prebond treatment. See *surface preparation.*

precure. A partial or full state of cure existing in an elastomer or thermosetting resin prior to its use as an adhesive or in a forming operation.

preform. (1) A preshaped fibrous reinforcement in which cut strands of roving are drawn by suction on to a shaped screen, sprayed with a binder, and cured. (2) A preshaped fibrous reinforcement of mat or cloth formed to the desired shape on a mandrel or mock-up before being placed in a mold press. (3) A compact "pill" formed by compressing premixed material to facilitate handling and control of uniformity of charges for mold loading.

preform binder. The resin applied to a preform for a reinforced plastic structure.

pregel. An unintentional extra layer of cured resin on part of the surface of a reinforced plastic (not relating to *gel coat*).

preimpregnation. The practice of mixing resin and reinforcement and effecting partial curing before use or shipment to the user. See also *prepreg.*

premix. Molding compound of glass fibers or roving impregnated with uncured resin and catalyst, ready for molding by application of heat and pressure. See also *BMC.*

prepreg. A combination of mat, fabric, nonwoven material, or roving with resin; usually in the B-stage, ready for molding.

preproduction test. A test or series of tests conducted by (1) an adhesive manufacturer, to determine the conformity of an adhesive batch to established production standards; (2) a fabricator, to determine the quality of an adhesive before parts are produced; or (3) an adhesive specification custodian, to determine the conformance of an adhesive to the requirements of a specification not requiring qualification tests.

pressure. Force measured per unit area. Absolute pressure is measured with respect to zero. Gage pressure is measured with respect to atmospheric pressure.

pressure bag molding. A process for molding reinforced plastics, in which a tailored flexible bag is placed over the contact lay-up on the mold, sealed, and clamped in place. Compressed air forces the bag against the part to apply pressure while the part cures.

pressure-sensitive adhesive. A viscoelastic material which in solvent-free form remains permanently tacky. Such materials will adhere instantaneously to most solid surfaces with the application of very slight pressure.

primary bond. Connection between a semi-cured lay-up and an uncured lay-up.

primary structure. A structure that is critical to flight safety.

primer. A coating applied to a surface prior to the application of an adhesive, lacquer, or enamel to improve the performance of the bond.

programmable controller. A control system often used to operate machinery in place of the standard electromechanical relays. The controls are programmed rather than permanently wired as in standard control methods.

promoter. See *accelerator.*

proportional limit. The greatest stress which a material is capable of sustaining without deviation from proportionality of stress and strain (Hooke's law); it is expressed in force per unit area.

prototype. A model suitable for use in complete evaluation of form, design, and performance.

pultrusion. A continuous process for manufacturing composites with a constant cross-sectional shape. The process consists of pulling a fiber reinforcing material through a resin-impregnation bath and into a shaping die where the resin is subsequently cured.

Q

quasi-isotropic laminate. A laminate approximating isotropy by orienting plies in several directions.

R

RIM. See *reaction injection molding.*

RPP. Reinforced pyrolyzed plastic.

RRIM. Reinforced reaction injection molding.

RTM. See *resin-transfer molding.*

radio frequency interference (RFI). The interference in electronic equipment caused by radio frequencies. These frequencies can range from 10 kHz to 1.0 GHz.

Ram travel. The distance the injection ram moves in filling the mold, in either injection or transfer molding.

reaction injection molding. A process which is applied to polyurethane, epoxy, and other liquid chemical systems. Mixing of two to four components in the proper chemical ratio is accomplished with a high-pressure impingement-type mixing head, from which the mixed material is delivered into the mold at low pressure.

reflection. The amount of electromagnetic energy reflected from the surface of a shield. Reflection depends on the imped-ance of the shield and the medium from which the signal originates.

reinforced molding compound. Compound supplied by raw-material producer in the form of ready-to-use materials, as distinguished from premix. See also *premix.*

reinforced plastics. Molded, formed, filament-wound, or shaped plastic parts consisting of resins to which reinforcing fibers, mats, fabrics, etc. have been added before the forming operation. Strength properties are improved.

reinforcement. A material used to reinforce, strengthen, or give dimensional stability to another material.

release agent. A material which is placed on a mold to prevent the resin from adhering to it during a bonding operation.

residual stress. In plastic (polymer) composites, residual stress results from cooldown after curing and from moisture content. On the micromechanical level, stress is tensile in the resin and compressive in the fiber. On the macromechanical level, it is tensile in the transverse direction to the unidirectional fibers, and compressive in the longitudinal direction, resulting in a lowered first-ply-failure load. Moisture absorption offsets this detrimental thermal effect on both micro and macro levels.

resilience. The ratio of energy returned on recovery from deformation to the work input to produce the deformation (usually expressed as a percentage); the ability to quickly regain an original shape after being stretched or distorted.

resin. Any of a class of solid, semisolid, or pseudosolid organic materials of natural or synthetic origin, generally of high molecular weight, insoluble in water, having no tendency to crystallize, exhibiting a tendency to flow when subjected to stress, usually having a softening or melting range, and usually fracturing conchoidally. See *gum.* Most resins are polymers, and in reinforced plastics a resin is the material (matrix) used to bind together the reinforcement material. See also *polymer.*

resin applicator. A device for depositing a liquid resin system. In filament winding, it applies the resin onto the reinforcement band.

resin content. The amount of resin in a laminate, expressed as a percent of either total weight or total volume.

resin pocket. An apparent accumulation of excess resin in a small, localized area that is visible on the cut edges of molded surfaces.

resin transfer molding. The transfer of catalyzed resin into an enclosed mold in which the fiber glass reinforcement has been placed.

resin-rich area. Localized area filled with resin and lacking reinforcing material.

resin-starved area. Localized area containing excess reinforcement and insufficient resin.

resistivity. The ability of a material to resist the passage of electric current through its bulk or along its surface.

resite. See *C-stage.*

resitol. See *B-stage.*

resol. See *A-stage.*

retarder. See *inhibitor.*

reverse helical winding. As the fiber-delivery arm traverses one circuit, a continuous helix is laid down, reversing direction at the polar ends, as contrasted with biaxial, compact, or sequential winding in that the fibers cross each other at definite equators, the number depending on the helix. The minimum crossover would be 3.

reverse impact test. A test for sheet material in which one side of the specimen is struck by a pendulum or falling object and the reverse side is inspected for damage.

rib. Configuration designed into a plastic part to provide lateral, horizontal, or other structural support.

ribbon. A fiber having essentially a rectangular cross section, where the width-to-thickness ratio is at least 4:1.

rigid plastic. A plastic with a stiffness or apparent modulus of elasticity greater than 100 ksi (690 MPa) at 23 °C (73 °F).

rise time. In urethane molding, the time between pouring of the urethane mix and the completion of foaming.

Rockwell hardness number. The hardness of a material expressed as a number derived from the net increase in depth of impression as the load on an indentor is increased from a fixed minor amount to a major load and then returned to the minor load.

room temperature curing adhesive. An adhesive that can reach full strength without heating and can be set to handling strength within 1 h at temperatures from 20 to 30 °C (68 to 86 °F).

rosin. A resin obtained as a residue in the distillation of crude turpentine from the sap of the pine tree (gum rosin) or from an extract of the stumps and other parts of the tree (wood rosin).

roving. (1) A multiplicity of single ends of continuous filament with no applied twist drawn together as parallel strands. (2) In filament winding, a collection of bundles of continuous filaments, either as untwisted strands or as twisted yarns.

roving ball. A term used to describe the supply package offered to the winder. It consists of a number of ends or strands wound to a given outside diameter onto a length of cardboard tubing.

roving cloth. A textile fabric, coarse in nature, woven from rovings.

rule-of-mixtures. Linear volume fraction relationship between the composite and the corresponding constituent properties.

runner. The channel in an injection or transfer mold which connects the sprue with the cavity gate.

S

S glass. A magnesia-alumina-silicate glass which provides very-high-tensile-strength reinforcement.

S-N diagram. A plot showing the relationship between stress, *S,* and the number of cycles, *N,* before fracture in fatigue testing.

SAP. Sintered-aluminum powder.

SPF/DB. Superplastic-forming diffusion bonding.

SMC. See *sheet molding compound;* see also *bulk molding compound.*

SPMC. Solid polyester molding compound.

safety hardener. A curing agent which causes only a minimum toxic effect on the human body, either on contact with the skin or as concentrated vapor in the air.

sagging. Term applied to the tendency of a wet paint film to flow downward and become thicker in some areas.

sandwich construction. A structure consisting of relatively dense, high-strength facings bonded to a less dense, lower-strength intermediate material or core.

sandwich heating. A method of heating a thermoplastic sheet before forming by heating both sides of the sheet simultaneously.

sandwich panel. Panel consisting of two thin face sheets bonded to a thick, lightweight, honeycomb or foam core.

scarf joint. A joint made by cutting the ends of two pieces at the same angle and fitting the two cut areas together.

scratch. Shallow mark, groove, furrow, or channel normally caused by improper handling or storage.

scrim. A low-cost reinforcing fabric made from continuous filament yarn in an open-mesh construction. Used in the processing of tape or other B-stage material to facilitate handing.

secant modulus. Idealized Young's modulus derived from a secant drawn between the origin and any point on a nonlinear stress-strain curve. The tangent modulus is the other idealized Young's modulus derived from the tangent to the stress-strain curve.

secondary bond. Connection between a cured lay-up and an uncured lay-up.

secondary structure. A structure that is not critical to flight safety.

self-extinguishing. A somewhat loosely used term describing the ability of a material to cease burning once the source of flame has been removed. PVC, vinyl chloride-acetate copolymers, polyvinylidene chloride, nylon, and casein plastics are examples of self-extinguishing materials.

self-skinning foam. A urethane foam which produces a molded part with a tough outer surface over a foam core upon curing.

self-vulcanizing. Pertaining to an adhesive that undergoes vulcanization without the application of heat.

selvage. The edge of a woven fabric finished off so as to prevent the yarns from raveling.

semipositive mold. A mold which allows a small amount of excess material to escape when it is closed (used where close tolerances are required).

semirigid plastic. For purposes of general classification, a plastic that has a modulus of elasticity, either in flexure or in tension, of between 700 and 7000 kg/cm^2 (10 and 100 ksi) at 23 °C (73 °F) and 50% relative humidity.

separate application adhesive. A term used to describe an adhesive consisting of two parts, one part being applied to one adherend and the other part to the other adherend and the two brought together to form a joint.

sequential winding. See *biaxial winding.*

serving. Wrapping a yarn such as rayon around a roving or yarn for protection.

set. (1) To convert an adhesive into a fixed or hardened state by chemical or physical action, such as condensation, polymerization, oxidation, vulcanization, gelation, hydration, or evaporation of volatile constituents; see *cure* and *dry.* (2) The irrecoverable deformation or creep usually measured by a prescribed test procedure and expressed as a percentage of original dimension.

set at break. Elongation measured ten minutes after rupture on reassembled tension specimen.

set up. To harden, as in curing.

setting time. The period of time during which an assembly is subjected to heat or pressure, or both, to set the adhesive.

shear. An action or stress resulting from applied forces, causing contiguous parts of a body to slide with respect to each other in a direction parallel to their plane of contact.

shear compliance. The reciprocal of *shear modulus.*

shear coupling. Induced shear strain from normal stress. This coupling is unique with anisotropic materials.

shear edge. The cutoff edge of the mold.

shear modulus (G). The ratio of shearing stress τ to shearing strain γ within the proportional limit of a material.

shear strength. The maximum shearing stress a material can develop, based on the original cross-sectional area.

sheet molding compound (SMC). A composite of glass fibers, polyester resins, and pigments, fillers, and other additives which have been compounded and processed into sheet form to facilitate handling in the molding operation.

shelf life. The period of time during which a packaged material may be stored under specific temperature conditions and remain suitable for use.

shielding effectiveness. A measure of the effectiveness of an EMI/RFI shield based on the logarithmic ratio of the energy passing between a signal source and the receiver with and without the shield placed between them. Shielding effectiveness is expressed in decibels.

shoe. A device for gathering filaments into a strand in glass-fiber forming. See *chase.*

Shore hardness. A test to determine a material's hardness using an indentation durometer or scleroscope.

short beam shear strength. The interlaminar shear strength of a parallel-fiber-reinforced plastic material as determined by three-point flexural loading of a short segment cut from a ring-type specimen.

short shot. Injection of insufficient material to fill the mold.

shot capacity. The maximum weight of material a machine can produce from one forward motion of the ram, screw, or plunger.

shrinkage. The relative change in dimension between the length measured on the mold when it is cold and the length on a molded object 24 h after it has been taken out of the mold.

silica glass. Noncrystalline SiO_2. It has short-range order but no long-range order. Sometimes inappropriately called quartz or *fused quartz.*

silicones. Resinous materials derived from organosiloxane polymers, furnished in different molecular weights including liquids and solid resins and elastomers. Silicones exhibit outstanding heat resistance (from −100 to +500 °F, or −73 to +260 °C), electrical properties, and compatibility with body tissues. They cure by a variety of mechanisms; are relatively expensive; and are available in many forms — e.g., laminating resins, molding resins, coatings, casting or potting resins, and sealants.

single spread. See *spread.*

single-circuit winding. A winding in which the filament path makes a complete traverse of the chamber, after which the following traverse lies immediately adjacent to the previous one.

singles. A yarn made from one or more strands twisted together but not plied. Examples: 1/0, 2/0, 4/0. Single-strand construction is a singles yarn made from one strand (1/0).

sink mark. A shallow depression or dimple on the surface of an injection molded part due to collapsing of the surface following local internal shrinkage after the gate seals; an incipient short shot.

size. A chemical substance, such as starch, gelatin, oil, or wax, that is coated on an adherend surface to reduce water absorption, scuffing, oil penetration, etc. In fiber manufacturing, sizes are often applied to yarn or fibers at the time of formation to protect the surface and facilitate handling and fabrication, or to control fiber characteristics. Before final fabrication of the fibers into a composite, the size is usually removed by heat cleaning, and a finish is applied.

sizing. The process of applying a material to a surface in order to fill pores and thus reduce the absorption of the subsequently applied adhesive or coating; also, the process of otherwise modifying the surface properties of the substrate to improve adhesion; also, the material used for this purpose. The latter is sometimes called *size.*

sizing content. The percentage of the total strand weight made up by the sizing, usually determined by burning off the organic sizing (*loss on ignition*).

skein. A continuous filament, strand, yarn, roving, etc., wound up to some measurable length and generally used to measure various physical properties.

skin. The relatively dense material that may form the surface of a cellular plastic or sandwich.

sliver. Overlapping, parallel staple fibers that have been gathered into a continuous bundle.

slurry preforming. Method of preparing reinforced plastic preforms by wet processing techniques similar to those used in the pulp molding industry. For example, glass fibers suspended in water are passed through a screen which passes the water but retains the fibers in the form of a mat.

soft flow. The behavior of a material which flows freely under conventional conditions of molding and which, under such conditions, will fill all the interstices of a deep mold where a considerable distance of flow can be demanded.

solidus. The locus of temperatures below which there is only solid. Also, solid-solubility limit, with the second plase being a liquid.

solution. A homogeneous mixture the proportions of which can be varied within certain limits. A solid is said to be in solution when the molecules of the liquid have exceeded the attraction of those of a solid.

solvent-activated adhesive. A dry adhesive film that is rendered tacky just prior to use by application of a solvent.

solvent adhesive. An adhesive having a volatile organic liquid as a vehicle.

solvent resistance. Ability of a plastic to resist swelling and dissolving in a solvent.

specific adhesion. Adhesion between surfaces which are held together by valence forces of the same type as those which give rise to cohesion.

specific gravity. The ratio of the weight of any volume of a substance to the weight of an equal volume of another substance taken as standard at a constant or stated temperature.

specific heat. The amount of heat required to raise the temperature of a unit mass of a substance one degree under specific conditions.

specific insulation resistance. See *volume resistance.*

specification. A detailed description of the characteristics of a product and of the criteria which must be used to determine whether the product is in conformity with the description.

spinneret. A type of extrusion die; that is, a metal plate with many tiny holes, through which a plastic melt is forced to make fine fibers and filaments which are hardened by cooling in air, water, etc., or by chemical action.

spinning. The process of forming fibers by extruding the material through spinnerets.

spiral. In glass-fiber forming, the device that is used to make the strand traverse back and forth across the forming tube.

splice line. The boundary between core details that are bonded together with a structural adhesive.

spline. (1) To prepare a surface to its desired contour by working a paste material with a flat-edged tool; the procedure is similar to screeding of concrete. (2) The tool itself.

split mold. A mold in which the cavity is formed from two or more components, known as splits, held together by an outer chase.

split-ring mold. A mold in which a split-cavity block is assembled in a chase to permit formation of undercuts in a molded piece. The parts are ejected from the mold and then separated from the piece.

spool. A term sometimes used to identify a *roving ball* (the preferred term).

spray. A complete impression of an injection mold, including the molded parts with their gates and runners attached.

spray-up. A general term covering several processes using a spray gun. In reinforced plastics, the term applies to the simultaneous spraying of resin and chopped reinforcing fibers onto the mold or mandrel. In the foamed plastics field, the term refers to the spraying of fast-reacting polyurethane or epoxy resin systems onto a surface where they react to foam and cure. In both processes, resins and catalysts are usually sprayed through separate nozzles so that they become mixed externally, thus avoiding pot life problems in the spray equipment and tanks.

sprayed-metal molds. Molds made by spraying molten metal onto a master until a shell of predetermined thickness is achieved. The shell is then removed and backed up with plaster, cement, casting resin, or other suitable material. Such molds are used primarily as molds in the sheet-forming process.

spread. The quantity of adhesive per unit joint area applied to an adherend, usually expressed in points of adhesive per thousand square feet of joint area; *single spread* refers to application of adhesive to only one adherend of a joint, whereas *double spread* refers to application of adhesive to both adherends of a joint.

spreading rate. The area of a surface over which a unit volume of paint will spread.

sprue. (1) The mold channel that connects the pouring basin with the runner or, in the absence of a pouring basin, directly into which liquid material is poured. Sometimes referred to as "downsprue" or "downgate." (2) Sometimes used to mean all gates, risers, runners, and similar scrap that are removed from castings after shakeout.

spun roving. A heavy, low-cost glass-fiber strand consisting of filaments that are continuous but doubled back on each other.

stabilization. In carbon-fiber forming, the process used to render the carbon fiber precursor infusible prior to carbonization.

standard deviation. A measure of dispersion of data from the average; the root mean square of the individual deviation from the average.

standard laboratory atmosphere. An atmosphere having a relative humidity of 50 ± 2% at a temperature of 23 ± 1 °C (73.4 ± 1.8 °F). Also: average room conditions — 40% relative humidity at 77 °F (25 °C); dry room conditions — 15% relative humidity at 85 °F (29 °C); moist room conditions — 75% relative humidity at 77 °F (25 °C).

staple fibers. Fibers produced from filaments in short lengths from the bushing (usually less than 17 in., or 43 cm), to be gathered into strands or slivers. See *continuous filament.*

starved area. An area in a plastic part which has an insufficient amount of resin to wet out the reinforcement completely. This condition may be due to improper wetting or impregation or excessive molding pressure.

starved joint. A joint which has an insufficient amount of adhesive to produce a satisfactory bond.

static decay rate. The time required for a material to dissipate induced surface charges of static electricity. This specification is used to compare the ESD capabilities of various materials.

static stress. A stress in which the force is constant or slowly increasing with time.

stiffness. Ratio between the applied stress and the resulting strain. Young's modulus is the stiffness of a material subjected to uni-axial stress; shear modulus, to shear stress. For composite materials, stiffness and other properties are dependent on the orientation of the material. They must be further identified with a direction, usually designated by subscripts such as x, y, s, or 1, 2, 6.

stoichiometry. The control of tolerances, levels, and amounts in a chemical mix. When exact amounts necessary for reaction are present without excess reactants, the reaction is said to be stoichiometric.

stoke. See *viscosity.*

storage life. See *shelf life.*

strain. Elastic deformation due to stress. Measured as the change in length per unit of length in a given direction, and expressed in percentage or inches per inch, etc.

strain relaxation. See *creep.*

strand. A primary bundle of continuous filaments (or slivers) combined in a single compact unit without twist.

strand count. U.S. Yardage System: the length, in hundreds of yards, of a single strand having a mass of 1 lb. European TEX System: the mass, in grams, of a strand 1000 m in length.

strand integrity. The degree to which the individual filaments making up the strand or end are held together by the sizing applied.

strength in compression. The maximum load sustained by the specimen divided by the original cross-sectional area of the specimen.

stress. The intensity of internal forces resisting deformation by external loads, generally expressed in pounds per square inch.

stress concentration. Increased ratio of a local stress over the average stress. On the micromechanical level, concentration occurs at the fiber/matrix interface. On the macromechanical level, concentration occurs at notches, ply termination points, joints, etc.

stress concentration factor. The ratio of the maximum stress in the region of a stress concentrator to the stress in a similar strained area without a stress concentrator. Compare with *fatigue notch factor.*

stress corrosion. Preferential attack of areas under stress in a corrosive environment, where this factor alone would not have caused corrosion.

stress crack. External or internal cracks in a plastic caused by tensile stresses less than its short-time mechanical strength. The stresses which cause cracking may be present internally or externally or may be combinations of these stresses. See also *crazing.*

stress decay. See *stress relaxation.*

stress relaxation. Time-dependent decreases in stress for a specimen constrained in a constant-strain condition. Also called *stress decay.*

stress-strain curve. Simultaneous readings of load and deformation, converted to stress and strain, are plotted as ordinates and abscissae to obtain a stress-strain curve.

striae. See *flow lines.*

stringiness. The property of an adhesive that results in the formation of filaments or threads when adhesive transfer surfaces are separated. See also *webbing.*

structural bond. A bond that joins basic load-bearing parts of an assembly. The load may be either static, as in a fixture bonded to a wall, or dynamic, as in a wing tip subject to aerodynamic loads.

styrene-butadiene. A copolymer of styrene and butadiene produced by emulsion polymerization.

sublaminate. A repeating multidirectional assemblage within a laminate.

successive ply failure. Sequential failures of plies in a multidirectional laminate due to increasing loads.

surface preparation. The procedure required with respect to a foundation surface or the materials to be adhered which will promote optimum performance of an adhesive, coating, or sealant. For example, if higher bond strength is required, abrading and/or acid etching of the surface can be the means of improving the adhesion of the bonding material to the mating surfaces. Common methods of surface preparation are solvent washing, sandblasting, and vapor degreasing. Also called *prebond treatment.*

surface resistivity. The ratio of dc voltage to the current that passes across the surface of the substrate. Surface resistivity expressed in ohms per square inch is an indication of a material's conductivity.

surface tension. The tension exhibited by the free surface of liquids measured in dynes per centimeter.

surface treatment. A material applied to fibrous glass during the forming operation or in subsequent processes — i.e., size or finish.

surfacing mat. Thin veil of fiber glass used to produce a smooth surface on reinforced plastic parts. See *overlay sheet.*

surfactant. A chemical compound which causes variations in the surface forces of a liquid or a solid in relation to other liquids, gases, or solids.

surging. An unstable pressure buildup leading to variable output and wavy extrudate surface. Surging may cause flow to stop momentarily at intervals.

symmetric laminate. A laminate possessing midplane symmetry. This is the most common construction because the curing stresses are also symmetric. The laminate does not twist when the temperature and moisture content change. An unsymmetric laminate, on the other hand, twists upon cooldown and untwists after moisture absorption.

syntactic foam. A composite made by mixing hollow microspheres of glass, epoxy, phenolic, etc. into fluid resins (with additives and curing agents) to form a moldable, curable, lightweight fluid mass.

synthetic resin. A complex, substantially amorphous, organic semisolid or solid material (usually a mixture) built by chemical reaction of comparatively simple compounds, approximating the natural resins in luster, fracture, comparative brittleness, insolubility in water, fusibility, or plasticity, and some degree of rubberlike extensibility, but commonly deviating widely from natural resins in chemical constitution and behavior with reagents.

T

TD. Thoria-dispersed.

TFRS. Tungsten-fiber-reinforced superalloy.

tack. Stickiness of a reinforced plastic prepreg or adhesive material.

tack range. The period of time during which an adhesive will remain in the tacky-dry condition after application to the adherend and under specified conditions of temperature and humidity.

tangent line. In a filament-wound bottle, any diameter at the equator.

tangent modulus. Idealized Young's modulus derived from the tangent drawn at the origin or any point on a nonlinear stress-strain curve. The secant modulus is the other idealized Young's modulus between the origin and the same point on the stress-strain curve.

tape. A composite ribbon consisting of continuous or discontinuous fibers that are aligned along the tape axis parallel to each other and bonded together by a continuous matrix phase.

tear resistance. Resistance of a material to a force acting to initiate and then propagate a failure at the edge of a test specimen.

template. A pattern used as a guide for cutting and laying plies.

tenacity. The strength of a yarn or of a filament of a given size; equals breaking strength divided by denier.

tensile compliance. The reciprocal of the *modulus of elasticity.*

tensile modulus. See *modulus of elasticity.*

tensile strength. The maximum tensile stress sustained by the specimen before failure in a tension test. Usually expressed in psi. The cross-sectional area is that of the original specimen at the point of rupture, not reduced by the break.

thermal conductivity. Ability of a material to conduct heat; the physical constant for quantity of heat that passes through a unit cube of a substance in unit time when the difference in temperature of two faces is one degree.

thermal decomposition. Decomposition resulting from action by heat. It occurs at a temperature for which some components of the material are separating or associating together, with a modification of the macro- or microstructure.

thermal endurance. The time required at a selected temperature for a material or system of materials to deteriorate to some predetermined level of electrical, mechanical, or chemical performance under prescribed conditions of test.

thermal load. One component of the hygrothermal load. The difference between the cure and use temperatures gives rise to in-plane thermal load for symmetric laminates; and both in-plane and flexural thermal loads for unsymmetric laminates. The presence of the flexural load causes twisting of unsymmetric laminates after curing.

thermal stress cracking. Cracking of the surface of a plastic part caused by exposure to excessive heat.

thermocouple. A device which uses a circuit of two wires of dissimilar metals or alloys, the two junctions of which are at different temperatures. A net electromotive force (emf) occurs as a result of this temperature difference. The minute electromotive force, or current, is sufficient to drive a galvanometer or potentiometer.

thermoplastic. Organic material that can reversibly change its stiffness by temperature change. One unique property of such a material is its large strain capability. However, processing requires high temperature and pressure, compared with thermosetting plastics.

thermosetting plastic. Organic material that can be converted to a solid body by crosslinking, accelerated by heat, catalyst, ultraviolet light, and others. This is the most popular matrix material for plastic (polymer) composite materials.

thinner. See *extender*.

thixotropic. The property of becoming a gel at rest, but liquifying again on agitation; to loose viscosity under stress.

thread. See *fiber*.

thread count. The number of yarns (threads) per inch (millimeter) in either the lengthwise (warp) or the crosswise (fill) direction of woven fabrics.

threshold concentration. See *critical concentration*.

throwing. Twisting and/or plying of strands or singles.

toggle. A mechanism that exerts pressure developed by applying force on a knee joint. It is used to close and exert pressure on a mold in a press.

tolerance. The guaranteed maximum deviation from the specified nominal value of a component characteristic at standard or stated environmental conditions.

tooling resins. Resins that have applications as tooling aids, coreboxes, prototypes, hammer forms, stretch forms, foundry patterns, etc. Epoxy and silicone are common examples.

torsion. Twisting stress.

torsional rigidity (fibers). The resistance of a fiber to twisting.

tow. A loose, untwisted bundle of filaments.

transducer. A force measuring device. It has the characteristics of providing an output, usually electrical, which serves as the measurement of load, force, compression, pressure, etc. when placed along the sensitive axis of the force cell.

transfer molding. Process in which molding material is placed into a well or pot and then forced into a runner system of a closed mold by means of an advancing hydraulic cylinder ram. Frictional heat, which develops during this transfer of material from the pot to the closed cavity, hastens the chemical reaction to substantially reduce the amount of curing time required. To facilitate handling of the raw material, the powder is usually consolidated into briquettes or preforms which contain the charge weight required for each shot.

transfer molding pressure. The pressure applied to the cross-sectional area of the material pot or cylinder.

transition temperature. The temperature at which the properties of a material change.

turns per inch (TPI). A measure of the amount of twist produced in a yarn during its conversion from a strand.

twist. The turns about its axis per unit of length observed in a yarn or other textile strand. Twist may be expressed as turns per inch (TPI). "S" and "Z" refer to direction of twist.

U

U-V stabilizer. A chemical compound additive to a thermoplastic resin which selectively absorbs U-V rays.

UDC. See *unidirectional composite*.

UHM. Ultrahigh-modulus.

UMC. Unidirectional molding compound.

ultimate elongation. Elongation at rupture.

ultimate tensile strength. The ultimate or final stress sustained by a specimen in a tension test; the stress at moment of rupture.

ultrasonic testing. A nondestructive test applied to sound-conductive materials having elastic properties for the purpose of locating inhomogeneities or structural discontinuities within a material by means of an ultrasonic beam.

ultraviolet. Zone of invisible radiations beyond the violet end of the spectrum of visible radiations. Since ultraviolet wavelengths are shorter than the visible, their photons have more energy — enough to initiate some chemical reactions and to degrade most plastics.

undercut. Reverse or negative draft in a mold, necessitating inserts or a split mold for removal of the part.

unidirectional composite. A composite having parallel fibers.

unidirectional laminate. A laminate with nonwoven reinforcements which are all laid up in the same direction.

unidirectional roving. Heavy parallel rovings with a smaller number of light rovings at right angles to them, so that highly directional strength properties result.

unsymmetric laminate. A laminate without mid-plane symmetry.

urethane plastics. Plastics generally reacted with polyols (e.g., polyesters or polyethers) when the reactants are joined by formation of urethane linkage.

V

VIM. Vibrational microlamination.

vacuum bag molding. A process for molding composites in which a film material is placed over the lay-up on the mold and sealed so that a vacuum can be applied to allow atmospheric pressure to help form the composite.

vacuum metalizing. A process in which metal, usually aluminum, is boiled in a vacuum chamber. The metal then condenses on the surface of a nonconductive substrate to form an EMI/RFI shield. The substrate surface is usually primed with a base paint.

vehicle. The film-forming or pigment-binding portion of a coating.

veil. An ultrathin fabric used as the surface layer in reinforced plastics.

vent. A small hole or shallow channel in a mold which allows air or gas to exit as the stock enters.

vermiculite. A granular material mixed with resin to form a filler of relatively high compressive strength.

vinyl. A general term applied to a class of resins containing many different materials such as polyvinyl chloride, polyvinyl acetate, etc.

virgin filament. An individual filament which has not been in contact with any other fiber or any other hard material.

viscosity. The measure of the resistance of a fluid to flow (either through a specific orifice or in a rotational viscometer). The absolute unit of viscosity measurement is the *poise* (or centipoise). Kinematic viscosity is expressed in *stokes*.

viscosity coefficient. The shearing stress tangentially applied that will induce a velocity gradient. A material has a viscosity of one poise when a shearing stress of one dyne per square centimeter produces a velocity gradient of (1 cm/s)/cm. See also *viscosity*.

void content. Volume percentage of voids, usually less than 1%. The experimental determination is, however, only indirect — i.e., calculated from the measured density of a cured composite and the "theoretical" density of the starting material. Such determination also implies that voids are uniformly distributed throughout the body.

void. Air or gas trapped in a composite material during curing.

volatile content. The percentage of volatiles which are driven off as vapor from a plastic or impregnated reinforcement.

volatile loss. Weight loss by vaporization.

volatiles. Materials which are capable of being driven off as vapors during molding.

volume fraction. Fraction of a constituent material based on its volume.

volume resistance. The volume resistance between two electrodes that are in contact with or embedded in a specimen is the ratio of the direct voltage applied to the electrodes, to that portion of the current between them that is distributed through the volume of the specimen. Also, the electrical resistance between opposite faces of a 1-cm cube of insulating material, commonly expressed in ohm-centimeters. Also called *specific insulation resistance*.

vulcanization. A chemical reaction in which the physical properties of a rubber are changed in the direction of decreased plastic flow, less surface tackiness, and increased tensile strength by reacting it with sulfur or other suitable agents. See also *self-vulcanizing*.

vulcanize. To subject to vulcanization.

W

warm-setting adhesive. See *intermediate temperature setting adhesive*.

warp. The lengthwise direction of the weave in cloth or roving; also the dimensional distortion of a plastic object. See *fill*.

water absorption. Ratio of the weight of water absorbed by a material to the weight of the same material in a dry condition. See also *moisture absorption*.

weathering. The exposure of plastics outdoors. Compare with *artificial weathering*.

weave. The particular manner in which a fabric is formed by interlacing yarns and usually assigned a style number.

web. A textile fabric, paper, or thin metal sheet of continuous length handled in roll form, as contrasted with the same material cut into sheets.

webbing. Filaments or threads that may form when adhesive transfer surfaces are separated. See also *stringiness*.

weeping. Slow leakage manifested by the appearance of water on a surface.

weft. The transverse threads or fibers in a woven fabric; those fibers running perpendicular to the warp; also called *fill, filler, yarn, woof*.

weld lines. The marks visible on a finished part made by the meeting of two flow fronts of resin during molding.

weld marks. See *flow lines*.

wet flexural strength (WFS). The flexural strength after water immersion, usually after boiling a test specimen for 2 h in water.

wet lay-up. A reinforced plastic which has liquid resin applied as the reinforcement is being laid up.

wet out. This condition occurs during the soaking of porous materials with resin when all voids between strands and filaments become filled with the resin.

wet strength. (1) The strength of paper when saturated with water, especially used in discussions of processes whereby the strength of paper is increased by the addition, in manufacture, of plastic resins. (2) The strength of an adhesive joint determined immediately after removal from a liquid in which it has been immersed under specified conditions of time, temperature, and pressure. (3) The strength of a laminate which has been submerged in water for a predetermined period to simulate the effect of contact with water in service. Wet strength should always be used in stress calculations for boat laminations.

wet winding. In filament winding, the process of winding glass on a mandrel where the strand is impregnated with resin just before contact with the mandrel. See also *dry winding*.

wet-out rate. Time required for the plastic resin to fill the interstices of the reinforcement material and wet the surface of the fibers. This is usually determined by optical or light-transmission means.

wetting. A condition in which the interfacial tension between a liquid and a solid is such that the contact angle is 0 to 90°.

wetting agent. A surface-active agent that produces wetting by decreasing the cohesion within the liquid.

whisker. A very short-fiber form of reinforcement, usually of crystalline material.

winding pattern. (1) The total number of individual circuits required for a winding path to begin repeating by laying down immediately adjacent to the initial circuit. (2) A regularly recurring pattern of the filament path after a certain number of mandrel revolutions, leading to the eventual complete coverage of the mandrel.

winding tension. In filament winding, the amount of tension on the reinforcement as it makes contact with the mandrel.

wire. A metallic filament.

woof. See *weft*.

working life. The period of time during which a liquid resin or adhesive, after mixing with catalyst, solvent, or other compounding ingredients, remains suitable for use. See also *gelatin time, pot life*.

woven fabrics. Fabrics produced by interlacing strands at more or less right angles.

woven roving. A heavy fabric made by the weaving of fiber glass roving.

wrinkle. A crease in one or more plies of a laminate.

Y

yarn. Generic term for strands of fibers or filaments in a form suitable for weaving. See also *continuous filament*.

yield point. The first point on the stress-strain curve at which an increase in strain occurs without an increase in stress.

yield strength. The lowest stress at which a material undergoes plastic deformation. Below this stress, the material is elastic; above it, viscous.

yield value. The stress (either normal or shear) at which a marked increase in deformation occurs without an increase in load.

Young's modulus. The ratio of normal stress to corresponding strain for tensile or compressive stresses less than the proportional limit of the material. See *modulus of elasticity*.

Bibliography

Composite Materials Handbook, by M. M. Schwartz, McGraw-Hill Book Co., New York, 1984

Handbook of Composites, edited by G. Lubin, Van Nostrand Reinhold, New York, 1982

"Composites Design," by S. S. Tsai and T. N. Massard, Report AFWAL-TR-84-4183, Materials Laboratory, AFWAL, Wright-Patterson Air Force Base, Dayton, OH, June–November, 1984

"Electrafil, Electrically Conductive Thermoplastics," Bulletin Elec 0385 10M Rev, Wilson-Fiberfil International, Evansville, IN

Physical Ceramics for Engineers, by L. H. Van Vlack, Addison-Wesley Publishing Co., Reading, MA, 1964

"Composites Glossary," Society of Manufacturing Engineers Composites Tutorial, 1985

Handbook of Adhesive Bonding, edited by C. V. Cagle, McGraw-Hill Book Co., New York, 1973

Handbook of Adhesives, 2nd Edition, edited by I. Skeist, Van Nostrand Reinhold, New York, 1977

Machine Design, April 15, 1982

"Key Terms–Advanced Composites," Du Pont technical literature, Du Pont Company, Wilmington, DE, April 17, 1985

"Properties of 3M Nextel 312 Ceramic Fibers," product data bulletin, Ceramic Fiber Products, 3M, St. Paul, MN

Metals Handbook Desk Edition, edited by H. E. Boyer and T. L. Gall, American Society for Metals, Metals Park, OH, 1985

Emulsion and Water-Soluble Paints and Coatings, by C. R. Martens, Reinhold Publishing Corp., New York, 1964

"Glossary of Plastic Terms, A Consensus," Plastec Note 14

"Glossary to the Science of Composites," Joint Report #HPC 66-11 by Monsanto Co./Washington University and Advanced Research Projects Agency (ARPA) of the Department of Defense

"Non-Metallic Materials" Grumman technical manual (unpublished)

Adhesive Bonding of Reinforced Plastics, by H. A. Perry, McGraw-Hill Book Co., New York, 1959

"Textile Fibers for Industry," technical bulletin, Owens-Corning Fiberglas Corp., 1979

Adhesive Bonding, by C. V. Cagle, McGraw-Hill Book Co., New York, 1968

Whittington's Dictionary of Plastics, by L. R. Whittington, sponsored by the Society of Plastics Engineers, Inc., Technomic Publishing Co., Inc., Stamford, CN, 1968

INDEX

Compiled by Marjorie R. Hyslop
Consultant — Information Science and Technology

Before compiling the index, several nomenclature decisions had to be made. The policies adopted fall in the following categories:

1. Use of trade names vs. generic designations. From the standpoint of greatest user convenience, generic designations are favored. However, all major trade names are cross-referenced to the appropriate generic term — for example, "Nicalon. *See* Silicon carbide fibers", "Celion. *See* Carbon fibers". A couple of notable exceptions are "Kevlar" (given main heading status with cross-reference to Aramid and vice versa) and "Spectra 900" (with cross-reference from "Polyethylene fibers"). This was done to conform to tabular data which more often than not gives only the trade name.

2. Index only by class of material or index by specific members of those classes. The practice is mixed. Specific members of classes have been indexed fairly profusely, particularly the metals and many inorganic compounds. Among the polymers and organics, frequency of mentions in the text was a guiding principle. Both broad classes of polymers and individual members of those classes will be found in fair proportion to text emphasis and the amount of detail provided.

3. Use chemical symbols and formulas or spell out the words? It was decided to spell out all terms used for main entries, but to allow abbreviations and formulas for second- and third-order entries. A prime example is "Silicon carbide", which is spelled out when used as a main heading but is indexed as SiC at second- and third-order levels. This was done to conform to text terminology, where SiC is used almost invariably in tabular matter and graphics. Since it is out of order alphabetically, the term SiC carries a "*see*" reference to Silicon carbide.

4. Property entries: The indexer well appreciates the convenience to a user of looking for such terms as, say, tensile strength, flexural strength, thermal expansion, etc., and finding page references for all materials in which he or she is interested. Unfortunately, because the major portion of the book is devoted to such data, the bulk of the index would have been inordinately increased — and its usefulness inversely affected — by the necessity for strings of page numbers and/or subheadings. However, such properties are amply covered by subheadings under the major materials classes as well as under individual members of those classes.